Sustainable Resource Management

Sustainable Resource Management

Modern Approaches and Contexts

Edited by

Chaudhery Mustansar Hussain

Department of Chemistry and Environmental Science, New Jersey Institute of Technology, University Heights, Newark, NJ, United States

Juan F. Velasco-Muñoz

Department of Economy and Business, Research Centre CIAIMBITAL, University of Almería, Almería, Spain

Elsevier
Radarweg 29, PO Box 211, 1000 AE Amsterdam, Netherlands
The Boulevard, Langford Lane, Kidlington, Oxford OX5 1GB, United Kingdom
50 Hampshire Street, 5th Floor, Cambridge, MA 02139, United States

Library of Congress Cataloging-in-Publication Data
A catalog record for this book is available from the Library of Congress

British Library Cataloguing-in-Publication Data
A catalogue record for this book is available from the British Library

ISBN: 978-0-12-824342-8

For information on all Elsevier publications visit our website at https://www.elsevier.com/books-and-journals

Publisher: Candice Janco
Acquisitions Editor: Marisa LaFleur
Editorial Project Manager: Grace Lander
Production Project Manager: Joy Christel Neumarin Honest Thangiah
Cover Designer: Mark Rogers

Typeset by TNQ Technologies

I would like to dedicate this book to my beloved GOD
"Mera Pyarey Allah"
Chaudhery Mustansar Hussain
I would like to dedicate this book to my partner and my parents
Juan F. Velasco-Muñoz

Contents

Contributors

Adeyemi Adesina Department of Civil and Environmental Engineering, University of Windsor, Windsor, ON, Canada

N. Ali Energy & Building Research Centre, Kuwait Institute for Scientific Research, Al-Shwaikh Educational Area, State of Kuwait, Kuwait

S.M. Al–Salem Environment & Life Sciences Research Centre, Kuwait Institute for Scientific Research, Al-Shwaikh Educational Area, State of Kuwait, Kuwait

Ignacio Amate-Fortes Department of Economics and Business, University of Almería, La Cañada de San Urbano, Almería, Spain

José A. Aznar-Sánchez Department of Economy and Business, Research Centre CIAIMBITAL, University of Almería, Almería, Spain

Lucia Bednárová Technical University of Košice, Faculty of Mining, Ecology, Process Control and Geotechnologies, Košice, Slovak Republic

Helen S.Y. Chen Faculty of Business, The Hong Kong Polytechnic University, Hong Kong SAR, PR China

T.C.E. Cheng Faculty of Business, The Hong Kong Polytechnic University, Hong Kong SAR, PR China

Shaswat Kumar Das Grøn Tek Concrete and Research, Bhubaneswar, Odisha, India; E&S Department, CSIR-Institute of Minerals and Materials Technology, Bhubaneswar, Odisha, India

Peter Drábik University of Economics in Bratislava, Research Institute of Trade and Sustainable Business, Bratislava, Slovak Republic

M.S. El-Eskanadarny Energy & Building Research Centre, Kuwait Institute for Scientific Research, Al-Shwaikh Educational Area, State of Kuwait, Kuwait

Daniel García-Arca Department of Economy and Business, Research Centre CIAIMBITAL, University of Almería, Almería, Spain

Almudena Guarnido-Rueda Department of Economics and Business, University of Almería, La Cañada de San Urbano, Almería, Spain

Mohammadhadi Hajian Department of Economics, Tarbiat Modares University, Tehran, Iran

Chaudhery Mustansar Hussain Department of Chemistry and Environmental Sciences, New Jersey Institute of Technology, University Heights, Newark, NJ, United States

Somayeh Jangchi Kashani Department of Agriculture Development, Islamic Azad University, Sciences and Researches Branch, Tehran, Iran

Vladimír Š. Kremsa Landscape Ecology Institute, Tábor, Czech Republic

Ana Labella-Fernández University of Almería, Ctra. Sacramento s/n, Almería, Spain

Belén López-Felices Department of Economy and Business, Research Centre CIAIMBITAL, University of Almería, Almería, Spain

Milan Majerník University of Economics in Bratislava, Research Institute of Trade and Sustainable Business, Bratislava, Slovak Republic

Marcela Malindžáková Technical University of Košice, Faculty of Mining, Ecology, Process Control and Geotechnologies, Košice, Slovak Republic

Diego Martínez-Navarro Department of Economics and Business (Applied Economy), University of Almería, Almería, Spain

Satya Prakash Maurya Indian Institute of Technology (BHU) Varanasi, Varanasi, Uttar Pradesh, India

Subhabrata Mishra Grøn Tek Concrete and Research, Bhubaneswar, Odisha, India

Jyotirmoy Mishra Department of Civil Engineering, VSSUT, Burla, Odisha, India

Syed Mohammed Mustakim E&S Department, CSIR-Institute of Minerals and Materials Technology, Bhubaneswar, Odisha, India

Jana Naščáková University of Economics in Bratislava, Faculty of Business Economy of the University of Economics in Bratislava with the Seat in Košice, Košice, Slovak Republic

Francisco J. Oliver-Márquez Department of Economics and Business (Applied Economy), University of Almería, Almería, Spain

Sukanchan Palit Department of Chemical Engineering, University of Petroleum and Energy Studies, Energy Acres, Dehradun, Uttarakhand, India

Belén Payán-Sánchez University of Almería, Ctra. Sacramento s/n, Almería, Spain

Pragati Priyadarshinee Chaitanya Bharathi Institute of Technology, CBIT(A), Hyderabad, Telangana, India

M. Mar Serrano-Arcos University of Almería, Ctra. Sacramento s/n, Almería, Spain

Ramesh Singh Indian Institute of Technology (BHU) Varanasi, Varanasi, Uttar Pradesh, India

Juan F. Velasco-Muñoz Department of Economy and Business, Research Centre CIAIMBITAL, University of Almería, Almería, Spain

Editors' biography

Chaudhery Mustansar Hussain, PhD, is an Adjunct Professor and Director of Labs in the Department of Chemistry and Environmental Sciences at the New Jersey Institute of Technology (NJIT), Newark, New Jersey, United States. His research is focused on the applications of nanotechnology and advanced technologies and materials, analytical chemistry, environmental management, and various industries. Dr. Hussain is the author of numerous papers in peer-reviewed journals as well as prolific author and editor of several scientific monographs and handbooks in his research areas published with Elsevier, Royal Society of Chemistry, John Wiley & sons, CRC, and Springer.

Juan F. Velasco-Muñoz, PhD, is an Assistant Professor in the Department of Economics and Business at the University of Almeria, Almeria, Spain. He is a member of the research group "Environmental and Natural Resources Economics." His research is focused on the sustainable management of resources, especially natural resources, socioeconomic valuation of ecosystem services, and development of circular economy systems. Dr. Velasco-Muñoz has numerous publications on these topics in peer-reviewed journals as well as book chapters and contributions to conferences. In addition, he is the editor of several scientific monographs, and he directs several doctoral theses on resource management and circular economy.

Preface

Sustainability is currently a transversal discipline, linked to both the social sciences, such as economics, sociology, or law, and the pure sciences, such as engineering, environmental sciences, or agronomy. The concept of sustainability and its application in the management of scarce resources are fundamental questions for the survival of organizations and society as a whole today. There are many documents dealing with partial topics on sustainable resource management; but, in general, they offer incomplete and disjointed information. There is no single document where you can find knowledge on the application of sustainability to the management of many types of resources. This book is intended as a reference guide in applying the concept of sustainability to the management of different types of resources.

This book defines what the concept of sustainability is and shows how it has evolved since its introduction in 1987 and how it has been adapted to each of the contexts in which it is used. Furthermore, sustainability is made up of three areas—economic, social, and environmental—which are rarely considered together. This handbook shows how these three scopes are implemented in different contexts. Specifically, this book collects, from the theoretical field as well as illustrative examples, the application of the current concept of sustainability to the management of natural resources, both renewable and nonrenewable, such as water, land, minerals, and metals. Productive ecosystems such as agricultural ecosystems, aquatic ecosystems, and forest ecosystems are also included. Finally, the recovered resources are also included.

This book is divided into four parts. The first part contains the theoretical framework within which the sustainable management of resources is framed, to link the theory of sustainable management with current reality and with different types of resources. In the second part, we contextualize the sustainable management of each type of resource. Each chapter includes the characteristics, specificities, problems, and alternatives of each type of resource. In the third part, we include the specific tools for each area of sustainability, economic sustainability, environmental sustainability, and social sustainability, as well as the tools to assess sustainability as a whole. The final part shows different questions related to the new types of resources, the newest practices, and innovative solutions to present and future challenges.

This book condenses all the relevant information on sustainable resource management in one book. It will be an amalgamation of the latest trends in this field. This book is a compilation where the state of the art of technologies in the field of sustainability is reviewed. Readers will get a book in a compact form with various aspects of sustainable resource management. The editors and contributors are prominent researchers, scientists, and professionals from academia and industry. On behalf of Elsevier, we are very pleased with all contributors for their exceptional and passionate hard work in the making of this book. Extraordinary acknowledgments to Peter Llewellyn (acquisition editors) and Grace Lander (editorial project manager) at Elsevier, for their dedicated support and help during this project. In the end, altogether appreciations to Elsevier for publishing this book.

Chaudhery Mustansar Hussain, PhD, and Juan F. Velasco-Muñoz, PhD
Editors

Evolution of the concept of sustainability. From Brundtland Report to sustainable development goals

1

Mohammadhadi Hajian[1], *Somayeh Jangchi Kashani*[2]
[1]Department of Economics, Tarbiat Modares University, Tehran, Iran; [2]Department of Agriculture Development, Islamic Azad University, Sciences and Researches Branch, Tehran, Iran

1. Introduction

The notion of sustainable development shaped the foundation of the Environment and Development Conference held by the United Nations (UN) in 1992 in Rio de Janeiro. The conference manifest the first worldwide effort to design strategies and action plans to a form development that is environmentally friendly and more sustainable. Almost all countries of the world sent delegates to the conference, more than 100 of which were leaders of their countries. Besides the delegates of countries, some international organizations representing civil societies sent representatives to the summit. In 1987, in its report "Our Common Future" the Brundtland Commission stated that the key to solving environmental problems was sustainable development.

The Brundtland Report considered the various serious problems regarding environmental degradation in previous decades: explicitly, that human activity had intense negative influences on the earth, and that development and growth would not be sustainable if the process of environment exploitation continued. Important tasks focusing this philosophy involved Rachel Carson's *Silent Spring* (1962), Garret Hardin's *Tragedy of the Commons* (1968), the ecologist magazine *Blueprint for Survival* (1972), and the report of Club of Rome's *The Limits to Growth* (1972).

At the Conference of UN on the Human Environment, held in Stockholm in 1972, experts presented a unique recognition of sustainable development for the first time. Sustainable development was referred to as development that considers environment conservation especially for future generations.

Fifteen years later, the term "sustainable development" was disseminated in the report of the World Commission on Environment and Development, entitled "Our Common Future" which involved the classic definition of sustainable development: "the development which guarantees meeting the needs of the present generation without reducing the ability of future generations to meet their own needs" Nevertheless, it was not considered to be the main global problem until the main leaders of

Sustainable Resource Management. https://doi.org/10.1016/B978-0-12-824342-8.00018-3

the world identified sustainable development as the most important challenge of humanity.

In 2002, 191 countries and international institutions attended the Johannesburg World Summit on Sustainable Development, which had three important consequences: a political statement, the Johannesburg Plan of Implementation, and a variety of partnership creativities. The main promises made in the summit involved those sustainability in production and consumption, energy, and water.

2. The concept of sustainable development

Sustainable development is a notion that at its center is groundbreaking, yet extremely hard to describe practically. The history behind sustainable development is not long. Pressures that may exist within the notion of sustainable development are various, extending from its unclear and imprecise description to the failure to reach a worldwide practical and functional context. The big problem the concept of sustainable development encounters is the needy to teach it to society and express it in words that individuals will understand. The definition of sustainable development is one that debatably is multifield, complicated, and regular; yet, describing the notion is undoubtedly an important challenge. The term "sustainable development" was invented in 1980, when the concept of the notion was purely rudimentary. It was at the World Conservation Strategy, an association between three leading environmental organizations the international union for conservation of nature (IUCN), world wide fund for nature (WWF), and the united nations environment programme (UNEP), where sustainable development was defined as "preserving the world's natural resources" (World Conservation Strategy, 2011). What the World Conservation Strategy had comprehended is that many countries increasingly exploit natural resources to achieve higher economic growth. The unique goal of the World Conservation Strategy was to ask countries to refuse to deplete natural resources because it was harming the environment. Therefore, sustainable development was considered only in general, not in detail, at the time it was invented; even after a decade, the definition remained almost the same. In 1987, the World Commission on Environment and Development published a paper entitled "Our Common Future" The article presented a commonly recited definition of sustainable development: development that guarantees meeting the needs of the current generation without reducing the ability of future generations to meet their own needs. Many worldwide organizations, institutions, and nongovernmental organizations have since applied the paper, repeatedly identified as the Brundtland Report, that became the first manuscript to support sustainable development as a multidisciplinary field, because it clarified that the economy, society, and the environment were vital to sustainable development. In 1992, the UN Conference on Environment and Development planned a program named Agenda 21, which supposedly was the primary design for sustainability in the 21st century. The UN Commission on Sustainable Development (CSD) observed countries that gave their agreement to Agenda 21. This commission is responsible for revising advancement in the enactment of Agenda 21 and the Rio Declaration on Environment and Development. The commission also provides policy guidance for applying the Johannesburg Plan of

Implementation at the domestic, local, and worldwide levels. Both Agenda 21 and the Brundtland Report are broadly used bases for countries, institutions, and organizations in current times, yet a perfect description still is not evident even though they are acknowledged.

The major problem with sustainable development is the lack of a perfect explanation of it. Although the Brundtland Report's definition is frequently used, one can see precisely where it fails. When looking again at the sentence defining sustainable development, two clear subjects come to mind. First, the definition of "needs" is not clear. There is a large difference between the needs of people living in an underdeveloped country and those of the residents of developed countries. Second, there is no type of time period in the definition, so "generations" is not clear. Because there are many undefined terms in the definition of sustainable development, it is hard to understand. An interesting way to look at subjects available with sustainable development is to observe the next quotation, which expresses a mixture of uncertainty about what to do and a sense of guilt about what is not being completed. In the other words, many individuals appear to be frightened to show what they feel is a lack of recognition of sustainable development. Thus, it is always easier to say that there is no need to talk about it. It is obvious that because of the ambiguous definition of sustainable development, governments and people seem to perform it differently. So far, for those who agree on a definition, the problem of application arises.

Nations that have made a conscious attempt to recognize sustainable development and are eager to make some changes, both countrywide and worldwide, encounter the challenge of fulfilling policies linked to sustainable development. Agenda 21 is a document that provides a framework for sustainable development; however, there is no enforcement of the document. In the other words, nations may pretend that they know about sustainable development and agree to change their policies, but no one really puts them into action. As the Brundtland report stated, our failure to endorse public attention in sustainable development is often the result of overlooking economic and social fairness within and among countries, which demonstrates another important subject in sustainable development: something more than a simple definition of sustainable development. The difficulty with putting into action policies related to sustainable development is that countries that need them the most have no motive to observe the policies. Because a gap between developing and developed countries undoubtedly exists, developed countries principally are telling underdeveloped ones an appropriate way to develop. This evidently proper way of telling what underdeveloped countries should do has roots in the dominant nature of western countries, but is also focuses on common benefits between both developing and developed countries. At the institutional level, the interreliant targets of economic growth, social development, and environmental preservation are achieved by organizations that are inclined to be disjointed and independent. Sustainable development concentrates on the reputation of organizations that are ready to participate in economic, social, and environmental tasks at the levels of both decision-making and policy development.

It is in the nature of sustainable development to outline the issues that not only a nation has with its own policies, but also the world has with its environmental problems.

Therefore, sustainable development is a notion that is groundbreaking yet imperfect within the scope of its perceiver.

Sustainable development has the capacity to be a revolutionary notion that can transform the methods that countries apply on a countrywide level, and further so on a worldwide level. However, because of its multidisciplinary nature, desirable targets, and pliable clarifications, an obvious explanation of the notion has yet to be realized. Furthermore, its imprecise explanation and uncertainty further enhance the strains found within this concept, because any nation could express that it is obeying policies of sustainable development. The difficulty is first to present a succinct definition and then to be able to apply it to any country practically across the world. Because sustainable development can divide the dialogue between developed and developing countries, any nation should make the required changes to its own policies to offer a cleaner, harmless, and more effective environment, economy, and society. What sustainable development really demands is transparency and collaborative countries that are enthusiastic to work together to improve the world. Perhaps for this single reason, sustainable development is hard to explain, because each country has a different viewpoint of what a better world may be.

3. The definition of sustainable development

Sustainable development is a notion that almost everyone has heard of but few recognize. That so many individuals are familiar with the concept of sustainable development is noteworthy considering that it was practically unidentified until the issue of the Brundtland Report by the World Commission on Environment and Development (1987). Certainly, it was not until the 1992 Earth Summit in Rio de Janeiro and the worldwide extension of Agenda 21 of the UN that sustainable development became definitely founded as an acceptable policy goal (UN, 1993). Since that time, many governments have presented a variety of new policy criteria in an effort to direct their economies to a more sustainable path. This seems to be a positive trend. However, we should clarify upon which conditions we can call a country successful in taking steps to achieve the goals of sustainable development. Maybe we have concentrated too deeply on policy criteria and have neglected to focus on tools for attaining sustainable development with an appropriate range of measures to evaluate sustainable development performance in a country. Or perhaps we do not have appropriate sustainable development measures. Costanza (1987) calls this situation a “social trap” because of a dependance on confusing signs or a failure to notice alarming signals exposed by currently developed measures.

4. The trend of sustainable development

The notion of sustainable development has become an extensively known purpose for human civilization in the 21st century. The concept of sustainable development was invented in 1987 with the publication of "Our Common Future," which definitely recognized sustainable development as a vital section of global development. Because irregularities were causing inequalities within and among countries, growing poverty, especially in underdeveloped countries; depleting the ozone layer; causing global warming; depleting natural resources and endangering species of animals and plants; polluting water and air pollution, and so forth, sustainable development was proposed as an attempt to revolutionize the way of thinking about the globe. Behind this was the reason for why today the idea of development is favored instead of growth. Growth is alleged to reflect only a quantifiable feature of nations without considering some other qualitative items such as equality, health, and education.

5. The evolution of sustainable development concept

The term "sustainable development" was used in 1980 by IUCN for the first time, yet the Brundtland Commission Report entitled "Our Common Future" made it important. The definition presented by the commission is still the most frequently used description across the globe. It describes sustainable development as that which meets the needs of the current generation without conciliating the needs of future generations (World Commission on Environment and Development, 1987). According to Gro Harlen Brundtland, sustainability can be guaranteed only when it stresses (a) the mitigation of deprivation and poverty; (b) the protection and improvement of the resources base, which alone can guarantee the permanent alleviation of poverty; (c) expansion of the idea of development, so that it covers economic growth as well as cultural and social development; and (d) a consideration of both economics and ecology in making decisions at all levels (Pearce et al., 1989).

The commission report emphasized considering policy-makers as well as organizers in both developed and developing countries. Environmental preservation institutes have been established in many countries and have the legal authority to inspect developmental plans on various sustainability scopes before approving or disapproving them. In 1992, the Earth Summit in Rio de Janeiro was shown to be revolutionary in this way; a complex of 27 principles were assumed to lead nations in attaining sustainable development for both present and future generations. Besides recognition of the right and need of the current generation for development, it stressed its responsibility to protect the public environment to guarantee intragenerational and intergenerational equity. Since then, sustainable development as an issue has become popular across the world (Mebratu, 1998; Kori and Gondo, 2012).

Sustainable development is a multidisciplinary subject with three main dimensions: social, economic, and environmental (Goodland and Daly, 1996). Social sustainability considers alleviating poverty; economic sustainability is related to the long-run sustainability of both renewable and nonrenewable resources so that they are put into the system of production and make long-run economic revenue; and environmental sustainability is linked to the protection and conservation of life forms existing on earth (Goodland, 1995; Sutton, 2004; Kori and Gondo, 2012). Sustainable development has become a major question of discussion among environment specialists and economists (AtKisson, 2006; Ayres et al., 2001) and is widely evaluated in weak and strong terms. A weak definition of sustainability is based on neoclassical capital theory and economic value principles, and is human-centric in the natural environment. It investigates the capital value of natural resources but totally overlooks their value in terms of the natural goods and services it delivers. In addition, strong definitions of sustainability are based on biophysical principles and consider certain functions that the environment does for humans (Hediger, 2006; Nourry, 2008). For instance, a weak sustainability model may state that the value of a jungle is equal only to the total number of trees cut and the value it produces to make equipment or paper. Nevertheless, a strong sustainability index will evaluate the financial aids of a jungle not only in terms of the economic value of the trees but also according to its environmental and social value. Trees produce food and shelter for animals and humans, help them by its rainfall, act as carbon sinks, and provide fresh air, all of which are considered in the weak index. Therefore, weak sustainability indexes are based on the viewpoint that man-made capital is more important than natural capital, which can be replaced by its manmade counterpart. In contrast, a strong sustainability index allots importance to natural and also human capital in contrast to man-made capital. Strong sustainability indexes represent the tendency to regulate resource consumption to the use of advanced technologies, and conserves natural capital for future generations (Pearce and Atkinson, 1993; Neumayer, 2003; Roberts, 2004; Barr, 2008 as citied in Davies, 2013). Thus, weak and strong sustainability indices are different from each other in terms of morality and logical standpoints.

Based on the strong sustainability concept model, the Brundtland Commission report recommended countries across the globe to pursue a development route calculated in terms of economic growth, but it also considered environmental and social features; therefore, all features should be measured and unified (Pope et al., 2004). Many governments pledged to take steps in sustainability in development by the early years of the third millennium. Agenda 21 at the Earth Summit in 1992, the Johannesburg Plan of Implementation (2002), and the CSD at its 11th (2003) and 13th (2005) meetings declared the need to develop and account for indicators that can describe sustainable development advancement by nations.

6. Indicator development

Indicators use frameworks to offer a communal language and standpoint on the question and its answer. This simplifies indicator development, chiefly when several performers are playing their part. The method in which subjects are outlined becomes significant in the clarification and profounder studies of the consequences, because the frameworks have assumptions for rationales on which the indicator is founded and must be made available to those willing to understand the indicators. Comprehending these assumptions and frameworks is vital to compare and deliberate indicators from various institutes because they may be based on different frameworks. For most users, nevertheless, presentation of the frameworks, or their classes, would only augment a needless degree of intricacy that may distract them from the outcomes. Institutes, working on sustainable development, must have a part in the policies of problem definition, responsive to matters of proper contribution and representation. Such institutes also should have their own role in science and technology, responsive to matters of quality control and expertise. Clark states that perhaps the most important message of the Johannesburg conference was that the research communities need to have a historic role in recognizing problems of sustainability with a larger tendency to link to other organizations in finding applied answers to those problems, and that institutes that spend most of their time doing pure science or pure politics are not expected to be as prosperous as those providing applied solutions that tend to be more effective than other organizations in generating information that can effect policy-making. The three measures of credibility, legitimacy, and salience are main features for depicting the effectiveness of sustainable development indicators in which credibility is related to the scientific and technical adequacy of the measurement system, legitimacy is related to the process of reasonable coping with the conflicting values and beliefs of stakeholders, and salience is related to the application of the indicator to decision-makers. The process of developing indicator guarantees at least the first two of these measures. Resources differ considerably between developing and developed countries. Socioeconomic, environmental, and knowledge contrasts between the two halves may be worsened by resource distribution. Finding ways to cover this resource gap is crucial for equitable representation, both geographically and in terms of identifying and bordering important subjects. Equitable representation raises the question of legitimacy and credibility of both the process and the final product. A capacity for activating and using science and technology is also a vital element of strategies endorsing sustainable development. Making scientific capacity and official support acceptable in developing countries is principally vital to improve resilience in regions that are vulnerable to numerous tensions arising from rapid, simultaneous changes in social and environmental systems.

Nevertheless, scientific capacity alone is inadequate for the goals of creating reliable sustainable development indicators. Instead, building capacity is required, with stress on supporting the broader processes that certify legitimacy and credibility of the process of developing indicators. Effective capacity building stresses the main elements of communication, translation, and mediation. Preparing for sufficient communication between stakeholders is necessary, as is confirming that common understanding is possible. Communication is often stuck by misunderstanding, experience differences, and beliefs about the basis of a convincing discussion. Mediation additionally increases transparency of the process by considering all viewpoints, defining the laws of behavior and process, and founding the criteria of decision-making.

An important pledge by institutes for managing the frontiers between expertise and decision making will help connect knowledge to action. Founding culpability to important actors across the frontier and using combined outputs to grow unity and pledge to the process are also useful in evolving capacity for sustainable development. Indicator legitimacy and acceptability depend on identifying the multiplicity of legitimate standpoints. Where there are complicated subjects, the excellence of the process of decision-making is itself vital, and processes planned to encourage the discourse between stakeholders rather than weakening the authority of science are important to making a comprehensive base of consensus. The role of indicators is to serve as contributors to this discourse and decision-making process. The value of a special indicator set diverges between users and situations. Users should be able to have an impact on selecting the indicators they will have to use. Sometimes this regional selection will lead to a lack of comparability as various groups and processes elect to use various indicators. This can be suitable when the key goal of indicators is to advance effective decision-making. On the other hand, when the key goal is comparability, more emphasis should be given to standardization. It is not usually probable to have both.

7. Environmental sustainability

The standard definition of sustainable development from Bruntland uses a three-pivot method to state, briefly, the connections among economic activity, quality of life, and the endlessness of ecosystems and natural resources. A society without operative life support systems cannot prosper; lack of supportive social constructions and institutes avoid economies from thriving. Furthermore, sustainable development has often been understood as social and economic development that should be sustainable from the environmental viewpoint as well. After the Bruntland definition was presented, it was gradually accepted that environmental sustainability has its own advantages. Goodland declares that environmental sustainability tends to advance human welfare by preserving the sources of raw materials used for human needs and guaranteeing that the volume of human waste will not be increased, to avoid damage to humans. This notion identifies that environmental sustainability worries about the insertion of restrictions on resource consumption, which is also a chief principle of the ecological economics framework of restrictions to growth. Moreover, Goodland describes environmental sustainability as the enforcement of constrictions on four key actions influencing the scale of the human economy: the use of renewable and nonrenewable resources on one side, and waste and pollution of resources on the other. Moldan, Janouskova, and Hak argue that the Organization of Monetary Cooperation and Development (OECD) similarly improves the idea of environmental sustainability within their environmental strategy for the first decade of the 21st century, published in 2001. The OECD's strategy defined four precise standards for environmental sustainability: (1) Regeneration: renewable assets will be used efficiently and their use should not be approved to exceed long-term quotes of natural regeneration. (2) Substitutability: nonrenewable resources will be used efficaciously and their use should be limited to ranges that may be offset through substitution with renewable assets or different sorts of capital. (3) Assimilation: release of unsafe or polluting materials into the environment should no longer exceed their waste assimilative capability. (4) The avoidance of irreversibility. The OECD used these four criteria for environmental sustainability as a way to consider five interrelated targets for advancing environmental policies in a sustainable development context (OECD, 2001):

(1) Preserving ecosystem integrity via the efficient management of natural resources
(2) Decoupling environmental pressures from economic growth
(3) Improving quality of life
(4) Improving international environmental connections by improving governance and cooperation
(5) Measuring progress, specifically using environmental indicators and indices

The concept of environmental sustainability can be further advanced through an ecosystem services viewpoint, because this reinforces the price referring to nonfinancial ecological features and capabilities, all of which can be vital for the OECD's five interrelated targets. Daily discusses "nature's services" to be produced from a worldwide life-assist structure (for example the climate system or hydrological cycle), items supplied via the geosphere (which includes mineral sources), and open space (inclusive of land on the surface of the earth, plus the distance above and beneath it). In meeting the OECD's five objectives for environmental sustainability, human welfare is preserved or enhanced. On this basis, ecosystem services can be considered a fundamental element of human welfare (Chaudhery et al., 2019). Therefore, environmental sustainability can be well defined as the preservation of nature's services at an appropriate level.

This needs environmental services on a local, countrywide and global scale to be reserved in a healthy form, and by definition needs governance systems to have a responsibility of maintenance and supervisory effect on environmental infrastructure.

To present an obvious and objective tool of assessing and representing a country's environmental sustainability, it is usually helpful to use environmental indicators and indices.

No set of countrywide environmental indicators is similar to the standard set of measures used to measure economic performance. In economic policy, countries are normally compared according to their gross domestic product (GDP) and overall performance. In the environmental discipline, the most comprehensive set of indicators is integrated within the environmental performance index (EPI), which attempts but largely fails to provide an overarching description.

The depiction, which captures the idea of environmental sustainability in preference to its specifics, consists of pollution, energy consumption, and soil degradation. A comparable loss of comprehensiveness can be found in sustainable development indicators, with the resulting effect that economic activities undertaken as sustainable development frequently threaten environmental integrity in a particular locality.

Moreover, the subjective method of normalizing and weighting indicators of environmental sustainability is consistently vulnerable to an excessive degree of arbitrariness and a shortage of attention, shown in the consistency and meaningfulness of results, which lowers their relevance in phrases of policy practice.

Theoretically, the use of environmental sustainability indices to delineate sustainable development is vital for explaining to policy-makers and whole society the relationships and exchanges among its three dimensions.

Although because of uncertainty, critics are skeptical about how well environmental indicators and composite indices represent environmental sustainability in practice, they can be beneficial instruments for evaluating the condition of the environment and observing trends over time, as well as describing conditions under which resource consumptions are sustainable.

Assessments of sustainability, especially environmental sustainability, have four specifications: (1) subject recognition regarding the relations among human activity and nature; (2) orientation to the long term and an unsure future; (3) a normative foundation regarding the idea of justice between people of current and future generations as well as among humans and nature; and (4) concern for economic efficiency in the allocation of goods and services, in addition to their man-made substitutes and enhancements. Apprehending the full dynamics of environmental sustainability elements as elaborated in Refs and presenting them in terms of easily explicable quantity indicators is a difficult function. The United Nations (UN) declaires that environmental indices need to have four components as below:

(1) Effects of economic activity on the environment (e.g., resource consumption, pollutant emissions, waste management)
(2) Impacts of resource productivity on the economic system (e.g., economic efficiency)
(3) Effects of environmental degradation on economic productivity (e.g., drop in absorptive ability, lack of forest cover)
(4) Impacts of environmental development on society (e.g., reduction of congestion costs, improvement in welfare and reduction of societal costs)

Regardless of obvious cohesions in the numerous definitions of environmental sustainability, there are no indicator sets or composite indices that satisfactorily and comprehensively measure the concept on a countrywide and global scale. Parris and Kates set out three reasons why this is the case: (1) the vague nature of sustainable development; (2) the variety of cause in characterization, and (3) measurement of sustainable development.

Confusion regarding the terminology, appropriate facts, and strategies of measurement is another reason why environmenal indicators are not satisfactory. Therefore, the number of indicators and indices needed to quantify any country's environmental sustainability credentials does not permit a uniform, well-described procedure; for this reason, professional judgment is crucial.

Several different international indices has been considered for assessment of environment such as the living planet index, satisfaction with life index, human development index (HDI) and sustainable society index which have been rejected because of some reasons as below:

- Environmental sustainability index: a composite index assessing 21 factors of environmental sustainability; however, in 2005 it was replaced through the more comprehensive EPI.
- Barometer of sustainability: includes all three sustainable development dimensions, but the restricted scope of the environmental component and lack of an approach, proposing approaches for achieving targets, rendered the EPI a more comprehensive instrument for evaluation.

- Surplus biocapacity index: this index lists nations according to the balance or deficit among their ecological footprint (EF) and countrywide biocapacity. Although the surplus or deficit in biocapacity is a critical criterion for evaluation and is considered, a global rating list gives nothing further in phrases of analytical value.
- Satisfaction with life index: measures subjective welfare across countries in compare to wealth, health, and accessto basic education. On the basis that it fails to comprise any type of ecological focus, the happy planet index (HPI) is more desirable.
- Human development index: although it is a comprehensive socioeconomic metric widely linked to the gratification with life index, it lacks attention to environmental problems and thus was not favored for the chosen indices.
  - Living planet index: concentrates especially on the problem of biodiversity stocks and in an international context; as a result, it misses enough scope to evaluate environmental sustainability troubles in a countrywide context.

7.1 Social sustainability

Social sustainability is the least determined and least apprehended of the various methods of achieving sustainability and sustainable development. Social sustainability has had significantly less interest in public discourse than economic and environmental sustainability.

There are numerous procedures for sustainability. The primary, which postulates a trio of environmental sustainability, economic sustainability, and social sustainability, is the most broadly accepted as a model for addressing sustainability. The idea of "social sustainability" in this method includes subjects such as social equity, livability, health equity, community development, social capital, social support, human rights, labor rights, place-making, social responsibility, social justice, cultural competence, community resilience, and human adaptation.

A second, newer one is the method suggesting that all fields of sustainability are social, consisting of ecological, economic, political, and cultural sustainability. Those fields of social sustainability are based on the connection between the social and the natural, with the ecological field defined as human embeddedness in the environment. In these items, social sustainability includes all human activities (Atkinson et al., 2003). It is not just relevant to the focused intersection of economics, the environment, and the social.

7.2 Economic sustainable development

Starting from the 21st century, increases in GDP are no longer considered the main macroeconomic indicator and the economic growth is not the major purpose of the economic system. Since the second world war, the concentration on economic development policy has changed through different paradigms up to the idea of sustainable development. The Keynesian concept of the postwar 1970s, in which economic policy was built on sturdy governmental interventions, to the monetarist thought of the 1980s, when initiatives were taken to reduce social disparities by including deprived groups into the mainstream economic system. In the late 1980s to 1990s, the emphasis of economic development policy shifted to a rationalist concept and projects to enhance the environmental and total quality of life with absorbing particularly talented individuals and enterprises started.

The simple standards of the UN specify sustainable development to be considered as the principal goal of the economic system from the current point of view. Economic growth calculated by increase in GDP represents a rise in economic activity and correlates with well-being. Rapid economic growth is rarely sustainable. There are some countries, particularly developing countries, in which excessive discrepancy exists between the level of economic activity and the people's well-being: precisely, high growth with low development. Development is broader than economic growth. Well-being cannot be best measured with financial phrases. Economic growth is an external notion, whereas development is a broader inner one that includes an enhancement in the standards of living and poverty reduction. Economic growth may also bring about an enhancement in the standards of living related to a small share of the population whereas most people are yet poor. It is the distribution of economic growth among the population as an index determining the level of economic development. Economic growth is measured through the growth of GDP, whereas economic development is an extra complex procedure that needs a couple of indicators.

The analysis of economic procedures in many nations indicates that fast economic growth has precipitated extreme issues from the point of view of sustainable development, such as social and local inequality, lack of adequate infrastructure and rural environment, lack of national capital, and many others. Sustainable development is an instant enhancement in well-being for all populations, not compromising the well-being within the near of far future. Economic growth is essential, but it is not enough for development.

7.3 Context of sustainable development goals

Sustainable Development Goals (SDGs) were emerged at the United Nations Conference on Sustainable Development in Rio de Janeiro in 2012. The aim was supplying a set of global goals that met urgent environmental, political, and economic problems encountering our world.

The SDGs update the Millennium Development Goals (MDGs), which commenced a worldwide attempt in 2000 to challenge the indignity of poverty. The MDGs created measurable, universally agreed-upon targets for tackling excessive poverty and starvation, preventing lethal sicknesses, and expanding primary schooling to all young people, among other development priorities.

For 15 years, the MDGs led development in several important regions: decreasing poverty, providing accees to water and sanitation, lowering infant mortality, and considerably enhancing maternal health. In addition, they began worldwide movement for free primary education, inspiring nations to invest for future generations. Most significantly, the MDGs made massive strides in combatting HIV/AIDS and other treatable diseases, including malaria and tuberculosis.

7.4 Key Millennium Development Goal achievements

More than one billion people have been lifted out of extreme poverty (since 1990):

- Child mortality dropped by more than half (since 1990)
- The number of youth who were out of college dropped by more than half (compared with 1990)
- HIV/AIDS infections fell by nearly 40% (compared with 2000)

The legacy and achievements of the MDGs offer us precious lessons and new goals. However, for tens of millions of humans, the job remains unfinished. We want to take steps towards ending hunger, attaining full gender equality, enhancing health services, and getting all children into education higher than primary level. The SDGs are also pressing to shift the sector onto a greater sustainable path.

The SDGs are an ambitious attempt to finish what we commenced and address a number of pressing demanding situations. All 17 goals interconnect, so accomplishment in one area affects success in others. Coping with the danger of climate change affects how we manage our fragile natural sources, achieving gender equality or better health facilitates, getting rid of poverty, and fostering peace and inclusive societies will lessen inequalities and help economies prosper. In brief, this is the finest field we should improve in life for future generations.

The SDGs was concurrent with another historic agreement reached in 2015 at the COP21 Paris climate change conference. Together with the Sendai Framework for Disaster Risk Reduction, signed in March 2015 in Japan, these agreements provide a collection of common standards and attainable aims to lessen carbon emissions, manage the risks of climate change and natural disasters, and to build back better after a catastrophe.

The SDGs are unique in that they cover subjects that have an effect on us all. They reaffirm our worldwide dedication to cease poverty permanently, anywhere. They are formidable in making sure nobody is left back. More importantly, they include all of us in building a more sustainable, safer, and more wealthy planet.

8. Sustainable development goals

8.1 Goal 1. End poverty everywhere in all of its forms

Through 2030, eliminate severe poverty for all of us anywhere, currently measured as humans livelihood on less than $5 a day. By 2030, reduce by at the least half the proportion of men, women, and children living in poverty in all its dimensions, in regard to national definitions, and enforce nationally suitable social protection structures and measures for all. By 2030, reach considerable coverage of vulnerable and poor people, assuring that everyone, specifically vulnerable and poor people, have the same rights to economic resources, in addition to access to elementary services, possession and control over land and different kinds of property, inheritance, natural resources, appropriate new technology, and economic services, such as microfinance by 2030, construct the resilience of the poor and those in vulnerable conditions and decrease their exposure and vulnerability to climate-related severe events and different financial, social and environmental shocks and failures (Sustainable Development Goals Report, 2020).

8.2 Goal 2. End hunger, achieve food security, improve nutrition, and promote sustainable agriculture

By 2030, eradicate hunger and ensure the right of access to resources for all people, mainly the poor and people in vulnerable conditions, including newborns, to safe, nutritious, and enough food year-round by 2030. Cease all types of malnutrition. The world has agreed to targets for improving the life qulaity of children under age 5 years, and to address the nutritional needs of adolescent girls, pregnant and lactating women, and older people by 2025. Moreover, it is agreed to double the productivity in agriculture sector and earning of small-scale food producers, specifically women, aboriginal peoples, family farmers, herders and fishers, including through safe and identical access to land, other efficient resources and inputs, education, monetary services, markets and opportunities for value addition and non-agriculture employment by 2030. Make sure sustainable food production structures and put in force resilient farming tasks that grow productiveness and production, that help preserve ecosystems, that enhance ability for variation to climate change, extreme climate, drought, flooding and different catastrophes and that step by step enhance land and soil quality.

8.3 Goal 3. Ensure healthy lives and promote welfare for all people

By 2030, reduce the worldwide maternal mortality ratio to less than seven in ten thousand live births. By 2030, eradicate avoidable deaths of infants and kids under age 5 years, with all international locations aiming to reduce neonatal mortality to less than 12 in 1000 live births and under-5 mortality to less than 25 in 1000 live births. By 2030, end the epidemics of AIDS, tuberculosis, malaria and out-left tropical diseases and fight hepatitis, water-borne illnesses and different communicable sicknesses. By 2030, reduce by one third premature mortality from non-communicable diseases thru prevention and remedy and promote intellectual health and well-being. Empower the prevention and remedy of substance abuse, including narcotic drug abuse and dangerous use of alcohol. By 2020, halve the number of global deaths and damages from road accidents. By 2030, ensure global access to sexual and reproductive health-care services, including for family planning, information and training, and the incorporation of reproductive health into countrywide strategies and programs. Obtain global health coverage, such as financial risk protection, access to high-quality vital health-care services and access to safe, effective and low-cost critical drug treatments and vaccines for all.

8.4 Goal 4. Ensure inclusive and equitable quality education and promote lifelong learning opportunities for all

By 2030, ensure that any girl and boy have access to free, and equitable primary and secondary education leading to relevant and effectivelearning outcomes. By 2030, ensure that early childhood improvement, care and pre-primary education are available for any child so that they get prepared for primary education. By 2030, ensure equal availability for all women and men to cheap and satisfactory technical, vocational, and tertiary education. By 2030, notably increase the number of children and adults who've relevant abilities, along with technical and vocational skills, for employment, respectable jobs and entrepreneurship. By 2030, readicate gender disparities in education, and make sure equal access to all stages of training and vocational education for the vulnerable people, such as individuals with disabilities, aboriginal peoples and youngsters in vulnerable situations. By 2030, make sure that all young people, and a substantial share of adults, each males and females, attain literacy and numeracy.

8.5 Goal 5. Achieve gender equality and empower all women and girls

Eradicate all kinds of discrimination against all women anywhere. Eradicate all kinds of violence against girls and women in public and personal fields, including trafficking and sexual and other kinds of exploitation. remove all harmful tasks, which include child marriage and female genital mutilation. Recognize and value unpaid care and homework through the availability of public services, infrastructure, and social safety regulations and promoting of shared responsibility within the family as nationally suitable. Ensure women's complete and powerful participation and identical opportunities for leadership at all levels of decision-making in political, economic and public life. Make certain common availability of sexual and reproductive health and reproductive rights as agreed in accordance with the Program of Action of the International Conference on Population and Development.

8.6 Goal 6. Ensure availability and sustainable management of water and sanitation for all

By 2030, ensure common and equitable access to safe and low-priced potable water, gurantee access to proper and equitable sanitation and hygiene for all and quit open defecation, paying special attention to the needs of women and those in vulnerable situations. By 2030, enhance water quality through reducing pollutants, removing of dumping and minimizing release of risky chemical compounds and materials, halving the share of untreated wastewater and significantly developing recycling and safe reuse globally. By 2030, significantly increase water-use efficiency for all sectors and ensure sustainable withdrawals and supply of freshwater to cope with water scarcity and noticeably reduce the number of people affected by water scarcity. By 2030, enforce integrated water resource management at all levels, including through transboundary cooperation as appropriate.

8.7 Goal 7. Ensure access to affordable, reliable, sustainable and modern energy for all

By 2030, gurantee access to low-priced, reliable and modern energy services. By 2030, increase significantly the share of renewable energy in the global energy combination. By 2030, double the worldwide rate of improvement in electricity efficiency. By 2030, improve worldwide cooperation to facilitate access to clean power research and technology, which include renewable power, increase efficiency and purify fossil-fuel technologies, and encourage investment in electricity infrastructure and clean energy technology. By 2030, promote infrastructure and upgrade production for imparting modern and sustainable technologies for all developing countries, particularly least developed countries, and land-locked developing countries, in accordance with their respective programs of support.

8.8 Goal 8. Promote sustained, inclusive, and sustainable economic growth, full and productive employment, and decent work for all

Make economic growth sustainable according to country conditions, especially at least seven percent annual growth of gross domestic product (GDP) in under-developed countries. Achieve higher levels of economic growth through diversification, technological upgrading and innovation, involving via an emphasis on high added value and labor-intensive sectors. Enhance development-orientated policies that assist productive activities, decent job creation, entrepreneurship, creativity and innovation, and inspire the formalization and increase of micro-, small- and medium-sized firms, including thru access to economic services. Enhance gradually, through 2030, worldwide efficient resource performance in consumption and production and endeavor to decouple economic growth from environmental degradation, in accordance with the 10—year framework of programmes on sustainable consumption and production, with developed countries taking the lead. By 2030, gain full and effective employment and decent job for all males and females, including for young persons and also people with disabilities, and equal pay for work of equal value.

8.9 Goal 9. Build resilient infrastructure, promote inclusive and sustainable industrialization, and foster innovation

Expand reliable, sustainable, and strong infrastructure, involving local and transboundary infrastructure, to aid economic development and human welfare, with a concentration on cheap and equitable availability for all people. Improve inclusive and sustainable industrialization and, by 2030, extensively boost industry's proportion of employment and gross domestic product (GDP), in line with country's circumstances, and double its proportion in under-developed countries. Promote the access of small-scale businesses and industries, mainly in developing countries, to financial services, including affordable credit, and their integration into value chains and markets. By 2030, improve infrastructure and retrofit industries to make them sustainable, and more efficient and lead them to extra adoption of environmental-friendly technologies and industrial processes, with all countries acting based on their respective skills. Enhance scientific research, promote the technological abilities of enterprises in all countries, specifically developing countries. Ensure halving the proportion of untreated wastewater and appreciably growing recycling and safe reuse globally by 2030. Significantly increase water-use efficiency throughout all sectors and ensure sustainable withdrawals and supply of freshwater to deal with water shortage and appreciably lessen the number of humans affected by water shortage by 2030. Implement incorporated water resource management in any regions, including via transboundary cooperation as appropriate.

8.10 Goal 10. Reduce inequality within and among countries

By 2030, steadily gain and maintain income growth of the bottom 40 percent of the population at a rate higher than the national average. By 2030, strengthen and improve the social, economic and political inclusion of all, no matter age, sex, disability, race, ethnicity, origin, religion or financial or different reputations. Ensure equal opportunity and reduce inequalities of outcome, involving through abolishing discriminatory laws, regulations and tasks and developing appropriate regulations and movement towards this regard. Enact policies, in particular financial, salary and social protection rules, and steadily achieve greater equality, Enhance the legislation and observation of the world financial markets and institutions and empower the fulfillment of such policies. Ensure greater illustration and louder voice for developing countries in making decisions in worldwide monetary and economic institutions in order to deliver extra powerful, credible, accountable and legally-valid establishments. Facilitate orderly, secure, well-organized and answerable migration and mobility of humans, including through fulfillment of planned and properly-managed migration policies. Enforce the principle of special and differential treatment for developing countries.

8.11 Goal 11. Make cities and human settlements inclusive, safe, resilient, and sustainable

By 2030, ensure for all people the availability of adequate, secure, and low-cost housing and essential services and improve the life quality of people living in slums. By 2030, provide access to secure, cheap, approachable, and sustainable transport systems for all, improving road security, significantly by growing public transport, with specific attention to the needs of those in vulnerable conditions, women, children, disabled persons, and elders. By 2030, improve inclusive and sustainable urbanization and ability for participatory, united and sustainable human settlement planning and management in all countries. Fortify attempts to guard and protect the global cultural and natural legacy. By 2030, noticeably lessen the rate of deaths and number of people affected and considerably reduce the direct economic losses relative to international gross domestic product (GDP) resulting from catastrophes, including water-related calamities, with an emphasis on supporting the poor and vulnerable people.

8.12 Goal 12. Ensure sustainable consumption and production patterns

Execute the 10−year framework of programs on sustainable consumption and manufacturing, all nations undertake, with developed countries taking the lead, considering the development and capacities of developing countries. By 2030, attain the sustainable management and efficient use of natural resources. By 2030, halve per capita global food waste at the retail and consumer levels and decrease food losses along production and supply chains, which includes postharvest losses. By 2020, attain the environmentally sound control of chemical fertilizers and all wastes throughout their life cycle, in accordance with agreed global frameworks, and appreciably lessen their release to air, water, and soil to minimize their destructive impacts on human health and the environment.

8.13 Goal 13. Take urgent action to combat climate change and its impacts

Increase resilience and adaptive ability to climate-related dangers and natural disasters in all countries. Merge climate change measures into local policies, tactics and planning. Expand education, and knowledge and improve human and institutional capacity on climate change mitigation, impact reduction and early caution. Put in force the dedication commenced with the aid of parties of developed countries to the UN Framework Convention on climate change to meet the needs of developing countries inside the context of significant mitigation actions and transparency on implementation and absolutely operationalize the Green Climate Fund via its capitalization as soon as possible. Promote mechanisms for raising potential for effective planning and management, related to climate change, in under-developed countries, including concentrating on women, youth, and local and marginalized groups.

8.14 Goal 14. Conserve and sustainably use oceans, seas, and marine resources for sustainable development

By 2025, avoid and extensively diminish all types of marine pollution, specially from land-based activities, such as marine debris and nutrient pollutants through 2020, sustainably control and guard marine and coastal ecosystems to prevent extensive detrimental impacts, via increasing their resilience, and acting for their renovation to achieve healthy and efficient oceans. Restrict and address the influences of ocean acidification, including via more advantageous clinical cooperation in any level. By 2020, effectively regulate harvesting and cease overfishing, illegal, unreported and unregulated fishing and negative fishing practices and put in force technological know-how-based management plans, to be able to restore fish stocks inside the shortest time possible, at least to volumes that could produce most sustainable yield as determined by their biological characteristics.

8.15 Goal 15. Protect, restore, and promote sustainable use of terrestrial ecosystems, sustainably manage forests, combat desertification, halt and reverse land degradation, and halt biodiversity loss

By 2020, gurantee preservation, renovation and sustainable use of terrestrial and inland freshwater ecosystems and their services, specifically forests, wetlands, mountains and drylands, in line with commitments under global agreements. By 2020, proceed the fulfillment of sustainable management of all kinds of jungles, prevent deforestation, restore degraded jungles and drastically boom afforestation and reforestation internationally. By 2030, fight desertification, renovate degraded land and soil, alongside land influenced by desertification, drought and floods, and try to acquire a land degradation-independent globe. By 2030, ensure the preservation of mountain ecosystems, including their biodiversity, on the way to improve their potential to make profits that are important for sustainable development. Take urgent and extraordinary actions to lessen the degradation of natural habitats, stop the loss of biodiversity and, by 2020, avoid the extinction of threatened species, Develop equitable sharing of the benefits of the utilization of genetic assets and increase suitable access to such sources, as the world over agreed. Take instant and substantial action to avoid poaching and trafficking of blanketed species of flora and fauna and cope with both supply and demand of illicit trade of wildlife.

8.16 Goal 16. Promote peaceful and inclusive societies for sustainable development, provide access to justice for all, and build effective, accountable, and inclusive institutions at all levels

Meaningfully lessen all sorts of violence and related death rates anywhere. Eradicate abuse, exploitation, trafficking and all forms of violence against and torture of children. Enact the rule of law at countrywide and worldwide levels and ensure equal access to justice for all. By 2030, notably decrease smuggling flows, reinforce the restoration and return of stolen belongings and fight all forms of organized crimes. Significantly lessen corruption and bribery in all their kinds. Expand powerful, responsible and transparent institutions at all levels. Ensure responsive, inclusive, participatory and representative decision-making in any level. Develop and fortify the participation of developing countries in the institutions of world governance. By 2030, provide legal identity for all, such as birth registration. Ensure public access to information and defend basic freedoms, in accordance with national law and international agreements. Develop applicable national establishments, consisting of through international cooperation, for constructing ability in any level, mainly in developing countries, to barricade violence and combat terrorism and crime. Develop and put in force nondiscriminatory laws and guidelines for sustainable development.

8.17 Goal 17. Strengthen the means of implementation and revitalize the global partnership for sustainable development

8.17.1 Finance

Improve national resource mobilization, which includes global aid to developing countries, to improve domestic ability for tax and different income collection. Developed countries to fulfill completely their legitimate development assistance commitments, which includes the dedication by many advanced economies to catch the target of 0.7 percent of official development assistance/gross national income (ODA/GNI) to developing countries and 0.15 to 0.20 percent of ODA/GNI to under-developed countries; ODA providers are advocated to remember setting a target to provide as a minimum 0.20 percent of ODA/GNI to least developed countries. Mobilize extra financial sources for developing countries from different sources. Aid developing countries in accomplishing long-run debt sustainability via coordinated policies aimed at fostering debt financing, debt relief and debt restructuring, as suitable, and cope with the foreign debt of rather indebted low-income countries to reduce debt distress. Assume and perform investment promotion regimes for least developed countries.

8.17.2 Technology

Improve cooperation among developed and developing countries and triangular regional and worldwide collaborations on for accessing to technology, knowledge, and innovation and collaborate information sharing via mutually agreed terms, consisting advanced coordination among current mechanisms, especially on the UN level, and thru a global mechanism of technology facilitation. Improve the development, transfer, dissemination, and diffusion of environmental-friendly technologies to under-developed countries on beneficial terms, which include on concessional and preferential terms, as mutually agreed. Completely operationalize banks for technology, make mechanisms for science and innovation for under-developed countries. By 2017 improve the usage of empowering technology, specifically information and communications technology (ICT).

8.17.3 Capacity-building

Improve worldwide aid for fulfilling effective and targeted capacity-building in developing countries to assist countrywide plans to put into effect all of the sustainable development goals, which includes via cooperations of developing and developed countries in regional and international levels and also intra developing countries as well.

8.17.4 Trade

Develop a global, law-based, open, nondiscriminatory, and equitable multilateral trading framework under the World Trade Organization (WTO), consisting through the outcomes of negotiations under Doha Development Agenda. Remarkably the exports of developing countries, especially so that it will double the under-developed countries' share of world exports by 2020. Recognize the long-term fulfillment of duty-free and quota-free market access on a permanent basis for all under-developed countries, consistent with the decisions of WTO, including through ensuring that preferential policies of origin applicable to imports from least-developed countries are transparent and simple, and contribute to facilitating marketplace access.

8.17.5 Systemic issues

Expand worldwide macroeconomic stability, such as through policy coordination and policy coherence. Develop policy coherence for sustainable development. REgard every country's policy space and management to set up and enforce rules for poverty eradication and sustainable development.

Develop the worldwide partnership for sustainable development, complemented thru the usage of multi-stakeholder partnerships that mobilize and share the knowledge, skills, technology and financial resources, to assist the accomplishment of the sustainable development goals in all countries, mainly developing countries. Inspire and improve effective public, public-private, and civil society partnerships, constructing at the practice and resourcing strategies of partnerships.

8.18 Sustainable development goals and targets 2030

According to an inclusive process of inter-governmental discussions, and based on the proposal of the Open Working Group on Sustainable Development Goals (SDGs), the agreed-upon goals are described. The SDGs and objectives are incorporated and inseparatable, worldwide in nature, and universally appropriate, considering various national realities, capacities, and levels of development and respecting countrywide regulations and priorities. Targets are described as ambitious and worldwide, with every government placing its own countrywide targets directed by the global level of ambition but thinking of countrywide circumstances. Each government may even determine how those ambitious and worldwide purposes ought to be included in national plans, tactics, policies, and strategies. It is crucial to understand the connection between sustainable development and other relevant ongoing tactics within the economic, social, and environmental fields. In determining these desires and goals, we realize that every country encounters special demanding situations to achieve sustainable development, and we underestimate the special difficulties fronting the vulnerable countries, especially African countries, least developed countries, landlocked developing countries, and small island developing states, as well as particular difficulties fronting middle-income countries. Countries in conditions of battle also require specific attention. We realize that baseline data for numerous targets are unavailable, and we require increased aid for strengthening data collection and capacity building in Member States (Sustainable Development Goals Report, 2020).

9. Conclusions

Sustainable development notion was primarily believed to be a resolution for ecological disaster arised from vast industrial exploitation of resources and deterioration of the environment and the initiative concentration was to maintain environmental quality. Nowadays, the idea of sustainable development has extended by considering quality of life in its intricacy, environmentally, economically and socially. Sustainable Development Goals (SDGs) were emerged at the United Nations Conference on Sustainable Development in Rio de Janeiro in 2012. The aim was supplying a set of global goals that met urgent environmental, political, and economic problems encountering our world. Sustainability has three main dimensions: economic, social and environmental aspects. The analysis of economic progression in many countries has indicated that the rapid economic growth led to severe difficulties from the viewpoint of sustainable development. It is the distribution of economic growth among the population as an index determining the level of economic development. Economic growth is measured through the growth of GDP, whereas economic development is an extra complex procedure that needs a couple of indicators. Economics is defined as the science of allocation of scarce resources to satisfy the unrestricted needs of the people of a given society so there is the need about new directions for economic policy if sustainable development is to be achieved including: difference between developing and developed countries, and justice in the international economy, and participatory development. Sustainable development is a continuous improvement of welfare for all people, without reducing the ability of future generations to meet their own needs. So, in order to satisfy the idea of sustainable development, it should be studied the best possible use of all available economic resources for producing maximum feasible output of goods and services that are required for the society now and in the future and the fair distribution of this output. Although economic sustainability is very important, it is essential to consider environmental and social sustainability as well. It is crucial to understand the connection between sustainable development and other relevant ongoing tactics within the economic, social, and environmental fields. Nowadays, the idea is extended in its intricacy, environmentally, economically and socially. The evaluation of the economic sustainability in many countries has proven that economic growth, without considering other aspects of development, precipitated critical issues from the point of view of sustainable development consisting of social, and environmental sustainability. The standard definition of sustainable development from Bruntland uses a three-pivot method to state, briefly, the connections among economic activity, quality of life, and the endlessness of ecosystems and natural resources. A society without operative life support systems cannot prosper; lack of supportive social constructions and institutes avoid economies from thriving. Furthermore, sustainable development has often been understood as social and economic development that should be sustainable from the environmental viewpoint as well. Environmental sustainability tends to improve human welfare by preserving the sources used for human needs and guaranteeing that the volume of human waste will not be increased, to avoid damage to humans. This notion identifies that environmental sustainability worries about the insertion of restrictions on resource consumption, which is also a chief principle of the ecological economics framework of restrictions to growth. Social sustainability is the least determined and least apprehended of the various methods of achieving sustainability and sustainable development. Social sustainability has had significantly less interest in public discourse than economic and environmental sustainability. There are numerous procedures for sustainability. The primary, which postulates a trio of environmental sustainability, economic sustainability, and social sustainability, is the most broadly accepted as a model for addressing sustainability. The idea of "social sustainability" in this method includes subjects such as social equity, livability, health equity, community development, social capital, social support, human rights, labor rights, placemaking, social responsibility, social justice, cultural competence, community resilience, and human adaptation. A second, newer one is the method suggesting that all fields of sustainability are social, consisting of ecological, economic, political, and cultural sustainability. Those fields of social sustainability are based on the connection between the social and the natural, with the ecological field defined as human embeddedness in the environment. In these items, social sustainability includes all human activities. It is not just relevant to the focused intersection of economics, the environment, and the social.

References

Ayres, R., Van den Berrgh, J., Gowdy, J., 2001. Strong versus weak sustainability: economics, natural sciences, and consilience. Environ. Ethics 23 (2), 155−168.

Atkinson, T., et al., 2003. Social Indicators: The EU and Social Inclusion. Oxford University Press, Oxford.

AtKisson, A., 2006. Sustainability is dead − long live sustainability. In: Keiner, M. (Ed.), The Future of Sustainability. Springer, Dordrecht: Netherlands, pp. 231−243.

Barr, S., 2008. Environment and Society Sustainability Ashgate, Policy and the Citizen. Aldershot.

Costanza, R., 1987. Social traps and environmental policy. Bioscience 37 (6), 407−412.

Chaudhery, M., Ajay, H., Mi, K., 2019. Modern Environmental Analysis Techniques for Pollutants. www.elsevier.com/books/modern-environmental-analysis-techniques-for-pollutants.

Davies, G.R., 2013. Appraising weak and strong sustainability. Searching for a middle ground. Consilience 10, 111−124.

Goodland, R., 1995. The concept of environmental sustainability. Annu. Rev. Ecol. Systemat. 26 (1), 1−24.

Goodland, R., Daly, H., 1996. Environmental sustainability: universal and non-negotiable. Ecol. Appl. 6 (4), 1002−1017.

Hediger, W., 2006. Weak and strong sustainability. environmental conservation and economic growth. Nat. Resour. Model. 19 (3), 359−394.

Kori, E., Gondo, T., 2012. Environmental sustainability. Reality fantasy or fallacy?. In: 2nd International Conference on Environment and Bioscience. IPCBEE, vol. 44. IACSIT Press, Singapore. https://doi.org/10.7763/IPCBEE.2012.V44.22.

Mebratu, D., 1998. Sustainability and sustainable development: historical and conceptual review. Environ. Impact Assess. Rev. 18 (6), 493–520.
Nourry, M., 2008. Measuring sustainable development: some empirical evidence for France from eight alternative indicators. Ecol. Econ. 67, 441–456.
Neumayer, E., 2003. Weak versus Strong Sustainability: Exploring the Limits of Two Opposing Paradigms. Edward Elgar Publishing.
OECD, 2001. Environmental Strategy for the First Decade of the 21st Century. OECD, Paris.
Pearce, D., Atkinson, G., 1993. Capital theory and the measurement of sustainable development: an indicator of "weak" sustainability. Ecological 8 (2), 103–108.
Pearce, D.W., Markandya, A., Barbier, E.B., 1989. Blueprint for a Green Economy. Earthscan, London, UK.
Pope, J., Annandale, D., Morrison-Saunders, A., 2004. Conceptualising sustainability assessment. Environ. Impact Assess. Rev. 24 (6), 595–616.
Roberts, J.A., 2004. Environmental Policy. Routledge, London.
Sustainable Development Goals Report, 2020. www.un.org/sustainabledevelopment/progress-report/.
Sutton, P., 2004. A perspective on environmental sustainability. Paper Vic. Commission. Environmen. Sustain. 1–32.
United Nations, 1993. Agenda 21 (New York, UN).
WCED, 1987. World Commission on Environment and Development. Report Our Common Future. United Nations, New York. https://sustainabledevelopment.un.org/content/documents/5987our-common-future.pdf.
World Commission on Environment and Development (WCED), 1987. Our Common Future. Oxford University Press, Oxford.
World Conservation Strategy, 2011. Living Resource Conservation for Sustainable Development, World Conservation Strategy. URL. http://data.iucn.org/dbtw-wpd/edocs/WCS-004.pdf.

Further reading

Chaudhery, M., Ajay, H., Mi, K., 2018. Nanotechnology in Environmental Science, 2 Olumes, ISBN 978-3-527-34294-5.
Commission of the European Communities, Eurostat, 1993. International Monetary Fund. Organization for Economic Co-operation and Development. United Nations and World BankNew York, Paris, Washington DC.
Nyman, M., 2003. Sustainable Development Indicators for Sweden. Concepts and Framework, Technical Report, Statistics Sweden. http://www.scb.se/eng/omscb/eu/eu.asp.
Pigou, A.C., 1952. Essays in Economics. London: Macmillan & Co. Ltd, London.
Sustainable Development Goals Report, 2013. www.un.org/sustainabledevelopment/progress-report/.
United Nations, 2015. Transforming Our World: The 2030 Agenda for Sustainable Development.
World Economic Forum, 2002. Global Leaders of Tomorrow Environmental Task Force: 2002 Environmental Sustainability Index. http://www.ciesin.columbia.edu/indicators/ESI.

Areas of sustainability: environment, economy, and society

2

Milan Majerník [1], Jana Naščáková [2], Marcela Malindžáková [3], Peter Drábik [1], Lucia Bednárová [3]
[1]University of Economics in Bratislava, Research Institute of Trade and Sustainable Business, Bratislava, Slovak Republic; [2]University of Economics in Bratislava, Faculty of Business Economy of the University of Economics in Bratislava with the Seat in Košice, Košice, Slovak Republic; [3]Technical University of Košice, Faculty of Mining, Ecology, Process Control and Geotechnologies, Košice, Slovak Republic

1. Areas of sustainability: environment, economy, and society

The current priority of each organization is to maintain prosperity, performance, and satisfaction of customers and all involved subjects operating in its area. In the world, the term socially responsible enterprise has been used for several decades. Socially responsible enterprises incorporate three basic areas, namely social, economic, and environmental. Within the framework of European legislation Corporate Social Responsibility (CSR) uses the standard EN ISO 26000:2010 Guidance on Social Responsibility. Under this standard EN ISO 26000:2010, CSR is defined as the responsibility of an organization for the impact of its decisions and activities on both society and environment, and simultaneously transparent and ethical behavior which contributes to sustainable development, including the health and welfare of society. It takes into consideration the expectations of all involved parties in compliance with legislation and international standards of conduct and it is integrated within the whole organization and applied in relations.

1.1 Pillars of social responsibility—social, economic, and environmental

Defining CSR should be a part of the plan of every organization concerned with increasing the rates of customer satisfaction. The organization must set its priorities, taking into account not only the resources available to facilitate achieving desirable results but also satisfying the needs of other involved parties. The enterprise needs to consider results it intends to achieve. However, it is necessary to identify critical factors which may occur too. Social responsibility aims at respecting not only social but also environmental sustainability in a way that allows for the goals of both local

Sustainable Resource Management. https://doi.org/10.1016/B978-0-12-824342-8.00006-7

and global communities to be reached. Social responsibility stems from the idea of volunteering and philanthropism. For that reason, in order to create their image modern companies meticulously work out a strategy of responsible behavior to both society and environment.

The European Union started to address the issue of social responsibility and its importance in 1996 when Jacques Delors, the President of the European Union, took the agenda under his auspices. In the same year, the European Centre for Social Responsibility—Corporate Social Responsibility Europe was established.

In 2001, the European Union published a document named Green Paper which defines social responsibility as "voluntary integration of social and environmental aspects into day-to-day business operations and interactions with corporate stakeholders (individuals, groups)".

With respect to these requirements it is necessary to invest more in human capital, the environment, but especially in green technologies as well as relations with stakeholders.

The analysis of social responsibility includes internal and external parties. The internal side includes employees, investment in human capital, health and safety, and environmentally responsible activities. The external side consists of business partners, suppliers, customers, competitors, communities, media, and public authorities. The analysis covers all entities which in some way affect activities and operation of an enterprise. Through social responsibility it is possible to achieve smooth functioning and long-term gain, while company's strategy must contain tools to identify and meet the needs of all stakeholders.

From the viewpoint of marketing social responsibility is perceived complexly, within the context of 3P "People, Planet, Profit" in the professional sense of "triple bottom line" business.

Corporate benefits are also part of CSR. There are three types of CSR:

1. ethical—involves a minimum level of responsibility, the enterprise avoids negative impact of its activities on the society,
2. altruistic—includes philanthropic activities through which the enterprise financially supports stakeholders,
3. strategic—focuses on voluntary activities from the perspective of benefits for the company.

1.1.1 Social pillar

The core of this pillar is represented by the development of human capital, employee care, health and safety, respect for human rights in a work place, ban on child labor, and support of charitable events. Within this pillar there are two areas granting opportunities to carry out many other activities.

The first of the mentioned areas is the working environment in which employees are constantly motivated to increase their efficiency and loyalty because, ultimately, they are high-quality specialists whose work results in long-term success of the company. It is up to the enterprise to develop and maintain quality and qualified employees. This area also includes realization of socially responsible activities. Based on these assumptions the enterprise should identify and determine which activities are beneficial for their employees and will consequently increase their performance and efficiency.

The second area is the local community within which the enterprise tries to gain trust participating in solving local problems and at the same time the enterprise makes efforts to eliminate negative effects of its own activities on the surroundings. Strengthening the competitive advantage is linked to maintaining good relations with stakeholders in order to gain their trust. Within this area it is also necessary to focus on and maintain cooperation with broader surroundings—participating entities/stakeholders.

1.1.2 Economic pillar

The implementation of social responsibility in an enterprise requires sufficient financial resources. Components of this pillar include creation of a document relating to the conduct of the company, such as the code of ethics, the principles of transparent conduct, the protection of intellectual property, and specification of relations to customers, investors, and suppliers. Most companies present their CSR-related activities through their annual reports as well as the Internet. In addition to these activities, CSR includes building relationships with stakeholders, transparent behavior of the enterprise, offering quality products, and responsible approach to customers. There are various forms of implementation of these activities, namely satisfaction surveys, effective handling of complaints, and product quality, which is assessed through EN ISO 9001:2015 Quality Management Systems. Requirements.

Transparency of a company is ensured by regular disclosure of financial as well as nonfinancial information, aiming at informing the environment about social responsibility and economic situation. Compliance with the terms and conditions and their constant monitoring vis-à-vis suppliers are also essential. The comparison of the information is also an important element in the return of other managerial and economic methods that have a significant impact on society (Bednárová et al., 2013). Within a company, it is important to set rules and implement CSR strategy through individual levels of the organization into all processes and activities.

1.1.3 Environmental pillar

The orientation of this pillar is related to organic production, green products and services, reducing the impact on the environment, protecting natural resources, and developing an overall green policy. Rational and sustainable use of energy, waste reduction, separation and recycling aimed at minimizing costs and minimizing benefits are at the forefront. Emphasis must be placed on identifying sustainable environmental aspects of business activity in order to bring a significant impact on the environment. Following the identification of environmental aspects and impacts, it is inevitable to monitor legal and other requirements. The priority of green business is to bring business opportunities to the company as well as financial savings. The company must specify the environmental objectives and values it implements at each level of the enterprise, monitoring and measuring them, and subsequently taking preventive and corrective measures if negative deviations are detected. As a rule, it is a priority to educate own workers in the field of environmentalism too.

1.1.4 Corporate philanthropy

Other benefits—corporate benefits, often difficult to measure but all the more monitored in the corporate environment—are also expected from an established CSR. The term corporate philanthropy means various voluntary activities oriented toward a public-benefit purpose, without expectation of added value. Corporate philanthropy is connected to the social environment in which the company does business. It is an opportunity to demonstrate one's own social responsibility by creating higher ethical standards in addition to making a profit and supporting the development of society outside its core business. The guiding principles of philanthropy include transparency, efficiency, innovation, flexibility, and sustainability. Both monetary and nonmonetary donations can be included among the instruments of philanthropy. These activities include charitable support, investments in projects of companies and communities, as well as commercial projects in cooperation with nonprofit organizations. Philanthropic/voluntary activities do not expect any reward/any consideration. The subject of social responsibility is the linking of relations between business and society, business and nature, entrepreneurship, and other social groups in local and global society, which have been greatly intensified by the globalization of the economy and new communication technologies. On the one hand, what is happening in businesses affects people's quality of life and, on the other hand, it affects the economic, political, and environmental sectors. Within the framework of global ethical responsibility, businesses need to address, on the one hand, social and environmental necessity. On the other hand, many multinationals must take into account the impact of global economic processes on the earth's humanity and nature. Social responsibility includes the following statements:

- enterprise belongs to social entities that are created and operate in the society,
- enterprise, as a collective entity, voluntarily claims responsibility for the consequences of its own activities,
- social responsibility includes economic, legislative, ethical, environmental, social, and civic responsibility,
- social responsibility is related to all corporate activities,
- social responsibility is accepted in relation to each interest group.

Social responsibility of an enterprise is therefore a complex structured system of relations between an enterprise and groups with which it exists in a particular social system. An enterprise, that has objective knowledge of its operations and integrity is its value, is consciously and voluntarily claiming responsibility for its activities. Social responsibility presupposes the following activities on the part of the enterprise (Fig. 2.1):

- adoption of the ethical principle of responsibility WHO?, WHAT?, TO WHOM?, WHY?, WHEN? and its application to itself,

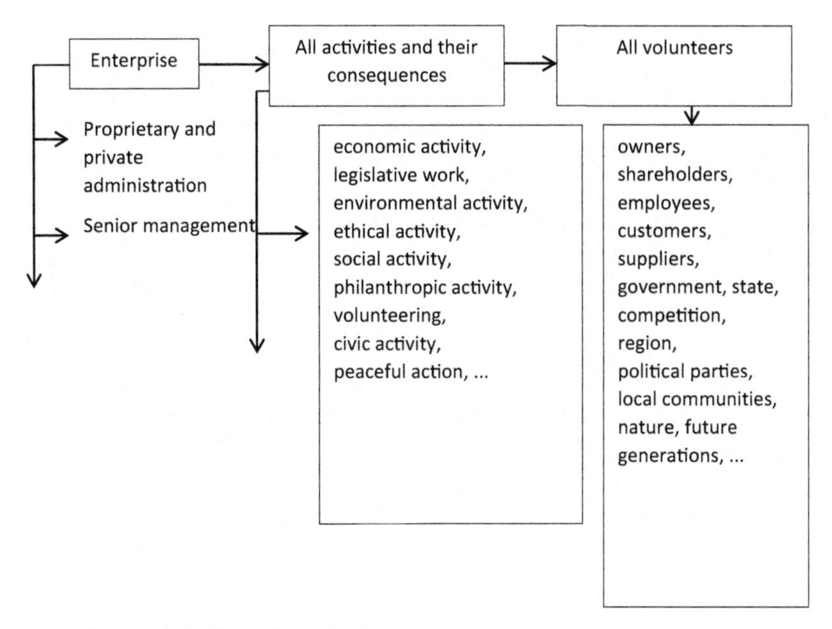

Figure 2.1 Links between business activities and volunteering.

- identification of specific groups or individuals—what the enterprise influences through its activities and at the same time what influences its activities, i.e., identify its own volunteers and create a database of relevant interests/expectations in economic, social, and ecological spheres,
- an analysis of the foreseeable consequences of all its activities (short, medium, long term) on individual volunteers.

1.2 Interconnections between economic growth, energy consumption, social welfare, and sustainable quality of life

In the past, people used renewable energy sources (sun, biomass, water, and wind energy) to meet their needs, and this development took place from the beginning of the Industrial Revolution. In the ancient past, people probably had low consumption of natural energy sources and energy (about 3 GJ per year). The discovery of fire and burning of wood increased annual human consumption per year to about 6 GJ. The emergence of economic activity and the use of the power of towing animals increased consumption to 20–30 GJ per person per year. At the turn of the 18th century, the steam engine was invented which triggered the Industrial Revolution when James Watt increased its efficiency by using coal to drive it. The ability to convert thermal energy into motion energy (useable for machine propulsion) meant that energy consumption and economic growth began to accelerate. Energy consumption thus increased on average up to 100 GJ per person per year.

The Industrial Revolution was a revolution of fossil fuel–based energy technologies. This development took place gradually from the use of coal, through oil to uranium and natural gas. Local and/or regional supply has been turned into a global transport of fuels across the Earth and economic growth has been supported by a steady increase in energy consumption.

The scenario of the Industrial Revolution was influenced by the idea of getting the most out of the minimum and in the early 20th century this idea was creatively developed further by Frederick W. Taylor. This founder of scientific management divided work tasks into specific steps and used time analysis to get maximum effectiveness from his workers. He was thinking about how to apply efficiency from the factory to every area of life in order to increase productivity of the whole society. Taylor was successful and the efficiency from factories penetrated even into households. It has become typical of modern life, in which it seems reasonable to use every minute to the maximum. However, efficiency also entails unexpected and unintended consequences. It leads to ever-increasing consumption and the associated economic and environmental consequences it entails. In the last two centuries of uninterrupted industrial and economic growth the so-called "*culture of exponential growth*" has been reflected.

According to Johanisová (2018), dependence on growth in general as well as economic growth may also be related to historical events. When fossil fuel machines were put into operation, people were losing their jobs and moving to other sectors. This created additional plants, new products and services, which were exported in case of demand and favorable price on world markets—the growth of production was motivated not only by domestic consumption but also by the possibility of exports to other countries.

Along with the production and consumption growth, the monetary system evolved as well. Banks lent entrepreneurs more money than they had in their vaults. The funds had to be repaid with interest and the banks lent them on to firms, consumers, and governments. Both the financial power of bankers and the society's indebtedness grew, and the need to repay interest contributed to the pressure on labor efficiency and productivity, which is reflected in—among other things—further job losses.

Economic growth is currently linked not only with employment but also with the lasting interest in credit. Most of our funds are incurred as debt, and if interest in loans fell, there would also be a risk of a fall in funds in circulation, which can lead to further feedback and destabilization of the whole system. However, increasing the indebtedness of individuals and countries does not solve the problem, but only postpones the solutions. According to Bartlett (2012), we are living in a time when exponential growth related to the economy as a whole, production and consumption, consumption of natural resources, population, indebtedness, poverty, environmental pollution, and global warming is hitting its limits and therefore such unlimited growth cannot be further promoted.

There are a number of innovations that make it possible to consume less material and energy per product or service, therefore many scientists and experts see increasing efficiency as a solution to environmental problems. However, the benefits of reducing environmental impact and pollution are nullified by an increase in traffic, construction activity, overall needs, and overall consumption. The prevailing views maintain that the increase in energy efficiency of products does not affect the intensity of their consumption—but the so-called *macroeconomic feedback* (*increase in* consumption of a product or service due to reduction in its cost price) confirms that efficiency and technological progress can cause an increase in consumption.

Already back in 1865 the English logician, methodologist, and economist William Stanley Jevons noticed that if technological improvements increase the efficiency with which the given resource is consumed, it will lead to an increase in its use, not a decline, as might appear at first glance. The economic explanation of the Jevons paradox is that the increase in efficiency is equal to the reduction in price (or increase in the efficiency of work) and, as a result, the demand for and consumption of the resource will increase—provided that the resource is available. Jevons observed this effect when James Watt significantly increased the efficiency of steam engine and this resulted in a sharp increase in coal consumption. At that time, many scientists predicted that increasing efficiency of steam engine would lead to a decrease in coal consumption. However, a decrease in coal consumption would have been observed only if the number of steam machines had not increased or if it increased by less than the saving of coal would bring by increasing efficiency.

Steadily increasing consumption is also one of the main causes of environmental problems. The problem is even more pronounced when environmental problems are combined with social problems linked with factual inequality. According to the Universal Declaration of Human *Rights, "all people are born free and equal in both dignity and rights"* Every person should therefore be entitled to the same amount of resources for his or her life. Currently, "rich" countries are estimated to consume about 80% of resources, while they make up only 20% of the world's population—consumption per capita is 16 times higher than that of "poor" countries.

According to some research results, while economic growth increases the living standard of individuals and countries, the feelings of happiness and satisfaction in individuals do not increase proportionally to rising living standards. Despite that, many countries do not intend to give up their own consumption, as they consider consumption growth to be an effective means of maintaining employment, economic growth, and also of solving environmental problems.

The "Kuznetsov curve" theory points out that environmental problems will be solved precisely owing to economic growth. Kuznetsov assumed that some local problems, such as atmospheric pollution of urban areas or river pollution, would seemingly be solved when the country got rich enough. He believed that, once the threshold of EUR 8000 GDP per capita had been reached, economic growth would correspond to improvement of the quality of the environment. An example of this type of model is a rich city with little pollution which mainly uses electric automobiles for transport. However, Carson (2010) notes that wealth allows to reduce local atmospheric pollution but does not reduce it at the global level. Carson questioned the assumption that greater efficiency and wealth can solve all environmental problems.

According to available surveys and scientific analyses, the process of uncontrollable economic growth leads to a reduction in the well-being of society, with the exception of a small minority within the society. The reason may be that, in a market economy, individual enterprises take decisions that lead to their own growth and thus to the growth of the country. One important way to achieve the growth of a company is introducing innovative technologies. When a company develops a process that will allow 10 workers at its highly automated plant to carry out the work of thousands in independent workshops across the country, and at the same time it will be financially profitable, the company will put it into practice. The technology the company chooses to use will make one group richer (its shareholders) but this happens at the expense of the loss of its redundancies and their families.

According to the Human Development Report, the "gap" between rich and poor countries is also widening. Over the past 3 decades, 20% of the poorest countries have seen their share of global income fall from 2.3% to 1.4%. As a result, the income ratio of the richest 20% of countries against the income of the poorest fifth has more than doubled, from 30:1 to 61:1. The gap between the rich and the poor is widening in the so-called developing countries. In Thailand, for example, where there was rapid economic growth 2 decades ago, the income ratio of the richest 10% to the poorest 10% rose from 17 times to 38 times (Naščáková and Červenka, 2009).

The disparity in income, which has been gaining considerable attention in American public and professional discourse as well as in the International Monetary Fund, is being pointed out by Frankel (2014), a Harvard University professor. The reason for the expert debate is the concern in the United States that income inequality is steadily increasing, not only in the United States but also in many other parts of the world. The given debate is interesting because it focuses to a large extent on the consequences of inequality beyond the adverse effects on the poor. One direction of discussion begins with the hypothesis that inequality is bad for overall economic growth. Another represents the opinion that inequality leads to volatility and instability. Inequality could have caused both the 2007 mortgage crisis and the global financial crisis in 2008. The third view is that inequality provokes envy and a feeling of unhappiness. A person, who might be satisfied with a certain income, may become unhappy if s/he finds out that other people have a higher income. The fourth concern in a way outclasses the first three hypotheses. It is a concern that, because of the huge amount of money in politics, the rich are managing to persuade governments to prioritize them as a class (www.1).

Globalization—a process that undermines the state's ability to maintain a healthy balance between the interests of transnational corporations and the well-being of its citizens—is also the cause of inequalities. Economic growth, which ensures a "material" increase and the desire of individual enterprises to produce and sell an ever-increasing number of goods and services, is linked with the reduction of the number of people employed in companies (productivity of labor), with the reduction of real wages and other benefits to employees (cost savings), with the constant increase in the amount of energy and raw materials consumed (the use of natural resources), with the uneven distribution of income in favor of the rich, the use of technologies that make work draconian and less satisfactory, increasing the amount of waste and chemicals that end up in landfills or the environment, and the decline of smaller and less competitive firms and local economies (Naščáková and Červenka, 2009).

1.3 Limitations of stable economic growth

Most economic models are based on the philosophy of continuous production growth. The fact that a number of the above conditions need to be met in order for economic growth to also have an effect on social well-being and quality of life may explain why the process of economic growth as we now know it has already known social and economic negative effects. Economic growth has always been linked with intense use of natural resources because it is based on "material" increase.

In a simplified manner economic growth can be characterized as growth of both production and consumption which requires consumption of natural resources, energy, and causes various kinds of pollution and interventions in the natural environment. Innovation technologies and recycling may, to a certain degree, restrain the negatives but cannot fully eliminate them. Recycling cannot reach 100% efficiency because it requires material and energy inputs and itself produces a particular level of pollution. The links between GDP growth and the growth in both natural resources consumption and some types of pollution are being referred to by many research projects which will be examined in the following parts of this work.

The ideology of economic growth and "the need of continuous economic growth" have become dogmas of government programs. In this respect the professor of physics Al Bartlett (2012) speaks of other aspects relating rules of arithmetic, growth of natural and energy resources consumption, and population growth.

1.3.1 Growth and the rules of arithmetic

Bartlett (2012) focuses his research on the general nature and consequences of any stable growth in a confined (limited) space (environment of the Earth). He explains valid laws of arithmetic—on the one hand independent from power, political, and economic interests and on the other hand uncompromisingly limiting the future stability of every society on every level. Some of these problems are local, other—national or global. All problems are interrelated and they are linked by the rules of arithmetic.

According to Bartlett, the weak point of the humankind is the lack of concern with understanding of the exponential function used to identify the size of "something" that is steadily growing.

If there is something that grows every year by 5%, we will use the exponential function to determine how big the annual increase is. This is a situation when the "something" always grows by a fixed measure within the same period of time, e.g., by 5% annually, where 5% is that fixed measure and 1 year is strictly defined period of time.

As the growth by 5% requires the specific strictly defined time, for the increase by 100% a longer fixed period of time will be necessary. The mentioned longer period of time can be labeled as the period of *doubling* and it can be calculated by splitting the number 70 by the percentage of growth per unit of time. Using the example with 5%

growth per year, we divide the number 70 by 5 (i.e., 5% growth) and find out that the growing amount will double every 14 years.[1]

$$T_2 = \frac{70}{\%growthpertimeunit}$$

1.3.2 Stable growth in a confined space

"Increase within every period of doubling is larger than the sum of all previous increases." In his speech on energy the former US President James Earl Carter talked about the issue. Apart from other things he said: *"In each of these decades (the 1950s and 1960s) more oil was consumed than during the entire previous history of mankind."* It was a shocking statement, and yet the President only explained the simple consequence of the arithmetic of the 7% annual growth in global oil consumption—that is exactly what growth was until the 1970s.

Entrepreneurs, government officials, politicians at local, regional, state, and international level are trying to maintain a society in which all values of indicators of material consumption would grow steadily each year.

1.3.3 Economic growth, increase in population, and energy resources consumption increase

It is estimated that about 50 million people lived on Earth before the agrarian (agricultural) revolution. About 10,000 years ago the agrarian revolution radically changed their way of life. It significantly improved the conditions for survival and life itself. It was the first significant impulse to a population explosion and from that point it took about 10,000 years for the number of people on Earth to reach 1 billion (1800 AD). The second impetus for the growth of human population was the discovery, improvement, and mass manufacturing of steam machines, which triggered technical development accompanied by exponential population growth. The second billion of inhabitants on Earth was reached as early as 30 years later (1910), and in the next 90 years (2000) the population on Earth grew up to six billion. 13 years later, in 2013, the population exceeded 7.3 billion. Today, about 200,000 people add up on Earth every day (i.e., the number of births minus the number of those who die as a result of old age, disease, or violence). The annual increase of the number of inhabitants is approximately 1.1%. If this growth rate is maintained, the doubling of the world's population (7.3 billion in 2013 x 2) to 14.6 billion will occur in 63 years−70/1.1 = 63 (2076).

The growth of the world's population is a global problem. However, the situation in developed and developing countries is inherently so different that we can talk about two problems. In developed countries, the number of inhabitants is decreasing or stagnating and the population is aging. In developing countries, the number of inhabitants

[1] Number 70 is the rounded result of multiplying 100 (100% increase) by the natural logarithm of two—100 x ln2 = 69,3. In case we wanted to find out the time it takes to triple, we would use the natural logarithm of three.

is increasing and the population structure contains a high proportion of the prereproductive component which will ensure an increase in the population in the coming decades, even with a possible significant reduction in fertility. The population is growing fastest in the world's poorest countries, which are least able to meet basic needs. These differences result from the humankind's response to the progress and modernization of society. With the change in living conditions, the reproductive behavior of the population changes too. Just as there are big differences in social development (tradition, culture, standard of living) between the countries of the world, there are also large differences in population development.

High population increases in developing countries also exacerbate existing environmental problems as well as problems with standard of living. In many cases, the situation has already crossed a viable threshold, and a realistic solution to the current and expected behavior of the humankind is out of sight. Critical problems include lack of agricultural land and drinking water, environmental pollution, climate changes, socioeconomic problems (access to education, housing, lack of employment opportunities, low standard of living).

Bartlett (2012) points out that we are facing a dilemma. Everything that is considered good from an economic and social points of view only worsens the population problem and everything we find bad helps to solve it. If we want to increase the rate of population growth and thus exacerbate the whole problem, we should support everything that serves humane objectives and reduces mortality (immigration, medicine, public health, hygiene, peace, rule of law and order, clean air, modern agriculture ...). If we want to reduce population growth and thus help solve the population problem, we should promote sexual restraint, contraception, abortion, small families, blocking off immigration, disease, wars, murders, famines, accidents, unhealthy way of life, air pollution.

Migration is a factor that can contribute to a partial reduction of disparities in population developments in the world and a more even distribution of population (and consequently wealth). Migration establishes a link between vastly different population developments in the developed and the developing parts of the world. Immigrants from less developed countries tend to move to developed prosperous countries. This way they solve their own economic and often existential problems. In the country of arrival, they compensate for the natural decline of the population, thereby increasing the potential of parents and the workforce. At the same time, immigrants create contact between developed and developing countries, helping to bring different worlds closer together.

According to Paul R. Ehrlich, an American biologist and professor of population studies and biological sciences at Stanford University, efforts to control birth rates need to be addressed because, as a result of population growth, energy demand will increase twice as much by 2050, according to experts. Not only is the population (number of energy consumers) growing but the average energy consumption of each individual is increasing too.

Increase in population is closely linked with the growth of energy consumption. The current condition of the stable increase in both population and natural resources consumption is not sustainable, according to Bartlett (2012), with regard to the limited natural resources of the Earth. Sustainable development is a kind of development which satiates the needs of the current generation without restricting the options of future generations to meet their needs. It is vital to know and be aware of the rules of arithmetic and consequences of growth, especially in respect of the size of population, the limited natural resources, protection of the environment, and sustainability.

1.3.4 Natural and environmental limitations of economic growth

Limitations of economic growth and the existence of physical boundaries of economic growth, defined by the planet Earth, have been and still are subject of numerous scientific and professional works.

In the updated version of *The Limits to Growth* from 1992, Meadows, D. and colleagues pointed out that one of the reasons why economy is nearing the physical boundaries of the planet is precisely the *exponential growth* which is being referred to in Professor Bartlett's lectures too. According to the authors of *The Limits to Growth,* the global increase of industrial production, consumption of natural resources, environment pollution, food production, and population is becoming ever faster. This growth has a character of stable exponential growth which, they believe, is unsustainable.

The issue of natural limitations of growth was in 1996 addressed also by Julian Lincoln Simon in his book *The Ultimate Resource 2*. Simon takes a different view from Bartlett and does not see problems related to the rarity and exhaustibility of natural resources or to the concept of stable economic growth. His views are also shared by other scientists/e.g., Musil (2009)/and, like Simon (2006), they question the existence of natural limits on economic growth and argue that natural resources are not limited, the consumption of natural resources brings new discoveries and more resources, natural resources are not natural, but are the result of human ingenuity, and the nonrenewable, ultimately exhaustible energy sources maintain economic growth.

One of the new scientific approaches to the issue of natural limitations of economic growth is a systemic approach to the processes on Earth defined as *planetary boundaries* which should not be crossed, provided we wish to avoid global environmental changes. Planetary boundaries are delimited by 28 internationally recognized scientists from across the world, working under the leadership of Stockholm Resilience Centre, established in 2007, which develops a disciplinary scientific research for socioecological systems management. The Centre involves scientists from the Stockholm University, the international Beijer Institute of Ecological Economics of the Royal Swedish Academy of Sciences, and the Stockholm Environmental Institute. The Centre's scientific team carried out a comprehensive research based on which they identified nine processes for which limits should be determined in order to minimize the risks of their exceeding:

1. Climate change—the first exceeded limit,
2. Extinction of biological species—the second exceeded limit,
3. Anthropogenic disruption of the natural cycle of nitrogen and phosphorus,

4. Disruption of stratospheric ozone layer,
5. Global use of freshwater,
6. Acidification of oceans,
7. Change in land use,
8. Aerosol burden,
9. Chemical pollution.

(www.2).

Vlčková (2014) maintains that one of the interesting aspects of the new *planetary boundaries* approach is that it handles *planetary boundaries* from the perspective of Earth and the nature as well as the things which the Earth and the nature allow people to do. The previous approaches were mostly based on societal factors and the needs of human society.

Exponential growth of anthropogenic activities (originating through the activities of people) burden the Earth to the degree which could gradually destabilize the decisive biophysical systems and trigger irreversible changes with negative impacts on human society. *All human activities require energy consumption so all human interventions into the living environment may be classified as consequences of energy use.* The work on quantification of the planetary boundaries was hindered by many uncertainties caused by insufficient scientific understanding of the properties of biophysical thresholds, uncertainties about how these complex systems behave, also those caused by insufficient scientific knowledge on how other biophysical processes (e.g., feedback mechanism) interact, etc. (Rockström et al., 2009).

The mentioned uncertainties then create further uncertainties by each proposed boundary. The team of scientists thus presents this new approach as their first proposal and draws attention to the inevitability of carrying out further research. According to the authors the benefit of defining planetary boundaries is that it concentrates on the biophysical processes of Earth which delimitate self-regulatory ability of the planet. The limits and boundaries of the main processes of Earth's systems exist regardless of human preferences, wishes, values or compromises and they are based on political and socioeconomic possibilities (Hegyi, 2012).

The concept of already mentioned *ecological footprint* also derives from the discovered discrepancy between existing natural resources and growing tempo of their use (due to the population increase and its anthropogenic activities). A very simplified analysis of ecological footprint can bring approximate results accounting for about 1.5 ha of ecologically productive area and 0.5 ha of seas per person, i.e., a total of 2 ha. After removing the ecological footprint of plants and animals (preserving biodiversity), 1.76 ha of biologically productive land per inhabitant of the Earth remains available. Along with the increase of population the amount of ecologically productive land per person is decreasing (Gajdoš et al., 2011).

In respect of sustainable forestry, it is necessary to ensure synergy between three pillars, i.e., ecological, economic, and sociocultural. It is significant to focus on the integration of advantages related to the local food sources protection down to the protection of biodiversity as well as forest ecosystems. On the other hand, the consequences of climate changes and poverty too need to be mitigated. Forests and forest stands are helpful in the global climate regulation via carbon storage thereby absorbing approximately 40% emissions of fossil fuels which are being produced on daily basis and simultaneously contributing to mitigation of flood risks. For the needs of the sustainable management of forest, it is necessary to ensure the good health of forest ecosystems as well as to provide sanctuary to people and wildlife in a long-term perspective. The basis for forest protection is to create intervention-free areas thus preserving genuine natural processes (www.7). Forest is characterized by functions which can be divided into three categories: economic, protective, and special-purpose forests. At present the share of economic forests represents 67.0% compared to the condition in the past. This share of forests has a falling tendency. While ensuring the other functions of forests, the priority of economic forests is production of quality wood mass. In the context of the concept of sustainable forestry, the integration of forest production and public service functions is necessary. Protective function of forests is specific for extreme habitat and climatic conditions, especially those focused on agricultural, water economic, and antiavalanche functions. Currently, the share of such forests is 16.9%. The share of special-purpose forests was gradually increasing up to 16.1%. This type of forest is focused on a special societal purpose which is related to carrying out public-benefit functions, specifically water protection, recreation, health restoration, hunting, nature protection, antiemission as well as education research, etc.

In 1998, the PEFC system (Programme for the Endorsement of Forest Certification schemes) was established by a group of private forest owners from six European countries, first as a procedure for forest certification according to the pan-European system. In 1999, a PEFC council was officially created in Paris, registered as an international organization in Luxembourg, where it is based to this day. Companies managing the certification system with help of the national certification procedures of the member states are permanent members of PEFC. Important international companies of forest and industry owners as well as trade and processing companies are special members of PEFC (www.8).

The General Assembly is the highest PEFC body that sets and recognizes standards for the establishment of independent national certification systems. The main part of the PEFC system consists of decisions taken in regard of the protection of forests held in Strasbourg, Helsinki, Lisbon, and, most recently, Vienna. Sustainable forest management is governed by pan-European criteria and indicators. The pan-European operational guidelines explained in the Pan-European Criteria documents constitute indicators and standards of implementation for sustainable forest management (Resolution L2 III of the Ministerial Conference on the Protection of Forests in Europe), which take into account national certification procedures. Based on the sustainable forest management practices, an agreement was drawn up on respecting independent national certification systems that goes beyond the borders of the European Union. Currently, PEFC manages 5 continents and 35 countries. At the moment 233 million hectares of forest are certified and 7618 consumer chain PEFC certificates have been issued in 30 national certification systems for sustainable management (www.8).

Accessibility and plausibility of the certification carried out shall be the responsibility of independent certification bodies which are not formally linked to the PEFC system and its structures. The conduct according to the indicators and pan-European criteria for sustainable management fully show the state and level of forest management in the area, as well as the accessibility of the system, which is determined by the emergence of national certification systems.

1.3.4.1 European criteria for sustainable forest management

With help of European standards which are used for forest qualification, forest management can be tackled in accordance with their sustainable management practices, the fulfillment of which is an essential condition of issuing the certificate. Forest plants must be governed by six basic European criteria for sustainable management (www.10):

Criterion 1: To conserve and adequately develop forest resources and their contribution to the global carbon cycle. The objective of this criterion is to focus on general preparedness, land area and use, and carbon balance.

Criterion 2: To maintain the vitality and health status of forest ecosystems. The point of this criterion is the care of the ecosystem.

Criterion 3: To preserve and promote productive forest functions (wood and non-wood products). This criterion focuses on forest production.

Criterion 4: To maintain, adequately develop, and protect the diversity of ecosystems in the forest. This area concentrates on general conditions, rare and easily vulnerable forest ecosystems, endangered species, and biological diversity in economic forests.

Criterion 5: To maintain and evenly develop the protection of forestry. This area is focused on general protection, soil erosion, and water protection in forests.

Criterion 6: To maintain other economic and social functions and requirements. Within this criterion, priority is given to orientation at the importance of the forestry sector, at the provision of work, vocational education and research, public awareness, and cultural values.

1.3.5 Change of climate conditions—global warming

The causes of the impact of climate change are the subject of many scientific and technical discussions which will be referred to below. It is proven now that the use of fossil resources influences climatic conditions, resulting in an increase in carbon dioxide concentration (CO_2). Carbon dioxide is a common part of the Earth's atmosphere, with fluctuating concentration in the air, depending on local conditions, altitude, and relative humidity of the air. The forecast of carbon dioxide concentration (CO_2) for the future can be realized through process modeling (Malindzak et al., 2015). As a result of industry emissions, its average concentration in the air continues to grow, which is one of the main causes of Earth's warming. Part of the solar energy is captured in the atmosphere and that effects warming of the Earth.

A hallmark of globalization is the creation of global or planetary structures in different areas of social reality and in different fields of human activity. The process of globalization is geared toward increasing the internal complexity of these structures as well as gradual interconnecting of subglobal structures. The dynamics of the

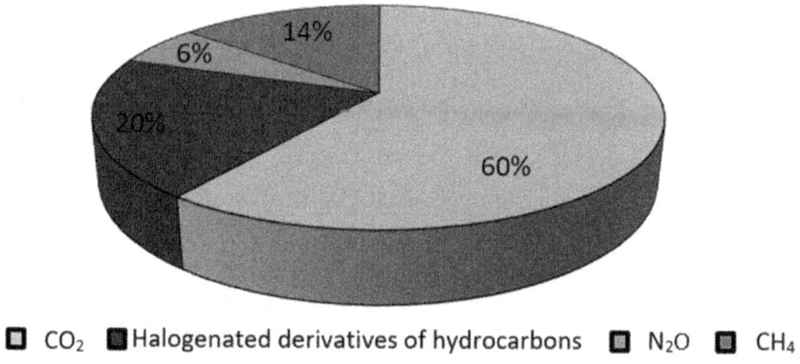

Figure 2.2 Influence of produced greenhouse gases on temperature changes. Source: author's own image.

development of individual partial globalization processes is realized by a procedure in which each subprocess affects all the others with its consequences, and the same applies to feedback.

Global warming needs to be addressed in terms of economic and environmental policy. Global warming is mainly caused by greenhouse gases, of which CO_2 is essential (Fig. 2.2).

Not all greenhouse gases have the same greenhouse effect. The same volume of methane is 23–25 times more effective, nitrous oxide (N_2O) 290–310 times, and freons are thousands to tens of thousands times more effective than CO_2. Some gases may capture or reflect heat more easily than others. Examples include water vapor, CO_2, methane (CH_4), nitrous oxide (N_2O), freons, and ozone. Other variables in the greenhouse effect influence are the amount of aerosols in the atmosphere and intensity of sunrays.

An important variable within the greenhouse effect is water vapor. Research results show that the natural greenhouse effect is made up to 60% of water vapor. Lomborg claims that "Strong water vapour feedback, however, is not primarily dependent on surface temperature, but mainly on temperature in the troposphere, which is the lowest part of the atmosphere, which spreads from the Earth's surface to the stratosphere to a height of 10–13 km." (Lomborg, 2006).

The intensity of sunrays has a direct effect on the amount of water vapor. Two-thirds of the Earth—of the blue planet—are covered by oceans and, if the solar activity is higher, the amount of water vapor in the atmosphere increases and thus amplifies the greenhouse effect. We cannot influence the vapors from the oceans and the intensity of sunshine.

The third influential variable is cloudcover. Amount of clouds affect the temperature on Earth—depending on their height and density, clouds contribute to both increase and fall of temperature.

Greenhouse gases make the climate warmer and, without the greenhouse effect, the average temperature on Earth would be lower than it currently is. The level of greenhouse gases is changing. A large part of the addition of CO of human origin comes from the combustion of carbohydrates (coal, oil, natural gas), the rest is due to deforestation (forests, which absorb CO_2, are being eliminated). It is necessary to note that not all CO_2 ends up in the atmosphere. If that were the case, the temperature on Earth would probably be significantly higher than it is today. More than half of the released CO_2 ends up in the oceans, the next part is absorbed by vegetation. CO_2 functions as a fertilizer for plants—at higher concentrations, their growth is faster and thus consumed faster. The cultivation of rice and wheat can be given as an example. These are two crops that feed humanity—they grow faster with increased CO_2 concentrations and offer higher yields too. The share of CO_2 emissions on Earth per person per year is about 5428 kg which makes 15 kg of emissions per day. Global CO_2 production amounts to 38 billion tonnes (Hodač and Kotrba, 2011).

Research shows that a person produces between 0.7 and 1.3 kg of CO_2 per day through his/her breathing. With an average mileage of 50 km per day, a car produces between 4 and 20 kg of CO_2 depending on consumption. For example, Skoda Fabia Greenline (a car developed to save not only financial resources but also the environment) produces 89 g of CO_2 for each kilometer. A comparison can be made between the CO_2 production of Skoda Fabia and BMW with an eight-cylinder engine, passing the same distance. While the Skoda Fabia produces 4.5 kg of CO_2, BMW with an eight-cylinder engine produces 12.8 kg CO_2. On planet Earth, humans produce about 38 Gton of CO_2 per year. The total volume of CO_2 in the atmosphere is higher by an order of magnitude, i.e., reaches approximately 3000 Gton, which is 80 times more. Research shows that only 1.2% of the total amount of CO_2 is of human origin. The remaining amount circulates in the atmosphere and biosphere, absorbed by oceans and various vegetation.

At present, the professional public is concerned with the issue of carbon footprint. A fuel-efficient car can drive more than 3 times more ecologically, i.e., with lower CO_2 production. Volcanoes around the world can be used as examples, which produced an estimated 200 million tonnes of CO_2 in 2010. Iceland's volcano Eyjafjallajökull itself spewed 300,000 tonnes of CO_2 in its explosions. For the sake of comparison, all European airlines release approximately 440 1000 tonnes of CO_2 per year (Hodač and Kotrba, 2011).

The best-known document aimed at stabilizing CO_2 emissions is the Kyoto Protocol. The Kyoto Protocol was adopted in 1997 in Kyoto. The purpose of the Kyoto Protocol is to specify new flexible instruments whose common objective is to achieve the maximum reduction potential in the most economically efficient way. The Kyoto Protocol also includes the issue of eliminating the production of another greenhouse gas, namely freon. In amendment to the Kyoto Protocol from December 2012, adopted in Qatar's Doha, the decision to continue the Protocol was taken and a second binding 8-year term (2013–20) was set. For the second period of the Kyoto Protocol, the 2020 emission reduction targets have been adopted according to the climate and energy package, i.e., a 20% reduction in greenhouse gas emissions compared to 1990 levels.

The Kyoto Protocol requires especially:

1. **Increase in energy efficiency**—investing in new technologies. In areas with capital shortage, it is a rather improbable solution.

2. **Promoting a sustainable way of forestry, afforestation, and the restoration of forest stands**—an idea feasible without major problems.
3. **Promoting sustainable forms of agriculture**—minimizing large-scale monocultures.
4. **Support for research and increased use of new and renewable forms of energy**, CO_2 separation technologies, and modern and innovative environmentally friendly technologies.
5. **Limiting and gradually exclusion of tax and tariff concessions and subsidies in sectors that emit greenhouse gases**—it can lead to the growth of prices of "traditional fuels" (mainly coal), which can create problems for the socially weaker sections of the population.
6. **Limiting or reducing methane emissions in waste management by capturing and using it, in production, transport, and energy distribution**—the use of landfill gas, that is common in some EU countries.

Freons are hydrocarbon halogen derivatives. In addition to carbon, they also contain two halogens, one of which is fluorine. Freons have no taste or color, they are not even flammable or toxic. Due to their excellent insulation properties, for many years they were used in cooling equipment, i.e., refrigerators, freezers, air conditioners, etc., which had a negative impact on the environment. In addition to the greenhouse effect, they contributed to the disruption of the ozone layer in the atmosphere.

In the mid-1980s, the objective of limiting the amount of freons discharged into the atmosphere was defined under the auspices of the UN, specifically UN Environment Programme. This objective was specified in 1987 in the famous Montreal Protocol. A total of 180 states have committed to limit the production of both "hard freons" (CFC-freons) and soft freons (HCFC). The Montreal Protocol was ratified by 196 states, with the UN having 193 members. States that have not been de jure recognized are also part of the UN.

The main sources of greenhouse gases from human activity are

- combustion of fossil fuels (coal, oil, and gas) in electricity generating, transport, industry, and households (CO_2);
- agriculture (CH_4) and changes in land use, such as deforestation (CO_2);
- waste dumps (CH_4);
- the use of fluorine-containing industrial gases.

(www.3).

There is a strong link between economic growth, rising energy consumption, pollution, and global warming. The mining, processing, and consumption of energy sources causes environmental interference, be it pollution itself or various other geomorphological and environmental interventions. Pecho (2012) claims that the links between these factors can be reduced, for example, by switching to better fuels or by applying technological changes aimed at improving economic performance as well as reducing pollution.

Any form of extraction of mineral resources cannot take place without interference with the environment. However, the environmental impact is not only due to the mining activity itself but also due to the subsequent processing, especially the waste from the treatment in both solid and liquid forms. Mineral extraction and the related development of industry, construction, transport, and agriculture constitute human intervention in the lithosphere, which, apart from the relocation of about 10 km^3 lithospheres, results in disruption of the water regime and geochemical and geomorphological changes. Under their influence the landscape changes in height, climate, and hydrological character, thus subjecting the emergence of new landscape habitats. The mining industry is constantly changing the original environment because every mineral deposit is nonreproducible and irreplaceable after extraction. Every extraction of mineral resources results in a change in the environment which (contrary to the influence of the processing industry) may not always be permanently negative. In part, the extraction of mineral resources also reflects in air pollution by various gases (CO_2, CO, SO_2, SO_3, N_2, NO, NO_2, NO_3). Dust pollution arises from blasts, especially chamber blasts, stone processing (shredders, sorters), and also in traffic on dusty roads. Gaseous air pollution is negligible in the extraction of raw materials and is mainly caused by gases released from used explosives, from the transporting vehicles, and from the oxidation of mined ore. Each of these methods of pollution, as well as their combination, has a negative impact not only on vegetation but also on human health.

Research in many scientific disciplines shows what the negative consequences of the planet warming can be (climate change), whether through greenhouse gases or thermodynamics and its laws (see Section 10 in Chapter 19). This applies not only to the ecological area (e.g., loss of biodiversity) but also to the social and economic spheres—the damage from natural disasters, changes in agriculture and industry, etc. The increase in quantified damage caused by natural disasters logically results from an increase in property damage (an amount of damaged material goods). In turn, the large amount of property which natural disasters can damage (or completely destroy it) is linked to population growth and increasing material consumption (construction of new residential areas, shopping centers, public institutions, infrastructure, etc.).

The second part of the Report of the Intergovernmental Panel on Climate Change, under the auspices of the United Nations, is one of the most complex reports examining the impact of climate change. The report is compiled from several scientific studies by renowned scientists and highlights the following problems:

***Coastlines and* low-lying areas**—hundreds of millions of people will be forced to move out of coastal areas which will become hit by rising sea and river levels. This will particularly affect parts of South-East Asia. A rise by a few centimeters may mean a disaster for countries such as Netherlands.

Food security—an increase in temperature by 1°C will have a negative impact on agricultural crop yields. Harvests of rice, grain, or corn may decline. The reduction in harvest caused by climate change and the growing demand for food will make the impact noticeable.

Human health—climate change on Earth will increase susceptibility to diseases. Heat waves and fires will increase the risk of malnutrition, diseases that the youngest children in particular will experience. The number of malnourished children under the age of five is expected to rise to between 20 and 25 million in the coming years.

***Lack* of drinking water**—greenhouse gases and the consequent rise in temperatures will result in the drying up of vitally important water. This will have devastating consequences in dry regions. The fight for water will escalate and its price will rise. Animal species directly dependent on water will face an existential risk.

Global security—migration will have a profound impact on the security of humanity. It will trigger political divisions leading to conflicts. Antimigrant sentiments and opinions will gradate and then the social tensions may occur. The situation will ultimately result in "economic shocks" and imbalances.

Unique ecosystems—because of weather fluctuations rare parts of the ecosystem may lose their unique character and gradually disappear.

Global economy—an increase in global temperatures by 2.5 will lead to an overall economic loss of between 0.2% and 2% of gross domestic product.

(Climate Change, 2014).

According to a new study by Professor Garrett from the University of Utah, an effort to reduce greenhouse gases emissions, which cause warming, of the planet does not have much sense. Scientists are of opinion that warming will not stop if economies all over the world do not refrain from industrial production. This scientific study in question, entitled *The Importance of Oil in the Global Economy*, is published in the Climatic Change journal, edited by leading climatologist Stephen Schneider of Stanford University. "*When asked whether and how to limit ever-increasing carbon dioxide emissions, we have a choice between a collapse of the economy and unrealistically large-scale building of energy capacity. It doesn't look likely to be a significant departure from the observed increase in carbon dioxide emissions for the foreseeable future*," Garrett (2012) says in the study.

Energy saving and the growth of energy efficiency in Garrett's view (2012) do not reduce the energy demand but rather stimulate economic growth and larger energy consumption. Garrett based his theory on the knowledge that energy savings do not actually lead to overall decrease of energy consumption but, to the contrary, promote economic growth and subsequently the energy consumption rises, corresponding to the accumulated economic activity.

The research of Nátr (1998) confirms that the short-term positive effects of technological progress in a long run bring negative consequences, and the later the negative consequences of short-term positive phenomena show the more persistent or even irreversible they can be. Like Pecho (2012) he argues that the impact of energy use on the living environment may change over time with regard to technological changes and innovations which may reduce emissions of individual polluting substances as well as energy requirements of production. Of course, in the case of restrictions on substitution or technological change, the potential reduction of negative environmental impact is limited and innovations that reduce one type of emissions (CO_2) can generate a different type of waste. The problem of tackling climate change is also challenging because almost all energy consumption and social welfare of modern civilization is now linked to greenhouse gas production and the subsequent global warming of the Earth. Changes in the orientation at other ways of energy use are costly so far or only applicable to a limited part of population (e.g., geothermal energy in Iceland).

The diagram produced by Donner (2007), confirms Garrett's (2012) and Nátr's (1998) research projects and already mentioned Jevons' paradox and its manifestations—as energy efficiency increases, CO_2 emissions increase as well.

Despite the link between energy consumption and economic growth, there certainly are opportunities to reduce the negative impact on the environment, and tackling this issue is a subject matter of many scientific, professional discussions, and the proposals of solutions resulting from them.

Not all the effects of energy consumption must be equally harmful to the environment and people's health. The shift from lower to higher quality of energy sources not only reduces the total energy needed to produce a GDP unit, but can also reduce the environmental impact by the remaining unused energy. An example is the use of natural gas, which has a demonstrably cleaner combustion and produces less CO_2 per unit of energy obtained compared to coal. Nuclear energy is also more cost-effective than coal, but its long-term environmental impact may not be cleaner than that of coal.

Against this background, it is evident that the problem-solving proposals require a comprehensive approach and also a change in the energy infrastructure to date.

2. Conclusion

In accordance with the available scientific analyses and the results of the scientific and professional projects, which were the subject of examination, comparison, and synthesis in the present work, it can be concluded that the currently preferred and supported stable economic growth of exponential nature in the confined environment of the Earth is not possible, not only from a mathematical but also from a physical point of view. Economic growth is making entropy increase, so economy should respect thermodynamic laws and make efforts to reduce entropy by effectively assessing the energy entering and operating in economic processes. Back in 1986, Georgescu-Roegen proposed the concept of bioeconomy, which understands operation of the economic system as part of the nature system and considers the second law of thermodynamics to be the basic law of functioning of the economy.

On the basis of the abovementioned results of research, it can be stated that the current main objective of economics, which is stable economic growth (as measured by the GDP indicator), needs to be changed, as it is not only contrary to natural laws but also makes it difficult to ensure economic growth, which would result in an increase in the social well-being and quality of the population's life in all its aspects. Following the change in the main objective of economics, it is also necessary to replace the current concept of the economic process (according to which demand stimulates supply/production/and production provides the income needed to feed further demand in an endless process) with a sustainable concept in which the economic process is set in the biophysical environment that sustains this economic process.

In the future, it will be necessary to adapt and reflect on the changing economic, demographic, social, and environmental aspects, resulting from the increasing consumption of natural and energy resources due to the preference for economic growth.

This chapter is part of the solution of scientific project KEGA 026EU-4/2018 and it was issued with ist support.

References

Bartlett, A., 2012. Arithmetic, population and energy, Odborná prednáška amerického profesora Alberta Bartletta Expert lecture. Dostupné na: http://www.priateliazeme.sk/cepa/sk/informacie/temy/922-aritmetika-populacia-a-energia.

Bednárová, L., Liberko, I., Rovňák, M., 2013. Environmental benchmarking in small and medium sized enterprises and there impact on environment. International Multidisciplinary Scientific GeoConference-SGEM 141–146. https://doi.org/10.33543/1001.

Carson, R.T., 2010. The environmental kuznets curve: seeking empirical regularity and theoretical structure oxford journals. Rev. Environ. Econ. Pol. ISSN: 1750-6824 4 (1), 3–23 published online December 22, 2009.

Climate Change, 2014. Impacts, Adaptation, and Vulnerability, Summary for Policymakers. Dostupné na: http://ipcc-wg2.gov/AR5/images/uploads/WG2AR5_SPM_FINAL.pdf.

Donner, S., 2007. Emissions Intensity: Declining for Decades. Dostupné na: http://simondonner.blogspot.sk/2007/05/emissions-intensity-declining-for.html.

Frankel, J., 2014. Omyl-o-oligarchii. Dostupné na: http://openiazoch.zoznam.sk/cl/144373/Omyl-o-oligarchii.

Garrett, T.J., 2012. The Physics of Long-run Global Economic Growth. Dostupné na: http://www.inscc.utah.edu/~tgarrett/Economics/Economics.html.

Gajdoš, J., Naščáková, J., Andrejovský, A., Ručinský, R., 2011. Ener supply - dôvody zapojenia sa do projektu. In: Nekonferenčný recenzovaný zborník v rámci riešenia projektov VEGA, KEGA, APVV, ENER SUPPLY. Vydavateľstvo EKONÓM, Bratislava, ISBN 978-80-225-3207-5, pp. 33–40.

Hegyi, L., 2012. Návrh modelu vzťahov prírodných limitov a trvaloudržateľného rozvoja spoločnosti (Design of a model of the relationship between natural limits and sustainable development of society), diploma thesis. Diplomová práca, Technická univerzita v Košiciach, Strojnícka fakulta, Katedra environmentalistiky, Košice.

Hodač, J., Kotrba, T., 2011. Globalizace. Barrister & Principal, Brno.

Johanisová, N. (2018). Je čas ukončit závislost Evropy na růstu. Otevřený dopis vědců a vědkyň. Available at: https://www.novinky.cz/kultura/salon/clanek/je-cas-ukoncit-zavislost-evropy-na-rustu-otevreny-dopis-vedcu-a-vedkyn-40208206.

Malindzak, D., Saderova, J., Vitko, D., Malindzakova, M., Gazda, A., 2015. The methodology and model for in-process inventories calculation in the conditions of metallurgy production. Metalurgija 54 (1), 227–230.

Musil, P., 2009. Globální energetický problém a hospodářská politika: se zaměřením na obnovitelné zdroje. C.H. Beck, Praha.

Naščáková, J., Červenka, P., 2009. Spoločenský blahobyt a limity ekonomického rastu. In: SEMAFOR 2009 - Medzinárodna vedecká konferencia. Michalovce, ISBN 978-80-225-2841-2.

Norma EN Norma EN ISO 26 000:2010 Guidance on Social Responsibility, n.d.

Nátr, L., 1998. Rostliny, lidé a trvale udržitelný život člověka na Zemi. Skripta. Nakladatelství Karolinum, Praha, ISBN 80-7184-681-3.

Lomborg, B., 2006. Skeptický ekolog. In: Jaký je skutečný stav světa? Liberální institute, Vydavateľstvo Dokořán.

Pecho, J., 2012. Zmena klímy - globálny problém s lokálnymi dopadmi. In: Klimatická zmena a lokálny rozvoj — výzva pre samosprávy: Príručka pre samosprávy. Karpatský rozvojový institut, Košice.

Rockström, J., et al., 2009. Planetary boundaries: exploring the safe operating space for humanity. Ecol. & Soc. 14 (2), 32. Dostupné na: http://www.ecologyandsociety.org/vol14/iss2/art32.

Simon, J.L., 2006. Největší bohatství. Centrum pro studium demokracie a kultury, Brno, p. 666. ISBN 80−7325-082-9.

Stockholm Resilience Centre, 2007. About Stockholm Resilience Centre. Dostupné na: http://www.stockholmresilience.org/aboutus.4.aeea46911a3127427980003326.html.

Vlčková, E., 2014. S uhlím vypouštíme džina z lahve, říká švédský ekolog. Lidové noviny, Praha. Dostupné na: http://www.lidovky.cz/s-uhlim-vypoustime-dzina-z-lahve-rika-svedsky-ekolog-pcs-/ln_veda.asp?c=A100510_114257_ln_veda_ev.

www.1, n.d. http://openiazoch.zoznam.sk/cl/144373/Omyl-o-oligarchii.

www.2, n.d. http://www.stockholmresilience.org/aboutus.4.aeea46911a3127427980003326.html.

www.3. n.d. http://www.eea.europa.eu/sk/themes/climate/intro.

www.7. n.d. https://zelenazeleni.sk/klimaticka-zmena-riesenia-udrzatelne-lesy/.

www.8. n.d. https://www.pefc.sk/obhospodarovanie-lesov/trvalo-udrzatelne-obhospodarovanie-lesov/kriteria-a-indikatory.

www.10. n.d. https://www.lesy.sk/o-nas/eticky-kodex/.

Further reading

Abulfotuh, F., 2007. Energy efficiency and renewable technologies: the way to sustainable energy future. In: Desalination the International Journal on the Science and Technology of Desalting and Water Purification. ISSN: 0011-9164, vol. 209. Elsevier B.V., pp. 275−282

Ayres, R.U., Kneese, A.V., 1996. Production, consumption, and externalities. Am. Econ. Rev. 59 (3), 282−297. Pub ID: 103-352-001.

Baek, J., Kim, H.S., 2013. Is economic growth good or bad for the environment? Empirical evidence from Korea. Energy Econ. 36, 744−749. Dostupné na: http://www.sciencedirect.com/science/article/pii/S0140988312003180.

Baranzini, A., Weber, S., Bareit, M., Mathys, N.A., 2013. The causal relationship between energy use and economic growth in Switzerland. Energy Econ. 36, 464−470. Dostupné na: http://www.sciencedirect.com/science/article/pii/S0140988312002290?np=y.

Bartlett, A., 2006. A depletion protocol for non-renewable natural resources: Australia as an example. In: Natural Resources Research, vol. 15, pp. 151−164. https://doi.org/10.1007/s11053-006-9018-1 (3).

Bednárová, L., Witek, L., et al., 2016. Assessment methods of the influence on environment in the context of eco-design process. In: Majerník, M., et al. (Eds.), Production Management and Engineering Sciences, 1st. CRC Press.

British Petrol, 2011. BP Statistical Review of World Energy. Dostupné na: http://www.bp.com/liveassets/bp_internet/globalbp/globalbp_uk_english/reports_and_publications/statistical_energy_review_2011/STAGING/local_assets/pdf/coal_section_2011.pdf.

Daly, H.E., Cobb JR., J.B., 1989. For the Common Good: Redirecting the Economy toward Community, the Environment, and a Sustainable Future. Beacon Press, Boston.

Heinberg, R., 2011. The End of Growth: Adapting to Our New Economic Reality. New Society Publishers, Gabriola Island, Kanada, ISBN 978-0-86571-695-7, 336s.

Howarth, R.B., 1997. Energy efficiency and economic growth. Contemp. Econ. Pol. 25, 1–9.

Islam, S.M.N., Clarke, M., 2002. The relationship between economic development and social welfare: a new adjusted GDP measure of welfare. In: Social Indicators Research, pp. 201–216.

Judson, R.A., Schmalensee, R., Stoker, T.M., 1999. Economic development and the structure of demand for commercial energy. Energy J. 20 (2).

Kenny, C., 2005. Does development make you happy? Subjective wellbeing and economic growth in developing countries. In: Social Indicators Research, vol. 73, pp. 199–219. Roč, 29–57.

Klinec, I., 2005a. Zelené myslenie, zelená budúcnosť. Alternatívne ekonomické a sociálne teórie podporujúce smerovnaie k udržateľnému rozvoju, vol. 258. Univerzita Palackého Olomouc, Olomouc, Česká republika.

Klinec, I., 2005b. Alternatívne ekonomické teórie podporujúce smerovanie k trvalo udržatelnému rozvoju. In: Sborník z projektu „K uržitelnému rozvoji Ceské republiky: vytvárení podmínek". Univerzita Karlova v Praze, Centrum pro otázky životního prostředí, Praha, pp. 52–119. Svazek 2: Teoretická východiska, souvislosti, instituce.

Lacy, P., Rutqvist, J., 2016. Waste to Wealth: The Circular Economy Advantage. Springer, ISBN 1137530707, pp. 2016–2264.

Lequiller, F., 2005. Is GDP a satisfactory measure of growth?. In: The OECD Observer, 246/247, pp. 30–31.

Lisý, J., 1999. Výkonnosť ekonomiky a ekonomický rast. Iura Edition, Bratislava.

Malindžáková, M., 2018. Kvalita logistických systémov. Edičné stredisko F BERG, Košice, ISBN 978-80-553-2664-1.

Mazanec, P., 2011. Odklon od jaderné energie zvyšuje poptávku po dřevu. Praha, Dostupné na: http://ekolist.cz/cz/zpravodajstvi/zpravy/odklon-od-jaderne-energie-zvysuje-poptavku-po-drevu.

Meadows, D., et al., 1972. The Limits to Growth. Universe Books, New York.

Mojžiš, M., 2012. Nízky uhlík. Dostupné na. http://blog.jetotak.sk/kriticka-ekonomia/2012/08/20/miroslav-mojzis-nizky-uhlik/.

Naščáková, J., Pčolinská, L., Gajdoš, J., 2011a. Kvalita života a jej dimenzie. In: Nekonferenčný recenzovaný zborník čiastkové výsledky riešených projektov „VEGA, KEGA, APVV, ENER SUPPLY". Ekonomická univerzita v Bratislave Podnikovohospodárska fakulta v Košiciach, Košice, ISBN 978-80-225-3207-5, pp. 56–65.

Naščáková, J., Pčolinská, L., Gajdoš, J., 2011b. Vybrané spôsoby merania ekonomického rastu a spoločenského blahobytu. In: Nekonferenčný recenzovaný zborník v rámci riešenia projektov VEGA, KEGA, APVV, ENER SUPPLY. Vydavateľstvo EKONÓM, Bratislava, ISBN 978-80-225-3207-5, pp. 66–70.

Naščáková, J., Pudło, P., 2011. Meranie kvality života z pohľadu ekonomicko-politického prostredníctvom ukazovateľov hrubého domáceho produktu, indexu udržateľného hospodárskeho blahobytu a reálneho pokroku. In: Nekonferenčný recenzovaný zborník v rámci riešenia projektov VEGA, KEGA, APVV, ENER SUPPLY. Vydavateľstvo EKONÓM, Bratislava, ISBN 978-80-225-3207-5, pp. 71–78.

Rifkin, J., 2011. The Third Industrial Revolution, How Lateral Power is Transforming Energy, the Economy and the World. Palgrave Macmillan, New York, 291 s.

Rowe, J., 2009. The Cult of GDP/Gross Domestic Product ignores Wealth Generated by the Commons/. Dostupné na: http://onthecommons.org/cult-gdp-0.

Sachs, J.D., 2014. Karbónová loby zabíja planétu. Dostupné na: http://komentare.hnonline.sk/komentare-167/karbonova-loby-zabija-planetu-603568.

Správa OECD, 2011. Towards Green Growth, ISBN 978-92-64-094970.

Stern, D.I., 2004. Economic growth and energy. In: Encyklopedia of Energy, vol. 2.

Steard, J.G., Steard, W.E., 2012. Manažment pre malú planétu - Prečo je dôležité meniť stratégie neobmedzeného rastu na stratégie udržateľnosti. Vydavateľstvo, Eastone, ISBN 9788081092169.

Willquist, K., 2012. Vodík: zdroj zelené energie v budoucnosti? Dostupné na: http://www.scienceinschool.org/2012/issue22/hydrogen/czech.

Záverečná správa za celú dobu riešenia projektu VEGA 1/0339/10, doba riešenia od 01/2010 do 12/2011 Ekonomický rast a jeho limitujúce faktory - návrh nových ekonomických cieľov a indikátorov kvality života potrebných pre vytvorenie všeobecne platnej, novej metodiky hodnotenia kvality života a trvalo udržateľného rozvoja na území Slovenskej republiky, 2011. Vedúci projektu Jana Naščáková. - Košice, 11s.

Evolution and trends of sustainable approaches

M. Mar Serrano-Arcos, Belén Payán-Sánchez, Ana Labella-Fernández
University of Almería, Ctra. Sacramento s/n, Almería, Spain

1. Introduction

Sustainable development has emerged as an influential concept for business and policy and is considered the solution to create a promising and prosperous future for humankind. There is growing awareness of the fact that a fundamental transformation in the way societies behave may be needed if we are to make progress toward sustainable development.

Environmental degradation is progressively becoming a significant threat to all humankind. What is more, social inequalities still prevail as a result of the "success" of the dominant industrialized society (Beck, 2008, p. 78). From the publication of the Brundtland Commission's Report, "Our Common Future" by the World Commission on Environment and Development (WCED) in 1987 for the publication of the Sustainable Development Goals (SDGs) by the United Nations in 2015, many initiatives have been taken to address environmental, economic, and social challenges in order to implement sustainable development agendas. An increasing number of organizations have realized that economic development must consider social and ecological issues simultaneously. Although many efforts have been made thus far, a great deal more remains to be done. Solving environmental and social issues requires changes in conventional policies that have proven insufficient to fight major sustainability-related problems.

The motivation of scholars, companies, and entrepreneurs to reach sustainability is closely linked to the economic objectives of firms, such as obtain a competitive advantage, be market leaders, and generate fewer negative effects on society (De Freitas et al., 2017). Bebbington et al. (2007) highlight the desires of companies and individuals to develop appropriate approaches according to initiatives in favor of sustainability. Additionally, in several business domains there is a substantial need for innovative approaches to achieve sustainability (Yang et al., 2017). In this sense, products, services, and technologies as well as business models are among the approaches for sustainable solutions discussed in this chapter. To attain sustainability, innovative sustainable products, services, technologies, and business models could be formulated taking into consideration not only environmental and social issues but also institutional and economic necessities. In this chapter, Product-Service System (PSS), Circular Economy (CE), and Industrial Symbiosis (IS) are among the sustainable business models explored.

Sustainable Resource Management. https://doi.org/10.1016/B978-0-12-824342-8.00013-4

Sustainability assessment and reporting has become of great concern among scholars and businesses since they were determined to be key to supporting all levels of firm decision-making and policy processes with the ultimate purpose of improving the management of human and natural systems, as well as obtaining more sustainable results that have fewer harmful impacts on the environment and society (Bond et al., 2013; Gibson, 2013; Ramos, 2019). In this sense, several approaches and tools have been developed to assess sustainability, among the most notable of which are sustainability indicators (SIs). These indicators are one of the most widely used approaches and are crucial for evaluating sustainability within decision-making processes and specially for communicating firm sustainability performance to stakeholders and the progress toward the achievement of sustainable development (Pope et al., 2017; Ramos, 2019; Sala et al., 2015). However, practitioners have at their disposal a wide variety of indicators and approaches, each one developed for a specific goal, thus there is no "one sustainability indicator to rule them all" (Bell and Morse, 2018b, p. 1). Consequently, several authors have recognized the need to define clear, straightforward, and robust frameworks to introduce the indicators (Bell and Morse, 2013; Gibson, 2013; Ramos, 2019).

Additionally, one methodology has appeared in literature to measure sustainability based on life cycles, i.e., Life Cycle Sustainability Assessment (LCSA), which combines the three dimensions of sustainability. Therefore, LCSA is composed of three methods: Life Cycle Assessment (LCA), social Life Cycle Assessment (S-LCA), and Life Cycle Costing (LCC), following the ISO 14040 framework (ISO, 2006a, b; Sala et al., 2013; Wulf et al., 2019).

Thus, the main objective of this chapter is to review the evolution and trends of sustainable development. Furthermore, this research aims to find approaches for firm sustainability assessment, as well as several sustainable solutions. For this purpose, the different types of approaches to sustainability evaluation will be discussed in terms of SIs and life cycle–based methodologies, as well as the types of approaches that exist for sustainable solutions.

The remainder of this chapter is structured as follows: Section 2 outlines the emergence and evolution of the sustainable development concept and agendas. Section 3 presents an overview on SIs and their relevance for sustainability assessment as well as a specific sustainability assessment approach based on life cycles—LCSA—which encompasses the three dimensions of sustainability. Section 4 discusses various sustainability solutions to be applied in firms regarding product, services, and technologies as well as novel business models. Finally, the main conclusions are detailed in Section 5.

2. Emergence of sustainable development

The idea of sustainable development gained widespread recognition with the publication of the Brundtland Commission's Report, "Our Common Future" in the 1980s. In the report, the WCED defines sustainable development as "development that meets the needs of the present without compromising the ability of future generations to meet their own needs" (WCED, 1987, p. 43). Two main components can be drawn from this definition. The first is the fulfillment of human needs. The report contends that economic growth is necessary as it is the main strategy that nations have to develop in order to avoid poverty and inequality and, in addition, to build fair opportunities and increase society's production capacity for all to satisfy their basic needs. The second component is the limited capacity of the natural environment to satisfy the basic needs and aspirations of future generations. The report argues that economic growth must be environmentally and socially sustainable to ensure the satisfaction and aspirations of humankind in the future (Akamani, 2020).

Although the concept of sustainable development proposed by the Brundtland Commission has had a great impact on policies concerning development and the environment, it has been subject to some criticism and has been regarded as a "weak sustainability" (Hediger, 1999). The inherent tensions in the concept of sustainable development are a lack of conceptual clarity; an absence of theoretical foundations; addressing the three pillars (social, environmental, and economic) independently; difficulty in practical application; a lack of recognition of the interactions between humans and nature; and a lack of consideration for the diversity of cultures and needs (Akamani, 2020; Burns and Witoszek, 2012; Fergus and Rowney, 2005; Giddings et al., 2002; Lélé, 1991; Stafford-Smith et al., 2017; Woods, 2006).

The Earth Summit, the United Nations Conference on Environment and Development was held in Rio de Janeiro in 1992 with the resulting "Agenda 21, Rio Declaration" (United Nations, 1992). The Agenda 21 was aimed to propose a comprehensive global action plan to promote participation and cooperation as one of the key strategies for achieving sustainable development. This plan was gradually adopted by multiple organizations of the United Nations System, governments, and other groups.

In 1997, Elkington introduced the concept of the triple bottom line with the objective to operationalize sustainability. The three pillars include environmental, social, and economic criteria (Elkington, 1997). Environmental sustainability focuses on human activity within the carrying capacity of the natural environment (in relation to, for example, water, land, energy, and materials). Environmental issues include climate change, biodiversity, water pollution, ozone layer depletion, waste management, etc. Economic sustainability focuses on the efficient use of resources to improve operational profit and maximize market value (Olawumi and Chan, 2018). Social sustainability considers the social well-being of human life. It is aimed to balance the needs of individuals with the needs of groups (Olawumi and Chan, 2018). Social issues include labor conditions, child labor, human rights, health and safety, minority development, the inclusion of disabled/marginalized people, and gender inequality (Yawar and Seuring, 2017).

As a major step toward sustainable development, the United Nations Millennium Declaration, signed in September 2000, adopted by the 191 United Nations members, IMPY, OECD, and the World Bank, led to the adoption of the Millennium Development Goals (MDGs). This declaration is an international framework consisting of 8 goals and 18 targets, complemented by 48 technical indicators to measure progress toward the MDGs. The eight MDGs include the eradication of extreme poverty and hunger, the promotion of universal primary education, improvement in maternal health, and the promotion of environmental sustainability, among others. The MDGs were developed at different levels by international, national, regional, municipal, and individual organizations.

At the Rio +20 Earth Summit conference, which was held in Rio de Janeiro in 2012, it was determined that significant progress had been made in the attempt to achieve MDGs, but this progress varied across goals, countries, and regions. Members States therefore highlighted the need to replace the MDGs.

In this regard, the United Nations Sustainable Development Summit in September of 2015 was designed to renew the MDGs. The document entitled "Transforming our world: the 2030 Agenda for Sustainable Development" defines 17 goals and 169 tasks, covering environmental, social, and economic issues which include a wide range of challenges such as poverty, food supply, education, gender, equality, energy, climate change, biodiversity, and economic growth (United Nations, 2015). These SDGs renew the sustainable development agenda and require that nations, communities, businesses, and organizations rethink and reconsider their sustainable development policies.

3. Approaches to sustainability assessment: sustainability indicators and assessment based on life cycles

3.1 Sustainability assessment and indicators

One of the main aims of sustainability assessment and monitoring and reporting approaches is to support firms' decision-making and policy processes (Ramos, 2019), ultimately improving management of socioecological systems and obtaining more sustainable results with fewer negative consequences (Bond et al., 2013; Gibson et al., 2013; Gibson, 2013). In this sense, sustainability assessments play a key role in planning and project processes at strategic and operational levels, with projects, programs, policies, plans, and activities/operations that address different goals and targets to reach sustainable development, regardless of their specific context and objective (Ramos, 2019).

Although a great variety of tools and methods have been conceived to evaluate and report sustainable development, SIs appear as one of the most used approaches, proving crucial for the assessment of sustainability within decision-making processes and especially for the communication of firm sustainability performance to stakeholders and progress toward sustainable development (Pope et al., 2017; Ramos, 2019; Sala et al., 2015; Singh et al., 2012). In fact, as Bell and Morse (1999, p. 11) expressed, "SIs were spawned from an understandable desire to 'do' sustainable development." As these indicators are mostly promoted at regional and supranational levels, in order to be well accepted by the stakeholders of each national government, these indicators ought to completely consider the different conditions and concerns of every nation (Tasaki and Kameyama, 2015).

Although some of the first important references to indicators from the 1970s mainly focused on environmental aspects (Ramos, 2019), in recent decades, there has been an expansion of SI initiatives internationally (e.g., Bell and Morse, 2018b; Hezri and Hasan, 2004; Ramos, 2009; Wilson et al., 2007). All these initiatives vary between transnational, national, regional, and local levels, as well as organizations (companies, universities, or NGOs), economic sectors, ecosystems, communities, families, and individuals (Ramos, 2019). More specifically, some of the most widely known approaches and indicators that have become popular are the Human Development Index (HDI) and a derivative called the Human Sustainable Development Index, the Ecological Footprint and the Environmental Performance Index, or its precursor Environmental Sustainability Index. At the same time, several targets and indicators have emerged which are linked to the concept of Planetary Boundaries (Steffen et al., 2015) or the SDGs (Bell and Morse, 2018b).

Ramos (2019) highlights some specific challenges for SIs, namely the widespread use of different definitions in literature and among practitioners for sustainable development, sustainability or even the indicators. This may be due to the uncertainties and complexities associated with these concepts, in addition to the need to balance several different dimensions (Bolis et al., 2014). Although SIs reflect the most studied issues and paradigms in practice, some authors such as Ramos (2019) and Viegas et al. (2018) express the need for a more transdisciplinary and flexible approach to deal with the real-world concerns and the goals of sustainability science (Lang et al., 2012).

Additionally, several studies in recent decades have also addressed the most relevant challenges that SIs face, some of which are related to the use itself of these indicators (Bell and Morse, 2013; Gibson, 2013). This translates into the need to define clear, straightforward, and robust frameworks to introduce the indicators, supported by the commitment of the individuals who are engaged in the indicator process (Ramos, 2019). Actually, as Bell and Morse (2018b, p. 1) asserted, "there is 'no one SI to rule them all' but a wide diversity of approaches and indicators, each emerging in their own time and space and designed to meet a defined set of objectives."

To address the need previously mentioned, in 2015, the United Nations Statistical Commission created an Inter-agency and Expert Group on Sustainable Development Goal Indicators (IAEG-SDGs), including Member States and regional and international agencies as observers, to give a proposal to create an SDG indicator framework (e.g., UNSC, 2015). With the adoption of the SDGs, the United Nations General Assembly signaled the journey toward a sustainable future (Tasaki and Kameyama, 2015). These targets and indicators demand taking into account the needs and desires of all countries, both developed and developing. In fact, it was thought that the dominant view of sustainability, and the SIs related, would arise out of a negotiation among the visions of many groups and individuals (Bell & Morse, 1999, 2018a).

To help in the identification of SIs and contribute to provide a clearer framework for their use, Tasaki and Kameyama (2015) conduct a survey to detect the sustainable development indicators adopted by international organizations, regions, and national governments. Together with the results of two other previous surveys (Tasaki et al., 2010), 1848 indicators were identified from 28 countries and international organizations, and then categorized into the 77 subcategories placed under the UNCSD economic, environmental, social, and institutional categories (UNCSD, 2001).

Based on the results of the analysis, Tasaki and Kameyama (2015) discovered the following:

- Within the economic category, the most common subcategories were "energy use," "transportation," and "economic performance." "Energy use" indicators focused on energy consumption, supply and intensity, as well as energy resources, energy efficiency, and energy price. In "transportation," indicators focused on travel distance, accessibility, share of specific transportation mode, traffic congestion, noise, emissions, and traffic energy consumption. For "economic performance," the most commonly used indicator is GDP.
- For the environmental category, "climate change," "ecosystems," and "agriculture and livestock" were found to be the most common subcategories. Most of the indicators for "climate change" focused on greenhouse gas and CO_2 emissions, together with some indicators of temperature and emissions per GDP or per capita to a lesser extent. Indicators within the "Ecosystem" subcategory are related to protected regions, the number of inhabitants of specific species, and diversity. Lastly, for "agriculture and livestock," many indicators focused on agricultural land use, organic agriculture, and the material balance of nitrogen and phosphorus.
- In the social category, the most commonly observed subcategories were "mortality, life expectancy, and health," "education," and "work." In the first subcategory, many indicators included life expectancy, mortality rate, and death causes from specific sicknesses. In "education," compulsory and lifetime education together with labor training were covered. Finally, "work" indicators focused on unemployment rate, long-term unemployment rate, and working poor.
- Finally, within the institutional category, "Environmental management and policy" and "Science and technology" were the most common subcategories found. In the former, indicators were mostly focused on several management systems (i.e., ISO 14000 series), while in the latter indicators such as "expenditure on research and development" and "patent registrations" were observed.

3.2 Sustainability assessment approach based on life cycles (LCSA)

As previously seen, sustainability assessment has been widely discussed in literature and industry, with many different approaches developed for its application (Zijp et al., 2015). Therefore, concepts able to integrate the three dimensions of sustainability (i.e., environmental, social, and economic) are needed (Wulf et al., 2019), and LCSA appears as one methodology with the possibility to evaluate sustainability in these circumstances. This approach stems from Life Cycle Thinking, thus conceiving the system (either a product, service, or the organization itself) from cradle to grave.

In this sense, LCSA can be defined as the "evaluation of all environmental, social and economic negative impacts and benefits in decision-making processes towards more sustainable products throughout their life cycle" (Valdivia et al., 2011, p. 3). Given that, according to this concept, sustainability dimensions have already been evaluated with the help of LCA, S-LCA, and LCC, all based on the ISO 14040 series (ISO, 2006a, b; Sala et al., 2013), LCSA follows this framework as well. Therefore, this methodology needs to pass through the phases of Goal and Scope Definition, Life Cycle Inventory, and Life Cycle Impact Assessment and Interpretation (Wulf et al., 2019).

Over time, the literature has applied LCSA to different products, services, sectors, and technologies. Tarne et al. (2017), for example, studied the existing approaches of LCSA in the automotive industry. As for De Luca et al. (2017), these authors focused attention on agricultural management, using a multicriteria decision analysis (MCDA) during LCSA process. Onat et al. (2017) also addressed LCSA literature concentrating on multiregional models and seeking tools, methods, and disciplines harmonization for LCSA.

The first example of what would commonly be named LCSA appeared in 2007 (Zhou et al., 2007), taking into account only economic and environmental indicators. However, the original idea was introduced by the Öko-Institut in 1987, with its analysis of the three sustainability dimensions based on life cycle, called Produktlinienanalyse. It was subsequently revised for an English version released in 2007, in which it was called Product Sustainability Assessment (PROSA) (Grießhammer et al., 2007). In 1995, the Social and Environmental Life Cycle Assessment (O'Brien et al., 1996) was proposed, which combined the social and environmental dimension of sustainability. The economic dimension was not included until 1998 (Andersson et al., 1998).

Notwithstanding, the LCSA concept did not emerge until Klöpffer released his publications which suggested linking LCA, S-LCA, and LCC (Klöpffer, 2003, 2008; Klöpffer and Grahl, 2014; Klöpffer and Renner, 2007). More specifically, he and his coauthors argue that this approach makes it possible to identify and avoid the impacts of the trade-offs between the different systems that are part of the product's life cycle, i.e., extraction of raw materials, production, use, and product disposal (Klöpffer, 2003). For this purpose, they demand a combined assessment of the three sustainability dimensions, leading to the approach LCSA = LCA + LCC + S-LCA (Wulf et al., 2019).

In order to delve deeper into LCSA and its characteristics, the main features and indicators chosen for each one of the three main methodologies in LCSA are now discussed, following the work of Wulf et al. (2019).

3.2.1 Life cycle assessment

LCA is the most consistent and standardized methodology. Besides its classical use for assessing a specific process in a product or service value chain, it also conducts economic input–output tables, concentrating on a sector or technology's supply chains.

Regarding the chosen indicators (or impact categories), life cycle impact assessment (LCIA) can address the cause–effect chain of environmental impacts using both midpoint and endpoint impact categories (Wulf et al., 2019). For this environmental domain, the categories considered are human health, resource availability, and ecosystem quality, for which the most common LCIA methods used are

Ecoindicator 99 and ReCiPe. Although most of the studies focused on ecological sustainability (Sadamichi et al., 2012; Zhou et al., 2007), other popular indicators not included in conventional LCIA methods can be, among others, cumulative energy demand (Ostermeyer et al., 2013; Valente et al., 2013), the latest version of climate change assessment (Valente, 2014), or some case study−specific approaches, such as product recyclability (Stamford, 2012), indoor air quality (Sims, 2014), or soil erosion (Rettenmaier et al., 2014).

3.2.2 Social life cycle assessment

The UNEP/SETAC established the guidelines for the LCSA as well as for the S-LCA (Andrews, 2009), also providing the Methodological Sheets (Benoît Norris et al., 2013), conceived as the most comprehensive and worldwide accepted methodological baseline. These guidelines provide five stakeholder groups (i.e., workers, local community, society, consumers, and value chain actors) that can be affected by social and socioeconomic impacts, providing possible indicators for each subcategory in each group although "a commonly accepted set of indicators has not yet been established by the scientific community and is still a controversial topic" (Traverso et al., 2012, p. 2).

Perhaps the greatest challenge when choosing appropriate indicators to evaluate social impacts is scaling them to the functional unit (Wu et al., 2014), for which it is recommended using activity variables, such as the number of working hours, to scale in a proper way (Norris, 2006). When selecting indicators, using UNEP/SETAC guidelines and the complementary Methodological Sheets is recommended by several authors, as well as stakeholder dialog and extensive literature reviews (Wulf et al., 2019), since data availability has been found to be a key variable (Martínez-Blanco et al., 2014; Moslehi and Arababadi, 2016; Traverso et al., 2012). Another option for selecting suitable indicators is to follow political frameworks or industry guidelines, such as those of Corporate Social Responsibility (Hu et al., 2013), the Global Reporting Initiative, or the UN SDGs (Wulf et al., 2018).

The most frequently used indicators are those related to health and safety, as well as those with employment implications, such as fatal and nonfatal accidents/injuries in the workplace, employment provision (Stamford and Azapagic, 2011), creation of employment opportunities (Menikpura et al., 2012), employment generation (Halog and Manik, 2016), and local employment (Dong, 2014). As finding purely social indicators is a challenging task, many studies tend to combine the social indicators with others related to economic, political, or even sociotechnical dimensions (Ibáñez-Forés et al., 2013; Kucukvar and Tatari, 2013; Ren et al., 2016).

3.2.3 Life cycle costing

LCC addresses the "economic dimension" of LCSA and is the oldest form of evaluation, although no standard is established (Wulf et al., 2019). To assess economic sustainability of a product, service, or technology, total costs are assessed from the viewpoint of all actors directly involved in the process (Swarr et al., 2011), thus LCC depends strongly on the actor's point of view.

LCC indicators refer to the product/technology's direct costs and the wider macroeconomic impacts. The most common practice is to discount the value of the life cycle components from their present value with a discount factor, using as indicators several investment analysis techniques, such as the internal rate of return, cost–benefit analysis, payback period, return on equity and return on investment, or the value added from activities, profit, and annualized costs (Wulf et al., 2019). Regarding macroeconomic impacts (e.g., employment, taxes, imports, exports, or R&D affecting the GDP growth), the indicators used are related to gross operating profit (Onat et al., 2014a, 2016), total value added, wages (Hu et al., 2013), tax revenue (Onat et al., 2014b), imports (Onat et al., 2014a), and the contribution to GDP (Noori et al., 2015; Tatari et al., 2015).

Over recent years, different approaches to LCSA have been developed in an attempt to streamline LCSA, ultimately improving its manageability, gathered in the work of Wulf et al. (2019).

- Product Sustainability Assessment (PROSA): this is one of the oldest concepts in LCSA literature (Grießhammer et al., 2007), concentrating on products, product portfolios, and services, although also offering the possibility of assessing technologies, large infrastructural projects, or even geographical units. It adds to the common LCSA analyses, risk assessment, benefit analysis (with consumer study), eco-efficiency analysis, and MCDA. 170 indicators are proposed which are clustered into the groups of employees, local and regional community, society, and users.
- New Energy Externalities Development for Sustainability (NEEDS): this project was carried out from 2004 until 2009 under the sixth Framework Programme of the EU, involving 80 different partners (universities, research institutions, industry, and NGOs) from 30 countries. It conducted a sustainability evaluation based on life cycle with MCDA and a total cost technique (all methodologies included in LCSA). Its final objective is "to evaluate the full costs and benefits (i.e., direct + external) of energy policies and of future energy systems, both for individual countries and for the enlarged EU as a whole" (European Commission, 2009).
- Co-Ordination Action for Innovation in Life Cycle Analysis for Sustainability (CALCAS): also developed under the sixth Framework Programme of the EU, this project involved scientists from France, Germany, Italy, The Netherlands, Portugal, Sweden, and the United Kingdom (Zamagni et al., 2009), who called their approach "Life Cycle Sustainability Analysis." This approach might be "broader and deeper" than LCS Assessment as it suggests a joint modeling stage including all synergies, linkages, and side effects among the three components of sustainability (Heijungs, 2010). Further developments have been made toward achieving this concept of Life Cycle Sustainability Unified Analysis (LiCSUA), including stakeholder integration, rebound effects, vulnerabilities, and resilience, together with data uncertainty and risk aversion (Kua, 2016).
- Prospective Sustainability Assessment of Technologies (Prosuite): this project was developed under the seventh Framework Programme of the EU from 2009 to 2013, involving scientists from more than 10 European countries (European Commission, 2019), with the main objective of developing an assessment method on which actors can take action for new

technologies. It is one of the few methods using endpoint indicators, and five impact categories are suggested regarding human health, social well-being, the natural environment, limited resources, and prosperity.

Integrated Life Cycle Sustainability Assessment (ILCSA): IFEU (Institut für Energie-und Umweltforschung Heidelberg) developed its own, new methodology called ILCSA, integrating other topics to the LCSA, such as local environmental effects (Wulf et al., 2019). As in the case of Prosuite, ILCSA also aims to evaluate new technologies, products, and processes including both quantitative and qualitative indicators.

- Tier Approach: this approach introduced by Neugebauer et al. (2015) defines three different layers (tiers) for indicators selection: (1) "sustainability footprint" including globally pertinent and practical indicators (only one for each sustainability dimension, i.e., climate change, fair salary, and added value); (2) "best practice" with robust and more numerous indicators, six environmental, four social, and three economic indicators; and (3) "comprehensive assessment" with 10 environmental, 7 social, and 6 economic indicators, as well as some new topics such as cultural heritage.
- SEEbalance: BASF developed its SocioEcoEfficiency Analysis (SEEbalance), publishing the methods and results involved in the process in scientific journals (Saling et al., 2005) with the help of several research organizations. The main objective is directly comparing product and process alternatives based on the three sustainability dimensions. One of the groups of stakeholders identified is "future generations," including indicators such as the number of apprentices (Ausberg et al., 2015).

Besides these approaches, some programs and projects, mostly supported by the EU, have been carried out to foster LCSA. For example, the project "Sustainable Process Industry through Resource and Energy Efficiency—SPIRE" (European Union, 2019) was conducted under the Horizon 2020 framework. This initiative is a public–private partnership with the aim of identifying "methodologies, tools and indicators for cross-sectorial sustainability assessment of energy and resource efficient solutions in the process industry." Furthermore, the project "Sustainability Toolkit for easy Life-cycle Evaluation—STYLE" was conducted between 2015 and 2016 to develop an ideal toolkit framework for technologies' sustainability evaluation (STYLE, 2017). During the same period, the project "Sustainability assessment methods and tools to support decision-making in the process industries—SAMT" was conceived to review and cluster different methods for assessing sustainability in the process industry (Pihkola et al., 2017), also including complete and exhaustive sustainability evaluation methods. Finally, the project "Metrics for Sustainability Assessment in European Process Industries—MEASURE" was created to examine the current state of sustainability assessment methods based on life cycle, while simultaneously developing a roadmap (Kralisch et al., 2016). In this last case, authors were in need of life cycle–based tools to facilitate sustainability evaluation, databases harmonization, methodological options, and results communication to industry and research community, while simultaneously supporting data exchange and collaborations across sectors.

Both projects, SAMT and MEASURE, greatly contributed to LCSA development in literature, providing helpful insights by integrating the industrial perspective into the dialog (Wulf et al., 2019).

4. Approaches to sustainable solutions

According to business activities, the concept of sustainable development has been described by the International Institute for Sustainable Development (1992, p. 11) as "the adaption of business strategies and activities that meet the needs of the enterprise and its stakeholders today, while protecting, sustaining, and enhancing the human and natural resources that will be needed in the future." In this sense, to achieve sustainability, this chapter reviews sustainable solutions through the creation of sustainable products, services, and technologies, as well as business models (i.e., PSS, CE, and IS).

4.1 Products, services, and technology

Green strategic partnerships have been defined as "voluntary arrangements between two or more organizations for the purpose of exchanging, sharing or co-developing environmentally friendly, or 'green', products, technologies or services to pursue a set of strategic green goals or address critical business needs" by Sadovnikova and Pujari (2017, p. 252). In this sense, the purpose of this collaboration between different organizations is to develop and implement innovative sustainable technology to offer eco-friendly products and sustainable services (McCormick et al., 2016). Consequently, there are different sustainable solutions that are fundamental to confront the deep effect of sustainability on society, namely technology, services, products, and a business model, such as PSS, CE, and IS (Nasiri et al., 2018).

As mentioned by Garetti and Taisch (2012), the scope of sustainability in the world is associated with three relevant aspects: innovation in technology, lifestyle, and business models. Compared to traditional manufacturing, advanced manufacturing consists of carrying out the entire process in a sustainable way (e.g., production processes activities, services, customer). Advanced manufacturing has been recognized as an essential element in the implementation of sustainability which improves the flexibility and quality of the production process, and in turn, the reduction of costs (BüLbüL et al., 2013). Following Nasiri et al. (2018, p. 358), technology can be described as "an influential item to achieve sustainability in advanced manufacturing." The reason is that during the production process technology allows the reduction of material consumption, toxic emissions, and energy (Zott et al., 2011).

Another sustainable solution that fosters the growth of value networks is to provide sustainable products and services (Legani et al., 2009). This solution is the reason why "green" products improve both the development rates of companies and the quality of life of individuals (Dangelico and Pujari, 2010). Thus, sustainable competitive advantages are not only provided by sustainable products but also by the construction of the service due to the rapid changes that occur in the marketplace (Rauch et al., 2016). Additionally, in keeping with the social dimension of sustainability, technology plays a crucial role in sustainable development for improving individuals' well-being, in terms of energy, agricultural industry, health, among other aspects.

Technology is among the most commonly discussed types of sustainable solutions in the literature, involving a group of processes, practices, and methods (Edmondson et al., 2019; Shrivastava et al., 2016). In this sense, technology can enhance the three dimensions of sustainability by providing a favorable relationship to the requirements of individuals and environmental restrictions (Saberi et al., 2019). For instance, technology can make it possible to achieve a sustainable environment by offering the smart consumption of energy sources (Park et al., 2018). Economic and social dimensions of sustainability can be provided by improvements in Information and Communication Technologies (Nasiri et al., 2018).

Technology has been recognized as the parameter that leads to sustainability by focusing on a resource's preservation, renewable energy consumption, waste management technology, and new methods of pollution control (Shrivastava et al., 2016). For this reason, advances in technology must be provided to overcome the challenges facing sustainable development (Caiado et al., 2018). The technologies employed for sustainable development in terms of the reduction of waste from air pollution, the effective use of water, land, and energy are environmental technology, clean technology, eco-technology, and end-of-pipe technology (Yusuf et al., 2018). Thus, it is necessary for organizations to focus on these elements to reach a sustainable society. However, barriers still exist due to the fact that technologies are limited to energy resources, the reduction of waste and emissions, and the implementation of water and soil (Pan et al., 2015).

4.2 Sustainable business models

While the sustainable solutions mentioned above are critical to achieving the commercial viability of the company, business models also play an important role in this domain (Geissdoerfer et al., 2018). Consequently, the construction of new business models could offer another potential solution for economic, social, and environmental sustainability (Evans et al., 2017). There are several business models that provide a set of sustainability features and support the sustainability of companies (e.g., closed-loop models) (Yang et al., 2017). In this chapter, we highlight the following business models: PSS (Reim et al., 2015), CE (Geissdoerfer et al., 2017), and IS (Baldasarre et al., 2019).

4.2.1 Product-service system

At present, sustainable consumption and production methods are attracting widespread interest due to the need to provide sustainable solutions (Morelli, 2006). Following these methods, there is a need to adapt the business model to offer more services that enhance sustainability of companies (Nasiri et al., 2018). In this sense, the notion of PSS is considered a method that makes reference to sustainable consumption and production (Geum and Park, 2011). According to Mont (2002), the principal objective of PSS is to decrease the environmental impact and, consequently, lead to sustainability by providing appropriate consumption of product/services. Additionally, an increasing amount of manufacturing companies are shifting their business models from the traditional product-based to the PSS business models, where manufacturers offer more than simply a product—they offer a product and service integration (Beuren et al., 2013). The development of this transition is named servitization (Neely, 2008).

The term PSS is considered an innovative business model including both products and services that satisfy the demand of the customer. Furthermore, many studies consider this method to be a sustainable business model (Sousa-Zomer and Miguel, 2018). In this regard, the implementation of customer demands is based on sustainable and innovative solutions throughout the entire production process (Vezzoli et al., 2015). The PSS concept intends to provide products and services using less pollution and waste, as well as reducing the amounts of natural resources used. For example, this can be achieved through regular maintenance that allows more efficient use of resources with less waste (Sousa-Zomer and Miguel, 2018).

More and more companies create sustainable solutions for their businesses as a differentiating factor and competitive advantage. The PSS is considered an effective "strategic alternative for sustainable development of firms" (Park and Yoon, 2015) due to the fact that the limits between the offer of products and services are not clear. This is in line with Morelli (2006), who noted that "the epochal shift from product-centred mass consumption to individual behaviours and highly personalized needs is now driving firms to rethink their industrial offerings."

4.2.2 Circular economy

More and more business models of linear systems are becoming circular (Morseletto, 2020), and this transition requires actions and policies. To obtain the CE, the government, state and private companies, and consumers all play important roles (Witjes and Lozano, 2016). The fundamental levels of action are to (Lowe, 2005) (i) seek a much higher effectiveness according to all the Rs of Cleaner Production (i.e., reduce resources-use and emission of pollutants and waste, reuse resources, and recycle by-products); (ii) reuse and recycle resources within grouped or chained industries and industrial parks so resources may fully circulate in the local production system; and (iii) integrate diverse consumption and production systems in a region so resources circulate between urban systems and industries. This last level of action requires the implementation of regional or municipal systems for the processing, storage, collection, and distribution of products.

The concept of CE is an innovative business model that allows achieving sustainable development in accordance with economic, environmental, institutional, and social issues (Manninen et al., 2018). The term CE was presented in the mid-2000s and focuses on waste recycling (Sakai et al., 2011). CE has been described as "a regenerative system in which resource input and waste, emission, and energy leakage are minimized by slowing, closing, and narrowing material and energy loops. This can be achieved through long-lasting design, maintenance, repair, reuse, remanufacturing, refurbishing, and recycling" (Geissdoerfer et al., 2017, p. 759). The principle objective of CE is not only to use fewer energy resources and raw materials but also to provide opportunities and generate competitive value for the advanced welfare of society (Ghisellini et al., 2016). The concept of CE is receiving increasing attention because of the need for a different approach to reach both environmental protection and economic development. The CE approach to resource efficiency involves industrial ecology and cleaner production within a broader system that includes industrial companies, eco-industrial parks, regional infrastructure, and business networks or chains to support optimization of resource use (Lowe, 2005).

The CE encompasses activities of economic, social, and environmental sustainability (Su et al., 2013). These sustainable activities give rise to the promotion of competitive profits both locally and nationally (Nasiri et al., 2018). According to the social dimension, this approach focuses on the general welfare of society while also facilitating the allocation of resources for economic development fair allocation. Regarding the environmental dimension, CE is considered an eco-friendly business model that promotes the use of sustainable methods with fewer negative effects on the environment through changes in the structure of industries. In terms of the economic dimension, CE contributes to the efficient consumption of resources and energy, as well as the development of resource distribution.

4.2.3 Industrial symbiosis

At present, the greatest task is to successfully achieve the proposal of sustainable and economically competitive industries, which can be achieved through the IS model.

The terms IS, sustainable development, and eco-industrial networks have been used interchangeably in accordance with the concept of CE (Winans et al., 2017). In this sense, IS is also related to the concept of eco-industrial parks, and, moreover, both the former and latter are critical terms in industry ecology (Lehtoranta et al., 2011). The Industrial Ecology approach is recognized as a sociotechnical process related to the cooperative interaction of independent company entities that exchange energy, materials, water, by-products, infrastructures, and services to achieve competitive advantage (Baldassarre et al., 2019).

According to Massard et al. (2014), the Industrial Ecology perspective focuses on the quantitative assessment of IS's positive environmental effects through LCA and Material Flow analysis, among others. According to Allenby (2006), Industrial Ecology is a systems-based, multidisciplinary discourse that seeks to understand emergent behavior of complex integrated human/natural systems. A more recent definition of Industrial Ecology has been proposed by Nasiri et al. (2018, p. 6) as "a policy with the aim of reducing the amount of waste creation by imitating natural ecosystems in industrial systems." The purpose of industrial ecology is to adapt the industrial system from a circular point of view.

Therefore, IS is considered as a subset of Industrial Ecology according to the incorporation of industries and exchange of waste, water, energy, and by-products between societies, industries, and companies (Pakarinen et al., 2010). IS networks are local systems that attempt to expand awareness of the principles of industrial ecology through the use of nearby industrial initiatives for the purpose of solving environmental issues (Chertow, 2000). Mirata and Emtairah (2005, p. 995) considered this concept as "a collection of long-term, symbiotic relationships between and among regional activities involving physical exchanges or materials and energy carriers as well as the exchange of knowledge, human or technical resources, concurrently providing environmental and competitive benefits." Thus, IS provides both economic and environmental sustainability through adequate energy consumption, resulting in the economic growth of local organizations and the reduction of their environmental impact. In this sense, Karlsson and Wolf (2008) proposed the implementation of IS in the forest industry and highlighted that additional waste and heat may be used in diverse processes and industries.

5. Conclusions

The main objective of this chapter was to review the notion of sustainable development and find approaches for organizations to achieve both sustainability assessment and several sustainable solutions.

On the one hand, SIs appear as one of the most used approaches to evaluate and report sustainable development, helping in decision-making processes and serving to communicate a firm's improvements regarding its sustainability performance to stakeholders. However, SIs face different challenges due to the complexity of different definitions of sustainability and sustainable development and the multitude of applications and interpretations associated with them, not to mention the need to define clear, straightforward, and robust frameworks to introduce the indicators, as there is no one single instruction or model to guide all the rest of the indicators. As for LCSA, it was designed to measure sustainability by taking into account its three dimensions, thus compiling the three methodologies of LCA, S-LCA, and LCC. This approach has been applied to different products, services, technologies, and industries and has received great attention by scholars and policy makers, resulting in the development of many different methodologies, programs, and tools based on this concept. This is the case, for example, of several projects promoted by the EU, such as NEEDS, CALCAS, Prosuite, SPIRE, STYLE, SAMT, and MEASURE. More specifically, the two latter projects have greatly contributed to LCSA development by integrating the industrial perspective into the discussion.

On the other hand, this chapter enhances sustainability research by investigating approaches for firms to achieve sustainable solutions. Four different approaches have been identified for sustainable solutions: products, services, and technologies and the business models PSS, CE, and IS. At a conceptual level, the transition of business models is a significant topic that attracts the attention of an increasing number of researchers, who are precisely the individuals who understand how to achieve sustainability. At a practical level, it is still necessary to address the way that these new business models are developed and implemented by businesses. According to the literature on business models, various streams of research have emerged through viable forms of business model innovation related to sustainability concerns, such as CE and ISAs stated above, these themes are closely related to PSS and can be seen as a subset of the PSS (or servitization) research topic. PSS, CE, and IS are research streams closely connected to sustainability and business model innovation, which is a subfield that is raising more and more awareness of emerging streams.

References

Akamani, K., 2020. Integrating deep ecology and adaptive governance for sustainable development: implications for protected areas management. Sustainability 12 (14), 5757.

Allenby, B., 2006. The ontologies of industrial ecology? Prog. Ind. Ecol. Int. J. 3 (1–2), 28–40. https://doi.org/10.1504/PIE.2006.010039.

Andersson, K., Eide, M.H., Lundqvist, U., Mattsson, B., 1998. The feasibility of including sustainability in LCA for product development. J. Clean. Prod. 6, 289–298.

Andrews, E.S., 2009. Guidelines for Social Life Cycle Assessment of Products: Social and Socio-Economic LCA Guidelines Complementing Environmental LCA and Life Cycle Costing, Contributing to the Full Assessment of Goods and Services within the Context of Sustainable Development. United Nations Environment Programme, Paris, France.

Ausberg, L., Ciroth, A., Feifel, S., Franze, J., Kaltschmitt, M., Klemmayer, I., Meyer, K., Saling, P., Schebek, L., Weinberg, J., et al., 2015. Lebenszyklusanalysen. In: Kaltschmitt, M., Schebek, L. (Eds.), Umweltbewertung für Ingenieure. Springer, Berlin, Germany, pp. 203–314.

Baldassarre, B., Schepers, M., Bocken, N., Cuppen, E., Korevaar, G., Calabretta, G., 2019. Industrial symbiosis: towards a design process for eco-industrial clusters by integrating circular economy and industrial ecology perspectives. J. Clean. Prod. 216, 446–460.

Bebbington, J., Brown, J., Frame, B., 2007. Accounting technologies and sustainability assessment models. Ecol. Econ. 61 (2–3), 224–236. https://doi.org/10.1016/j.ecolecon.2006.10.021.

Beck, U., 2008. Climate change and globalisation are reinforcing global inequalities: high time for a new social democratic era. Globalizations 5 (1), 78–80.

Bell, S., Morse, S., 1999. Sustainability Indicators, Measuring the Immeasurable. Earthscan, London, UK.

Bell, S., Morse, S., 2013. Measuring Sustainability: Learning by Doing. Routledge, London, UK.

Introduction: indicators and post truth. In: Bell, S., Morse, S. (Eds.), 2018a. Routledge Handbook of Sustainability Indicators. Routledge, New York, USA, pp. 1–17.

Bell, S., Morse, S., 2018b. Sustainability indicators past and present: what next? Sustainability 10, 1688.

Benoît Norris, C., Traverso, M., Valdivia, S., Vickery-Niedermann, G., Franze, J., Azuero, L., Ciroth, A., Mazijn, B., Aulisio, D., 2013. The Methodological Sheets for Subcategories in Social Life Cycle Assessment (S-LCA). UNEP, Nairobi, Kenya (SETAC, Pensacola, FL, USA).

Beuren, F.H., Ferreira, M.G.G., Miguel, P.A.C., 2013. Product-service systems: a literature review on integrated products and services. J. Clean. Prod. 47, 222–231.

Bolis, I., Morioka, S.N., Sznelwar, L.I., 2014. When sustainable development risks losing its meaning. Delimiting the concept with a comprehensive literature review and a conceptual model. J. Clean. Prod. 83, 7–20.

Bond, A., Morrison-Saunders, A., Howitt, R., 2013. Sustainability Assessment: Pluralism, Practice and Progress. Routledge, New York, USA.

BüLbüL, H., ÖMüRbek, N., Paksoy, T., Bektaş, T., 2013. An empirical investigation of advanced manufacturing technology investment patterns: evidence from a developing country. J. Eng. Technol. Manag. 30 (2), 136–156.

Burns, T.R., Witoszek, N., 2012. Brundtland report revisited: toward a new humanist agenda. J. Hum. Ecol. 39 (2), 155–170.

Caiado, R.G.G., Leal Filho, W., Quelhas, O.L.G., de Mattos Nascimento, D.L., Ávila, L.V., 2018. A literature-based review on potentials and constraints in the implementation of the sustainable development goals. J. Clean. Prod. 198, 1276–1288.

Chertow, M.R., 2000. Industrial symbiosis: literature and taxonomy. Annu. Rev. Energy Environ. 25 (1), 313–337.

Dangelico, R.M., Pujari, D., 2010. Mainstreaming green product innovation: why and how companies integrate environmental sustainability. J. Bus. Ethics 95 (3), 471–486.

De Freitas, J.G., Costa, H.G., Ferraz, F.T., 2017. Impacts of lean six sigma over organizational sustainability: a survey study. J. Clean. Prod. 156, 262–275.

De Luca, A.I., Iofrida, N., Leskinen, P., Stillitano, T., Falcone, G., Strano, A., Gulisano, G., 2017. Life cycle tools combined with multi-criteria and participatory methods for agricultural sustainability: insights from a systematic and critical review. Sci. Total Environ. 595, 352–370.

Dong, Y., March 2014. Life Cycle Sustainability Assessment Modeling of Building Construction (Ph.D. thesis). The University of Hong Kong, Hong Kong.

Edmondson, D.L., Kern, F., Rogge, K.S., 2019. The co-evolution of policy mixes and socio-technical systems: towards a conceptual framework of policy mix feedback in sustainability transitions. Res. Pol. 48 (10), 103555.

Elkington, J., 1997. Cannibals with Forks: The Triple Bottom Line of 21st Century. Capstone, Oxford.

European Commission, 2009. New Energy Externalities Development for Sustainability. Available from: https://cordis.europa.eu/project/rcn/73947/factsheet/en. (Accessed 20 July 2020).

European Commission, 2019. Development and Application of Standardized Methodology for the Prospective Sustaınability Assessment of Technologies. Available from: https://cordis.europa.eu/project/id/227078/es. (Accessed 17 July 2020).

European Union, 2019. SPIRE Trio Map Out Road to Sustainability Measurement. Available from: https://www.spire2030.eu/projects/casestudies/spire-trio-map-out-road-sustainability-measurement. (Accessed 20 July 2020).

Evans, S., Vladimirova, D., Holgado, M., Van Fossen, K., Yang, M., Silva, E.A., Barlow, C.Y., 2017. Business model innovation for sustainability: towards a unified perspective for creation of sustainable business models. Bus. Strat. Environ. 26 (5), 597–608.

Fergus, A.H., Rowney, J.I., 2005. Sustainable development: lost meaning and opportunity? J. Bus. Ethics 60 (1), 17–27.

Garetti, M., Taisch, M., 2012. Sustainable manufacturing: trends and research challenges. Prod. Plan. Control 23 (2–3), 83–104. https://doi.org/10.1080/09537287.2011.591619.

Geissdoerfer, M., Savaget, P., Bocken, N.M., Hultink, E.J., 2017. The circular economy–A new sustainability paradigm? J. Clean. Prod. 143, 757–768.

Geissdoerfer, M., Vladimirova, D., Evans, S., 2018. Sustainable business model innovation: a review. J. Clean. Prod. 198, 401–416.

Geum, Y., Park, Y., 2011. Designing the sustainable product-service integration: a product-service blueprint approach. J. Clean. Prod. 19 (14), 1601–1614.

Ghisellini, P., Cialani, C., Ulgiati, S., 2016. A review on circular economy: the expected transition to a balanced interplay of environmental and economic systems. J. Clean. Prod. 114, 11–32.

Why sustainability assessment? In: Gibson, R.B. (Ed.), 2013. Sustainability Assessment: Pluralism, Practice and Progress. Routledge, New York, USA, pp. 3–17.

Gibson, R.B., Hassan, S., Holtz, S., Tansey, J., Whitelaw, G., 2013. Sustainability Assessment: Criteria and Processes. Routledge, New York, USA.

Giddings, B., Hopwood, B., O'brien, G., 2002. Environment, economy and society: fitting them together into sustainable development. Sustain. Dev. 10 (4), 187–196.

Grießhammer, R., Buchert, M., Gensch, C.O., Hochfeld, C., Manhart, A., Rüdenauer, I., 2007. PROSA—Product Sustainability Assessment. Öko-Institut, Freiburg, Germany.

Halog, A., Manik, Y., 2016. Life cycle sustainability assessments. In: Encyclopedia of Inorganic and Bioinorganic Chemistry. JohnWiley & Sons, Hoboken, NJ, USA.

Hediger, W., 1999. Reconciling "weak" and "strong" sustainability. Int. J. Soc. Econ. 26, 1120–1143.

Heijungs, R., 2010. Ecodesign—carbon footprint—life cycle assessment—life cycle sustainability analysis. A flexible framework for a continuum of tools. Environ. Clim. Technol. 4, 42–46.

Hezri, A.A., Hasan, M.N., 2004. Management framework for sustainable development indicators in the State of Selangor, Malaysia. Ecol. Indicat. 4, 287–304.

Hu, M., Kleijn, R., Bozhilova-Kisheva, K.P., Di Maio, F., 2013. An approach to LCSA: the case of concrete recycling. Int. J. Life Cycle Assess. 18, 1793–1803.

Ibáñez-Forés, V., Bovea, M.D., Azapagic, A., 2013. Assessing the sustainability of best available techniques (BAT): methodology and application in the ceramic tiles industry. J. Clean. Prod. 51, 162–176.

International Institute for Sustainable Development, Deloitte & Touche, & Business Council for Sustainable Development, 1992. Business Strategy for Sustainable Development: Leadership and Accountability for the 90s. Diane Publishing, Collingdale, PA, USA.

ISO, 2006a. DIN EN ISO 14040:2006 Environmental Management—Life Cycle Assessment—Principles and Framework. Beuth Verlag, Berlin, Germany.

ISO, 2006b. DIN EN ISO 14044:2006 Environmental Management—Life Cycle Assessment—Requirements and Guidelines. Beuth Verlag, Berlin, Germany.

Karlsson, M., Wolf, A., 2008. Using an optimization model to evaluate the economic benefits of industrial symbiosis in the forest industry. J. Clean. Prod. 16 (14), 1536–1544.

Klöpffer, W., 2003. Life-cycle based methods for sustainable product development. Int. J. Life Cycle Assess. 8, 157–159.

Klöpffer, W., 2008. Life cycle sustainability assessment of products. Int. J. Life Cycle Assess. 13, 89–95.

From LCA to sustainability assessment. In: Klöpffer, W., Grahl, B. (Eds.), 2014. Life Cycle Assessment (LCA): A Guide to Best Practice. Wiley-VCH, Weinheim, Gremany, pp. 357–374.

Klöpffer, W., Renner, I., 2007. Lebenszyklusbasierte nachhaltigkeitsbewertung von Produkten. Technikfolgenabschätzung – Theorie und Praxis 3 (16), 32–38.

Kralisch, D., Minkov, N., Manent, A., Rother, E., Mohr, L., Schowanek, D., Sfez, S., Lapkin, A.A., Jones, M., 2016. Roadmap for Sustainability Assessment in the European Process Industries. Friedrich-Schiller-University Jena, Jena, Germany.

Kua, H.W., 2016. Toward a more integrated and holistic assessment framework for life cycle modeling – life cycle sustainability unified analysis. In: Proceedings of the 22nd International Sustainable Development Research Society Conference, Lisbon, Portugal, 13–15 July 2016.

Kucukvar, M., Tatari, O., 2013. Towards a triple bottom-line sustainability assessment of the U.S. construction industry. Int. J. Life Cycle Assess. 18, 958–972.

Lang, D.J., Wiek, A., Bergmann, M., Stauffacher, M., Martens, P., Moll, P., Swilling, M., Thomas, C.J., 2012. Transdisciplinary research in sustainability science: practice, principles, and challenges. Sustainability Science 7, 25–43.

Legnani, E., Cavalieri, S., Ierace, S., 2009. A framework for the configuration of after-sales service processes. Prod. Plan. Control. 20 (2), 113–124.

Lehtoranta, S., Nissinen, A., Mattila, T., Melanen, M., 2011. Industrial symbiosis and the policy instruments of sustainable consumption and production. J. Clean. Prod. 19 (16), 1865–1875.

Lélé, S.M., 1991. Sustainable development: a critical review. World Dev. 19 (6), 607–621.

Economic solutions. In: Lowe, E. (Ed.), 2005. Environmental Solutions. Academic Press, pp. 61–114.

Manninen, K., Koskela, S., Antikainen, R., Bocken, N., Dahlbo, H., Aminoff, A., 2018. Do circular economy business models capture intended environmental value propositions? J. Clean. Prod. 171, 413–422.

Martínez-Blanco, J., Lehmann, A., Muñoz, P., Antón, A., Traverso, M., Rieradevall, J., Finkbeiner, M., 2014. Application challenges for the social Life Cycle Assessment of fertilizers within life cycle sustainability assessment. J. Clean. Prod. 69, 34–48.

International survey on ecoinnovation parks. In: Massard, G., Jacquat, O., Zürcher, D. (Eds.), 2014. Workshop on Eco-Innovation Parks, vol. 20, p. 12.

McCormick, K., Neij, L., Mont, O., Ryan, C., Rodhe, H., Orsato, R., 2016. Advancing sustainable solutions: an interdisciplinary and collaborative research agenda. J. Clean. Prod. 123, 1–4.

Menikpura, S.N.M., Gheewala, S.H., Bonnet, S., 2012. Framework for life cycle sustainability assessment of municipal solid waste management systems with an application to a case study in Thailand. Waste Manag. Res. 30, 708–719.

Mirata, M., Emtairah, T., 2005. Industrial symbiosis networks and the contribution to environmental innovation: the case of the Landskrona industrial symbiosis programme. J. Clean. Prod. 13 (10–11), 993–1002.

Mont, O.K., 2002. Clarifying the concept of product–service system. J. Clean. Prod. 10 (3), 237–245.

Morelli, N., 2006. Developing new product service systems (PSS): methodologies and operational tools. J. Clean. Prod. 14 (17), 1495–1501.

Morseletto, P., 2020. Targets for a circular economy. Resour. Conserv. Recycl. 153, 104553.

Moslehi, S., Arababadi, R., 2016. Sustainability assessment of complex energy systems using life cycle approach-case study: Arizona state university tempe campus. Proc. Eng. 145, 1096–1103.

Nasiri, M., Rantala, T., Saunila, M., Ukko, J., Rantanen, H., 2018. Transition towards sustainable solutions: product, service, technology, and business model. Sustainability 10 (2), 358.

Neely, A., 2008. Exploring the financial consequences of the servitization of manufacturing. Oper. Manag. Res. 1 (2), 103–118.

Neugebauer, S., Martínez-Blanco, J., Scheumann, R., Finkbeiner, M., 2015. Enhancing the practical implementation of life cycle sustainability assessment—proposal of a Tiered approach. J. Clean. Prod. 102, 165–176.

Noori, M., Kucukvar, M., Tatari, O., 2015. A macro-level decision analysis of wind power as a solution for sustainable energy in the USA. Int. J. Sustain. Energy 34, 629–644.

Norris, G.A., 2006. Social impacts in product life cycles—towards life cycle attribute assessment. Int. J. Life Cycle Assess. 11, 97–104.

O'Brien, M., Doig, A., Clift, R., 1996. Social and environmental life cycle assessment (SELCA). Int. J. Life Cycle Assess. 1 (4), 231–237. https://doi.org/10.1007/BF02978703.

Öko-Institut Projektgruppe Ökologische Wirtschaft, 1987. Produktlinienanalyse Bedürfnisse, Produkte und ihre Folgen. Kölner Volksbl, Cologne, Germany, p. 180 (In German).

Olawumi, T.O., Chan, D.W., 2018. A scientometric review of global research on sustainability and sustainable development. J. Clean. Prod. 183, 231–250.

Onat, N.C., Gumus, S., Kucukvar, M., Tatari, O., 2016. Application of the TOPSIS and intuitionistic fuzzy set approaches for ranking the life cycle sustainability performance of alternative vehicle technologies. Sustain. Prod. Consumption 6, 12–25.

Onat, N., Kucukvar, M., Halog, A., Cloutier, S., 2017. Systems thinking for life cycle sustainability assessment: a review of recent developments, applications, and future perspective. Sustainability 9, 706.

Onat, N., Kucukvar, M., Tatari, O., 2014a. Integrating triple bottom line input–output analysis into life cycle sustainability assessment framework: the case for US buildings. Int. J. Life Cycle Assess. 19, 1488–1505.

Onat, N., Kucukvar, M., Tatari, O., 2014b. Towards life cycle sustainability assessment of alternative passenger vehicles. Sustainability 6, 9305–9342.

Ostermeyer, Y., Wallbaum, H., Reuter, F., 2013. Multidimensional Pareto optimization as an approach for site-specific building refurbishment solutions applicable for life cycle sustainability assessment. Int. J. Life Cycle Assess. 18, 1762–1779.

Pakarinen, S., Mattila, T., Melanen, M., Nissinen, A., Sokka, L., 2010. Sustainability and industrial symbiosis—the evolution of a Finnish forest industry complex. Resour. Conserv. Recycl. 54 (12), 1393–1404.

Pan, S.Y., Du, M.A., Huang, I.T., Liu, I.H., Chang, E.E., Chiang, P.C., 2015. Strategies on implementation of waste-to-energy (WTE) supply chain for circular economy system: a review. J. Clean. Prod. 108, 409–421.

Park, H., Yoon, J., 2015. A chance discovery-based approach for new product–service system (PSS) concepts. Service Business 9 (1), 115–135.

Park, L.W., Lee, S., Chang, H., 2018. A sustainable home energy prosumer-chain methodology with energy tags over the blockchain. Sustainability 10 (3), 658.

Pihkola, H., Pajula, T., Federley, M., Myllyoja, J., 2017. Sustainability Assessment in the Process Industry—Future Actions and Development Needs. VTT Technical Research Centre of Finland, Brussels, Belgium.

Pope, J., Bond, A., Hugé, J., Morrison-Saunders, A., 2017. Reconceptualising sustainability assessment. Environ. Impact Assess. Rev. 62, 205–215.

Ramos, T.B., 2009. Development of regional sustainability indicators and the role of academia in this process: the Portuguese practice. J. Clean. Prod. 17 (12), 1101–1115.

Ramos, T.B., 2019. Sustainability assessment: exploring the frontiers and paradigms of indicator approaches. Sustainability 11, 824.

Rauch, E., Dallasega, P., Matt, D.T., 2016. Sustainable production in emerging markets through distributed manufacturing systems (DMS). J. Clean. Prod. 135, 127–138.

Reim, W., Parida, V., Örtqvist, D., 2015. Product–Service Systems (PSS) business models and tactics–a systematic literature review. J. Clean. Prod. 97, 61–75.

Ren, J., Xu, D., Cao, H., Wei, S.A., Dong, L., Goodsite, M.E., 2016. Sustainability decision support framework for industrial system prioritization. AIChE J. 62, 108–130.

Rettenmaier, N., Harter, R., Himmler, H., Keller, H., Kretschmer, W., Müller-Lindenlauf, M., Reinhardt, G.A., Scheurlen, K., Schröter, C., 2014. Integrated Sustainability Assessment of the BIOCORE Biorefinery Concept Report Prepared for the BIOCORE Project. Institut für Energie und Umweltforschung, Heidelberg, Germany.

Saberi, S., Kouhizadeh, M., Sarkis, J., Shen, L., 2019. Blockchain technology and its relationships to sustainable supply chain management. Int. J. Prod. Res. 57 (7), 2117–2135.

Sadamichi, Y., Kudoh, Y., Sagisaka, M., Chen, S.S., Elauria, J.C., Gheewala, S.H., Hasanudin, U., Romero, J., Shi, X., Sharma, V.K., 2012. Sustainability assessment methodology of biomass utilization for energy in East Asian countries. J. Jpn. Inst. Energy 91, 960–968.

Sadovnikova, A., Pujari, A., 2017. The effect of green partnerships on firm value. J. Acad. Market. Sci. 45, 251–267.

Sakai, S.I., Yoshida, H., Hirai, Y., Asari, M., Takigami, H., Takahashi, S., et al., 2011. International comparative study of 3R and waste management policy developments. J. Mater. Cycles Waste Manag. 13 (2), 86–102.

Sala, S., Ciuffo, B., Nijkamp, P., 2015. A systemic framework for sustainability assessment. Ecol. Econ. 119, 314–325.

Sala, S., Farioli, F., Zamagni, A., 2013. Progress in sustainability science: lessons learnt from current methodologies for sustainability assessment: part 1. Int. J. Life Cycle Assess. 18 (9), 1653–1672.

Saling, P., Maisch, R., Silvani, M., König, N., 2005. Assessing the environmental-hazard potential for life cycle assessment, eco-efficiency and SEEbalance. Int. J. Life Cycle Assess. 10, 364–371.

Shrivastava, P., Ivanaj, S., Ivanaj, V., 2016. Strategic technological innovation for sustainable development. Int. J. Technol. Manag. 70 (1), 76–107.

Sims, R.J., August 1, 2014. Life Cycle Sustainability Assessment of the Electrification of Residential Heat Supply in UK Cities (Ph.D. thesis). The University of Manchester, Manchester, UK.

Singh, R.K., Murty, H.R., Gupta, S.K., Dikshit, A.K., 2012. An overview of sustainability assessment methodologies. Ecol. Indicat. 15, 281–299.

Sousa-Zomer, T.T., Miguel, P.A.C., 2018. Sustainable business models as an innovation strategy in the water sector: an empirical investigation of a sustainable product-service system. J. Clean. Prod. 171, 119–129.

Stafford-Smith, M., Griggs, D., Gaffney, O., Ullah, F., Reyers, B., Kanie, N., O'Connell, D., 2017. Integration: the key to implementing the sustainable development goals. Sustain. Sci. 12 (6), 911–919.

Stamford, L., December 31, 2012. Life Cycle Sustainability Assessment of Electricity Generation: A Methodology and an Application in the UK Context (Ph.D. thesis). Universita of Manchester, Manchester, UK.

Stamford, L., Azapagic, A., 2011. Sustainability indicators for the assessment of nuclear power. Energy 36, 6037–6057.

Steffen, W., Richardson, K., Rockström, J., Cornell, S.E., Fetzer, I., Bennett, E.M., et al., 2015. Planetary boundaries: guiding human development on a changing planet. Science 347 (6223), 1259855.

STYLE, 2017. Ideal Toolkit Framework—A High-Level View of Featrues and Functions. SPIRE, Brussels, Belgium.

Su, B., Heshmati, A., Geng, Y., Yu, X., 2013. A review of the circular economy in China: moving from rhetoric to implementation. J. Clean. Prod. 42, 215–227.

Swarr, T.E., Hunkeler, D., Klöpffer, W., Pesonen, H.L., Ciroth, A., Brent, A.C., Pagan, R., 2011. Environmental life cycle costing: a code of practice. Int. J. Life Cycle Assess. 16, 389–391.

Tasaki, T., Kameyama, Y., 2015. Sustainability indicators: are we measuring what we ought to measure? Global Environ. Res. 19, 147–154.

Tasaki, T., Kameyama, Y., Hashimoto, S., Moriguchi, Y., Harasawa, H., 2010. A survey of national sustainable development indicators. Int. J. Sustain. Dev. 13 (4), 337–361.

Tarne, P., Traverso, M., Finkbeiner, M., 2017. Review of life cycle sustainability assessment and potential for its adoption at an automotive company. Sustainability 9, 670.

Tatari, O., Kucukvar, M., Onat, N., 2015. Towards a triple bottom line life cycle sustainability assessment of buildings. In: Ayyub, B.M., Galloway, G.E., Wright, R.N. (Eds.), Measurement Science for Sustainable Construction and Manufacturing, vol. I. University of Maryland Report to the National Institute of Standards and Technology, Office of Applied Economics, Gaithersburg, MD, USA, pp. 226–231.

Traverso, M., Finkbeiner, M., Jørgensen, A., Schneider, L., 2012. Life cycle sustainability dashboard. J. Ind. Ecol. 16 (5), 680–688.

UNCSD (United Nations Commission on Sustainable Development), 2001. Indicators of Sustainable Development: Guidelines and Methodologies, second ed.

UNSC (United Nations Statistical Commission), 2015. Technical Report on the Process of the Development of an Indicator Framework for the Goals and Targets of the Post-2015 Development Agenda.

United Nations, 1992. Declaración de Río sobre Medio Ambiente y Desarrollo. Río de Janeiro.

United Nations, 2015. Transforming Our World: The 2030 Agenda for Sustainable Development.

Valdivia, S., Ciroth, A., Finkbeiner, M., Hildenbrand, J., Klöpffer, W., Mazijn, B., Prakash, S., Sonnemann, G., Traverso, M., Ugaya, C.M.L., Vickery-Niederman, G., 2011. Towards a Life Cycle Sustainability Assessment: Making Informed Choices on Products. UNEP/SETAC Life Cycle Initiative, Paris, France.

Valente, C., 2014. Sustainability Assessment of Chestnut and Invaded Coppice Forests in Piedmont Region (Italy). Ostfold Research, Kråkerøy, Norway.

Valente, C., Modahl, I.S., Askham, C., 2013. Method Development for Life Cycle Sustainability Assessment (LCSA) of New Norwegian Biorefinery. Ostfold Research, Kråkerøy, Norway.

Vezzoli, C., Ceschin, F., Diehl, J.C., Kohtala, C., 2015. New design challenges to widely implement 'sustainable product–service systems'. J. Clean. Prod. 97, 1–12.

Viegas, O., Caeiro, S., Ramos, T.B., 2018. Conceptual model for the integration of non-material components in sustainability assessment. Ambiente Sociedade 21.

WCED, 1987. Our Common Future. World Commission on Environment and Development. Oxford University Press, Oxford.

Wilson, J., Tyedmers, P., Pelot, R., 2007. Contrasting and comparing sustainable development indicator metrics. Ecol. Indicat. 7, 299–314.

Winans, K., Kendall, A., Deng, H., 2017. The history and current applications of the circular economy concept. Renew. Sustain. Energy Rev. 68, 825–833.

Witjes, S., Lozano, R., 2016. Towards a more circular economy: proposing a framework linking sustainable public procurement and sustainable business models. Resour. Conserv. Recycl. 112, 37–44.

Woods, D., 2006. Sustainable Development: A Contested Paradigm. Foundation for Water Research, pp. 1–8.

Wu, R., Yang, D., Chen, J., 2014. Social life cycle assessment revisited. Sustainability 6, 4200–4226.

Wulf, C., Werker, J., Ball, C., Zapp, P., Kuckshinrichs, W., 2019. Review of sustainability assessment approaches based on life cycles. Sustainability 11, 5717.

Wulf, C., Werker, J., Zapp, P., Schreiber, A., Schlör, H., Kuckshinrichs, W., 2018. Sustainable development goals as a guideline for indicator selection in life cycle sustainability assessment. Proc. CIRP 69, 59–65.

Yang, M., Evans, S., Vladimirova, D., Rana, P., 2017. Value uncaptured perspective for sustainable business model innovation. J. Clean. Prod. 140, 1794–1804.

Yawar, S.A., Seuring, S., 2017. Management of social issues in supply chains: a literature review exploring social issues, actions and performance outcomes. J. Bus. Ethics 141 (3), 621–643.

Yusuf, M.F., Ashari, H., Razalli, M.R., 2018. Environmental technological innovation and its contribution to sustainable development. Int. J. Technol. 8, 1569–1578.

Zamagni, A., Buttol, P., Buonamici, R., Masoni, P., Guinée, J.B., Huppes, G., Heijungs, R., van der Voet, E., Ekvall, T., Rydberg, T., 2009. D20 Blue Paper on Life Cycle Sustainability Analysis. Institute of Environmental Sciences. Leiden University (CML), Leiden, The Netherlands.

Zhou, Z., Jiang, H., Qin, L., 2007. Life cycle sustainability assessment of fuels. Fuel 86 (1–2), 256–263.

Zijp, M., Heijungs, R., van der Voet, E., van de Meent, D., Huijbregts, M., Hollander, A., Posthuma, L., 2015. An identification key for selecting methods for sustainability assessments. Sustainability 7 (3), 2490–2512.

Zott, C., Amit, R., Massa, L., 2011. The business model: recent developments and future research. J. Manag. 37 (4), 1019–1042.

Modern age of sustainability: supply chain resource management

Belén Payán-Sánchez, Ana Labella-Fernández, M. Mar Serrano-Arcos
University of Almería, Ctra. Sacramento s/n, Almería, Spain

1. Introduction

Sustainability is a complex problem and it has become a great concern to all countries, industries, and citizens, and this situation is even more dire in the business field as a result of globalization. Global production networks and supply chains (SCs) are at the core of modern economic development (Yakovleva et al., 2019), and globalization has brought new perspective to the table regarding the limits and responsibilities of organizations in relation to the environment (Boström et al., 2015). Different stakeholders, such as customers, NGOs, other firms, and governments, exert pressures over organizations to reconsider their sense of responsibility toward their external environment, using their SCs as a means of managing their requirements at different levels and scales (Morana, 2013). Thus, these pressures have resulted in the development of proactive institutional and regulatory innovations to achieve "Sustainable Supply Chain Management" (SSCM), such as eco-labels, codes of conduct, inspection plans, and provisioning guidelines (Boström et al., 2015).

One of the areas of the organization that has received more attention by scholars is the SC, primarily for its connections with the environment throughout the process and the multiple contributions it offers to the environment, thereby potentially enhancing the sustainability of the firm. In fact, numerous businesses have achieved many improvements by considering sustainability implications and applying sustainability principles along their SCs (Feng et al., 2017). Research has demonstrated that supply relationships are an essential pillar for businesses in order to achieve the sustainability of products and services (Simpson and Power, 2005; Toubpulic et al., 2014). Therefore, organizations should make great efforts to cooperate and collaborate to adjust operations within each company and among companies to achieve the sustainability of their products and services and to respond to the customer's requirements in terms of quality, speed, and costs (Fritz, 2019). This demands that the company not only make a commitment to collaborating with suppliers and customers to further integrate environmental and social considerations in the SC but also the involvement of other stakeholders such as stockholders, NGOs, and governments.

There are several practices that can be associated with SSCM, namely Eco-Efficiency, Eco-Design, Green Purchasing, Reverse Logistics, and Investment Recovery. These practices have been categorized as advanced (e.g., Eco-Efficiency,

Sustainable Resource Management. https://doi.org/10.1016/B978-0-12-824342-8.00003-1

Eco-Design, Green Purchasing) and common (e.g., Reverse Logistics and Investment Recovery), which include both social and environmental aspects. All practices are oriented toward improving the environmental performance of companies and even reducing their negative environmental effect (Geng et al., 2017; Sarkis et al., 2011), while also promoting socially responsible behavior in relation to purchasing decisions made by firms (Mani et al., 2018). Additionally, Design for Sustainability (DfS) appears in literature as the point where the firm's SC and the environment converge (Arnette et al., 2014), and also as a response from the design field within the firm to sustainability issues. In this sense, DfS has been developed to combine and consolidate the different "Design for" approaches into a single taxonomy, including all the dimensions of sustainability (not only the environment) (Vogtlander et al., 2001), guiding managers and designers through the design process.

Besides these practices, the assessment of the environmental performance of products and services appears to be crucial to making the SC, and the activities it entails, more sustainable. In this sense, Life Cycle Assessment (LCA) appears in literature as one of the most influential and robust methodologies, along with Environmental Labeling, both of which play key roles in supporting and guiding the decision-making process (Ruiz-Méndez and Güereca, 2019).

Consequently, the main objective of this chapter is to examine the connection of one of the organization's processes that is most closely related to resource management, i.e., the SC, and sustainability issues. In this sense, we delve into the term SSCM and discuss the different manners, practices, and tools that allow firms to manage their resources in a more sustainable way, minimize waste, and contribute to the welfare of the planet.

With this aim, this chapter is organized as follows: Section 2 introduces the concept of SSCM, as the means to achieve firms' goals from all three dimensions of sustainability through the SC. Section 3 presents different practices to be implemented in the firm to improve environmental and social impacts. Furthermore, several sustainability evaluation tools are compiled in Section 4. Finally, the main conclusions of this chapter are discussed in Section 5.

2. Sustainable supply chain management

Supply chain management (SCM) is broadly defined as "the systemic, strategic coordination of the traditional business functions and the tactics across these business functions within a particular company and across businesses within the supply chain, for the purpose of improving the long-term performance of the individual companies and the supply chain as a whole" (Mentzer et al., 2001, p. 18). The incorporation of environmental concerns into SCM literature gave rise to the term "Green supply chain management" (GSCM). GSCM refers to the integration of environmental issues in all functions of the SC. In this regard, Srivastava (2007, pp. 54–55) defined GSCM as "integrating environmental thinking into supply-chain management, including product design, material sourcing and selection, manufacturing processes, delivery of the final product to the consumers as well as end-of-life management of the product after its useful life." However, the concept of GSCM limits sustainability to environmental practices (Yawar and Seuring, 2017), and researchers have also identified the relevance of social problems (Ashby et al., 2012; Martins and Pato, 2019) due to the growth of global trade activities.

The Brundtland definition of sustainability includes the social, environmental, and economic pillars (WCED, 1987). In the same line, the term "Triple Bottom Line" coined by Elkington (1998) also takes into consideration economic, environmental, and social criteria. John Elkington argued that organizations should take into account the traditional financial profit and loss bottom line as well as the social and the environmental dimensions. Moreover, Carter and Rogers (2008) argued that environmental, social, and economic goals are interdependent, and organizations should strive to combine them.

The concept of SSCM emerged with the objective of taking goals from all three dimensions of sustainability. Ahi and Searcy (2013) asserted that SSCM extends the concept of GSCM by integrating the social and economic dimensions along with the environmental dimension. Although more than 16 definitions of SSCM exist, definitional consensus has yet to be achieved (Dubey et al., 2017). Most authors define SSCM as the integration of the three pillars of sustainability into SCM (e.g., Pagell and Shevchenko, 2014; Seuring and Müller, 2008; Wittstruck and Teuteberg, 2012). In this respect, SSCM is conceived as "the creation of coordinated supply chains through the voluntary integration of economic, environmental, and social considerations with key inter-organizational business systems designed to efficiently and effectively manage the material, information, and capital flows associated with the procurement, production, and distribution of products or services in order to meet stakeholder requirements and improve the profitability, competitiveness, and resilience of the organization over the short and long-term" (Ahi and Searcy, 2013, p. 339).

SSCM integrates environmental and social issues in the product design, materials sourcing and selection, purchasing, manufacturing processes, packaging, warehousing, transport and disposal phases of products and services (Haake and Seuring, 2009).

Some scholars contend that sustainability must be implemented simultaneously from the top-down and bottom-up approaches (Meadows et al., 2004). In this regard, the integration of sustainability in SCs could take place simultaneously at different strategic levels (1) in governance mechanisms and top management; (2) in operations; (3) in products/services; and (4) through SC partners (Fritz, 2019).

First, sustainability could be acknowledged as a fundamental value by top management (Labella-Fernández and Martínez-del-Río, 2019). From a top-down perspective, organizations could introduce sustainability into their mission and vision statements with the objective of creating a truly sustainable chain (Pagell and Wu, 2009). However, achieving the sustainability of the SC requires much more effort by top management to increase the awareness of middle management and employees. This could be achieved through the formulation and implementation of corporate sustainability policies, goals, and practices. Top-level managers and entrepreneurs tend to be more proactive and innovative and take more risks in turbulent and uncertain environments (Rauch et al., 2009; Zhu et al., 2008). Thus, the leadership and commitment of top-level management with environmental and social issues are important aspects in the implementation of SSCM activities and programs (Lambert et al., 1998). In addition, according to Kaur and Sharma (2018), in order to maintain sustainability as a strategy within the organization not only is it necessary to establish complete organizational inclusion but is also important to ensure employee involvement in the decision-making process along the value chain.

Second, sustainably managing the SC also requires integrating sustainability at the operational level into the different departments or functions of the firm (e.g., marketing, R&D, accountancy, operations, human resource management). This step is of paramount importance in order to incorporate social, environmental, and economic criteria into the SC. In this regard, companies should devote all efforts to guaranteeing collaboration among all functional areas that take part in activities related to product design, sourcing, procurement and logistics, as well as use and postuse (Ahi and Searcy, 2013; Badurdeen et al., 2009). This collaboration could lead organizations to explore and implement environmental and social opportunities. For this purpose, it is essential to have a robust information management system to control all processes along the SC.

Third, at the product level, organizations should incorporate social and environmental criteria in the design of the product and service in order to reduce or even eliminate its negative environmental and social impacts. The integration of such criteria in the design of the product and service involves taking into consideration all stages of a product life cycle, from the extraction of raw materials to final product disposal. For this purpose, one of the most commonly used methods to evaluate the environmental and social impact of a product during its production, use, and end-of-life phase is the LCA, as will be seen below.

Finally, with the objective of achieving social and environmental sustainability along the SC, it is especially important to extend sustainability to other SC partners such as suppliers (Krause et al., 2009; Sancha et al., 2016). This is due to the fact that stakeholders, especially customers, are not in contact with all the different actors in the SC (Seuring and Gold, 2013) and, therefore, focal firms should be responsible for their suppliers (Hartmann and Moeller, 2014; Sancha et al., 2016). Extending sustainability to SC partners implies that suppliers can attain the same sustainability performance so that a firm's sustainability performance cannot be damaged by its suppliers (Faruk et al., 2001). Successfully achieving this endeavor involves the integration of sustainable criteria (environmental, social, and economic considerations) into supplier selections, collaboration, and assessment. On the one hand, collaboration with suppliers entails the cooperation between a focal firm and a supplier and is intended to enhance performance in collaboration (Gavronski et al., 2011; Klassen and Vereecke, 2012). This collaboration entails sharing knowledge and information with suppliers, coordinating resources with suppliers such as design, sourcing, and production, and providing assistance, for example, dedicating firm workforce temporarily to the supplier (Lee and Klassen, 2008; Rao, 2002; Sancha et al., 2016; Vachon and Klassen, 2008). In this light, Green et al. (2012) found empirical evidence that environmental collaboration not only improves environmental performance of the focal firm but also its organizational performance.

On the other hand, supplier evaluation is also crucial to secure a valuable relationship between a focal firm and a supplier which improves the social and environmental performance of the focal firm and, ultimately, the sustainability performance of the supplier. Moreover, supplier evaluation allows focal firms to measure progress, assess sustainability performance, and identify what suppliers need to improve (Fritz, 2019; Hahn et al., 1990; Sancha et al., 2016). In this regard, companies conduct audits of their suppliers to ensure they meet social and environmental requirements (Sancha et al., 2016).

Managing sustainability implies the adoption of socially and environmentally responsible practices throughout the SC to achieve sustainability outcomes: social and environmental performance. Social performance considers compliance with human rights, employment of minority groups, improved health and safety, fair labor practices, and impact on local communities (Koberg and Longoni, 2019; Yawar and Seuring, 2017). Environmental performance considers the reduction of resource consumption, pollution, waste and emissions, and recycling (Koberg and Longoni, 2019; Rao and Holt, 2005).

3. Practices for a greener and more sustainable supply chain

The extant literature has categorized certain SSCM activities into two main groups (De Sousa et al., 2013): advanced practices (e.g., Eco-efficiency, Eco-Design, Green Purchasing) and common adopted practices (e.g., Reverse Logistics and Investment Recovery), involving both social and environmental aspects. These practices are all focused on enhancing the firm's environmental performance as well as decreasing its harmful environmental impact (Geng et al., 2017; Payán-Sánchez et al., 2019), in addition to promoting socially responsible behavior by firms in purchasing decisions (Mani et al., 2018). Regarding DfS, it appears to be the response of firm SCM to global sustainability–related issues (Arnette et al., 2014).

3.1 Design for sustainability

Business's understanding of the behavior change required in human society (Ryan, 2013a, b) has evolved over time. This evolution, as Ceschin and Gaziulusoy (2016) have shown, can be observed in reports by the World Business Council for Sustainable Development (WBCSD): in 2000, with product innovation and efficiency promotion to address environmental problems (WBCSD, 2000); in 2004, drafting sustainability risks as intrinsic mega-risks that present exceptional difficulties to organizations and government (WBCSD, 2004); and, by 2010, suggesting a vision for change (WBCSD, 2010).

Current comprehension of sustainability suggests that it is not a property of the individual elements forming systems but rather a property of the system itself (Ceschin and Gaziulusoy, 2016). Therefore, sustainability would require a systemic, multiscale, and process-based approach to pursue goal-oriented sustainability rather than traditional approaches based on goal optimizations (Bagheri and Hjorth, 2007; Walker et al., 2004).

Design is considered one of the primary functions necessary for innovation in business, governments, and other social units, such as local communities (Design Council, 2007; Gruber et al., 2015). Consequently, it has been incorporated into different aspects of the discourse and practice of sustainability since the mid-20th century, more systematically so in the 1980s due to new, heightened interest by industry in environmental and social issues (Ceschin and Gaziulusoy, 2016).

Thus, the DfS appears as a response from the design field to sustainability issues, considered to be the intersection between the SC and the environment within a firm (Arnette et al., 2014). Over the years, literature has increasingly expanded DfS from a technical and product point of view to changes at the large-scale system level where sustainability "is understood as a sociotechnical challenge" (Ceschin and Gaziulusoy, 2016, p. 1).

DfS belongs to the line of "design for" literature in which product design is related to all aspects of product development (i.e., production, distribution, use, and end-of-life). The "Design for" studies are gathered under the umbrella term Design for X (DfX), given that the different techniques have focused on several topics such as manufacturing, SC, and the environment, among others (Arnette et al., 2014). The original approaches of "Design for" were created to improve the operations and production aspects of the product creation process, thereby making it more efficient and, eventually, reducing cost and errors.

A common error in previous "Design for" research, as well as in popular terminology (Vogtlander et al., 2001), is the interchanging of the terms *sustainability* and *environment*. The technique Design for the Environment (DfE) has typically been used as the equivalent of sustainability inside the firm. In this sense, Hart (1997) outlined the role that DfE plays in product stewardship, while simultaneously recognizing that it refers to only one component of sustainable business development (Arnette et al., 2014). The most common definition for sustainability requires the "three pillars" or three Es: ecology, equity (social equity), and economy (Elkington, 1998; United Nations General Assembly, 2005). Thus, focusing exclusively on environmental concerns while utilizing the term sustainability, or vice versa, is both deceiving and inappropriate as focusing on one of the sustainability pillars disregards the other two, possibly leading noneconomical designs to be produced or potentially generating negative social effects (Arnette et al., 2014).

In the work of Arnette et al. (2014), all the types of "Design for" existing in literature are gathered within the three Es dimensions. What this work demonstrates is that there is still a need to combine and consolidate the different DfX approaches into a single taxonomy (i.e., DfS) in order to guide managers and designers through the design process and ultimately provide insights for future research. In this regard, a number of studies have been carried out to produce a more extensive DfS approach based on other DfX studies.

One example is a more recent work by Ceschin and Gaziulusoy (2016) in which the different DfS are collected and categorized into four different innovation levels:

- Product innovation level: focusing on improving existing products or completely developing new ones. This level includes Green design and Eco-design, Emotionally durable design, Design for sustainable behavior, Cradle-to-cradle design and Biomimicry design (Nature-inspired design), and Design for the Base of the Pyramid (BoP).
- Product-service system (PSS) innovation level: focus placed beyond individual products and on integrated combinations of products and services, such as development of new business models. This level includes the PSS design for eco-efficiency, the PSS DfS, and the PSS design for the Bottom of the Pyramid (BoP).
- Spatio-Social innovation level: focusing on human settlements and their communities' spatio-social conditions, addressed on different scales, from neighborhoods to cities. This level consists of Design for social innovation and Systemic design.
- Socio-Technical System innovation level: focusing on promoting radical changes regarding the fulfillment of societal needs, such as nutrition and transport/mobility, thus supporting the shifts to new socio-technical systems. This level is related to Design for system innovations and transitions.

3.2 Eco-efficiency

The concept of eco-efficiency defines business practices that generate economic value while also reducing the use of resources and ecological effects (DeSimone and Popoff, 2000). Thus, eco-efficiency is a key objective in environmental management; it is also considered a noteworthy business approach by several institutions, such as the Organization for Economic Co-operation and Development and the European Environmental Agency (Angelakoglou and Gaidajis, 2015). On an abstract level, there are three approaches for applying efficiency to environmental projects, more specifically eco-efficiency, interpreted by the WBCSD (WBCSD, 2000) as (i) an *economic approach* which is strongly associated with cost efficiency and considers the conditions upon which a specific economic resource makes the maximum profit; (ii) an *ecological efficiency approach* with a purely environmental vision that defines the emissions and the service ratio; and (iii) an *economic ecological efficiency approach* that makes reference to a cross-efficiency that is responsible for addressing economic and environmental problems (Mickwitz et al., 2006). Consequently, there are substantial eco-efficiency benefits to the environment, such as increasing responsibility of businesses regarded environmental protection, reducing hazardous elements, and, as a result, creating a shift toward renewable natural resources (Grosse-Sommer et al., 2020).

The term eco-efficiency is generally understood to mean an essential element for assessing the economic as well as ecological aspects of the environmental issues pertaining to an organization (Carvalho et al., 2017; Tseng et al., 2014). According to Schaltegger and Burritt (2017), in the literature there seems to be no general definition of the term eco-efficiency, and, consequently, this lack of precision (see, e.g., Côté et al., 2006) impedes the evolution of the term's definitions. Many organizations (national, regional, and global) have also offered definitions of this term and they all coincide on the concept of producing "more with less" (e.g., European Environment Agency, Industry Canada and Australia Environmental Protection Agency, among others). However, one of the most widely used definitions was provided by the WBCSD, describing eco-efficiency as "achieving more value from lower inputs of material and energy and with reduced emissions" (Côté et al., 2006, p. 544). In addition, the Environmental Protection Agency argued that by being eco-efficient goods and services can be produced using less raw material and energy, resulting in less waste, pollution, and cost (Sorvari et al., 2009). According to the definition by the World Business Council, various factors are recognized as identifying the concept's main characteristics, namely (Oliveira et al., 2017) reduction of toxic dispersion and material, extension of product durability, increase of goods and services in service intensity, and sustainability of renewable resources.

The main problem of eco-efficiency is ruled by the "win-win" motto (Tregidga et al., 2013). It is widely thought that this philosophy implies a significant restriction on not only the scale of environmental action but also on its scope (Banerjee, 2012). Hence, it efficiently discards the strategies that have an unfavorable environmental effect (Figge and Hahn, 2013). Eco-efficiency is then questioned due to the excessive use of resources that an organization consumes in order to boost economic profits (Passetti and Tenucci, 2016). For this reason, other approaches have been developed to decrease this environmental effect on the management of SCs, such as "eco-effectiveness" (Niero et al., 2017). This concept is based on redesigning fewer environmentally destructive processes, and subsequently, enhancing the company's awareness of creating and being more efficient with less (Nieuwenhuis et al., 2019). It stands to reason that any business reaches a significant level of sustainability due to well-designed eco-efficiency strategies by offering a competitive position within the international market (Zhang et al., 2008). From a different perspective, sustainability is a matter of concern for researchers and practitioners (Carvalho et al., 2017). There is an "ecocentric" paradigm that forces us to examine the notion that supports eco-efficiency (Nieuwenhuis et al., 2019)—how corporations manage their benefits in the first place. In a deeper eco-efficiency analysis, the environmental impacts may be included in relation to both pollution emissions and resource use (Shao et al., 2019).

3.3 Eco-design

Eco-design, a term defined by Johansson (2002), makes reference to a proactive environmental management perspective on product development. Taking into account Deutz et al. (2013), the application of eco-design enhances environmental value of the related product process (e.g., functionality, quality, cost, and performance). Most recently, eco-design has been defined as "the integration of environmental aspects into product design and development with the aim of reducing adverse environmental impacts throughout the whole product's life cycle" (Lewandowska and

Matuszak-Flejszman, 2014, p. 1795). Eco-design approaches are becoming a research stream both in scientific and academic literature. Thus, national and international organizations have carried out sustainability policies as a response to eco-design methods (Favi et al., 2018). Within the framework of more eco-friendly product designs, environmental sustainability has been widely accepted (Marconi and Favi, 2020). There are several factors that include intrinsic ecological characteristics, for example, national and/or international legislation, economic and environmental outcomes, and a wide range of new opportunities (Iranmanesh et al., 2019).

Eco-design aims to improve products' environmental performance (Al-Sheyadi et al., 2019). The eco-design management practices are not product-related practices but process-related. Ciconi (2020) provides a list of eco-design practices related to materials which include, among others, reduced consumption of material, maximum use of recyclable material, use of returnable packaging, and extending the lifespan of materials. The Eco-design Maturity Model (EcoM2) features a number of eco-design management practices from eco-design management studies (see study by Pigoso et al. (2013) for a detailed description of eco-design management practices). Additionally, EcoM2 comprises the incorporation of eco-designs into the development of product processes and organizations, along with barriers and success factors for its implementation (Pigosso et al., 2013).

In general terms, any type of product is designed according to customers' expectations and the desire to increase its demand (Seo et al., 2016). On the one hand, it stands to reason that the main purpose of the new product development team is to focus on the design of an eco-friendly product that consequently creates a high demand (Ganji et al., 2018). On the other hand, the priority of the manufacturing and logistics team is to minimize the costs in the SC (Zhu and He, 2017). Thus, there is a clear conflict of interests between SC and product design issues. In short, it is necessary to align both objectives in order to obtain better results (Flynn et al., 2018). However, it is sometimes difficult to combine all these expectations due to the rapid technological advances of today, along with the awareness of customers concerning eco-friendly products (Tsai and Liao, 2017). As a result, the theory of design for manufacturability requires that every single step of product designs be taken into account to meet all or most of the market requirements (Thamsatitdej et al., 2017).

With regard to the literature, eco-design or DfE is referred to as "green" that incorporate ecological aspects into product development and design (Dangelico and Pontrandolfo, 2017). Green design involves two concepts: LCA and environmentally related design (Eksi and Karaosmanoglu, 2018). This design seeks to develop a less toxic and environmentally friendly product in order to increase customer collaboration in eco-design and minimize waste (Mohtashami et al., 2020). Technological development is carried out through the use of a wide range of integrated activities led by the continual development of the ecological product performance (Schöggl et al., 2017). Product and process development innovation mediates the relationship between sustainability performance and GSCM practices, highlighting the main operational life cycle stages (Silva et al., 2019).

3.4 Green purchasing

Environmentally preferable purchasing (also known as green purchasing) has been conceptually defined as "making environmentally conscious decisions throughout the purchasing process, beginning with product and process design, and through product disposal" by the Institute for Supply Management. There are three main characteristics involved in green purchasing: the reduction of waste, the ability to conserve water and energy, and the use of recycled materials (Dubey et al., 2013). Green purchasing activities cover all environmental considerations related to supply management decisions, along with traditional purchasing aspects, such as supplier location and product price (Yook et al., 2018). This approach considers social and environmental issues as well as several other factors, including performance, cost, and purchase decisions (Çankaya and Sezen, 2019).

Green supply development and its practices have been widely studied, encompassing three main factors: green knowledge communication, transfer of resources and investment, and corporate management activities (Ageron et al., 2012). In the context of SC, suppliers collaborate with this industry by providing the equipment and raw materials (Balon, 2020). According to the US National Institutes of Health, green purchasing is the favorable selection and acquisition of goods and services, which efficiently reduce negative environmental effects over the lifespan of products. Moreover, consumers are concerned with ecological problems (e.g., global warming, decreasing natural resources, pollution) and take them into consideration when conducting green purchasing (Shao and Ünal, 2019).

Green purchasing represents a key activity in the SC for purchasing managers seeking to improve corporate social responsibility (Blome and Paulraj, 2013). The purchasing function is an important stage in the value chain because its success will depend on the integration of the company's environmental efforts, purchasing activities and environmental objectives (Çankaya and Sezen, 2019). Therefore, the goal of green purchasing is considered to be a significant element of GSCM as it integrates environmental issues and concerns into the acquisition process. Green purchasing is mainly concerned with controlling the environmental performance of suppliers (Eltayeb et al., 2011); companies tend to focus on their initial green purchasing activities as less effort is made to build innovative capacities to support green purchasing practices (Foo et al., 2019). For this reason, choosing the most suitable supplier is very important, as the latter has substantial impact on whether a firm meets its environmental objectives. Therefore, the supply process must be conducted by adopting a collaborative and strategic point of view with suppliers. Once the supplier is selected and managed, it is also important to evaluate whether the supplier meets the firms' environmental standards (Luthra et al., 2017).

3.5 Reverse logistics

Reverse logistics has been described as "the process of planning, implementing and controlling the efficient, effective inbound flow and storage of secondary goods and related information opposite to the traditional SC direction for the purpose of recovering value and proper disposal" (Fleischmann, 2001). It is believed that reverse logistics is "unique enough to undergo specialized research" (Tibben-Lembke and Rogers, 2002, p. 271). This approach makes reference to several factors including collection, separation, transition processes, delivery, and integration (Sarkis, 2003). From an environmental point of view, the concept of reverse logistics places emphasis on using reusable materials or returning recyclable ones (Sahay et al., 2006). Thus, reverse logistics is considered a common adopted GSCM practice (De Sousa Jabbour et al., 2013), as it involves traditional transportation and inventory management. Additionally, this process has also been linked to circular economy, which encompasses recycling, reusing, and reducing the quantity of raw materials in production or post-consumption stage (De Oliveira et al., 2019).

The adoption of reverse logistics activities has become increasingly necessary for all organizations as a result of population growth and the depletion of natural resources (Agrawal et al., 2015). In recent years, this concept has gained considerable attention in the literature due to Government legislation, increased environmental awareness, and corporate social responsibility (Govindan and Bouzon, 2018). It offers a business advantage to obtain a competitive position, providing both environmental and economic benefits (Goudenege et al., 2013). Following Mimouni et al. (2015), product returns are collected and disassembled after the product has been used to convert products into three categories: (1) a product which needs minor corrections (i.e., finished goods), (2) a product that requires major corrections (i.e., raw material), and; (3) a final product which is of no use to the market and is finally rejected (i.e., inconvertible product). At management level, it is crucial to shorten the reverse processes and to provide visibility, and in turn, to obtain additional benefits from optimizing reverse logistics (Sciarrotta, 2018). This requires taking measures and monitoring significant Key Performance Indicators (Stuart, 2018).

Reverse logistics activities can generate profits by emphasizing the reduction of resources, that is, obtaining more value from the reduction of disposal costs and the recovery of products (Agrawal and Singh, 2019). Investment recovery is integral for the operations related to the reuse of products and materials (Zhu et al., 2011). Reverse logistics has been perceived by firms as an investment recovery rather than simply a way of minimizing the cost of waste management (Ravi et al., 2005). Investment recovery refers to the company's strategic use of recycling, redistribution, and reselling to obtain greater value from its materials and products (Chan et al., 2012). Its goal is to convert surplus assets into revenue by selling inactive assets, reducing storage space, and reallocating inactive assets for another, more productive use (Somjai and Jermsittiparsert, 2019). Furthermore, investment recovery is also a dimension of study investigated in GSCM (Aslam et al., 2019). This dimension is recognized as a traditional business practice by which products are resold (Zhu and Sarkis, 2004). The principal aim of investment recovery is to obtain the utmost value from obsolete products, attempting to include them in the reverse logistics process (Çankaya and Sezen, 2019). According to Masi et al. (2018), investment recovery may increase a product's life as the product can be recycled into other useable materials. Therefore, collaboration with suppliers is essential to ensure new products are designed in an environmentally sound way.

4. Sustainability evaluation methods

Sustainability performance assessment along the SC has awakened great interest among scholars due to three main points, in keeping with Bai and Sarkis (2010): sustainability understanding from science, regulatory demands from the political landscape, and firms' concern for the environment. These three viewpoints highlight the need to take action to guide organizations in the adoption of various methods to make SCs more sustainable (Ruiz-Méndez and Güereca, 2019).

LCA appears in literature as one of the most pertinent and robust methods used by firms to evaluate the sustainability of SCs and support the decision-making process (Ruiz-Méndez and Güereca, 2019). In this sense, the different typologies of LCA that have been studied will be presented: Full LCA, Direct LCA, and Streamlined LCA. In addition, environmental labeling is also studied, as it plays a key role in guiding decision-making within the firm as well (ISO, 2000).

4.1 Full LCA

LCA is a highly scientific robust methodology that provides a well-structured framework for the analysis of the environmental performance of products and services in a holistic way, either in the public or private sectors, while being controlled by international standards, such as ISO 14040 and ISO 14044 (UNEP/SETAC, 2015). LCA evaluation considers all the stages of the product life cycle, from raw materials extraction to the distribution of the final product (ISO, 2006a; Ruiz-Méndez and Güereca, 2019).

A full LCA process consists of four reiterative phases, as Ruiz-Méndez and Güereca (2019) outlined: target and scope definition, life cycle record analysis, life cycle impact evaluation, and interpretation (ISO, 2006a, b). The results obtained when performing a full LCA are considered of high quality, constituting one of the main advantages of the method. Normally, it is the first study performed for a product or group of products, when no other analysis has been conducted previously. Thus, it aids in recognizing the significance and order of the stages to perform, precisely to reduce the environmental impacts and improve the process within the firm (Ruiz-Méndez and Güereca, 2019).

Although full LCA entails two major challenges, i.e., the complexity of the method demanding specialized knowledge and the large amount of resources needed to perform LCA (Muralikrishna and Manickam, 2017; Scanlon et al., 2013), these problems do not detract merit from the method (Ruiz-Méndez and Güereca, 2019). The use of a detailed LCA can help with methods of standardization and harmonization to support policies, proving of utmost importance to practitioners and legislators (Bai and Sarkis, 2010). In fact, LCA is the approach suggested by the United Nations for the firm to be able to accomplish Sustainable Consumption and Production (SCP) patterns (UN/UNEP, 2007). Additionally, a full LCA is a common preliminary step for other life cycle approach instruments, communicating the results through Environmental Product Declarations or simplified models, such as eco-labels (European Commission, 2010; UNEP, 2015).

As Ruiz-Méndez and Güereca (2019) assert, academies face significant challenges in this respect, with the necessary improvements of life cycle databases and essential staff training on topics such as SCP, LCA, and environmental impacts, so they may learn how to manage the tool. Furthermore, it is important to expand the method so that the knowledge generated goes beyond the academic domain.

4.2 Direct life cycle assessment

Direct LCA consists of a more simplified LCA, with respect to communication. In this sense, it is a more straightforward method to communicate certain product's environmental specifications, being a more easily understandable eco-label for those who are nonspecialists in the matter (Ruiz-Méndez and Güereca, 2019).

This simplification might come from restraining the scope of the study regarding the impact categories evaluated (Finkbeiner, 2016) or from the exclusion of life cycle stages of the process (Todd and Curran, 1999), either upstream or downstream. In the first case, two concepts appear as primary drivers (Ruiz-Méndez and Güereca, 2019): water footprint and carbon footprint, both focused on only one impact category—water waste and climate change, respectively. As for the latter, it can involve evaluations from resources extraction to the factory gate (commonly known as cradle-to-gate) or from resources extraction to final disposal (i.e., cradle-to-grave) (ISO, 2014).

4.3 Streamlined life cycle assessment

As previously stated, applying the full LCA method demands a great deal of resources: information/data, funds, time, and expertise (Ruiz-Méndez, 2019), which ultimately limits its applicability. While seeking alternatives to easily implement LCA, especially in areas such as design, packaging, and materials and suppliers selection (Skaar and Fet, 2012), researchers have managed to develop a streamlined LCA. It attempts to maintain the technical attributes of the fixed structure of LCA while simultaneously making the method more achievable, efficient, and direct (Muralikrishna and Manickam, 2017), thereby helping during product assessment and final supplier selection.

There is an increasing need among businesses for technical skills to correctly evaluate the environmental impacts of their products and services, constantly seeking less expensive methods that provide practical and useful information in less time. Streamlined LCA tools appear to satisfy this necessity by providing results on products' environmental performance in less time and at a fractional cost (Ruiz-Méndez and Güereca, 2019), applied mostly in product development and procurement processes within the firm (Hochschorner and Finnveden, 2003).

Guinée et al. (2001, p. 95) described the streamlined LCA as "a simplified variety of detailed LCA conducted according to guidelines not in full compliance with the ISO 14040/44 standards and representative of studies typically requiring from 1 to 20 person-days of work." Life cycle inventory is one of the most time-consuming steps, which is where most efforts have been focused (Zamagni et al., 2008), and to which more robust and specially designed life cycle databases have contributed (Ruiz-Méndez and Güereca, 2019).

Streamlined LCA methodology consists of distinguishing omissible elements of an LCA or where database information can be utilized without significantly influencing the quality of the outcomes (Todd and Curran, 1999). Therefore, the complexity of a full LCA study can be diminished, simultaneously reducing aspects such as the costs, time, and efforts required for the LCA study. This streamlined LCA can be achieved by excluding some life cycle phases from the process, diminishing process's inputs and outputs, lessening the quantity of impact categories assessed, or utilizing generic information for the system under examination (Ruiz-Méndez and Güereca, 2019).

Although full LCA has been standardized within the ISO 14040 series (e.g., ISO 14040 and ISO 14044), there is no standard for streamlined LCA (Scanlon et al., 2013). Essentially, streamlined LCA is founded on the idea of the impossibility of a "one-size-fits-all" method for simplification (Finkbeiner, 2016; Todd and Curran, 1999), since "the choice of the most suitable simplified method, or combination of simplified methods, depends on the type of results users are looking for" (Zamagni et al., 2008, p. 114). In other words, the criteria will be selected depending on the type of results sought by the organization and the application field (Hochschorner and Finnveden, 2003).

This lack of standardization is a challenge for firms seeking a rapid evaluation of their supply, following LCA principles to fulfill the international challenges of SCP. Thus, as Ruiz-Méndez and Güereca (2019) highlight, some of the leading areas of streamlined LCA could be utilized as guidelines (i.e., packaging design, eco-design, and materials selection) to perform the evaluation—first, to avoid repeating actions every time an assessment is required, and second, to allow nonspecialized staff to apply the method, thanks to how clear the technique is and that it can be utilized with merely a fundamental understanding of LCA.

4.4 Environmental labeling

Labels and environmentally related declarations are environmental management tools providing information about the environmental attributes or characteristics of products or services offered by the firm (ISO, 2000), rather than performance qualifications. Since these labels are not explained to buyers, they may draw their own conclusions, meaning each of their decisions can be different. In this sense, ISO has been working to regulate and normalize eco-labels to address this issue, eventually resulting in three types of environmental labeling specifically regulated by the ISO 14020 series, as Ruiz-Méndez and Güereca (2019) reflect

- Type I (regulated by ISO 14024). Through a voluntary labeling program, a third party provides a license approving the utilization of environmental labels on items that demonstrate an item's normal environmental inclination over another dependent on life cycle contemplations (ISO, 1999).
- Type II (regulated by ISO 14021). Here the label includes self-declarations by manufacturers, importers, distributors, sellers, or anyone benefitting from the label. Since the information comes from interested parties in the process (with no third-party verification), reliability must be guaranteed to avoid negative consequences on the market (ISO, 2016), with clear, transparent, and scientifically based information.

- Type III (regulated by ISO 14025). The label provides quantified environmental information on the product's life cycle which makes it possible to compare products with the same functions as well as carry out an LCA for the products of the same group (UNEP, 2015). This information is given by one or more organizations, based on an independent certification of LCA data, life cycle record analysis, or information modules (according to ISO 14040), using deliberate parameters, which is all controlled by a program manager (ISO, 2006b).

The standardization of these three types of eco-labels allows a greater understanding by users of the technical concepts of the products, making labeling more robust and reliable, and avoiding the confusion generated with general environmental labeling (Ruiz-Méndez and Güereca, 2019).

5. Conclusions

The aim of this chapter was to explain that, in addition to establishing a change in the organization's areas that have the most influence on resource management, it is important to adopt innovative approaches to SCM. The review conducted shows that the approach used to achieve firm goals can prove to be a crucial factor for changing the current structure of corporate SC to a more sustainable and proactive environmental and social management system. It stands to reason that any business achieves a substantial level of sustainability due to well-designed advanced and common implemented SSCM strategies, by offering a competitive position within the international market.

In this chapter, we first presented several definitions which lay the ground for the subsequent sections of this chapter. We introduced and compared the concepts of GSCM and SSCM. SSCM is conceived as a means to achieve sustainability targets through the SC. At present, organizations must strive to become more environmentally and socially responsible while maintaining their economic competitiveness. To successfully achieve this goal, several strategies have been outlined to integrate sustainability into the SC of organizations.

On the one hand, certain business practices have been detailed (i.e., eco-efficiency, eco-design, green purchasing, reverse logistics, investment recovery) which would lead to a decrease in the harmful environmental impact of companies and the improvement of their environmental performance, while also revealing the importance of a firm's socially responsible behavior. The key advantages of these practices are appraising the economic as well as ecological aspects of an organization's environmental problems, reducing the use of resources and addressing ecological issues. It is based on product development and design, the purchasing process, reuse and recyclable materials for recovery products.

In the same vein, DfS has appeared in literature as the point where the firm's SC and the environment converge, constituting the response from the design field to sustainability issues. Over time, each one of the "Design for" techniques has been used to refer to only one of the dimensions of sustainability (i.e., the environment), although the literature has treated these terms as interchangeable. As a solution, DfS has been developed to combine and consolidate the different "Design for" approaches into a single taxonomy, guiding managers and designers through the design process and providing insights for future research.

On the other hand, besides these practices, certain sustainability evaluation methods have been presented, given that sustainability performance assessment along the SC has become crucial to make organizations and their activities more sustainable. In this sense, LCA appears in literature as one of the most influential and robust methodologies to assess the environmental performance of products and services, which along with environmental labeling plays key roles supporting and guiding the decision-making process within the firm. Although full LCA and eco-labeling have been found to be standardized (ISO 14040 series and ISO 14020 series, respectively) to allow a greater understanding by users of the procedures, for Streamlined LCA there is no standard. In contrast, this tool contributes by providing results of products' environmental performance in less time and at a fractional cost, saving more resources than full LCA.

References

Ageron, B., Gunasekaran, A., Spalanzani, A., 2012. Sustainable supply management: an empirical study. Int. J. Prod. Econ. 140 (1), 168–182.

Agrawal, S., Singh, R.K., 2019. Analyzing disposition decisions for sustainable reverse logistics: triple Bottom Line approach. Resour. Conserv. Recycl. 150, 104448.

Agrawal, S., Singh, R.K., Murtaza, Q., 2015. A literature review and perspectives in reverse logistics. Resour. Conserv. Recycl. 97, 76–92.

Ahi, P., Searcy, C., 2013. A comparative literature analysis of definitions for green and sustainable supply chain management. J. Clean. Prod. 52, 329–341.

Al-Sheyadi, A., Muyldermans, L., Kauppi, K., 2019. The complementarity of green supply chain management practices and the impact on environmental performance. J. Environ. Manag. 242, 186–198.

Angelakoglou, K., Gaidajis, G., 2015. A review of methods contributing to the assessment of the environmental sustainability of industrial systems. J. Clean. Prod. 108, 725–747.

Arnette, A.N., Brewer, B.L., Choal, T., 2014. Design for sustainability (DFS): the intersection of supply chain and environment. J. Clean. Prod. 83, 374–390.

Ashby, A., Leat, M., Hudson-Smith, M., 2012. Making connections: a review of supply chain management and sustainability literature. Supply Chain Manag. 17 (5), 497–516. https://doi.org/10.1108/13598541211258573.

Aslam, M.M.H., Waseem, M., Khurram, M., 2019. Impact of green supply chain management practices on corporate image: mediating role of green communications. Pakistan J. Commerce Soc. Sci. 13 (3), 581–598.

Badurdeen, F., Iyengar, D., Goldsby, T.J., Metta, H., Gupta, S., Jawahir, I.S., 2009. Extending total life-cycle thinking to sustainable supply chain design. Int. J. Prod. Lifecycle Manag. 4 (1–3), 49–67.

Bagheri, A., Hjorth, P., 2007. Planning for sustainable development: a paradigm shift towards a process-based approach. Sustain. Dev. 15 (2), 83–96.

Bai, C., Sarkis, J., 2010. Integrating sustainability into supplier selection with grey system and rough set methodologies. Int. J. Prod. Econ. 124 (1), 252–264.

Balon, V., 2020. Green supply chain management: pressures, practices, and performance-an integrative literature review. Business Strategy Dev. 3 (2), 226–244.

Banerjee, S.B., 2012. Critical perspectives on business and the natural environment. In: Bansal, P., Hoffman, A.J. (Eds.), The Oxford Handbook of Business and the Natural Environment. Oxford University Press, Oxford.

Blome, C., Paulraj, A., 2013. Ethical climate and purchasing social responsibility: a benevolence focus. J. Bus. Ethics 116 (3), 567–585.

Boström, M., Jönsson, A.M., Lockie, S., Mol, A.P., Oosterveer, P., 2015. Sustainable and responsible supply chain governance: challenges and opportunities. J. Clean. Prod. 107, 1–7.

Çankaya, S.Y., Sezen, B., 2019. Effects of green supply chain management practices on sustainability performance. J. Manuf. Technol. Manag. 30 (1), 98–121.

Carter, C.R., Rogers, D.S., 2008. A framework of sustainable supply chain management: moving toward new theory. Int. J. Phys. Distrib. Logist. Manag. 35 (5), 360–387.

Carvalho, H., Govindan, K., Azevedo, S.G., Cruz-Machado, V., 2017. Modelling green and lean supply chains: an eco-efficiency perspective. Resour. Conserv. Recycl. 120, 75–87.

Ceschin, F., Gaziulusoy, I., 2016. Evolution of design for sustainability: from product design to design for system innovations and transitions. Des. Stud. 47, 118–163.

Chan, R.Y., He, H., Chan, H.K., Wang, W.Y., 2012. Environmental orientation and corporate performance: the mediation mechanism of green supply chain management and moderating effect of competitive intensity. Ind. Market. Manag. 41 (4), 621–630.

Cicconi, P., 2020. Eco-design and Eco-materials: an interactive and collaborative approach. Sustain. Mater. Technol. 23, e00135.

Côté, R., Booth, A., Louis, B., 2006. Eco-efficiency and SMEs in nova scotia, Canada. J. Clean. Prod. 14 (6–7), 542–550.

Dangelico, R.M., Pujari, D., Pontrandolfo, P., 2017. Green product innovation in manufacturing firms: a sustainability-oriented dynamic capability perspective. Bus. Strat. Environ. 26 (4), 490–506.

Design Council, 2007. The Value of Design: Factfinder Report. Design Council, UK.

De Oliveira, C.T., Luna, M.M., Campos, L.M., 2019. Understanding the Brazilian expanded polystyrene supply chain and its reverse logistics towards circular economy. J. Clean. Prod. 235, 562–573.

De Sousa Jabbour, A.B.L., de Souza Azevedo, F., Arantes, A.F., Jabbour, C.J.C., 2013. Green supply chain management in local and multinational high-tech companies located in Brazil. Int. J. Adv. Manuf. Technol. 68 (1–4), 807–815.

DeSimone, L.D., Popoff, F., 2000. Eco-efficiency: The Business Link to Sustainable Development. MIT Press.

Deutz, P., McGuire, M., Neighbour, G., 2013. Eco-design practice in the context of a structured design process: an interdisciplinary empirical study of UK manufacturers. J. Clean. Prod. 39, 117–128.

Dubey, R., Bag, S., Ali, S.S., Venkatesh, V.G., 2013. Green purchasing is key to superior performance: an empirical study. Int. J. Procure. Manag. 6 (2), 187–210.

Dubey, R., Gunasekaran, A., Papadopoulos, T., Childe, S.J., Shibin, K.T., Wamba, S.F., 2017. Sustainable supply chain management: framework and further research directions. J. Clean. Prod. 142, 1119–1130.

Eksi, G., Karaosmanoglu, F., 2018. Life cycle assessment of combined bioheat and biopower production: an eco-design approach. J. Clean. Prod. 197, 264–279. https://doi.org/10.1016/j.jclepro.2018.06.151.

Elkington, J., 1998. Partnerships from cannibals with forks: the triple bottom line of 21st-century business. Environ. Qual. Manag. 8 (1), 37–51.

Eltayeb, T.K., Zailani, S., Ramayah, T., 2011. Green supply chain initiatives among certified companies in Malaysia and environmental sustainability: investigating the outcomes. Resour. Conserv. Recycl. 55 (5), 495–506.

European Commission, 2010. *Making Sustainable Consumption and Production a Reality*. European Union.

Faruk, A.C., Lamming, R.C., Cousins, P.D., Bowen, F.E., 2001. Analyzing, mapping, and managing environmental impacts along supply chains. J. Ind. Ecol. 5 (2), 13–36.

Favi, C., Germani, M., Mandolini, M., Marconi, M., 2018. Implementation of a software platform to support an eco-design methodology within a manufacturing firm. Int. J. Sustain. Eng. 11 (2), 79–96.

Feng, Y., Zhu, Q., Lai, K.H., 2017. Corporate social responsibility for supply chain management: a literature review and bibliometric analysis. J. Clean. Prod. 158, 296–307.

Figge, F., Hahn, T., 2013. Value drivers of corporate eco-efficiency: management accounting information for the efficient use of environmental resources. Manag. Account. Res. 24 (4), 387–400.

Finkbeiner, M., 2016. Special Types of Life Cycle Assessment. Springer, Berlin.

Fleischmann, M., Beullens, P., Bloemhof-Ruwaard, J.M., Van Wassenhove, L.N., 2001. The impact of product recovery on logistics network design. Prod. Oper. Manag. 10 (2), 156–173.

Flynn, B., Pagell, M., Fugate, B., 2018. Survey research design in supply chain management: the need for evolution in our expectations. J. Supply Chain Manag. 54 (1), 1–15.

Foo, M.Y., Kanapathy, K., Zailani, S., Shaharudin, M.R., 2019. Green purchasing capabilities, practices and institutional pressure. Manag. Environ. Qual. Int. J. 30 (5), 1171–1189.

Fritz, M.M.C., 2019. Sustainable supply chain management. In: Leal Filho, W., Azul, A., Brandli, L., Özuyar, P., Wall, T. (Eds.), Responsible Consumption and Production. Springer, Cham, pp. 1–14. Encyclopedia of the UN Sustainable Development Goals.

Ganji, E.N., Shah, S., Coutroubis, A., 2018. An examination of product development approaches within demand driven chains. Asia Pac. J. Market. Logist. 30 (5), 1183–1199.

Gavronski, I., Klassen, R.D., Vachon, S., do Nascimento, L.F.M., 2011. A resource-based view of green supply management. Transport. Res. E Logist. Transport. Rev. 47 (6), 872–885.

Geng, R., Mansouri, S.A., Aktas, E., 2017. The relationship between green supply chain management and performance: a meta-analysis of empirical evidences in Asian emerging economies. Int. J. Prod. Econ. 183, 245–258.

Goudenege, G., Chu, C., Jemai, Z., 2013. Reusable containers management: from a generic model to an industrial case study. Supply Chain Forum Int. J. 14 (2), 26–38.

Govindan, K., Bouzon, M., 2018. From a literature review to a multi-perspective framework for reverse logistics barriers and drivers. J. Clean. Prod. 187, 318–337.

Green, K.W., Zelbst, P.J., Bhadauria, V.S., Meacham, J., 2012. Do environmental collaboration and monitoring enhance organizational performance? Ind. Manag. Data Syst. 112 (2), 186–205.

Grosse-Sommer, A.P., Grünenwald, T.H., Paczkowski, N.S., van Gelder, R.N., Saling, P.R., 2020. Applied sustainability in industry: the BASF Eco-efficiency toolbox. J. Clean. Prod. 120792.

Gruber, M., de Leon, N., George, G., Thompson, P., 2015. Managing by design. Acad. Manag. J. 58 (1), 1–7.

Guinée, J.B., Gorré, M., Heijungs, R., Huppes, G., Klejn, R., de Koning, A., van Oers, L., Wegener, S.A., Suh, S., Udo de Haes, H.A., de Brujin, H., van Duin, R., Huijbregts, M.A.J., 2001. Life Cycle Assessment, an Operational Guide to the ISO Standards. Kluwer Academic Publishers, Dordrecht.

Haake, H., Seuring, S., 2009. Sustainable procurement of minor items—exploring limits to sustainability. Sustain. Dev. 17 (5), 284—294.

Hahn, C.K., Watts, C.A., Kim, K.Y., 1990. The supplier development program: a conceptual model. J. Purch. Mater. Manag. 26 (2), 2—7.

Hart, S.L., 1997. Beyond greening: strategies for a sustainable world. Harv. Bus. Rev. 75 (1), 66—77.

Hartmann, J., Moeller, S., 2014. Chain liability in multitier supply chains? Responsibility attributions for unsustainable supplier behavior. J. Oper. Manag. 32 (5), 281—294.

Hochschorner, E., Finnveden, G., 2003. Evaluation of two simplified life cycle assessment methods. Int. J. Life Cycle Assess. 8 (3), 119—128.

Iranmanesh, M., Fayezi, S., Hanim, S., Hyun, S.S., 2019. Drivers and outcomes of eco-design initiatives: a cross-country study of Malaysia and Australia. Rev. Manag. Sci. 13 (5), 1121—1142.

ISO, 1999. ISO 14024. Environmental Labels and Declarations — Type I Environmental Labelling — Principles and Procedures.

ISO, 2000. ISO 14020:2000. Environmental Labels and Declarations — General Principles.

ISO, 2006a. ISO 14040:2006. Environmental Management — Life Cycle Assessment — Principles and Framework.

ISO, 2006b. ISO 14025. Environmental Labels and Declarations — Type III Environmental Declarations — Principles and Procedures.

ISO, 2014. ISO/TS 14072:2014. Environmental Management — Life Cycle Assessment — Requirements and Guidelines for Organizational Life Cycle Assessment.

ISO, 2016. ISO 14021. Environmental Labels and Declarations — Self-Declared Environmental Claims (Type II Environmental Labelling).

Johansson, G., 2002. Success factors for integration of ecodesign in product development. Environ. Manag. Health 13 (1), 96—107.

Kaur, A., Sharma, P.C., 2018. Social sustainability in supply chain decisions: Indian manufacturers. Environ. Dev. Sustain. 20 (4), 1707—1721.

Klassen, R.D., Vereecke, A., 2012. Social issues in supply chains: capabilities link responsibility, risk (opportunity), and performance. Int. J. Prod. Econ. 140 (1), 103—115.

Krause, D.R., Vachon, S., Klassen, R.D., 2009. Special topic forum on sustainable supply chain management: introduction and reflections on the role of purchasing management. J. Supply Chain Manag. 45 (4), 18—25.

Koberg, E., Longoni, A., 2019. A systematic review of sustainable supply chain management in global supply chains. J. Clean. Prod. 207, 1084—1098.

Labella-Fernández, A., Martínez-del-Río, J., 2019. Green human resource management. In: Leal Filho, W., Azul, A., Brandli, L., Özuyar, P., Wall, T. (Eds.), Responsible Consumption and Production. Encyclopedia of the UN Sustainable Development Goals. Springer, Cham.

Lambert, D., Stock, J.R., Ellram, L.M., 1998. Fundamentals of Logistics Management. McGraw-Hill, Irwin.

Lee, S.Y., Klassen, R.D., 2008. Drivers and enablers that foster environmental management capabilities in small-and medium-sized suppliers in supply chains. Prod. Oper. Manag. 17 (6), 573—586.

Lewandowska, A., Matuszak-Flejszman, A., 2014. Eco-design as a normative element of environmental management systems-the context of the revised ISO 14001: 2015. Int. J. Life Cycle Assess. 19 (11), 1794–1798.

Luthra, S., Govindan, K., Kannan, D., Mangla, S.K., Garg, C.P., 2017. An integrated framework for sustainable supplier selection and evaluation in supply chains. J. Clean. Prod. 140, 1686–1698.

Martins, C.L., Pato, M.V., 2019. Supply chain sustainability: a tertiary literature review. J. Clean. Prod. 225, 995–1016.

Mani, V., Gunasekaran, A., Delgado, C., 2018. Enhancing supply chain performance through supplier social sustainability: an emerging economy perspective. Int. J. Prod. Econ. 195, 259–272. https://doi.org/10.1016/j.ijpe.2017.10.025.

Marconi, M., Favi, C., 2020. Eco-design teaching initiative within a manufacturing company based on LCA analysis of company product portfolio. J. Clean. Prod. 242, 118424.

Masi, F., Rizzo, A., Regelsberger, M., 2018. The role of constructed wetlands in a new circular economy, resource oriented, and ecosystem services paradigm. J. Environ. Manage. 216, 275–284. https://doi.org/10.1016/j.jenvman.2017.11.086.

Meadows, D., Randers, J., Meadows, D., 2004. Limits to Growth: The 30-Year Update. Chelsea Green Publishing.

Mentzer, J.T., DeWitt, W., Keebler, J.S., Min, S., Nix, N.W., Smith, C.D., Zacharia, Z.G., 2001. Defining supply chain management. J. Bus. Logist. 22 (2), 1–25.

Mickwitz, P., Melanen, M., Rosenström, U., Seppälä, J., 2006. Regional eco-efficiency indicators-a participatory approach. J. Clean. Prod. 14 (18), 1603–1611.

Mimouni, F., Abouabdellah, A., Mharzi, H., 2015. Study of the reverse logistics' break-even in a direct supply chain. Int. Rev. Model. Simul. 8 (2), 277–283.

Mohtashami, Z., Aghsami, A., Jolai, F., 2020. A green closed loop supply chain design using queuing system for reducing environmental impact and energy consumption. J. Clean. Prod. 242, 118452.

Morana, J., 2013. FOCUS Series: Sustainable Supply Chain Management. Wiley, Somerset.

Muralikrishna, I.V., Manickam, V., 2017. Life cycle assessment. In: Muralikrishna, I.V., Manickam, V. (Eds.), Environmental Management: Science and Engineering for Industry. Elsevier, Oxford, Butterworth-Heinemann, UK, pp. 57–75.

Niero, M., Hauschild, M.Z., Hoffmeyer, S.B., Olsen, S.I., 2017. Combining eco-efficiency and eco-effectiveness for continuous loop beverage packaging systems: lessons from the Carlsberg circular community. J. Ind. Ecol. 21 (3), 742–753.

Nieuwenhuis, P., Touboulic, A., Matthews, L., 2019. Sustainable supply chain management sustainable? In: Yakovleva, N., Frei, R., Murthy, S.R. (Eds.), Sustainable Development Goals and Sustainable Supply Chains in the Post-global Economy. Springer, Cham, pp. 13–30.

Oliveira, R., Camanho, A.S., Zanella, A., 2017. Expanded eco-efficiency assessment of large mining firms. J. Clean. Prod. 142, 2364–2373.

Pagell, M., Shevchenko, A., 2014. Why research in sustainable supply chain management should have no future. J. Supply Chain Manag. 50 (1), 44–55.

Pagell, M., Wu, Z., 2009. Building a more complete theory of sustainable supply chain management using case studies of 10 exemplars. J. Supply Chain Manag. 45 (2), 37–56.

Passetti, E., Tenucci, A., 2016. Eco-efficiency measurement and the influence of organisational factors: evidence from large Italian companies. J. Clean. Prod. 122, 228–239.

Payán-Sánchez, B., Pérez-Valls, M., Plaza-Úbeda, J.A., 2019. Supply chain management in a degrowth context: the potential contribution of stakeholders. Sustainable Development Goals and Sustainable Supply Chains in the Post-global Economy. In: Yakovleva, N., Frei, R., Murthy, S.R. (Eds.), In: Greening of Industry Networks Studies 7. Springer Nature Switzerland AG, pp. 31–45. https://doi.org/10.1007/978-3-030-15066-2_3.

Pigosso, D.C., Rozenfeld, H., McAloone, T.C., 2013. Ecodesign maturity model: a management framework to support ecodesign implementation into manufacturing companies. J. Clean. Prod. 59, 160–173.

Rao, P., 2002. Greening the supply chain: a new initiative in South East Asia. Int. J. Oper. Prod. Manag. 22 (6), 32–655.

Rao, P., Holt, D., 2005. Do green supply chains lead to competitiveness and economic performance? Int. J. Oper. Prod. Manag. 25 (9), 898–916.

Rauch, A., Wiklund, J., Lumpkin, G.T., Frese, M., 2009. Entrepreneurial orientation and business performance: an assessment of past research and suggestions for the future. Enterpren. Theor. Pract. 33 (3), 761–787.

Ravi, V., Shankar, R., Tiwari, M.K., 2005. Analyzing alternatives in reverse logistics for end-of-life computers: ANP and balanced scorecard approach. Comput. Ind. Eng. 48 (2), 327–356.

Ruiz-Méndez, D., Güereca, L.P., 2019. Streamlined life cycle assessment for the environmental evaluation of products in the supply chain. In: Yakovleva, N., Frei, R., Rama Murthy, S. (Eds.), Sustainable Development Goals and Sustainable Supply Chains in the Post-global Economy. Springer, Cham, pp. 115–131.

Ryan, C., 2013a. Critical agendas: designing for sustainability from products and systems. In: Walker, S., Giard, J. (Eds.), The Handbook of Design for Sustainability. Bloomsbury, London, UK, pp. 408–427.

Ryan, C., 2013b. Eco-acupuncture: designing and facilitating pathways for urban transformation, for a resilient low-carbon future. J. Clean. Prod. 50, 189–199.

Sahay, B.S., Srivastava, S.K., Srivastava, R.K., 2006. Managing product returns for reverse logistics. Int. J. Phys. Distrib. Logist. Manag. 36 (7), 524–546.

Sancha, C., Gimenez, C., Sierra, V., 2016. Achieving a socially responsible supply chain through assessment and collaboration. J. Clean. Prod. 112, 1934–1947.

Sarkis, J., 2003. A strategic decision framework for green supply chain management. J. Clean. Prod. 11 (4), 397–409.

Sarkis, J., Zhu, Q., Lai, K.H., 2011. An organizational theoretic review of green supply chain management literature. Int. J. Prod. Econ. 130 (1), 1–15. https://doi.org/10.1016/j.ijpe.2010.11.010.

Scanlon, K.A., Cammatata, C., Siart, S., 2013. Introducing a streamlined life cycle assessment approach for evaluating sustainability in defense acquisitions. Environ. Syst. Decisions 33, 209–223.

Schaltegger, S., Burritt, R., 2017. Contemporary Environmental Accounting: Issues, Concepts and Practice. Routledge.

Schöggl, J.P., Baumgartner, R.J., Hofer, D., 2017. Improving sustainability performance in early phases of product design: a checklist for sustainable product development tested in the automotive industry. J. Clean. Prod. 140, 1602–1617.

Sciarrotta, T., 2018. Directing reverse logistics – a corporate paradigm shift. Reverse Logist. Magazine 12 (4), 40–41.

Seo, S., Ahn, H.K., Jeong, J., Moon, J., 2016. Consumers' attitude toward sustainable food products: ingredients vs. packaging. Sustainability 8 (10), 1073.

Seuring, S., Gold, S., 2013. Sustainability management beyond corporate boundaries: from stakeholders to performance. J. Clean. Prod. 56, 1–6.

Seuring, S., Müller, M., 2008. From a literature review to a conceptual framework for sustainable supply chain management. J. Clean. Prod. 16 (15), 1699–1710.

Shao, J., Ünal, E., 2019. What do consumers value more in green purchasing? Assessing the sustainability practices from demand side of business. J. Clean. Prod. 209, 1473–1483.

Silva, G.M., Gomes, P.J., Sarkis, J., 2019. The role of innovation in the implementation of green supply chain management practices. Bus. Strat. Environ. 28 (5), 819–832.

Simpson, D.F., Power, D.J., 2005. Use the supply relationship to develop lean and green suppliers. Supply Chain Manag.: Int. J. 10 (1), 60–68.

Skaar, C., Fet, A.M., 2012. Accountability in the value chain: from environmental product declaration (EPD) to CSR product declaration. Corp. Soc. Responsib. Environ. Manag. 19 (4), 228–239.

Somjai, S., Jermsittiparsert, K., 2019. The trade-off between cost and environmental performance in the presence of sustainable supply chain. Int. J. Supply Chain Manag. 8 (4), 237–247.

Sorvari, J., Antikainen, R., Kosola, M.L., Hokkanen, P., Haavisto, T., 2009. Eco-efficiency in contaminated land management in Finland—barriers and development needs. J. Environ. Manag. 90 (5), 1715–1727.

Srivastava, S.K., 2007. Green supply-chain management: a state-of-the-art literature review. Int. J. Manag. Rev. 9 (1), 53–80.

Stewart, R., Niero, M., 2018. Circular economy in corporate sustainability strategies: a review of corporate sustainability reports in the fast-moving consumer goods sector. Bus. Strat. Environ. 27 (7), 1005–1022.

Thamsatitdej, P., Boon-Itt, S., Samaranayake, P., Wannakarn, M., Laosirihongthong, T., 2017. Eco-design practices towards sustainable supply chain management: interpretive structural modelling (ISM) approach. Int. J. Sustain. Eng. 10 (6), 326–337.

Tibben-Lembke, R.S., Rogers, D.S., 2002. Differences between forward and reverse logistics in a retail environment. Supply Chain Manag.: Int. J. 7 (5), 271–282.

Todd, J.A., Curran, M.A., 1999. Streamlined Life-Cycle Assessment: A Final Report from the SETAC North America Streamlined LCA Workgroup. Society of Environmental Toxicology and Chemistry (9. 31). SETAC Press, Pensacola, FL, 6/99.

Touboulic, A., Chicksand, D., Walker, H., 2014. Managing imbalanced supply chain relationships for sustainability: a power perspective. Decis. Sci. J. 45 (4), 577–619.

Tregidga, H., Kearins, K., Milne, M., 2013. The politics of knowing "organizational sustainable development". Organ. Environ. 26 (1), 102–129.

Tsai, K.H., Liao, Y.C., 2017. Innovation capacity and the implementation of eco-innovation: toward a contingency perspective. Bus. Strat. Environ. 26 (7), 1000–1013.

Tseng, M.L., Tan, K.H., Lim, M., Lin, R.J., Geng, Y., 2014. Benchmarking eco-efficiency in green supply chain practices in uncertainty. Prod. Plann. Contr. 25 (13–14), 1079–1090.

UN/UNEP, 2007. Proceso de Marrakech sobre Consumo y Producción Sustentable. New York, US.

UNEP, 2015. Product Sustainability Information: State of Play and Way Forward. United Nations Environment Programme Division of Technology, Industry and Economics, Paris.

UNEP/SETAC, 2015. Life Cycle Initiative. Available from: http://www.lifecycleinitiative.org/. (Accessed 10 June 2020).

United Nations General Assembly, 2005. 2005 World Summit Outcome, Resolution A/60/1, Adopted by the General Assembly on 15 September 2005.
Vachon, S., Klassen, R.D., 2008. Environmental management and manufacturing performance: the role of collaboration in the supply chain. Int. J. Prod. Econ. 111 (2), 299–315.
Vogtländer, J.G., Brezet, H., Hendriks, C.F., 2001. The virtual eco-costs '99 A single LCA-based indicator for sustainability and the eco-costs-value ratio (EVR) model for economic allocation. Int. J. Life Cycle Assess. 6 (3), 157–166.
Walker, B., Holling, C.S., Carpenter, S.R., Kinzig, A., 2004. Resilience, adaptability and transformability in social-ecological systems. Ecol. Soc. 9 (2), 5.
WBCSD, 2000. Eco-efficiency: Creating More Value with Less Impact. WBCSD.
WBCSD, 2004. Running the Risk e Risk and Sustainable Development: A Business Perspective. WBCSD.
WBCSD, 2010. Vision 2050: The New Agenda for Business. WBCSD.
WCED, 1987. Our Common Future. World Commission on Environment and Development, Oxford University Press, Oxford.
Wittstruck, D., Teuteberg, F., 2012. Understanding the success factors of sustainable supply chain management: empirical evidence from the electrics and electronics industry. Corp. Soc. Responsib. Environ. Manag. 19 (3), 141–158.
Editorial introduction: achieving sustainable development goals through sustainable supply chains in the post-global economy. In: Yakovleva, N., Frei, R., Murthy, S.R. (Eds.), 2019. Sustainable Development Goals and Sustainable Supply Chains in the Post-global Economy. Greening of Industry Networks Studies. Springer, Cham, pp. 1–9.
Yawar, S.A., Seuring, S., 2017. Management of social issues in supply chains: a literature review exploring social issues, actions and performance outcomes. J. Bus. Ethics 141 (3), 621–643.
Yook, K.H., Choi, J.H., Suresh, N.C., 2018. Linking green purchasing capabilities to environmental and economic performance: the moderating role of firm size. J. Purch. Supply Manag. 24 (4), 326–337.
Zamagni, A., Buttol, P., Porta, P.L., Buonamici, R., Masoni, P., Guinée, J., Heijungs, r., Ekvall, T., Bersani, R., Bienkowska, A., Pretato, U., 2008. Critical review of the current research needs and limitations related to ISO-LCA practice. In: Deliverable D7 of Work Package 5 of the CALCAS Project 106. ISBN: 88-8286-166-X Cover.
Zhu, Q., Sarkis, J., 2004. Relationships between operational practices and performance among early adopters of green supply chain management practices in Chinese manufacturing enterprises. J. Oper. Manag. 22 (3), 265–289. https://doi.org/10.1016/j.jom.2004.01.005.
Zhu, Q., Sarkis, J., Lai, K.H., 2008. Green supply chain management implications for "closing the loop". Transport. Res. E Logist. Transport. Rev. 44 (1), 1–18.
Zhu, W., He, Y., 2017. Green product design in supply chains under competition. Eur. J. Oper. Res. 258 (1), 165–180.
Zhang, B., Bi, J., Fan, Z., Yuan, Z., Ge, J., 2008. Eco-efficiency analysis of industrial system in China: a data envelopment analysis approach. Ecol. Econ. 68 (1–2), 306–316. https://doi.org/10.1016/j.ecolecon.2008.03.009.
Zhu, Q., Geng, Y., Sarkis, J., Lai, K.H., 2011. Evaluating green supply chain management among Chinese manufacturers from the ecological modernization perspective. Transport. Res. E Logist. Transport. Rev. 47 (6), 808–821.

Other references of interest

Hussain, C.M., Mishra, A.K. (Eds.), 2019. Nanotechnology in Environmental Science, vol. 2. John Wiley & Sons, United States.
Hussain, C.M., Keçili, R. (Eds.), 2019. Modern Environmental Analysis Techniques for Pollutants, first ed. Elsevier, United Kingdom.
Hussain, C.M. (Ed.), 2019. Handbook of Environmental Materials Management, first ed. Springer International Publishing, Switzerland.
Hussain, C.M. (Ed.), 2020. The Handbook of Environmental Remediation: Classic and Modern Techniques, first ed. Royal Society of Chemistry.

Relevant websites

International Organization for Standardization (ISO). https://www.iso.org.
United Nations Environment Programme (SETAC/UNEP). https://www.unenvironment.org.
UNEP/SETAC Life Cycle Initiative. https://www.lifecycleinitiative.org/.
World Business Council For Sustainable Development. https://www.wbcsd.org/.

Sustainable management of agricultural resources (agricultural crops and animals)

Vladimír Š. Kremsa
Landscape Ecology Institute, Tábor, Czech Republic

1. Introduction

The world's population is projected to grow from around 7.2 billion today to 9.3 billion in 2050 (United Nations, 2013). Food and agriculture production systems worldwide are facing unprecedented challenges from an increasing demand for food for a growing population, overexploitation of natural resources, loss of biodiversity, adverse climate change effects, rising hunger and malnutrition, and food loss and waste. In 2017, 821 million people were undernourished, or almost 11% of the world's population (World Bank, 2019). A growing number of people have had to reduce the quantity and quality of the food they consume. Published reports World Economic Outlook 2020 "A crisis like no other, an uncertain recovery" (IMF, 2020), World Economic Outlook Update, January 2021 "Policy Support and Vaccines Expected to Lift Activity" (IMF, 2021a) and World Economic Outlook 2021 "Managing Divergent Recoveries" (IMF, 2021). *The 2030 Agenda for Sustainable Development*, adopted by United Nations Member States in 2015, provides guidelines for peace and prosperity for people and the planet. The 17 *Sustainable Development Goals* are an urgent call for action by all countries in a global partnership. The United Nations Decade of Action on Nutrition (2016–2025) was officially announced. **The Food and Agriculture Organization of the United Nations (FAO)** is a specialized agency that leads international efforts to improve agriculture, food security, and nutrition, and defeat hunger. The World Food Programme is the food-assistance branch and the world's largest humanitarian organization addressing hunger and promoting food security. FAO launched Save and Grow: a policymaker's guide to the sustainable intensification of smallholder crop production, as a new paradigm for intensive crop production for that would enhance both productivity and sustainability through an ecosystem approach (FAO, 2011). ***The World Food Programme*** (WFP) is the food-asistance branch and the world's largest humanitarian organization addressing hunger and promoting food security. ***The Commission on Genetic Resources for Food and Agriculture*** **(CGRFA)** of the FAO UN published in 2019 *The State of the World's Biodiversity for Food and Agriculture* and *The state of the World's Aquatic Genetic Resources for Food and Agriculture*. **The World Bank** published *World Development Report* (WB, 2008), which addressed "*Agriculture for Development*," calling for greater investment in agriculture in developing countries, and warned that the sector

Sustainable Resource Management. https://doi.org/10.1016/B978-0-12-824342-8.00010-9

must be at the center of the development agenda if the goals of halving extreme poverty and hunger by 2015 are to be realized. *Enabling the business in Agriculture* (World Bank, 2019) measures how regulation affects the livelihood of domestic farmers, which helps policy-makers assess the regulatory environment in agriculture. *World Development Report* (WB, 2020) "Trading for Development in the Age of Global Value Chains" examines whether there is still a path to development through GVCs and trade (WB, 2020) The World Development Report for 2021 will focus on "Data for Development" *Earth's biodiversity* provides resources for increasing the range of food and other products suitable for human use. *Agricultural production* depends on natural resources and endowments (knowledge, management skills, production technologies, etc.), and it influences a wide range of natural resources. Use of agricultural resources depends on the decisions made by politicians and the farm operators. The great gains in agricultural productivity in the past century have been accompanied by substantial degradation to global ecosystems (Millennium Ecosystem Assessment, 2005). A new focus on social, environmental, and economic sustainability has increased demand for information on agricultural production methods, their social and environmental effects, and ways to measure them. Consumers are now concerned not just with the cost of food but also with long-term impacts on the environment and agricultural workers. Concise and accurate information about the current state of, and complex interactions between, public policies, economic conditions, farming practices, conservation, resources, and the environment can assist public and private decision-making. In 1980, the International Union for Conservation of Nature (IUCN) and Natural Resources published the "*World Conservation Strategy: Living Resource Conservation for Sustainable Development.*" Sachs (2015) published the book "*The Age of Sustainable Development.*" Norton (2002) published a book Searching for Sustainability: Interdisciplinary Essays in the Philosophy of Conservation Biology. The Convention on Biological Diversity (UN, 1993) defines the resources: **"*Biological resources*"** includes genetic resources, organisms or parts thereof, populations, or any other biotic component of ecosystems with actual or potential use or value for humanity. **"*Genetic resources*"** means genetic material of actual or potential value. The knowledge of *agricultural resources* (species, agroecosystems, agrolandscapes) is the basis for scientific sustainable management.

2. Sustainable management of agrobiodiversity

The Convention on Biological Diversity defines "*Sustainable use*" as the use of components of biological diversity in a way and at a rate that does not lead to the long-term decline of biological diversity, thereby maintaining its potential to meet the needs and aspirations of present and future generations (UN,1993).

2.1 Biodiversity

"*Biological diversity*" means the variability among living organisms from all sources including terrestrial, marine, and other aquatic ecosystems and the ecological complexes of which they are part; this includes diversity within species, between species and of ecosystems (UN, 1993). It is typically a measure of variation at the genetic, species, ecosystem, and landscape level. *Global Biodiversity Outlook* is a periodic report published by CBD that summarizes the latest data on the status and trends of biodiversity (CBD, 2014). Approximately nine million types of plants, animals, protists, and fungi inhabit the earth. So do seven billion people. Intergovernmental Science-Policy Platform on Biodiversity and Ecosystem Services presented *The Global Assessment Report on Biodiversity and Ecosystem Services* (IPBES, 2019). According to the report 25% of plant and animal species are threatened with extinction as the result of human activity. *The Cartagena Protocol on Biosafety* to the CBD is an international agreement on biosafety, which seeks to protect biological diversity from the potential risks posed by genetically modified organisms (GMOs) resulting from modern biotechnology. *The Nagoya Protocol on Access to Genetic Resources and the Fair and Equitable Sharing of Benefits Arising from their Utilization* is a supplementary agreement to the CBD.

Biodiversity loss is the extinction of species (plant, animal, microbes) worldwide, and also the local reduction or loss of species in a certain habitat. *Global extinction* has so far been proven to be irreversible. *Local extinction* can be temporary or permanent, depending on whether the environmental degradation that leads to the loss is reversible through ecological restoration/resilience or effectively permanent (e.g., through land loss). The *Red List of Threatened Species* (IUCN, 2020) is the world's most comprehensive inventory of the global conservation status of plant and animal species. More than 32,000 species are threatened with extinction! Biodiversity loss and its impact on humanity was analyzed by Cardinale et al. (2012).

Threats are broad categories of phenomena that directly harm biodiversity: (a) habitat loss and degradation; (b) overexploitation; (c) climate change; (d) pollution; and (e) invasive species. ***Drivers*** are activities (e.g., agricultural extensification, logging, etc.) that spur threats. ***Factors*** are activities and phenomena (e.g., population growth) that spur drivers.

2.2 Agrobiodiversity

Agricultural biodiversity is the part of biodiversity recognized as a resource by farmers for agricultural production. It corresponds to the diversity of living organisms consciously managed by the farmer. Humans use at least 40,000 species, but cca 80% of humans' food supply comes from just 20 plant species. Thrupp (2000) studied the role of agrobiodiversity for sustainable agriculture and the links with food security. Agricultural biodiversity is essential for a sustainable improvement in food and nutrition security (Frison et al., 2011).

In 1996, the Third Conference of Parties of the Convention on Biological Diversity established a program of work on **Agricultural Biological Diversity.** *Agricultural biodiversity* was defined to include all components of biological diversity of relevance to food and agriculture: genetic resources of harvested crop varieties, livestock breeds, fish species and nondomesticated ("wild") resources within field, forest, rangeland, and aquatic ecosystems; biological diversity that provides ecological services (nutrient cycling, pest and disease regulation, maintenance of local wildlife, watershed protection, erosion control, climate regulation, and carbon sequestration). The topic is addressed in national reports and in **National Biodiversity Strategies and Action Plans.**

Agricultural diversity can be divided into two categories: (a) *intraspecific* diversity (the genetic variation within a single species), (b) *interspecific* diversity (the number and types of different species). *Functional classification* of agricultural diversity can also be used: (a) *planned diversity* (the crops and animals which a farmer has encouraged, planted, or raised), and (b) *associated diversity* arrives among the crops, uninvited (e.g., weed, pathogens, herbivores, etc.). The control of associated biodiversity is one of the great agricultural challenges that farmers face. Agrobiodiversity is divided into four *levels* which interact with one another: genetic, specific, agroecosystem, and agrolandscape diversity. FAO (2020f) published "*Biodiversity for food and agriculture and ecosystem services* - Thematic Study for The State of the World's Biodiversity for Food and Agriculture."

2.2.1 Genetic agrobiodiversity

It includes, for each species (domesticated and used in agriculture), all *plant varieties* and *animal species* created by humans since Neolithic times, as well as their *wild relatives* which are an important diversity reservoir for the genetic improvement of these varieties and species.

2.2.2 Specific agrobiodiversity

It is the diversity of species involved in the agroecosystem (domestic or wild), but whose survival is dependent on agricultural practice. The development of agroecology has led all species which play a part in the ecosystem processes that support agricultural production to be considered a part of agrobiodiversity, for instance, soil fauna and flora.

2.2.3 Ecosystem agrobiodiversity

It considers the diversity of *ecological habitats* and how they fit together both in time and space at different scales (field, farm, landscape). It includes crops, *seminatural habitats* (permanent grasslands, hedges, etc.) and *natural habitats* (ponds, groves) which are part of the agricultural landscape. The importance and organization of this agrobiodiversity play a decisive role in providing *agroecosystem services*. It is a key component of food security, human health, and well-being. Agrobiodiversity has eroded in a very worrying way since the end of the 19th century. Its preservation and conservation is subject to specific measures (conservation of genetic resources, preservation of natural elements, etc.).

2.2.4 Landscape agrobiogeodiversity

lIt deals with the diversity of ecological landscapes (structure, function, changes) and how they fit together in spatiotemporal scale of a region and geobiome. Landscapes are open (natural, seminatural, semiagricultural, agricultural) and build-up (rural, suburban, urban, industrial). Agricultural, rural, suburban, urban, and industrial landscapes are *cultural landscapes.* Kremsa (1999) described the hierarchical organization of landscapes.

2.2.5 Loss of agrobiodiversity

The UN's *Global Biodiversity Outlook* estimates that 70% of the projected loss of terrestrial biodiversity are caused by agriculture use (United Nations, 2014). More than one-third of the planet's land surface is utilized for crops and grazing of livestock. Agriculture destroys biodiversity by converting natural habitats to intensely managed systems and by releasing pollutants, including greenhouse gases. Food value chains further amplify impacts including through energy use, transport, and waste. The original 25 biodiversity hotspots covered 11.8% of the land surface area of the Earth. The current hotspots cover more than 15.7% of the land surface area, but have lost around 85% of their habitat.

2.2.6 Conservation of agrobiodiversity

FAO (2019a,2019b) published report with the analysis of genetic resources and "*Strategic Plan for the Commission on Genetic Resources for Food and Agriculture" (2019–27)* and with following goals: 1. *Sustainable use:* Promote the sustainable use and development of genetic resources for food and agriculture and, more generally, all biodiversity relevant to food and agriculture, to increase production for world food security and sustainable development. 2. *Conservation:* Maintain the diversity of genetic resources for food and agriculture. 3. *Access and benefit-sharing*: Promote appropriate access to genetic resources for food and agriculture and fair and equitable sharing of benefits arising from their utilization. 4. *Participation:* Facilitate the participation of relevant stakeholders in decision-making. Work plan for the sustainable use and conservation of microorganism and invertebrate genetic resources for food and agriculture is included. "*Ex situ conservation*" means the conservation of components of biological diversity outside their natural habitats. "*In situ conditions*" means conditions where genetic resources exist within ecosystems and natural habitats, and, in the case of domesticated or cultivated species, in the surroundings where they have developed their distinctive properties (UN, 1993).

3. Sustainable management of agricultural species

"Domesticated or cultivated species" means species in which the evolutionary process has been influenced by humans to meet their needs. *"Habitat"* means the place or type of site where an organism or population naturally occurs. (UN, 1993).

The Commission on Genetic Resources for Food and Agriculture of the FAO UN is an intergovernmental body that addresses issues specifically related to the management of biodiversity of relevance to food and agriculture. In 2019, the Commission published *The State of the World's Biodiversity for Food* (FAO, 2019a) *and Agriculture* and *The state of the World's Aquatic Genetic Resources for Food and Agriculture* (FAO, 2019b).

A center of origin is a geographical area where a group of organisms, either domesticated or wild, first developed its distinctive properties. They are also considered *centers of diversity*. Centers of origin were first identified in 1924 by Nikolai Vavilov.

3.1 Agricultural plants (Crops)

Domesticated plants are plant species cultivated by humans in agriculture, horticulture, gardening, and forestry for consumption (food crops, herbs, spices), healing (medical drugs), esthetics (ornamental herbs and trees, floristry), and raw material (lumber, biofuel, etc.). Groups of plants and plant products used by man include the following: cereals, pulses, vegetables, fruits, nuts, oilseeds, spices, condiments, sugars and starches, fibers, beverages, rubber forages, green and green leaf manure, and narcotics.

Crop diversity is the variance in genetic and phenotypic characteristics of plants used in agriculture. Over the past 50 years, there has been a major decline in genetic diversity within each crop and the number of species commonly grown. Crop diversity is important for food security. Biodiversity International presents "*Crop Wild Relatives Global Portal.*"

Crop diversity loss threatens global food security, as the world's human population depends on a diminishing number of varieties of a diminishing number of crop species. Nautiyal and Kaechele (2007) presented conservation of crop diversity for sustainable landscape development. Khoury et al. (2016) deal with measuring the state of conservation of crop diversity as baseline for marking progress toward biodiversity conservation and sustainable development goals. The Commission on Genetic Resources for Food and Agriculture of the FA O published *Second Global Plan of Action for Genetic Resources for Food and Agriculture* (2011). *The International Plant Protection Convention (IPPC)* is a multilateral treaty overseen by the FAO that aims to secure coordinated, effective action to prevent and to control the introduction and spread of pests of plants and plant products. IPPC (2020) published Annual Report 2019—*Protecting the world's plant resources from pests.*

3.2 Agricultural animals

3.2.1 Livestock biodiversity

Livestock are domesticated animals (cattle, pigs, poultry, horses, sheep, bees, etc.) raised in agriculture to produce commodities (meat, milk, eggs, fur, leader, pool, etc.) and labor. *The Second Global Assessment of Animal Genetic Resources* (FAO, 2015) provides a comprehensive assessment of livestock biodiversity and its management. It sets out the latest available information on the state of livestock diversity, trends in the livestock sector, the state of capacity to manage animal genetic resources, the state of the art in animal genetic resources management, gaps and needs in animal genetic resources management, and country reports. *Key findings*: Livestock diversity facilitates the adaptation of production systems to future challenges and is a source of resilience in the face of greater climatic variability, The roles and values of animal genetic resources remain diverse, particularly in the livelihoods of poor people, The adaptations of specific species and breeds to specific environmental challenges need to be better understood, The impact of many livestock sector trends on animal genetic resources and their management is increasing, The world's livestock diversity remains at risk, The assessment of threats to animal genetic resources needs to be improved, Institutional frameworks for the management of animal genetic resources need to be strengthened, Establishing and sustaining effective livestock breeding programs remains challenging in many countries, particularly in the low-input production systems of the developing world, Conservation programs for animal genetic resources have become more widespread, but their coverage remains patchy, Emerging technologies are creating new opportunities and challenges in animal genetic resources management, and Livestock diversity and the sustainable management of animal genetic resources are acquiring a greater foothold on policy agendas.

FAO has a unit focused on Animal Genetic Resources, which are defined as "those animal species that are used, or may be used, for the production of food and agriculture, and the populations within each of them." These populations within each species can be classified as wild and feral populations, landraces and primary populations, standardized breeds, selected lines, varieties, strains, and any conserved genetic material; all of which are currently categorized as Breeds. FAO assists countries in implementation of the *Global Plan of Action for Animal Genetic Resources.* FAO supports a variety of ex situ and in situ conservation strategies including cryoconservation of animal genetic resources.

Globally, 51% of all sheep, 44% of goats, 38% of cattle, 21% of pigs, and 27% of chickens are assumed to occur in systems where predominantly *locally adapted breeds* thrive and grasses and roughages are the major feed resources. In most of these low-input extensity systems, small-scale livestock keepers predominate, with pastoralists widespread in arid rangelands. Globally, 49% of all sheep, 56% of goats, 62% of cattle, 79% of pigs, and 73% of chickens are assumed to occur in systems where both *locally adapted breeds, exotic breeds, and their cross-breeds* thrive and where feed quality tends to increase. Both small-scale and large-scale livestock keepers can be found in these higher-input systems. More than half of sheep and goats are found in hyperarid to semiarid systems, whereas the share is lower than 10% for pigs. The number of chicken is highest in humid, followed by temperate climatic zones. Pig numbers are low in arid climatic zones. Similar shares of cattle are kept in humid (39%) and arid (hyperarid to semiarid) climates (35%), whereas more than half of sheep and goat are kept in arid zones (Hoffmann et al., 2014). Breed diversity in dryland ecosystems was studied by FAO (2006).

4. Sustainable management of agroecosystems

"Ecosystem" means a dynamic complex of plant, animal, and microorganism communities and their nonliving environment interacting as a functional unit (UN, 1993). The philosophy of sustainable adaptive ecosystem management was presented by Norton (2005). Schulze & Mooney presented the book about biodiversity and ecosystem

Table 5.1 Properties of agroecosystems and their determinants (Lal et al., 2016).

Property	Indication	Controls
1 Productivity	Total output	Soil quality, micro- and mesoclimate
2 Stability	Consistency	Management, availability of inputs
3 Equitability	Distribution and availability	Economic, social, and cultural factors
4 Autonomy	Independence	Political, economic, and social factors
5 Perpetuity	Forever	Prudent management, education, strong institutions
6 Efficiency	Input	Intensification, production of more from less

function. Kremen (2005) explained what do we need to know about the ecology for ecosystem services management. Table 5.1 shows the properties of agroecosystems and their determinants (Lal et al., 2016).

"*In situ conditions*" means conditions where *genetic resources* exist within ecosystems and natural habitats, and, in the case of domesticated or cultivated species, in the surroundings where they have developed their distinctive properties. (UN, 1993).

4.1 Agroecology

Agroecology is scientific field, an approach integrating the concepts and methods of a variety of disciplines (agronomy, ecology, economics, sociology, etc.). Aiming to promote the services rendered by natural processes, it analyses at different levels (from field to territory, from individual to community, from short term to long term) the evolutionary relations which are created within these systems between the living, its management method, and the ecological, economic, and social context of this management. An *agroecosystem* is the basic unit of study in agroecology. *Agroecosystem analysis* is an analysis of an agricultural environment that considers ecology, sociology, economics, and politics with equal weight. Principles of agroecology, science of sustainable agriculture, were presented in books of Altieri (1995) and Gliesman (1997). *Agroecology* aims to promote *sustainable food systems*, respectful of people and the environment. These systems involve agricultural production methods and sectors that value the ecological, economic, and social potential of a territory. Their development relies on transdisciplinary approaches that bring together professionals from the agriculture, scientists, actors of agroecology and public policy social movements. Agroecology is an alternative to intensive agriculture based on the artificialization of crops through the use of synthetic inputs (fertilizers, pesticides, etc.) and fossil fuels. It promotes agricultural production systems that value biological diversity and natural processes (the cycles of N, C, water, the biological balances between pests and auxiliary crops ...). FAO (2018a) published The 10 elements of agroecology, Guiding the transition to sustainable food and agricultural systems. FAO, Rome. FAO (2018b) presented, "FAO's work on agroecology. A pathway to achieving the SDGs", which introduces the agroecology approach to linking food, livelihoods and natural resources. FAO (2018c) presented a report "Constructing Markets for Agroecology, An Analysis of Diverse Options for Marketing Products from Agroecology".

4.2 Agroecosystems services

Ecosystem services are defined as the goods and services that humans can get from ecosystems, directly or indirectly, in order to ensure their well-being. *Agroecosystem services* are coproduced by nature and humans. This notion emerged from the interface between economics and ecology. It is based on the postulate that a value, often monetary, can be assigned to nature. Commission on Ecosystem Management of The IUCN published a book *Ecosystem Services*. Everard (2017) presents key issues of ecosystem services. The ecosystem services were formally categorized in 2005 by the United Nations *Millennium Ecosystem Assessment*, which recognized four broad categories: *provisioning* (food and water), *regulating* (controlling climate and disease), *supporting* (nutrient cycles, crop pollination), and *cultural* (recreational and spiritual) (Millennium Ecosystem Assessment, 2005).

Agroecosystem services provided by biodiversity**:** *Food production* (the portion of gross primary production extractable as raw food or for processing of food), *Genetic resources* (sources of unique biological materials and products), *Raw materials* (the portion of gross primary production extractable as raw material), *Biological control*: trophic (food web) dynamic regulations of populations, climate and gas regulation (global temperature, precipitation, other biologically mediated climatic processes at global/local levels; of atmospheric chemical composition), *Erosion control* and sediment retention (prevent loss of soil by wind, rain impact, runoff; storage of silt in ecosystem, in lakes and wetlands), *Nutrient cycling* (storage, cycling, processing, input of nutrients), *Pollination* (movement of floral gametes), *Refugee* (habitat for local/transient populations), *Resilience*/disturbance regulation (ecosystem response to environmental fluctuation, mainly controlled by vegetation structure), *Soil formation* (processes of weathering of rock; soil buildup), *Waste detoxification* (recovery of mobile nutrients, removal/breakdown of excess or toxic nutrients/compounds, pollution control), *Water regulation and supply* (hydrological flow/regimes; water retention, storage, provisioning in the watershed: water supply in aquifers, surface water bodies; availability for consumption, irrigated agriculture, industry, transport).

Provision of ESS may be stabilized or increased with increasing *biodiversity*, which also benefits the variety of ESS available to society. Understanding the relationship between biodiversity and the stability of an ecosystem/landscape is essential to the management of natural resources and their services. Estimation of *the functional structure of an ecosystem/landscape* and then combination with information about individual species' traits can help understanding of the resilience of an ecosystem as environmental change occurs, e.g., through explicit human intervention or as a byproduct of human actions, e.g., climate change. Agroecology strongly mobilizes ecosystem services, whether those contributing to agricultural production (pollination, soil fertility), thus enabling a reduction in the use of chemical inputs, or those provided by agriculture to society (landscapes, water quality). Altieri (1999) evaluated the ecological role of biodiversity in agroecosystems.

4.2.1 Agricultural crop services

Sustainable crop production refers to agricultural production in such a way that does not impose any harm to environment, biodiversity, and quality of agricultural crops. Producing crops sustainably increases the ability of the system to maintain stable levels of food production and quality for long term without increasing the demand and requirements of agricultural chemical inputs to control the system. Sustainable crop production deals with keeping the soil alive with organic matter, integrated pest management and reduction in usage of pesticides, protecting biodiversity, ensuring food safety and food quality, improving nutrient quality, and fertilizing the soil with organic fertilizers. Sustainable agricultural production leads to lowering of greenhouse gas emission and carbon footprint of overall world. Sustainably produced crops and food are more beneficial to consume by humans as compared to commercial crops. Sustainable usage of resources ensures the pollution-free environment for our future generations.

The concept of ecosystem services is particularly relevant in the context of permanent *grassland* systems. Indeed, by limiting inputs and adapting practices to the environment potential (altitude, exposure, type of soil, …), farmers sustain a diversified seminatural vegetation while providing herds with an annual grass-based resource. In addition to the production of quality products, these farming systems provide society with many ecosystem services (water quality, biodiversity, esthetic landscapes, carbon storage, …). It is nowadays possible, thanks to multifunctional typologies of permanent grasslands, to establish links between this plant biodiversity and the provision of ecosystem services.

4.2.2 Agricultural animals services

Over millennia, humans have selected and bred livestock species, creating a range of breeds with their own special traits. Different livestock species and breeds are able to provide ecosystem services as a result of their adaptation to different environments, production systems, societal needs, and cultural preferences.

Livestock plays a special role in the provision of ecosystem services and is an essential part of many agroecosystems. They do so by (1) *transforming* feeds unsuited for human consumption into nutritious foods and useful products (converting grass into milk or meat); (2) *interacting* directly with ecosystems through grazing, browsing, tramping, and the production of dung and urine; and (3) *moving around*, meaning they can respond to fluctuations in resource availability and climate. The roles of livestock species and breeds in providing ecosystem services depend strongly on how people manage them and the production systems of which they form a part. Depending on how they are managed, their impacts can be positive or negative. Hoffmann et al. (2014) elaborated the review of ecosystem services provided by livestock species and breeds (Table 5.2).

Table 5.2 Ecosystem services provided by livestock (Hoffmann et al., 2014).

Provisioning services (products obtained from ecosystems):
Food: Meat, milk, eggs, honey. *Fiber, skins,* and *related products*: Wool, fiber, leather, hides, skins, wax
Genetic resources: Basis for breed improvement and medicinal purposes
Biotechnical/medicinal resources: Laboratory animals, test-organisms, biochemical products
Fertilizer: Manure and urine for fertilizer. *Power:* Draught power
Fuel: Manure and methane for energy, biogas from manure, slaughterhouses, etc.
Regulating services (benefits from regulation of ecosystem processes):
Waste recycling and conversion of nonhuman edible feed, *recycling* of crop residues, household waste, swill, and primary vegetation consumption
Land degradation and erosion prevention: Maintenance of vegetation cover
Water quality regulation/purification, water purification/filtering in soils, regulation of water flows, natural drainage, drought prevention, influence of vegetation on rainfall, timing and magnitude of runoff and flooding. *Climate regulation*: Soil carbon sequestration, greenhouse gas mitigation
Pollination: Yield and seed quality in crops and natural vegetation; genetic diversity
Biological control and animal/human disease regulation: destruction of habitats of pest and disease vectors; yields. *Moderation of extreme events,* avalanche and fire control
Supporting services (necessary for the production of all other ES)
Maintenance of *soil structure and fertility*, soil formation
Nutrient cycling on farm and across landscapes. *Primary production:* Improving vegetation growth/cover
Habitat services
Maintenance of life cycles of species. Habitat for species, especially migratory, habitat connectivity, seed dispersal in guts and coats. Maintenance of genetic diversity. Gene pool protection and conservation. **Cultural services** (nonmaterial benefits from ecosystems)
Opportunities for recreation: Eco/agrotourism, sports, shows, and other recreational activities involving specific animal breeds.
Knowledge systems and educational values: Traditional and formal knowledge about the breed, the grazing and sociocultural systems, information for cognitive development, scientific discovery
Cultural and historic heritage: Breed in the area helps to maintain heritage of the region; cultural identity, especially for indigenous peoples, inspiration for culture, art, and design
Natural (landscape) heritage: Values associated with the landscape as shaped by the animals themselves or as a part of the landscape, e.g., esthetic values, sense of place, inspiration
Spiritual and religious experience: Values related to religious rituals, human life cycle such as religious ceremonies, funerals, or weddings.

Continued

Table 5.2 Ecosystem services provided by livestock (Hoffmann et al., 2014).—cont'd

Traditional art and handicraft; fashion; cultural, intellectual, and spiritual enrichment and inspiration; pet animals, advertising
Provisioning roles of livestock for rural families
Cattle: Milk, blood, and meat for food and income from sales; hides for shelter, bedding, clothing and footwear; savings, dowry, and bride-price; manure and animal traction.
Camels: Meat and milk for food and income; hides for shelter, bedding, and footwear; savings, dowry, and bride-price; manure, including for use in papermaking; transport.
Goats and sheep: Sales for cash, milk, blood, and meat for food; skins for clothing; savings; wool and wool products. *Pigs*: Meat for food and income; savings.
Donkeys: Transport of water and goods; milk for medicinal purposes.
Llamas and alpacas: Transport in mountainous terrain; meat and fiber; income from tourism.
Poultry: Eggs and meat for food and income; feathers for bedding

4.3 Agroecosystem health

The health of an agroecosystem is its ability to develop and self-sustain the production of a diversity of ecosystem services. This way of qualifying an agroecosystem is generally part of an agroecological perspective for the development of sustainable farming systems. A healthy agroecosystem is an agricultural system that has found a balance between stability and resilience; this means a system which, depending on the situation in which it finds itself, is capable of maintaining a form of continuity in the way it works, and to reorganize itself when facing important disturbances in order to restore its primary functions (productive, landscape, social functions ...). *International Plant Protection Convention (IPPC)* dedicated the annual theme to plant health and capacity development (IPPC, 2020). The qualities of a healthy agricultural system are its capacity to enhance its autonomy (decision-making autonomy, emancipation from the use of fertilizers, pesticides, and energy) and its ability to maximize ecological processes which are underlying to the production of ecosystem services. This concept has been differently declined by the scientific community since its birth in the 1990s. On one hand, it is a concept used to develop *indicators* to monitor and evaluate the health of agroecosystems, at the scale of the agricultural landscape, in order to model scenarios of change. Altieri (1994) studied the role of biodiversity and pest management in agroecosystems.

4.4 Agroecosystems conservation

Agroecosystem level conservation looks at landscape level, with landscapes managed by the group of stakeholders working together to achieve biodiversity, production, and livelihood goals. Land use mosaics combine "natural" areas, agricultural production areas, institutional mechanisms to coordinate initiatives to achieve production, conservation and livelihood objectives at landscape, farm and community scales, by exploiting synergies and managing trade-offs among them. There are limited initiatives that focus on conserving entire landscapes or agroecosystems. One is "*Globally Important*

Agricultural Heritage Systems," which are conserved and maintained as unique systems of agriculture, in order to sustainably provide multiple goods and services, food and livelihood security for millions of small-scale farmers.

5. Sustainable management of agricultural landscapes

Landscape science is a metadiscipline, evolving from synergistic theories, methods, and knowledge of different scientific disciplines applied to study landscapes. The disciplines include ecology, geography, agricultural science, forestry science, conservation science, geosciences, biology, social sciences, engineering sciences, and mathematics. *Agricultural landscapes* are *multifunctional* through their simultaneous support of habitat, productivity, regulatory, social, and economic functions. *Heterogeneity* is a basic characteristic of landscape, and implies the capacity of landscape to support various, sometimes contradictory, functions simultaneously. Many elements in cultural landscapes have a multifunctional character. *Land use* is the key activity that determines the *performance of landscapes* with respect to socioeconomic functions such as land-based production, infrastructure, and housing. Kremsa (1999b) presented the information system for monitoring of land use/cover changes. The degree of integration between these socioeconomic functions and environmental functions including natural resources protection depends on the patterns and intensities of land use. Biodiversity conservation in multifunctional, human-dominated landscapes needs a coherent large-scale spatial structure of ecosystems. The theory and empirical knowledge of *territorial ecological networks* provides a framework for the design of such structures. Principles of **Territorial system of landscape ecological stability** (TSLES) described Kremsa (1999a). Ecological networks are the most important guarantees of landscape multifunctionality. Depending on the spatial scale, the number of functions covered varies. Landscape types and high intensity agricultural and forestry production aimed at homogenizing the heterogeneous landscape features in order to facilitate trafficability and management procedures. This process was coupled with *environmental degradation* including soil erosion, nutrient losses, groundwater pollution, a decrease in biodiversity and landscape scenic values. Landscape types and human influence play an important role. Kremsa (1999c) presented Strategies for Sustainable Development of Rural Landscape in the Mountains for Sustainable Mountain Development.

5.1 Agricultural landscape ecology

Landscape ecology focuses on three *characteristics of the landscape* (Forman and Godron, 1986): (1) *structure*—the spatial relationships among the distinctive ecosystems (elements) present (the distribution of energy, materials, and species in relation to the sizes, shapes, numbers, kind, and configurations of the ecosystems); (2) *function*—the interactions among the spatial elements (the flow of energy, materials, and species among the component ecosystems); (3) *change*—the alternation in the structure and function of the ecological mosaic over time.

The central theme of landscape ecology is defined by Naveh and Lieberman (1994) as the study of the complex totality of all landscapes on earth and the safeguarding of their integrity, health, and natural and cultural diversity. In landscape ecology, attention is given not only to the natural dimensions but also to *historical, cultural, social, political*, and *economic aspects*. Human beings play several roles as elements of landscape, namely they appear as (a) existential (ontological) landscape elements (by the very fact of existence), (b) a landscape moderator (forming its structure, and creating urban, industrial, and cultural space), (c) a social element (collaboration within a group, while remaining under the influence of a group). Kremsa and Žigrai (2021a,b) investigated theoretical and meta-scientific aspects of Landscape Ecology in Mexico (evaluation, research, education and future) using meta-landscape ecological approach.

5.2 Sustainable management of agricultural landscapes

Principles *for landscape approaches* adopted by the Convention on Biological Diversity are as follows: Continuous learning and adaptation, Common concern entry point, Multiple scales, Multifunctionality, Multiple stakeholders, Negotiated and transparent change logic, Clarification of rights and responsibilities, Participatory and user-friendly monitoring, Resilience, Strengthened stakeholder capacity. Kremsa (1998) presented strategies for sustainable development of rural landscape and experience with Sustainable Landscape Management in Czech Republic Kremsa (2000).

5.2.1 Agrolandscape services

The term *landscape function* is used in analogy with the concept of ecosystem functions: it indicates the capacity of the landscape to provide goods and services to society. The reason for specifically addressing landscapes and not ecosystems is because landscapes consist of different systems, arranged in specific spatial patterns. This thesis addresses land systems that are strongly modified by humans, such as agricultural landscapes. Landscapes are considered holistic spatial systems in which humans interact with their environment, while ecosystems are often perceived as merely natural and seminatural systems. As a product of landscape functions, *landscape services* are defined as the flow of goods and services provided by the landscape to society. These landscape goods and services are the connection between the landscape and human benefits, the actual contributions to well-being.

Typology of landscape functions. The multiple benefits that are provided by ecosystems and landscapes have been described in a large number of studies which provided the basis for a recent global assessment of ecosystem goods and concept and valuation of landscape functions at different scales 17 services (Millennium Ecosystem Assessment, 2005). Four groups of functions (or **services**) are primarily distinguished by the Millennium Assessment: *provisioning, regulating, cultural*, and *supporting services*. The Millennium Ecosystem Assessment (2005) made a commendable attempt to bring order in the many definitions of the terms “functions,” “goods,” and “services.” Agreement was reached to define services as “the benefits people derive from ecosystems," and to avoid lengthy texts decided to use the term "services" for both goods and services, as well as the underlying functional processes and components of the ecosystems providing them. Landscape diversity and economic valuation of environmental functions was studied and presented by Kremsa (2002).

5.2.2 Sustainable landscape management

Integrated landscape management is a way of managing a landscape that brings together multiple stakeholders, who collaborate to integrate policy and practice for their different land use objectives, with the purpose of achieving sustainable landscapes. It recognizes that, for example, one river basin can supply water for towns and agriculture, timber and food crops for smallholders and industry, and habitat for biodiversity; the way in which each one of these sectors pursues its goals can have impacts on the others. The intention is to minimize conflict between these different land use objectives and ecosystem services. This approach draws on landscape ecology, as well as many related fields that also seek to integrate different land uses and users, such as watershed management. Methods of monitoring of rural landscapes are published by Kremsa (2002a).

Sustainable landscape management. We propose a platform for interdisciplinary and transdisciplinary research into sustainable landscape and ecosystem services to overcome the fragmentation of scientific knowledge. We will collect case studies of best practices and identify key factors of success supported by effective integration of scientific knowledge into planning and decision-making processes for sustainable land use development. The platform will integrate research results and ensure knowledge transfer and the platform's members will cooperate on preparation of regional and national strategic documents (e.g., national strategies on biodiversity conservation). To cover these goals, an interface between policy-making and science is needed. Public awareness of landscape issues will therefore be raised using communication tools (public campaigns, seminars, and excursions). Review of trends in advanced technologies for Landscape Monitoring is published by Kremsa (1993,1995).

6. Sustainable agriculture

Sustainable agriculture is an agricultural production and distribution system that (a) achieves the integration of natural biological cycles and controls, (b) minimizes adverse impacts on health, safety, wildlife, water quality, environment, (c) optimizes the management and use of on-farm resources, (d) promotes opportunity in family farming and farm communities, (e) protects and renews soil fertility and the natural resource base, (f) provides an adequate and dependable farm income, and (g) reduces the use of nonrenewable resources and purchased production inputs. *The State of Food and Agriculture* (FAO, 2019a,b) present balanced science-based assessments of important issues in the field of food and agriculture. Agrobiodiversity is central to sustainable food systems and sustainable diets. The use of agricultural biodiversity can contribute to food security, nutrition security, and livelihood security. Jamieson (1998) traced the concept of *sustainable development* from a 1980 report from the IUCN through the 1987 Brundtland Commission report to its current plethora of uses and applications. Thompson (2007) presents a philosophical analysis of concepts of *agricultural sustainability*. Programs in sustainable agriculture apply human, biological, and financial resources to the development of technology and social institutions. They generally draw upon agronomy and other agricultural sciences to research and disseminate tools and techniques that farmers can use, or they draw on the applied social sciences to support decision-making and social organization to meet the local problems of rural communities. The variety of views on what it means to be sustainable has multiplied since the early 1980s, when critics of conventional agriculture began to claim that it was "*unsustainable*." The debate is no longer confined to agriculture. Others now want to talk about "sustainable development," "sustainable land use," etc. "*Sustainable agriculture*" *is an integrated system of plant and animal production practices* having a site-specific application that will, over the long term, satisfy human food and fiber needs, enhance environmental quality and the natural resource base upon which the agriculture economy depends, make the most efficient use of nonrenewable resources and on-farm resources, and integrate, where appropriate, natural biological cycles and controls, sustain the economic viability of farm operations, enhance the quality of life for farmers and society as a whole. Pretty (2008) has stated several *key principles of sustainability of agriculture*: (a) the incorporation of biological and ecological processes (nutrient cycling, soil regeneration, and nitrogen fixation, etc.) into agricultural and food production practices, (b) using decreased amounts of nonrenewable and unsustainable inputs, (c) using the expertise of farmers to productively work the land and to promote the self-reliance and self-sufficiency of farmers, (d) solving agricultural and natural resource problems through the cooperation and collaboration of people with different skills, (e) the problems tackled include pest management and irrigation.

Indicators of Sustainable Agriculture: Sustainability cannot be measured directly; it is too elusive concept and it operates over too long time scale. We can identify *measurable phenomena* (indicators) that suggest how sustainable our system might be. Indicators are widely used as benchmarks to help gauge performance in a number of human endeavors. For example, the consumer price index and gross domestic product are indicators, albeit crude ones, of economic performance.

Regional and national indicators could be divided into four main areas: (a) profitability; (b) land and water quality to sustain production; (c) managerial skills; and (d) offsite environmental impacts. A number of criteria can be used to judge the usefulness of a given indicator: Is it relevant and easy to use? Does it provide a representative picture? Is it easy to interpret and does it show trends over time? Is it responsive to changes? Does it have a reference to compare it against so that users are able to assess the significance of its values? Can it be measured at a reasonable cost? Can it be updated?

6.1 Sustainable management of agricultural resources (sustainable management of crop and animal production)

Sustainable agriculture is the use of farming systems and practices that maintain or enhance (a) the economic viability of agricultural production; (b) the natural resource base; and (c) other ecosystems that are influenced by agricultural activities. *Fundamental principles of sustainable agriculture* are (a) Farm productivity is enhanced over the long term; (b) Adverse impacts on the natural resource base and associated ecosystems are ameliorated, minimized, or avoided; (c) Residues resulting from the use of chemicals in agriculture are minimized; (d) Net social benefit (in both monetary and nonmonetary terms) from agriculture is maximized; and (e) Farming systems are sufficiently flexible to manage risks associated with the vagaries of climate and markets.

6.1.1 Alternative agriculture

Alternative agriculture is defined as production systems that do not use conventional methods. They aim at following the concept of agroecology. These kinds of systems seek sustainable performances while optimizing all agroecosystem resources. Alternative agriculture gathers a lot of different systems (organic agriculture, sustainable agriculture, integrated agriculture, agroforestry, permanent agriculture, etc.). Despite their differences, these systems share common values. Their technical itineraries were actually firstly thought as ways to preserve the environment and more precisely soil and water. They also seek to reduce or suppress the use of chemicals and mineral fertilizers, thanks to, respectively, biological control and organic fertilizers and amendments. They intend to comply with natural cycles, by using crop rotations, cover crops or no-tillage, etc. These systems try to fit their territories. The farmers who practice alternative agriculture often seek the overall improvement of their living standards. They also aim at including themselves in the local social network and at selling quality products. Farmers who practice alternative agriculture are able to ensure a profit and respect the environment and people. To achieve these goals, they develop on-farm processing, producer-to-consumer schemes, agrotourism, etc. Alternative agriculture can be linked to agroecology, thanks to their common concepts and objectives as well as biotechnical and socioeconomical aspects.

6.1.2 Intensive agriculture

The principal goals of IA are to (1) increase and sustain high productivity, (2) decrease positive feedbacks to climate change; (3) reduce offsite and onsite impacts; (4) minimize trade-offs in ecosystem services; and (5) improve social and gender equity. *The principal objectives of ecological intensification* are to (1) increase and sustain productivity; (2) reduce gaseous emissions; (3) alleviate human drudgery; and (4) improve nutritional security. Notable benefits are higher production and income levels, the development of agroindustries, and significant increases in rural employment. *The principal challenges to sustainable intensification* are high inputs and high risks of environmental pollution, soil degradation, and gaseous emissions.

6.1.3 Biodynamic agriculture

Biodynamic agriculture is an alternative farm management mode, free from synthetic inputs. It has participated in the development of organic agriculture and is based on philosopher Rudolf Steiner's ideology, *anthroposophy*. Biodynamic agriculture differs from organic agriculture in as much as it involves specific practices aimed at improving plant vitality by strengthening plant, ground, and environmental interactions. Biodynamic agriculture is based on the concept of the farm as an autonomous and living organism and structure. It is commonly seen as a quest for a balance between the production system and its environment. Biodynamic agriculture and organic agriculture share the multiannual crop rotations, as well as the use of mixed plants with mutual benefits and the use of compost made from animal droppings.

6.1.4 Ecological agriculture

Ecologically intensive agriculture is an agricultural way of production based on the sustainable use of ecosystem services in agroecosystems and landscapes. The term "intensive" refers to getting higher productivity (concerning plants or cattle) while boosting the natural features of agroecosystems. Ecologically intensive agriculture tries to replace the use of chemicals and fossil energies with natural mechanisms such as biocontrol, pest control, and those that lead to the soil natural fertility. The use of natural services is not common in the management of conventional agriculture. In accordance with the pillars of sustainable development, the ecological intensity should follow the economic and social situation of the farm. It must be compatible with the farmer's lifestyle quality and provide him/her with decent income.

6.1.5 Conservation agriculture

Conservation Agriculture is a farming system that promotes minimum soil disturbance (i.e., no tillage), maintenance of a permanent soil cover, and diversification of plant species. It enhances biodiversity and natural biological processes above and below the ground surface, which contribute to increased water and nutrient use efficiency and to improved and sustained crop production. There are three principles of Conservation Agriculture: (a) Minimum mechanical soil disturbance (i.e., no tillage) through direct seed and/or fertilizer placement, (b) Permanent soil organic cover (at least 30%) with crop residues and/or cover crops, (c) Species diversification through varied crop sequences and associations involving at least three different crops.

6.1.6 Organic agriculture

Organic agriculture is a production method based on agricultural practices that exclude the use of synthetic biocides and GMOs or products derived from GMOs. OA also seeks to limit the environmental impact by reducing its consumption of inputs and fossil energy but also by promoting natural processes such as the recycling of organic materials. It aims to respect the living; from the soil microorganisms to human beings without forgetting the agroecosystems uses. Several preventive methods can be used in organic farming: crop rotation, prophylaxis, integrated crop protection; they make it possible to limit the pressures due to bioaggressors or yield loss. Concerning animals must be managed with a low stocking density per hectare, fed with an organic-based feedstock, treated without antibiotics, and their well-being must respect natural living conditions, for instance, grazing for cattle.

6.1.7 Permanent agriculture (permaculture)

Permanent agriculture is an integrated and progressive production system inspired by natural ecosystems. It is also an ethical way of thinking and a philosophy. It is built around "the triple-win solutions" which are taking care of the Earth, taking care of people and sharing resources fairly. Permaculture is usually mentioned together with vegetable cropping, gardening, and kitchen gardening. At first, permanent agriculture was thought as a resilient, stable, and sustainable production system. Permanent agriculture is mainly based on the use of biodiversity around and on the farm. Farmers also try to be as autonomous and self-sufficient as possible by implementing a low-energy system. Thus, with this kind of agriculture it is possible to be productive along with using new technologies. PA is usually linked to a sustainable society.

6.1.8 Regenerative agriculture

Regenerative agriculture is a conservation and rehabilitation approach to food and farming systems. It focuses on topsoil regeneration, increasing biodiversity, improving the water cycle, enhancing ecosystem services, supporting biosequestration, increasing resilience to climate change, and strengthening the health and vitality of farm soil. Practices include recycling as much farm waste as possible and adding composted material from sources outside the farm. *RA Principles* include the following: Increase soil fertility, Work with whole systems (holistically), not isolated parts, to make changes to specific parts, Improve whole agroecosystems (soil, water, and biodiversity), Connect the farm to its larger agroecosystem and region, Make holistic decisions that express the value of farm contributors, Each person and farm is significant, Make sure all stakeholders have equitable and reciprocal relationships. *Practices*: No-till farming and pasture cropping, Organic annual cropping, compost and compost tea, Biochar and terra preta, Holistically managed grazing, animal integration, biological aquaculture, perennial crops, Silvopasture, Agroforestry. Management of Environmental Materials and Classic and Modern Techniques of Environmental Remediation is presented by Hussain (2019, 2020).

6.1.9 Climate-smart agriculture

Climate-smart agriculture (CSA) covers all agricultural sectors and brings together practices, policies, and institutions that are not necessarily new but are used in the context of climatic changes. It is conceived to develop the technical, policy, and investment conditions to address the interlinked challenges of sustainably increasing food production, achieving food security and development targets while addressing the challenges of climate change. *Principles* that are more specific to CSA include

(a) The need to identify site-specific solutions to achieve food security under climate change; and (b) Increasing resilience in social as well as production systems and broad-based strategies to manage risk ex ante and ex-post. Its main themes are in line with the priorities of CSA: (1) Data and knowledge to assess impact and vulnerabilities; (2) Institutions, policies, and financing to strengthen capacities to adaptation; (3) Sustainable climate-smart management of land, water, and biodiversity; (4) Technology, practices, and processes for adaptation; and (5) Disaster risk management.

7. Sustainable rural development

FAO (2020) presented 20 interconnected sustainable food and agriculture (SFA) actions, which embrace the ***2030 Agenda's vision of sustainable development*** in which food and agriculture, people's livelihoods, and the management of natural resources are addressed as a whole. These actions integrate the three dimensions of *sustainable development* (environmental, economic, and social), and its five *principles* (increase productivity in food systems, protect and enhance natural resources, improve livelihoods, increase resilience of people and ecosystems, adapt governance to new challenges): 1. Facilitate access to productive resources, finance, and services, 2. Connect smallholders to markets, 3. Encourage diversification of production and income, 4. Build producers' knowledge and develop their capacities, 5. Enhance soil health and restore land, 6. Protect water and manage scarcity, 7. Mainstream biodiversity conservation and protect ecosystem functions, 8. Reduce losses, encourage reuse and recycle, and promote sustainable consumption, 9. Empower people and fight inequalities, 10. Promote secure tenure rights, 11. Use social protection tools to enhance productivity and income, 12. Improve nutrition and promote balanced diets, 13. Prevent and protect against shocks: enhance resilience, 14. Prepare for and respond to shocks, 15. Address and adapt to climate change, 16. Strengthen ecosystem resilience, 17. Enhance policy dialogue and coordination, 18. Strengthen innovation systems, 19. Adapt and improve investment and finance, 20. Strengthen the enabling environment and reform the institutional framework. In 2015, *The Sustainable Development Goals* were adopted as part of this agenda—a universal call to action to end poverty, protect the planet, and improve the lives and prospects of everyone, everywhere. Thrupp (1993) presented a book "Political Ecology of Sustainable Rural Development: Dynamics of Social and Natural Resource Degradation."

7.1 Sustainable rural systems

The report "*Building a common vision for sustainable food and agriculture: Principles and Approaches*" (FAO, 2014) sets out five *key principles* that balance the social, economic, and environmental dimensions of sustainability: (1) improving efficiency in the use of resources; (2) conserving, protecting, and enhancing natural ecosystems; (3) protecting and improving rural livelihoods and social well-being; (4) enhancing the resilience of people, communities, and ecosystems; and (5) promoting good governance of both natural and human systems.

7.1.1 Guidelines for sustainable rural systems

Integration of crop and animal production. Animal holdings should reduce the number of livestock or split up into smaller, more evenly distributed animal holdings.

Biodiversity: The number of ruminants should locally correspond to the amount of old permanent grazing land to preserve biodiversity. Select valuable grazing areas. Save natural biotopes. Conserve landraces and old species on-farm.

Organic farming: Promote OF wherever it contributes to sustainable development.

Less competitive farming: Support remote and less market-competitive farming and essential services and complementary employment, in order to preserve a viable countryside.

Cooperation within watersheds: Cooperation between neighboring farms to overcome negative effects of extensive specialization on individual farms by mutual care for the arable resources (permanent grazing land, exchange of feed and manure, etc.).

Water quality: Secure long-term water quality by suitable land use within pumping areas for high quality groundwater. This usually corresponds to less intensive forms of land use.

Soil fertility: Maintain and improve SF (soil organic matter, soil structure, nutrient status, nonbiotic elements and chemicals) by use of nonpolluted production, noncompacting machinery, and cultivation promoting soil organic matter. Soil monitoring, analysis, and nutrient balances programs for appropriate use of arable land.

Animal health and welfare: Animals should be fed a balanced diet, not long distance transportation, preferably have outdoor access and be kept in loose housing systems. Antibiotics use in animal medication should decrease. Terminate growth promoters.

Genetically modified organisms (GMO): Restrictive approval procedure for GMO in food production. Any increase in the use of plant protection products should not be allowed!

Bioenergy: Increase bioenergy production on excess arable land. Present land use must not jeopardize possibilities in the future to produce high quality food on the same land.

Recirculation: Recirculate nutrients and organic matter in urban biowaste to the production of biomass on arable land. Efficient administrative systems for waste quality assessment.

Plant protection: Reduce the use and risks of plant protection products. Select crops and cropping systems with less need for plant protection, improve spraying techniques, obligatory certificates after participation in courses on safe handling practices for all farmers. All plant protection products must be registered and approved by national or international authorities! Cropping systems increase the soil organic matter content (permanent grassland, perennial crops, and reduced soil tillage). Reduced use of mineral fertilizers and imported feed. CO_2 energy taxes on nonrenewable energy. Reduce ruminant livestock numbers and/or increase production level.

Transport: Minimize transport of feed, food, and wastes—promote local alternatives before centralized ones when this is deemed profitable by life cycle analysis.

Finances: Farmers' income should be sufficient to provide a fair standard of living and consist of reasonable compensation for products and other services.

Employment: New profitable services and products based on farm assets and production.

Expertise: Education, demonstrations, and advisory activities for sustainable agriculture. Research for sustainable agriculture and mitigate and adapt to climate change.

7.1.2 Sustainable crop production

Sustainable crop production does not impose any harm to environment, biodiversity, and quality of agricultural crops. It increases the ability of the system to maintain stable levels of food production and quality for long term without increasing the demand and requirements of agricultural chemical inputs. SCP deals with keeping the soil alive (organic matter, integrated pest management, reduced usage of pesticides), protecting biodiversity, ensuring food safety and food quality, improving nutrient quality, and improving the soil with organic fertilizers. SAP leads to lowering of greenhouse gas emission and carbon footprint. Sustainably produced crops and food are more beneficial to consume by humans as compared to commercial crops. Sustainable usage of resources ensures the pollution-free environment.

7.1.2.1 Assessment and implementation of sustainable crop production

Lewandowski et al. (1999) present *a method for assessing and implementing sustainable crop production*, in the context of "sustainable agriculture." The steps are as follows: (1) identify emissions and other releases linked to different crop production practices, (2) trace each different release from its source (the crop management practice) to its sinks (i.e., agroecosystems and other ecosystems or components of ecosystems directly or indirectly affected), (3) select indicators that adequately describe the ecosystem condition affected directly or indirectly by crop production practices, (4) determine threshold values for the selected ecosystem indicators (i.e., values which should not be exceeded to avoid irreversible changes in the affected ecosystems), (5) transpose the ecosystem threshold values to the farm level by retracing the impact pathways backward to crop production itself, (6) derive farm-level indicators that point to separate or combined agronomic practices that could cause irreversible changes in affected ecosystems, (7) determine farm-level threshold values for management-induced releases on the basis of ecosystem level threshold values, and (8) identify production schemes that adhere to the framework set by the farm-level thresholds. The farmer can select from these production schemes those most in line with his available resources and objectives.

7.1.2.2 Sustainable crop production intensification

Climate change, environmental degradation, and stagnating yields threaten crop production and world food security. It is now recognized that the enormous gains in agricultural production and productivity achieved through the green revolution were often accompanied by negative effects on agriculture's natural resource base, which jeopardize its productive potential in the future. Current food production and distributions systems are failing to feed the world. In most countries there is little space for arable land expansion. The declining quality and increased competition for the land and water resources for crop production has major implications for the future. Resource degradation reduces the effectiveness of inputs, such as fertilizer and irrigation. Given the current and increasing future challenges to our food supply and to the environment, sustainable intensification of agricultural production is emerging as a major priority for policy-makers and international development partners. Sustainable crop production intensification (SCPI) aims to produce more from the same area of land while reducing negative environmental impacts, conserving natural resources and enhancing healthy ecosystem services. This eco-friendly approach to farming combines farmers' traditional knowledge with modern technologies adapted to the needs of small-scale producers. This approach is founded on a set of science-based environmental, institutional, and social *principles*. It provides adoptable and adaptable farming practices and technologies that support the development of resilient crop production systems.

Environmental principles. SCPI is based on an *ecosystems approach* to production and incorporates management practices that include maintaining healthy soil to enhance crop nutrition; cultivating a wider range of species and varieties in association, rotations and sequences; using well-adapted, high-yielding varieties and good quality seeds; integrated management of insect pests, diseases, and weeds; and efficient water management. *Sustainable management practices*: minimum soil disturbance and permanent organic soil cover through conservation agriculture; increased on-farm species diversity through conservation and sustainable use of plant genetic resources; use of quality seeds of best-adapted varieties which are accessible to farmers; integrated pest management; and plant nutrition based on sustainable water and soil management (i.e., integration of nitrogen-fixing legumes and trees to cropping systems).

Institutional principles. Implementing these environmental principles into large-scale programs requires institutional support at both national and local levels. Strengthening institutional linkages is key to improving the formulation of policies and strategies for SCPI, and to enable scaling up of pilot studies, farmers' experiences, and local/traditional knowledge. Cooperation and integration among government and other key stakeholders to develop multisector and multistakeholder policy frameworks and strategies is needed.

Social principles. To support the adoption of SCPI by farmer's rural *participatory* advisory services, from both traditional and nontraditional sources, must be strengthened. Mobilizing social capital will require people's participation in local decision-making, ensuring decent and fair working conditions in agriculture, and the recognition of the critical role of women in agriculture. *Farmer-centered approach* incorporates sustainable natural resource management into their production systems. Promotes technological innovation through dissemination of information, tools, and capacity building on ecosystem-based approaches to crop production adapted to the specific agroecological and socioeconomic conditions of farmers.

7.1.2.3 Sustainable crop production techniques

Planting methods. Sustainable systems focus on *plant diversity* (polycultures) rather than monocultures. Repeatedly planting the same crop in the same place makes it vulnerable to a wide array of pests, exhausts the soil of nutrients, makes it more susceptible to erosion, and decreases yields, requiring use of fertilizers and pesticides. C*rop rotation* can replenish soil and prevent pests from getting established. *Partial* crop rotations are populár (strip cropping: planting alternating rows of a row crop (corn, soybeans with small grains, hay to reduce erosion)). Other methods of preventing runoff of water and nutrients (contour buffers and grassland strips with native grasses to retain water). Using a variety of plants increases the biodiversity of the farm ecosystem, attracting pollinators (butterflies, bees), as well as birds and insects that prey on pests. Pesticides eliminate plants' best natural defenses. Plants that attract pests can be grown as *trap crops* near the main crop, protecting it by luring away predators. Plants can benefit each other in many other ways (attracting beneficial insects, fixing soil nitrogen, providing shade or support, emitting growth-enhancing chemicals; intercropping places beneficial crops near each other).

Sustainable seeds. Sustainable crop farmers use a wide range of seeds. *Hybrids* are bred through conventional plant breeding, cross-pollinating two varieties of a plant to produce an offspring with the best traits of each parent. Hybrids are bred for disease resistance, uniformity, productivity, ease of growing, and other popular traits. Sustainably and independently produced seed stock is under *threat*, as seed companies have been bought up at an alarming rate. We worry about long-term food security if one company controls the vast majority of the genetic information for the world's major crops. Some independently owned seed companies still remain and are the chief suppliers of sustainable farmers.

Soil health. Managing and building the health of the soil is the most critical element of sustainable crop farming. The structure, organic matter, insects, and microbes of healthy soil retain water, deliver nutrients to roots, and keep plants healthier and more resilient under stress than those fed with chemical fertilizer. Healthy soil contains bacteria, which break down organic matter; pull nitrogen from the air and make it available to plant roots; help with the movement of water in soil; help with plant communication and defense; and perform many other functions. To restore depleted soil nutrients and microorganisms, farmers use *organic amendments* (manure, compost, worm castings, and seaweed). In the winter or off-season, it is common to plant *cover crops* (oats, rye, or clover). This "green manure" adds organic matter, fixes nitrogen, improves soil structure, reduces erosion, etc. *Low- or no-till practices* are popular, because soil can sequester a great deal of carbon. Plowing the land releases carbon into the air, disrupts microorganisms, compacts the soil, and can even hasten its erosion. *Reduced tillage* increases organic matter and the amount of carbon, as well as improves its structure and water retention capacity. *No-till farming* leaves crop residues to decompose into the soil (rather than removing weeds by mechanical tillage) and allows spray pesticides. Modern techniques of environmental analysis for pollutants are presented by Hussain and Keçili (2019).

Weed control. Without tilling or pesticides, weeds must be managed by farmers using sustainable methods. Cover crop; mulching, spreading a cover layer over unplanted soil can be used. Materials from straw and leaves to sheets of black plastic are used as mulch. *Flaming weeds* (e.g., burning them with specialized equipment) is another method of dealing with leeds.

7.1.2.4 Sustainable management of crop nutrients

Nitrogen input. Application rates should not exceed the crop nutrient requirements. National guidelines with fertilizing recommendations: (a) soil conditions, soil nutrient content, soil type, slope, (b) climatic conditions, precipitation, and irrigation, (c) land use and agricultural practices, including crop rotation systems, (d) all external nutrient sources. Nitrogen nutrient balances should be performed on the farm to show the nitrogen surplus when planning fertilization.

Phosphorus input. The available P content of arable topsoils should not exceed the requirements of acceptable crop production. The annual P input should be calculated in relation to the phosphorus content in the field and the crop requirements. Good monitoring data on the P status of arable land are needed, as well as nutrient balances to show whether the supply of P in the soils is increasing or being depleted. Phosphorus nutrient balances should be prepared for the farm to show the size of the phosphorus surplus when planning fertilization.

Livestock density and manure handling. In regions with high average livestock density, and on individual farms, the total number of animals should be reduced to a level consistent with efficient recycling of nitrogen and phosphorus. Efficient circulation of nutrients on animal farms in combination with a high degree of self-sufficiency in feed is a prerequisite for limited losses of plant nutrients. The utilization efficiency of the nutrient content in animal manure should be improved. That can be achieved by building sufficient storage capacity for manure for optimal timing of spreading, covering slurry and urine stores to reduce the odor and the emissions of ammonia nitrogen, improving manure spreading techniques, incorporating slurry, urine and solid manure into the soil immediately after spreading on open soils to minimize ammonia nitrogen losses. *Nutrient point sources* on the farm, such as from manure storage, milking parlors, silage storage, etc., should be identified and eliminated.

Crops and crop rotations. Choose crops and crop rotations with a minimum need for soil cultivation. Keep a high proportion of arable land covered by crops during autumn and winter. In areas with more than 50% annual crops, the proportion of perennial crops or green cover crops should be increased specially in areas with sandy soils, areas for drinking water, and on land sensitive to erosion.

New technologies. Development of new technologies that can reduce nutrient losses (precision farming with site-specific crop management by use of GPS, etc.).

Criteria for surplus land. The farmers should take environmental considerations into account when removing surplus land from food production: soils poor in phosphorus, organic soils on previously drained wetlands, soils sensitive to erosion, and soils sensitive to nitrate leaching.

Nutrient traps. Create buffer zones and wetlands to reduce nutrient losses and increase biodiversity.

7.1.3 Sustainable animal production

Animal husbandry is the breeding and raising of animals for meat, milk, eggs, or wool, and for work and transport. Working animals (horses, mules, oxen, water buffalo, camels, llamas, alpacas, donkeys, and dogs) have for centuries been used to help cultivate fields, harvest crops, wrangle other animals, and transport farm products to buyers. *Livestock production systems* can be defined based on feed source, as grassland-based, mixed, and landless.

7.1.3.1 Global agenda for sustainable livestock

The Agenda is a partnership of livestock sector stakeholders committed to the sustainable development of the sector. Livestock are critical to building sustainability. Sustainability is a process of continuous practice change that addresses social, economic and environmental objectives simultaneously. To be sustainable, livestock sector growth needs to support the livelihoods of an estimated one billion people, contribute to enhancing economic and social well-being, protect public health through balanced diets and the reduction of health threats from livestock, and protect the natural resources. The Agenda is working in three major *focus areas*: (a) global food security and health; (b) equity and poverty reduction; and (c) resources and climate. Stakeholders of the Agenda include the private sector, NGOs, and social movements, government partners, research institutions, international agencies, and foundations. The Agenda catalyzes policy dialogue into practice change. FAO supports the Agenda through dialogue, analysis, and policy advice, as well as through pilots and investment strategies.

7.1.3.2 Integrated crop-livestock production

Integrated crop-livestock production is a promising approach for enhancing overall productivity and profitability through recycling of the farm by-products and efficient use of available resources. This integrates natural resources into farming system to achieve maximum replacement of off-farm inputs. Crop residues are substantial component of livestock diets. It can generate rural employment opportunities to the farming sector and provide better economic and nutritional security. This can go a long way to uplift rural life through increased income. The *advantages* of integrated system include increased productivity, profitability, and soil fertility; provide balanced diet, recycling of nutrients, generation of employment, and money flow round the year; and solve energy crisis.

7.2 Sustainable intensification of rural systems

The objectives of sustainable intensification are as follows: (a) Production of more food with less environmental impact, (b) Input use efficiency of key production (nutrients, water, pesticides, labor, energy), (c) Use of natural biodiversity (cropping systems, agroforestry) to overcome biotic, abiotic, and economic stresses (Murray 2012).

Sustainable agriculture and rural systems is an agricultural production and distribution system that (a) achieves the integration of natural biological cycles and controls,

(b) minimizes adverse impacts on health, safety, wildlife, water quality, environment, (c) optimizes the management and use of on-farm resources, (d) promotes opportunity in family farming and farm communities, (e) protects and renews soil fertility and the natural resource base, (f) provides an adequate and dependable farm income, and (g) reduces the use of nonrenewable resources and purchased production inputs.

Sustainable intensification of rural systems aims to produce more food from the same piece of land with less environmental impact. Sustainable agriculture and rural systems are a profitable way to produce high-quality food and fiber that protects and renews the natural environment, builds local economies, and enhances the quality of life of farmers. New technologies and innovation will help to overcome these challenges and help to achieve the aim. Sustainable intensification utilizes *ecological processes* (biological N fixation, natural predators, etc.), uses integrated modern and traditional strategies, minimizes environmental hazards, reduces risk, maintains soil and water quality, and acknowledges local environmental and cultural conditions, resulting in increased productivity. Global food demand and the sustainable intensification of agriculture was studied by Tilman et al. (2011).

7.2.1 Sustainable intensification of crop and animal production

The sustainable intensification is founded on science-based *principles*: *Economic Sustainability.* Sustainable livelihoods and improved well-being through growth and poverty reduction. *Environmental Sustainability.* Target agricultural land, forests, water resources, protected areas, and biodiversity, so that opportunities and options of future generations are not degraded. *Fiscal and Institutional Sustainability.* Must be realistic about cost and institutional requirements of instruments.

7.2.1.1 Farming practices for sustainable intensification

Agricultural biodiversity. *The components* of agricultural biodiversity consist of vegetation, herbivores, predators and parasites, pollinators, soil mesofauna, soil microfauna, livestock, and earthworms. *The functions* of biodiversity include population regulation, biological control, predation, sources of natural enemies, disease suppression, pollination, genetic introgression, crop wild relatives, competition, allelopathy, biomass consumption, nutrient cycling, soil structure, and organic matter decomposition. Agrobiodiversity can be enhanced by agroforestry, intercropping, cover crops, crop rotations, green manures, composting, no-tillage, organic matter inputs, and windbreaks (Altieri, 1994). The soil management strategies can also be used to enhance soil quality through biodiversity.

Improved water management practices. The fundamental question is how to increase *water productivity in agriculture*. It is important to shift from water-intensive crops (wheat, sugar beets, citrus) to less water-intensive crops (barley, fruit trees, etc.). This applies particularly to rain-fed crops, where there is untapped potential to be developed through research. Many *technologies* are available for improving the water use efficiency in agriculture, both rain-fed and irrigated: Integrated water resources management: surface water, groundwater, catchment basin, and land-use planning, Conservation tillage and planting on raised beds (efficient use of irrigation water), Water-saving technologies in rain-fed agriculture, Improved small-scale and supplemental irrigation systems for rain-fed agriculture, reefficient crop sequencing and timely planting (savings in water use), New crops, requiring less water, are now available. Drip irrigation systems and center pivot irrigation, drought tolerant crops, for rain-fed agriculture. Precision irrigation systems for supplying water to plants only when needed. Wastewater treatment and use for irrigation (mainly for peri-urban agriculture).

Retention of crop cover. The retention of crop residues/cover crops can be achieved by leaving the crop residues after harvest, planting cover crops after harvest, and adding organic amendments (farm yard manure, compost, wood chips, straw, leaves, etc.) as soil mulch. This helps in reducing the soil erosion. Growing cover crops, such as cereals, legumes (hairy vetch, crimson clover, red clover, white clover, winter pea), and brassicas (mustard and radish) which provide quick soil cover, have an extensive root system, and preferably survive winter. If the cover crops are killed without tillage and the main crop established with no-till methods, additional erosion protection will be provided by the resulting mulch. The tine system results in short residue, well spread to maximize trash flow, while disk system produces long residue that is well spread. Expected changes due to *residue management systems* include improved soil aggregate stability, increased water-holding and infiltration capacity, reduced soil erosion, increased organic matter and C sequestration, soil structure improvement, nitrate recapture, atmospheric nitrogen fixation, reduced water evaporation, reduced soil temperature, weed smothering, nutrient release, and less surface ponding due to rainfall.

Selection of suitable cultivars. Select suitable crop cultivars which can perform well under different agroecosystems, socioeconomic circumstances, farming practices, and resilient to climate change. The following aspects should be considered while selecting and developing crop varieties suitable for sustainable intensification: Adapted to various agroecosystems, changing production practices, farming systems, and less favored areas and resilient to climate change, Elaborate root system to improve the root efficiency for mining the water and absorption of nutrients, High input, water, and nutrient use efficiency, High quality with good production potential, Resistant/tolerant to biotic (pests, diseases, and nematodes) and abiotic (drought, heat, flood, salinity, and frost) stresses, Varieties with higher nutritional value and desirable organoleptic properties, and Enhanced carbon sequestration.

Agroecological pest management. The ecological goal of pest management strategy is to use naturally based controls (farmscaping, beneficial overwintering sites, cover crops, no-till), in order to increase number of natural enemies for long-term pest control. Leaving a residual pest population is important for natural enemies to have food in order to survive. Pest residues for natural enemies should be lower than the economic threshold. The pests should be suppressed but not to be eradicated. The pesticide treatments should be based on the need—"treat when necessary" driven by data.

Ecological pest management practices include the following: Conservation tillage/residue management, Crop nutrient management, Conservation buffers, Filter strips and crop rotation, Other conservation practices. *Integrated pest management* is an approach to pest control that combines biological, cultural, and other alternatives to chemical control with the judicious use of pesticides. The objective of IPM is to maintain pest levels below economically damaging levels while minimizing harmful effects of pest control on human health and environmental resources by (a) Use of cultural methods, biological controls, and other alternatives to pesticides to help delay the pesticide resistance, (b) Use of field scouting, pest forecasting, and economic thresholds to ensure that pesticides are only used for real pest problems, and (c) Matching of pesticides with site characteristics to minimize offsite environmental risks. Approaches for managing pests are often grouped in the following categories:

Biological control. Biological control is the use of natural enemies—Predators (free-living animals that eat other animals), Parasites, Parasitoids (insect parasites of other insects), and Pathogens (disease-causing microorganisms) to control pests and their damage. Invertebrates, plant pathogens, nematodes, weeds, and vertebrates have many natural enemies.

Cultural controls. Cultural controls reduce pest establishment, reproduction, dispersal, and survival. Practices: Certified seed that is free of pathogens and weed seed, Crop rotation, Cover crops, Pest-resistant varieties, Altering planting and harvest dates, Altering seeding rates/crop spacing, Sanitation practices (cleaning tillage and harvesting equipment), Seedbed preparation, fertilizer application, and irrigation schedules that help plants outgrow pests, Trap crops, Changing irrigation practices, etc.

Mechanical and physical controls. Mechanical and physical controls kill a pest directly, block pests out, or make the environment unsuitable for it. Traps for rodents are examples of mechanical control. Physical controls include mulches for weed management, steam sterilization of the soil for disease management, or barriers such as screens to keep birds or insects out.

Chemical control. Chemical control is the use of pesticides. In IPM, pesticides are used only when needed and in combination with other approaches for more effective, long-term control. Pesticides are selected and applied in a way that minimizes their possible harm to people, nontarget organisms, and the environment. With IPM you will use the most selective pesticide that will do the job and be the safest for other organisms and for air, soil, and water quality; use pesticides in bait stations rather than sprays; or spot-spray a few weeds instead of an entire area.

Conservation tillage. CT includes many varied tillage systems that leave more than 30% crop residue cover after planting (no-till, minimum tillage, ridge tillage, mulch tillage). Crop rotations and cover crops are central to this system. On most soils, no-till yields should be similar to yields obtained with tillage. No-till should outyield tilled crops in areas where drought stress is a problem, due to the water conserved by the mulch cover. *Practices*: 1. No-Till/Strip-Till (>30% Residue), 2. Ridge-Till (>30% Residue), 3. Mulch-Till (>30% Residue), 4. Reduced-Till (15%–30% Residue).

Balanced crop nutrition. Today, nutrients are exported away from the soil. Loss of soil also occurs through erosion that comes as a result of extensive tillage. Nutrients should be managed in a way that meets crop needs without applying in excess which can lead to water quality problems. Plants need *macronutrients* such as N, P, K, Ca, Mg, and S (>50 mg/kg in plant) and *micronutrients* such B, Cl, Cu, Fe, Mn, and Zn (<50 mg/kg in plant). The sources of nutrients include soil, organic manures, and inorganic fertilizers. Nutrients can be applied to crop plants either in organic or inorganic form. Nutrients in inorganic fertilizers are dissolved into soil solution and are available immediately. Decomposition of organic matter releases nutrients, and hence they are slowly available. *Manure* utilization recycles nutrients back to the land. Nutrient management means integration of on-farm nutrient sources (soil reserves and organic manures), with commercial inorganic fertilizers to meet the crop need, minimizing nutrient loss. The development of nutrient management plan helps to decide on the suitable form, correct amount and proper timing of application of crop nutrients to obtain optimum yields, and less impact on quality of water. Best *nutrient management strategies* include soil testing, timing of applications, conservation tillage, buffer strips, cover crops, P-based nutrient management, and balance: inputs = exports. There is an urgent need for integrated management of all sources of nutrients for promoting efficient and balanced use of plant nutrients.

Careful management of farm machinery. Agriculture mechanization refers to usage of any improved tool, implemented or machined by the farm workers to improve the efficiency of any farm operation to increase crop productivity. The timeliness of operations has assumed greater significance in obtaining optimal yields from different crops. Comparison of agricultural, ecological and sustainable intensification shows Lal et al. (2016) (Table 5.3).

Table 5.3 Sustainable intensification of agriculture: merits and challenges (Lal et al., 2016), adapted.

Merits	Challenges
Agricultural intensification	
Low inputs	Increasing productivity
High productivity	Reducing positive feedbacks
Soil and resources conservation	Decreasing adverse impacts
High biodiversity	Minimizing trade-offs
High residence	Increasing equity
Ecological intensification	
Low by-products	Increasing productivity
Minimal material fluxes	Increasing emissions
Low waste levels	Reducing drudgery
Low risks	Improving nutritional security
Sustainable intensification	
High production	Reducing inputs
High income	Decreasing environmental pollution
Agroindustry	Reducing soil degradation
Rural employment	Decreasing emissions

8. Conclusions

8.1 The State of Food and Agriculture

The State of Food and Agriculture 2019: *Moving forward on food loss and waste reduction* provided new estimates of the percentage of the world's food lost from production up to the retail level. It is a comprehensive analysis of the critical loss points in specific supply chains. Guiding principles for interventions based on the objectives being pursued through food loss and waste reductions were elaborated, be they in improved economic efficiency, food security and nutrition, or environmental sustainability.

The State of Food and Agriculture 2020: *Overcoming water challenges in agriculture* presents new estimates on the pervasiveness of water scarcity in irrigated agriculture and of water shortages in rainfed agriculture, as well as on the number of people affected. The publication provides guidance on how countries can prioritize policies and interventions to overcome water constraints in agriculture, while ensuring efficient, sustainable, and equitable access to water.

The impact of disasters and crises on agriculture and food security was also studied (FAO, 2017a, 2017b, 2021). The FAO also presented a report, "Tracking progress on food and agriculture-related SDG indicators," which offers detailed analysis and trends on 22 indicators under six SDGs under its custodianship (FAO, 2020b). FAO (2017a) elaborated "Alternative pathways for future development of food and agriculture to 2050."

8.2 The State of Food Security and Nutrition in the World

The State of Food Security and Nutrition in the World 2019 analyzed ***Safeguarding against economic slowdowns and downturns*** (FAO, IFAD, UNICEF, WFP, WHO, 2019). The report provided new estimates of the percentage of the world's food lost from production up to the retail level. It is a comprehensive analysis of the critical loss points in specific supply chains. Guiding principles for interventions based on the objectives being pursued through food loss and waste reductions were elaborated, be they in improved economic efficiency, food security and nutrition, or environmental sustainability.

The State of Food Security and Nutrition in the World 2020: ***Transforming Food Systems for Affordable Healthy Diets*** informs on progress toward ending hunger, achieving food security, and improving nutrition and to provide analysis on key challenges for achieving this goal in the context of the 2030 Agenda for Sustainable Development (FAO, IFAD, UNICEF, WFP, WHO, 2020).

Food Security Information Network (FSIN) published "Global Report on Food Crises 2019" (FSIN, 2019). Global Panel on Agriculture and Food Systems for Nutrition published "Food systems and diets: facing the challenges of the 21st century" (GLOPAN, 2016). Guiding principles for sustainable healthy diets were elaborated by FAO and WHO (2019). FAO (2018) presented a report "The State of World Fisheries and Aquaculture 2018. Meeting the Sustainable Development Goals". FAO (2019c) published the study "The fisheries and aquaculture advantage. Fostering food security and nutrition, increasing incomes and empowerment". FAO (2019d) published The State of the World's Biodiversity for Food and Agriculture. FAO (2019e) presented a report "The state of the Worlds Aquatic Genetic Resources for

Food and Agriculture." FAO (2019f) reported about the 17th Session of he Commission on Genetic Resources for Food and Agriculture.

8.3 Industrialized agriculture

Although the industrialized model of agricultural production has been very successful at increasing agricultural yields, it is highly dependent on fossil fuel inputs and thus is vulnerable to increased energy costs and declining energy supplies. The great gains in agricultural productivity in the past century have been accompanied by substantial degradation to global ecosystems / landscapes (Millennium Ecosystem Assessment, 2005).

Sustaining increased agricultural production while preserving ecological integrity and environmental quality is one of the grand challenges for agriculture for the 21st century. To be sustainable, agriculture must meet the needs of present and future generations, while ensuring profitability, environmental health, and social and economic equity. Sustainable food and agriculture (SFA) contributes to all four pillars of food security (availability, access, utilization, and stability) and the dimensions of sustainability (environmental, social, and economic). Perspectives of sustainable agriculture from industry published Richgels et al. (1990).

8.4 Holistic agriculture

A holistic view on agriculture must be adopted. All inputs and resources should be used most effectively. Improved technology and combining the best from all farming systems is one of the keys. *Ecologically designed agrosystems* (intercropping, integrated crop-livestock systems, agroforestry, etc.) should be used. *The welfare and health* of both man and beast should also be high on the agenda. *Climate change* must be combated and it will most likely lead to new crops, pests, and diseases on plants, animals, and humans, as well as a longer growing season. *Good products* of high quality should also lead to *fair prices* that make it possible to farm with a reasonable standard of living. Minimize transport. *Traditional agricultural knowledge* should be integrated with more ecologically designed agricultural knowledge into the intensification process. *Integrated nutrient—organic matter management* and *pest management* approaches should be introduced. *Precision agriculture*, which facilitates application of inputs differentially across fields to match crop demands, allows increased efficiency. *Education, training,* as well as the advisory service, will play an important role. Arable land will be more and more important to feed the world population and produce renewable energy. *Capacity building* for farmers through technical support and providing hands-on learning opportunities through *farmer field schools* to improve the skills of all stakeholders are the need. This will require coordinated efforts of extension workers, researchers, and policy makers to enhance the adaptation of sustainable intensification.

8.5 Agrobiodiversity

Agricultural biodiversity is essential to satisfy basic human needs for food and livelihood security; many components of agrobiodiversity depend on human influence. Indigenous knowledge and culture are integral parts of the management of agricultural biodiversity. For crops and domestic animals, diversity within species is at least as important as diversity between species and has been greatly expanded through agriculture. Because of the degree of human management of agricultural biodiversity, its conservation in production systems is inherently linked to sustainable use. Nonetheless, much biological diversity is now conserved ex situ in gene banks or breeders' materials. The interaction between the environment, genetic resources, and management practices that occurs *in situ* within agroecosystems and agricultural and rural landscapes often contributes to maintaining a dynamic agricultural biodiversity. Genetic erosion narrows our ability to adapt our global food system to challenges such as population growth, emerging diseases, and climate change. Policy environments frequently disadvantage the traditional production systems that typically harbor adapted livestock species and breeds, thus fostering genetic erosion. The provisioning roles of livestock in poor rural households are often not fully captured in economic data. Production systems that overemphasize the provisioning roles of livestock often do so at the expense of other ecosystem services.

8.6 Agroecosystems and Agrolandscapes

The urgent need to preserve multiple ecosystem / landscape services is one of the key challenges in sustainable natural resource management globally.

Landscape scientists must find better ways of communicating their results to nonacademic users. Universal criteria for evaluating landscapes are not available. *Landscape multiple functions* are notalways compatible, and conflicts between landscape functions are common. One particular landscape typically has different functions for different people. Finding markets for ecosystem / landscape services (for instance, through the trading of permits for using particular services) might be a promising strategy to balance the various interests. *Nonagricultural landscape functions* provide new opportunities for rural entrepreneurs outside the agricultural and forestry sectors, and might thus help slow down the economic, sociocultural, and demographic decline in rural areas.

Conservation of agricultural ecosystem / landscape services:

Understand the concept that agricultural ecosystem services can sustain themselves with proper design (farmers, policymakers, etc.). (Agro) Ecosystem services have the potential to reduce both off-site inputs and on- and off-site pollution. Promoting identification and taxonomy is necessary. Assessment of risks over time, relative dependence, and sustainable livelihoods are critical issues for agricultural biodiversity, and need to be in appropriate balance. Policy makers are biased toward large-scale plans, whereas much of agrobiodiversity is fine-scaled. Costs and benefits of agrobiodiversity goods and services need to be identified. Costs and benefits need to be distributed on the basis of careful assessment of possible trade-offs, paying attention to incentives and subsidies, and making them appropriate. Creating popular awareness and education is necessary for change. It is necessary to enhance capacity for adaptation to change.

8.7 Sustainable agriculture

To be sustainable, agriculture must meet the needs of present and future generations, while ensuring profitability, environmental health, and social and economic equity. *Sustainable food and agriculture (SFA)* contributes to all four pillars of food security (availability, access, utilization, and stability) and the dimensions of sustainability (environmental, economic, and social). *The benefits of sustainable intensification of agriculture* are: a) resembles a natural ecosystem, b) soil is continuously covered, c) builds soil organic carbon, d) fertilizer requirement is reduced, e) no turning of the soil, f) less use of energy (30-40 %), g) helps in pest and weed control, h) yields are comparable to conventional agriculture, i) higher yields especially in drought years, j) adds up to 1 mm soil per year, k) organic matter increase (about 0.1-0.2 % per year), l) production costs can be reduced, m) more efficient use of soil nutrients, n) soil structure more stable, o) drudgery is reduced, p) planting area can be expanded with existing resources, q) soil erosion and degradation stopped/reversed, r) less labor requirement, s) machinery use efficiency is increased, t) mitigation through emission reductions (fuel, NOx, CH 4), u) mitigation through carbon sequestration (up to 0.2 t of C/ha/yr), v) adaptation through better drought tolerance, w) adaptation through better water infiltration (less flooding), etc.

Consider the ecosystem and landscape concepts and how all parts of the ecosystem / landscape are dependent on one another. Develop the conceptual basis and practical strategies for incorporating broader environmental, ecological, and social factors into agroecosystems / agrolandscapes management and science. Sustaining increased agricultural production while preserving ecological integrity and environmental quality is one of the grand challenges for agriculture for the 21st century. Production systems that sustain agricultural productivity, support natural processes and functions, as well as diversity, and provide economically viable and socially attractive opportunities for farmers should be developed. Integrated approaches for sustaining whole system productivity, diversifying agricultural operations, and managing agricultural systems for multiple purposes, including the ecological integrity of farming systems are needed.

8.8 Sustainable rural systems

The concept of *knowledge exchange* has shifted. Scientific information is part of a four-way process: from science to science, from science to citizens, from citizens to science, and from citizens to citizens. The growth of *citizen science* offers new approaches and opportunities for engagement between researchers, practitioners, and communities. Shaw et al. (2017) presented potentials and challenges of contributions of citizen science to landscape democracy.

Sustainable crop production: Survey and inventory of plant genetic resources for food and agriculture; Support of on-farm management and improvement of plant genetic resources for food and agriculture; In situ conservation and management of crop wild relatives and wild food plants; Assistance to farmers in disaster situations to restore crop system; Support for targeted collection of plant genetic resources for food and agriculture; Sustaining and expanding *ex situ* conservation of germplasm;

Regenerating and multiplying *ex situ* accessions; Characterization, evaluation, and development of specific subsets of collections to facilitate use; Support for plant breeding, genetic enhancement, and base-broadening efforts; Diversification of crop production and broadening of crop diversity for sustainable agriculture; Promotion of development and commercialization of all varieties, primarily farmers' varieties/landraces, and underutilized species; Support of seed production and distribution (local, regional, national varieties); Building and strengthening national programs; Promotion and strengthening of network; Construction and strengthening of comprehensive information systems for plant genetic resources; Systems for monitoring and safeguarding genetic diversity and minimizing genetic erosion; Building and strengthening human capacity; and Public awareness.

Sustainable animal production: Improvement of knowledge of the characteristics of different types of animal genetic resources, the production systems in which they are kept, and the trends affecting these production systems; Development of stronger institutional frameworks for animal genetic resources management; Improvement of awareness, education, training, and research in all areas of animal genetic resources management; Strengthening of breeding strategies and programs; Expanding and diversifying of conservation programs.

8.9 Landscape approaches for climate-smart agriculture

The process of applying landscape approaches for CSA interventions follows these steps: ***Designing the methodology:*** Development of the analytic framework and the selection of tools for the intervention. Raising awareness of climate risks and the need for adaptation and mitigation to promote the wider uptake of climate-smart agriculture. For country-level interventions, this is the step in which the strategy and action plan are prepared. The action plan is generally aimed at the implementation of the intervention, the expansion of activities on the ground, and the mainstreaming of CSA practices. ***Assessing and prioritizing:*** This step includes the preparation of assessments of land and other natural resources, and socioeconomic conditions. System-wide capacity needs assessment. Training materials are developed and disseminated. Climate change impact and vulnerability assessments are also carried out, which can also include climate analysis, agro-meteorology forecasts, and climate modelling. Participatory wide-scale assessment is undertaken to identify *hot spots* (e.g., areas where there is a severe degradation of ecosystem services or declining production) and *bright spots* (e.g., areas where the land is being managed sustainably). Based on the assessment's findings, the priority landscapes interventions are selected through a collaborative process involving all stakeholders and sectors. This process considers the livelihoods, ecosystem functions and services, and other agro-environmental factors in the landscape. A participatory and inclusive process is fundamental to ensure countryownership and commitment, which are key ingredients for making a transition to CSA. ***Analyzing and planning:*** The detailed biophysical characterization of the environment. The detailed assessment of selected areas in the landscape allow for the joint selection of the most suitable practices for CSA based on local livelihoods and natural resources. In building *climate change scenarios,* the assessment considers the impacts of climate change, determines mitigation benefits, and identifies options for adaptive management. Community or territorial management plans are then developed through a negotiated and collaborative multistakeholder right-based process.

Implementing, monitoring, and learning to scale up best practices: The implementation of plans is undertaken by using a variety of technologies and approaches based on both indigenous and scientific knowledge. Activities are "retrofitted" through endogenous monitoring, self-evaluation, and the sharing of lessons learned. The sustainability of climate-smart practices demands continued action and supportfrom all stakeholders. The mainstreaming of the best practices requires appropriate policy, planning, and institutional support and the establishment of sustainable financing for scaling up climate-smart practices and ensuring all stakeholders have adequate incomes. This should include financial and nonfinancial incentives for ecosystem services and should be negotiated between stakeholders from all sectors (Table 5.4).

Table 5.4 Innovative governance and technologies that sustainably increase agricultural production (FAO 2014).

Crops
1. Better practices for biodiversity (*insitu* and *exsitu* conservation of plant genetic resources, IPM…)
2. Better practices for soil: land rehabilitation, appropriate cropping systems.
3. Better practices for water management: deficit irrigation, preventing water pollution.
4. Payments for using and providing environmental services (pollinators, C sequestration.)
5. Policies, laws, incentives, and enforcement to promote the above
Livestock
1. Conserve animal genetics *insitu* and *exsitu*
2. Use grassland for biodiversity, carbon storage, and water services
3. Protect water from pollution through waste management
4. Use better practices for reduced emission intensity
5. Set payments for using and for providing environmental services, e.g., grazing fees
6. Set policies, laws, incentives, and enforcement to promote the above
Aquaculture
1. Conserve aquatic genetic resources
2. Promote aquaculture certification for environmental protection
3. Ensure biosecurity: pathogens, escapees, veterinary drugs, invasive species, biodiversity.
4. Use integrated aquaculture−agriculture systems
5. Implementing the Ecosystem Approach to Aquaculture (EAA)

Table 5.4 Innovative governance and technologies that sustainably increase agricultural production (FAO 2014).—cont'd

Forestry
1. Conserve biodiversity and forest genetic resources
2. Restore and rehabilitate degraded landscape
3. Enhance the role of forests in soil protection and conservation
4. Enhance the role of forests in the protection and conservation of water resources
5. Use reduced impact harvesting techniques
6. Certification of forest management
Fisheries
1. Assess nontarget resources
2. Develop and use low-impact fishing gears
3. Build fish passes in dams
4. Rebuild depleted stocks and protect critical habitats
5. Restock inland fisheries
6. Implement the Ecosystem Approach to Fisheries (EAF): protect vulnerable marine ecosystems (VMEs), use MPAs in fishery management, and implement ecolabelling
7. Code of Conduct for Responsible Fisheries (CCRF) and international action plans
8. Deter illegal (IUU) fishing

9. Metascientific approach to sustainable management of agricultural resources

1. **Nature of met-scientific approach to solve the sustainable management of agricultural resources.** One of the possible scientific approaches to solve the broad-spectrum issues of sustainable management of agricultural resources is also metascientific one. It is based on knowledge of science itself and its individual scientific disciplines, how they are organized and structured inside, how they communicate with the environment, and the extent to which they can generalize their rich spatiotemporal, complex and integrative entities and experiences into universally valid regularities that will help solve the issue of sustainable management of the agricultural resources. The metascientific approach has been used, for example, in landscape ecological research and in the establishment of metalandscape ecology as a new ecological discipline (Žigrai, 2003, 2007, 2016, 2020; Kremsa, Zigrai, 2021a, 2021b).
2. **Advantage of metascientific approach to solve the issue of sustainable management of agricultural resources.** A great advantage of the metascientific approach, whose main mission is the management of immaterial information resources, is the ability to significantly contribute to solving complicated problems in spatiotemporal and natural–social contextuality, complexity, and integrity. This can undoubtedly include a case study of the issue of sustainable management of the agricultural resources. Its structure reflects the combined nature of the issue, expressed by the cross-sectional intersection of ecological–environmental ideas and principles of sustainable development with the above types of anthropocentric and economically oriented management resources.

3. **Purpose of metascientific approach to solve the issue of sustainable management of agricultural resources.** Purpose of the metascientific approach is to increase the degree of generalization of existing empirical—methodical, theoretical, metascientific and applied-didactic knowledge, and research results of issue of sustainable management of agricultural resources, so that their generally valid regularities can be determined. This oriented approach will, among other things, contribute to increasing the nomotheticity and exactness of this research.
4. **Tools of metascientific approach to solve the issue of sustainable management of agricultural resources.** Among the most important tools of metascientific research needed to generalize the issue in our case of sustainable management of the agricultural resources is its *analysis, metaanalysis* (analysis of analyzes), *synthesis* and *metasynthesis* (synthesis of syntheses). Their results must then be compiled into a system that would represent the framework of future consistent science with its particular object, approach, subject, and goal of research.
5. **Creating a compact science of sustainable management of agricultural resources.** In solving such a complicated issue of sustainable management of agricultural resources, it is necessary in the future to design a compact scientific discipline with its research object, approach, and goal. Its content should reflect the empirical—methodical and theoretical—application aspects of the research of the abovementioned main structure of agricultural resources, such as agrobiodiversity, agricultural species, agricultural ecosystems, and agricultural landscape.
6. **Selection of scientific disciplines supporting sustainable management of agricultural resources.** One of the conditions for the creation of a compact science of sustainable management of agricultural resources is the selection of scientific disciplines that deal with the main structure of agricultural resources. → in *agrobiodiversity research*: (agroecology, synecology, aquatic ecology, etc.); → in *agricultural species research*: (plant ecology, animal ecology, grassland ecology, evolutionary ecology, physiological and conservation ecology, etc.); → in *agricultural ecosystems research*: (ecosystem agroecology, functional ecology, comparative ecology, agroecology, etc.); → in *agricultural landscape research*: (agricultural landscape ecology, historical landscape ecology and land use science, etc.), and → in *sustainable agriculture management research*: (sustainable science, management science, restaurant ecology, applied ecology, forest ecology, dynamic ecology, environmental ecology, theoretical ecology, urban ecology, restoration ecology, etc.).
7. **Creating clusters of selected scientific disciplines and their convergence and cooperation for the needs of sustainable management of agricultural resources.** The large number of these disciplines implies the need to gradually group them into scientific clusters within the convergence field. The aim of the convergence of sciences is to bring closer the theoretical bases, methodological instruments, empirical application, and educational experience of individual scientific disciplines in solving complex problems such as sustainable management of agricultural resources. The result and goal of this gradual convergence of individual scientific disciplines is the creation of integrated management of agricultural resources as part of integrated landscape management. From a metascientific point of view, *integrated landscape management* represents a synthesis of *the science of landscape use* and *the science of landscape protection.*
8. **Science of sustainable management of agricultural resources as a system.** The proposed science of sustainable management of agricultural resources creates a certain system that consists of an internal and external subsystem. The core of the *internal* subsystem consists of elements of sustainable development (ecological environmental, economic and social pillars, principles, indicators and limits), as well as elements of management of agricultural resources

(their management, organization, administration, decision-making, control, planning, protection and use). The *external* subsystem sustainable management of agricultural resources is formed by the cooperation of scientific disciplines examining its internal subsystem, such as sustainable science, as well as the cooperation of the abovementioned scientific disciplines studied the main structure of agricultural resources.

10. Covid-19 pandemic impacts on agriculture, food security and nutrition

10.1 COVID-19 pandemic

The coronavirus pandemic and social distancing measures to contain it have plunged the global economy into the worst recession since World War II.

The State of Food Security and Nutrition in the World called for action on two fronts (FAO, IFAD, UNICEF, WFP and WHO, 2019): (a) safeguarding food security and nutrition through economic and social policies that help counteract the effects of economic slowdowns or downturns, (b) tackling existing inequalities at all levels through multisectoral policies that make it possible to escape from food insecurity and malnutrition. These policy recommendations are more essential in the face of the COVID-19 pandemic. The expected economic impact of the COVID-19 pandemic is a severe global recession for 2020 and 2021. Massive lockdowns and household isolation measures are delivering a major fall in economic activity. According to projections from the Organisation for Economic Co-operation and Development (OECD), the lockdown will directly affect sectors that account for one-third of GDP in the major economies (OECD, 2020). Economist Intelligence Unit (EIU) predicted that COVID-19 will send almost all G20 countries into a recession. The International Labour Organization (ILO) initially predicted that global unemployment could increase by almost 25 million people (ILO, 2020).

10.2 COVID-19 pandemic impact on food and agriculture

The effects of the COVID-19 pandemic are crippling agriculture and food systems. Approximately 80 percent of the 734 million extreme poor live in rural areas and cca 70 percent of the Sustainable Development Goals (SDGs) targets relate to rural areas. Food systems, which directly employ over 1 billion people and provide livelihoods to another 3.5 billion, are experiencing disruptions that could at least temporarily disrupt the incomes and, by extension, food access of 1.5 billion people. Preliminary projections suggest that the COVID-19 pandemic may add an additional 83 to 132 million people to the ranks of the undernourished in 2020.

The Food and Agriculture Organization of the United Nations (FAO) has published report "COVID-19 global economic recession: avoiding hunger must be at the centre of the economic stimulus" (FAO, 2020) and the analysis of the impact of COVID-19 pandemic on food and agriculture (FAO, 2020a). The combined impacts of COVID-19, its suppression measures and subsequent global recession will make hunger and malnutrition worse, increasing the number of hungry and poor people, especially in low-income countries that rely on food import. It is likely to erase a decade of progress on poverty reduction. Economic stimulus must be focused on keeping the food supply chains functioning, while also protecting access to locally-, regionally- and globally-produced food. FAO Director-General urges G20 to ensure that food value chains are not disrupted during COVID-19 pandemic (FAO, 2020e). FAO & WHO (2020) presented report "COVID-19 and food safety: guidance for food businesses." FAO also published reports Responding to the impact of the COVID-19 outbreak on food value chains through efficient logistics (FAO, 2020c) and Social protection and COVID-19 response in rural areas (FAO, 2020d). International Food Information Council studied the impacts of COVID-19 on food purchasing, eating behaviors, and perceptions of food safety (IFIC, 2020).

10.3 COVID-19 pandemic development and solutions

FAO's response plan is part of the United Nations Office for the Coordination of Humanitarian Affairs-led **Global Humanitarian Response Plan for COVID-19**. Key activities include: (a) Rolling out data collection and analysis, (b) Ensuring availability of and stabilizing access to food for the most acutely food-insecure populations, (c) Ensuring continuity of the critical food supply chain for the most vulnerable populations is a key determinant of food security and nutrition (d) Ensuring food supply chain actors are not at risk of virus transmission is crucial to maintaining food supplies. 1.3.1

The FAO COVID-19 Response and Recovery Programme

FAO conducted a comprehensive assessment to identify the most likely or most dangerous threats food and agriculture production, food availability and accessibility and distribution systems, and published "data resources and analysis to understand the COVID-19 pandemic and its impacts" (FAO, 2020). The COVID-19 Response and Recovery Programme addresses seven **priority themes**:

(a) ***The Global Humanitarian Response Plan***: *Addressing the impacts of COVID-19 and safeguarding livelihoods in food-crisis contexts.*

Actions: (1) rolling out data collection and analysis; (2) ensuring availability of and stabilizing access to food for the most acutely food-insecure populations; (3) ensuring continuity of the critical food supply chain for the most vulnerable populations; and (4) ensuring food supply chain actors are not at risk of virus transmission.

(b) ***Data for Decision-making***: *Ensuring quality data and analysis for effective policy support to food-systems and Zero Hunger.*

FAO monitors trade and collects information on logistical issues, assesses how problems have been resolved and signals the market to reduce uncertainty.

Actions: (1) rapid, repeated assessments of the impact of COVID-19 on food insecurity, using the Food Insecurity Experience Scale (FIES); (2) leveraging innovative data sources to monitor the impact of COVID-19; (3) adapting agricultural data collection methods to meet new demands, while maintaining the continuity of technical assistance on agricultural surveys; and (4) evidence-based policy support for post-COVID-19 economic and social recovery.

(c) ***Economic Inclusion and Social Protection to Reduce Poverty***: *Pro-poor COVID-19 responses for an inclusive post-pandemic economic recovery.*

The socio-economic impacts of the COVID-19 pandemic add urgency to the call to eradicate poverty, particularly in rural areas. FAO is promoting pro-poor COVID-19 responses for an inclusive post-pandemic economic recovery.

Actions in this area include: (1) expanding social protection to better reach underserved groups, integrate rural areas into risk-informed and shock-responsive protection components, and scale up nutrition-sensitive social protection; (2) strengthening the sustainable economic inclusion of small-scale producers; (3) strengthening rural women's economic empowerment; and (4) protecting and empowering rural workers and entrepreneurs.

(d) ***Trade and Food Safety Standards***: *Facilitating and accelerating food and agricultural trade during COVID-19 and beyond.*

According to the World Trade Organization (WTO), world merchandise trade in 2020 could fall by as much as 32 percent. Labour shortages due to curtailed mobility are affecting all aspects of the food and agriculture supply chains, from production, to processing and retailing, leading to both immediate and longer-term risks for food production and availability.

Actions: (1) conducting country-specific agricultural trade and trade-policy assessments, comprehensive regional trade assessments, and "deep dive" analyses as needed for specific value chains and thematic areas; (2) strengthening regional multi-stakeholder trade networks and platforms to promote trade policy coordination, deter ad hoc policy responses, advance regulatory cooperation and foster private sector engagement; (3) facilitating trade through technical assistance to implement reforms and address obstacles to trade; (4) establishing or strengthening market intelligence and early warning systems based on regional and country specificities; and (5) supporting capacity development of national and regional institutions.

(e) ***Boosting Smallholder Resilience for Recovery***: *Protecting the most vulnerable, promoting economic recovery and enhancing risk management capacities.*

The effects of the COVID-19 pandemic and associated containment measures are eroding the livelihoods and resilience of vulnerable groups in both rural and urban areas. Pandemic has aggravated existing gender inequalities in terms of reducing access to basic services, increasing domestic and work responsibilities, escalating gender-based violence and the loss of working opportunities in the informal sector (on average women make up 43 percent of the agricultural labour force).

Actions: (1) safeguarding the most vulnerable in rural and urban settings; (2) promoting transformative and inclusive economic recovery; and (3) building capacities and institutions for resilience.

(f) ***Preventing the Next Zoonotic Pandemic***: *Strengthening and extending the One Health approach to avert animal-origin pandemics.*

The COVID-19 originated from an animal source. The risk is highest where there is close interaction between wildlife and intensifying livestock or agricultural production, and is often exacerbated where agriculture has encroached upon or put pressure on natural ecosystems.

Actions: (1) enhancing national and international preparedness and performance during the emergency response; (2) developing policies for spillover containment through the foresight approach; (3) mainstreaming a One Health approach in environment and natural resource agencies at every level; (4) improving national capacity to apply an extended One Health approach to prevent and manage spillovers; and (5) strengthening policy implementation.

(g) ***Food Systems Transformation***: *"Building to transform" during response and recovery.*
FAO will lead efforts to stimulate investments to improve market functioning, foster inclusive and sustainable recovery and accelerate progress toward the 2030 Agenda and the SDGs.
Actions: (1) fostering innovations for increased efficiency, inclusiveness and resilience of food supply chains; (2) ensuring food safety and nutritional quality of diets; (3) reducing food loss and waste; (4) sustaining and strengthening agri-food enterprises; and (5) fostering investment in the green recovery of food value chains.

FAO UN published Strategic Framework 2022–2031, in the context of recent global developments, global and regional trends and major challenges in the areas of FAO's mandate (FAO, 2021). The development of this report took place during a period of unprecedented challenges driven by the COVID-19 pandemic - a global crisis, which highlighted the critical mandate of FAO to ensure functioning and sustainable agri-food systems that allow for sufficient production and consumption of food.

Acknowledgment

I would like to thank to Professor RNDr. Florin Žigrai, DrSc, Dr.h.c., the founder of Meta-landscape ecology, for his valuable consultations and remarks.

References

Altieri, M.A., 1994. Biodiversity and Pest Management in Agro-Ecosystems. Haworth Press, New York.
Altieri, M.A., 1995. Agroecology: The Science of Sustainable Agriculture. Westview Press, Boulder, CO.
Altieri, M.A., 1999. The ecological role of biodiversity in agroecosystems. Agric. Ecosyst. Environ. 74 (1–3), 19–31.
Bryan, G., 2003. Searching for Sustainability. Cambridge University Press, Cambridge.
Cardinale, et al., 2012. Biodiversity loss and its impact on humanity. Nature 486 (7401), 59–67.
Convention on Biological Diversity (CBD), 2014. Global Biodiversity Outlook, 4. CBD, Montréal.
Everard, M., 2017. Ecosystem Services: Key Issues. Routledge, Abingdon, Oxon., UK.
FAO & WHO, 2019. Sustainable Healthy Diets: Guiding Principles. FAO, Rome.
FAO & WHO, 2020. COVID-19 and Food Safety: Guidance for Food Businesses. FAO, Rome.
FAO, 2006. Breed Diversity in Dryland Ecosystems. FAO, Rome.
FAO, 2011. Save and Grow: A Policymaker's Guide to the Sustainable Intensification of Smallholder Crop Production. FAO, Rome.
FAO, 2014. Building a Common Vision for Sustainable Food and Agriculture: Principles and Approaches. FAO, Rome.

FAO, 2015. The Second Global Assessment of Animal Genetic Resources. FAO, Rome.
FAO, 2017a. The Future of Food and Agriculture - Alternative Pathways to 2050. FAO, Rome.
FAO, 2017b. The Impact of Disasters and Crises on Agriculture and Food Security. FAO, Rome.
FAO, 2018. The State of World Fisheries and Aquaculture. Meeting the Sustainable Development Goals. FAO, Rome.
FAO, 2018a. The 10 Elements of Agroecology. Guiding the Transition to Sustainable Food and Agricultural Systems. FAO, Rome.
FAO, 2018b. FAO's Work on Agroecology. A Pathway to Achieving the SDGs. FAO, Rome.
FAO, 2018c. Constructing Markets for Agroecology – An Analysis of Diverse Options for Marketing Products from Agroecology. FAO, Rome.
FAO, 2019a. State of Food and Agriculture. Rome.
FAO, 2019b. Report of the 17th Session of the Commission on Genetic Resources for Food and Agriculture, Rome.
FAO, 2019c. The Fisheries and Aquaculture Advantage. Fostering Food Security and Nutrition, Increasing Incomes and Empowerment. FAO, Rome.
FAO, 2019d. The State of the World's Biodiversity for Food and Agriculture. FAO, Rome.
FAO, 2019e. The state of the Worlds Aquatic Genetic Resources for Food and Agriculture. FAO, Rome.
FAO, 2019f. Report of the 17th Session of he Commission on Genetic Resources for Food and Agriculture. FAO, Rome.
FAO, 2020. COVID-19 Global Economic Recession: Avoiding Hunger Must be at the Centre of the Economic Stimulus. FAO, Rome.
FAO, 2020a. The Impact of COVID-19 Pandemic on Food and Agriculture. FAO, Rome.
FAO, 2020b. Tracking progress on food and agriculture-related SDG indicators. FAO, Rome.
FAO, 2020c. Responding to the impact of the COVID-19 outbreak on food value chains through efficient logistics. FAO, Rome.
FAO, 2020d. Social Protection and COVID-19 Response in Rural Areas. FAO, Rome.
FAO, 2020e. FAO Director-General Urges G20 to Ensure that Food Value Chains are not Disrupted during COVID-19 Pandemic. Available from. http://www.fao.org/news/story/en/item/1268254/icode/. (Accessed 26 March 2020).
FAO, 2020f. Biodiversity for Food and Agriculture and Ecosystem Services – Thematic Study for The State of the World's Biodiversity for Food and Agriculture. Rome. https://doi.org/10.4060/cb0649en.
FAO, 2021. The Impact of Disasters and Crises on Agriculture and Food Security. FAO, Rome.
FAO, IFAD, UNICEF, WFP, WHO, 2019. The state of food security and nutrition in the world 2019. Safeguarding Against Economic Slowdowns and Downturns. FAO, Rome.
FAO, IFAD, UNICEF, WFP, WHO, 2020. The State of Food Security and Nutrition in the World 2020: Transforming Food Systems for Affordable Healthy Diets. FAO, Rome.
Food Security Information Network (FSIN), 2019. Global Report on Food Crises 2019. Rome.
Forman, R.T.T., Gordon, M., 1986. Landscape Ecology. Wiley, New York.
Frison, E.A., Cherfas, J., Hodgkin, T., 2011. Agricultural biodiversity is essential for a sustainable improvement in food and nutrition security. Sustainability 3, 238–253.
Gliessman, S.R., 1997. Agroecology: Ecological Processes in Sustainable Agriculture. Ann Arbor Press, Chelsea, MI.
Global Panel on Agriculture and Food Systems for Nutrition (GLOPAN) (2016) Food systems and diets: facing the challenges of the 21st century. London., 2016. Food Systems and Diets: Facing the Challenges of the 21st Century. GLOPAN, London.
Hoffmann, I., From, T., Boerma, D., 2014. Ecosystem Services provided by Livestock Species and Breeds. FAO Commission on Genetic Resources for Food and Agriculture (CGRFA).

Hussain, C.M., 2019. Handbook of Environmental Materials Management, first ed. Springer International Publishing.
Hussain, C.M., 2020. The Handbook of Environmental Remediation: Classic and Modern Techniques, first ed.
Hussain, C.M., Keçili, R., 2019. Modern Environmental Analysis Techniques for Pollutants, first ed. Elsevier Royal Society of Chemistry. March 25, 2020.
IFAD, 2019. The fisheries and aquaculture advantage. Fostering Food Security and Nutrition, Increasing Incomes and Empowerment. Rome.
IFIC, 2020. Food & Health Survey. International Food Information Council.
ILO, 2020. World Employment and Social Outlook: Trends 2020. ILO, Genève.
IMF, 2020. World Economic Outlook 2020 "A Crisis Like No Other, An Uncertain Recovery". International Monetary Fund, Washington, DC.
IMF, 2021. World Economic Outlook 2021 "Managing Divergent Recoveries". International Monetary Fund, Washington, DC.
IMF, 2021a. World Economic Outlook Update, January 2021 "Policy Support and Vaccines Expected to Lift Activity". International Monetary Fund, Washington, DC.
International Union for Conservation of Nature and Natural Resources., 1980. World conservation strategy: living resource conservation for sustainable development. IUCN.
IPBES, 2019. The Global Assessment Report on Biodiversity and Ecosystem Services. Intergovernmental Science Policy Platform on Biodiversity and Ecosystem Services, Bonn.
IPPC, 2020. Annual Report 2019 – Protecting the World's Plant Resources from Pests. Rome.
IUCN, 2020. Red List of Threatened Species. International Union for Conservation of Nature and Natural Resources.
Jamieson, D., 1998. Sustainability and beyond. Ecol. Econ. 24, 183–192.
Jules, N., 2008. Agricultural sustainability: concepts, principles and evidence. Philos. Trans. R. Soc. Lond., B, Biol. Sci. 363 (1491), 447–465.
Khoury, C.K., Castañeda-Álvarez, N.P., Dempewolf, H., Eastwood, R.J., Guarino, L., Jarvis, A., Struik, P.C., 2016. Measuring the State of Conservation of Crop Diversity: A Baseline for Marking Progress toward Biodiversity Conservation and Sustainable Development Goals. CGIAR.
Kremen, C., 2005. Managing ecosystem services: what do we need to know about their ecology? Ecol. Lett. 8, 468–479.
Kremsa, V., 1993. Advanced technologies for landscape ecology and management. In GIS for environment. In: Proceedings of 1st Conference on Geographical Information Systems in Environmental Studies. November 25–27, 1993, Krakow, Poland.
Kremsa, V., 1995. Landscape monitoring: trends in advanced technologies. In: Historical Changes of Ecological Situations in People-Environment Relationships. Symp. Proceedings of the Commission "Historical Monitoring of Environmental Change", International Geographical Union, 18–21 August. Prihrazy, Czech Republic.
Kremsa, V., 1998. Estrategias par el desarrollo sustentable del paisaje rural (Strategies for sustainable development of rural landscape). In: Sustainable Development of Built Environment. José R. Garcia Chavez (Compilator). Universidad Autónoma Metropolitana, pp. 287–291 (In Spanish).
Kremsa, V., 1999. Hierarchical landscape organization in Mexico. In: Landscape Ecology: The Science & the Action (Abstracts). 5th World Congress of International Association for Landscape Ecology (IALE), July 29–August 3, 1999, Snowmass, Colorado, USA.

Kremsa, V., 1999a. Territorial system of landscape ecological stability(TSLES) in Mexico. In: Landscape Ecology: The Science & the Action (Abstracts). 5th World Congress of International Association for Landscape Ecology (IALE), July 29—August 3, 1999, Snowmass, Colorado, USA.

Kremsa, V., July 10—13, 1999b. Information system for monitoring of land use/cover changes. In: Proceedings of IGU-LUCC'99 Open International Symposium on Land Use/Cover Change, East-West Center. University of Hawaii at Manoa, Honolulu, Hawaii, USA.

Kremsa, V., 1999c. Estrategias para desarrollo sustentable del paisaje rural en las montañas (Strategies for Sustainable Development of Rural Landscape in the Mountains). In: "Sustainable Mountain Development", Proceedings of III International Symposium. Quito, Ecuador, December 9—14, p. 1998.

Kremsa, V., 2000. Sustainable landscape management in Czech republic. In: "Living with Diversity", 29th International Geographical Congress 14—18 August, 2000, Seoul, Korea.

Kremsa, V., 2002. Landscape diversity: economic valuation of environmental functions. In: Seminario Internacional "Incentivos para la conservación de la biodiversidad y el uso sustentable". Organizado por Banco Mundial, OECD, CCA, SEMARNAT/INE. Oaxaca, México, 10—14 June 2002.

Kremsa, V., 2002a. Historical monitoring of Czech rural landscapes. In: 22nd EARSeL Symposium & General Assembly "Geoinformation for European-wide Integration" Prague, Czech Republic, June 4—6, 2002.

Kremsa, V., Žigrai, F., 2021a. Landscape ecology in Mexico: evaluation, research, education and future (selected theoretical and meta-scientific aspects). J. Landsc. Ecol. Submitted for publication.

Kremsa, V., Žigrai, F., 2021b. Meta-Landscape Ecology Approach to Determine the Nature of Mexican Landscape Ecology (Selected Meta-Scientific Aspects). IALE-NA 2021 Annual Meeting, University of Nevada, Reno, USA.

Lal, R., et al., 2016. Climate Change and Multi-Dimensional Sustainability in African Agriculture: Climate Change and Sustainability in Agriculture. Springer.

Lewandowski, et al., 1999. Sustainable crop production: definition and methodological approach for assessing and implementing sustainability. Crop Sci. 39 (Issue 1), 184—193.

Millennium Ecosystem Assessment, 2005. Ecosystems and Human Well-Being: Synthesis. Island Press, Washington D.C., USA.

Mishra, A.K., 2019. Nanotechnology in Environmental Science, 2. John Wiley & Sons.

Murray, W.J., 2012. Sustainable crop production intensification. In: Barney, S.T. (Ed.), Proceedings of the International Scientific Symposium on Biodiversity and Sustainable Diets United against Hunger. FAO Headquarters, Rome, Italy, pp. 66—74.

Nautiyal, S., Kaechele, H., 2007. Conservation of crop diversity for sustainable landscape development. Manag. Environ. Qual. 18 (5), 514—530.

Naveh, Z., Lieberman, A.S., 1994. Landscape Ecology. Theory and Application. Springer-Verlag, New York.

Norton, B.G., 2002. Searching for Sustainability: Interdisciplinary Essays in the Philosophy of Conservation Biology. Cambridge University Press, Cambridge.

Norton, B.G., 2005. Sustainability. University of Chicago Press, Chicago.

Norton, K.C.K., Castañeda-Álvarez, N.P., Dempewolf, H., Eastwood, R.J., Guarino, L., Jarvis, A., Struik, P.C., 2016. Measuring the State of Conservation of Crop Diversity: A Baseline for Marking Progress Toward Biodiversity Conservation and Sustainable Development Goals. CGIAR.

OECD, 2020. Economic Outlook "Turning Hope into Reality". OECD Publishing, Paris.

Pretty, J.N., 2008. Agricultural sustainability: concepts, principles and evidence. Philos. Trans. R. Soc. Lond. B Biol. Sci. 363 (1491), 447–465.
Richgels, C.E., et al., 1990. Sustainable agriculture, perspectives from industry. J. Soil Water Conserv. 45, 31–33.
Sachs, J.D., 2015. The Age of Sustainable Development. Columbia University Press, New York.
Schulze, E.D., Mooney, H.A., 1993. Biodiversity and Ecosystem Function. Springer-Verlag.
Shaw, B.J., et al., 2017. Contributions of citizen science to landscape democracy: potentials and challenges of current approaches. Landsc. Res. 42, 831–844.
Thrupp, L.A., 1993. Political ecology of sustainable rural development: dynamics of social and natural resource degradation. In: Allen, P. (Ed.), Food for the Future: Conditions and Contradictions of Sustainability. John Wiley and Sons, New York, pp. 47–73.
Thrupp, L.A., 2000. Linking agricultural biodiversity and food security: the valuable role of agrobiodiversity for sustainable agriculture. Int. Aff. 76, 283–297.
Tilman, D., Balzer, C., Hill, J., Befort, B.L., 2011. Global food demand and the sustainable intensification of agriculture. Proc. Natl. Acad. Sci. U.S.A. 108, 20260–20264.
United Nations (UN), 1993. Convention on Biological Diversity (CBD). United Nations.
United Nations (UN), 2013. World Population Prospects, the 2012 Revision. United Nations.
World Bank, 2008. World Development Report. Agriculture for Development, Washington, DC.
World Bank, 2019. Enabling the Business in Agriculture. Washington, DC.
World Bank, 2020. The World Development Report 2020: Trading for Development in the Age of Global Value Chains. Washington, DC.
Žigrai, F., 2003. The meaning of meta-landscape ecology for the development of the theory, methodology, application and education of the landscape ecology (selected aspects). In: Ekológia 22 (Suppl. 1), 20–33.
Žigrai, F., 2007. Contribution of metascience to the development of landscape ecology. In: Landscape Ecology in Slovakia (Development, Current State and Perspectives). Monograph, pp. 38–53. In: Contribution of the Slovak Landscape Ecologists. IALE World Congress Wageningen 2007.
Žigrai, F., 2016. Time-spatial contextuality, complexity, and integrity of the development and cognition of landscape ecology. (selected theoretical and metascientific aspects). J. Landscape Ecol. 9 (3), 14–32.
Žigrai, F., 2020. Meta-landscape Ecology as a New Ecological Science (Selected Meta-Scientific Aspects). Monograph. University of Prešov, p. 570.

Internet links

2030 Agenda for Sustainable Development www.un.org/sustainabledevelopment/development-agenda/.
Second Global Assessment of Animal Genetic Resources www.fao.org/publications/sowangr/en/.
The State of the World's Biodiversity for Food and Agriculture www.fao.org/3/CA3129EN/CA3129EN.pdf.
International Union for Conservation of Nature and Natural Resources www.iucn.org.
Consultative Group for International Agricultural Research (CGIAR) www.cgiar.org.
Convention of Biological Diversity www.cbd.int.
Millennium Ecosystem Assessment www.millenniumassessment.org.

Food and Agriculture Organization www.fao.org.
Strategic Plan for the Commission on Genetic Resources for Food and Agriculture (2019–2027) www.fao.org/3/ca8345en/ca8345en.
Global Biodiversity Outlook. (2014) www.cbd.int › gbo4.
Global Panel on Agriculture and Food Systems for Nutrition (GLOPAN) http://glopan.org.
High Level Panel of Experts on Food Security and Nutrition (HLPE) www.fao.org/right-to-food/areas-of-work/projects/rtf-global-regional-level/hlpe/en/.
International Food Information Council (IFIC) https://ific.org.
International Fund for Agricultural Development (IFAD) www.ifad.org.
International Union for Conservation of Nature and Natural Resources www.iucn.org.
IUCN Red List of Threatened Species www.iucnredlist.org.
IUCN Red List of Threatened Species www.iucnredlist.org.
International Plant Protection Convention www.ippc.int.
Sustainable Development www.un.org/sustainabledevelopment/.
United Nations Department of Economic and Social Affairs (UNDESA) www.un.org/development/desa/dpad/category/un-desa/.
United Nations Environment Programme www.unep.org.
United Nations System Standing Committee on Nutrition (UNSCN) www.unscn.org.
United Nations: www.un.org.
WORLD BANK www.worldbank.org.
World Food Program www.wfp.org.
World Health Organization (WHO) www.who.int.
World Development Report www.worldbank.org/en/publication/wdr2020.
International Monetary Fund (IMF) www.imf.org.

Sustainable water resources

Satya Prakash Maurya, Ramesh Singh
Indian Institute of Technology (BHU) Varanasi, Varanasi, Uttar Pradesh, India

"The earth, the air, the land, and the water are not an inheritance from our forefathers but on loan from our children. So we have to handover to them at least as it was handed over to us"

- M. K. Gandhi.

1. Introduction

A quotation of Aberts Szent-Gyoryi, M. D. Discoverer of vitamin C "Water is life's matter and matrix, mother and medium. There is no life without water" outlines the essence of water for plant, animal, and human. United Nations stated that water is not merely an economic commodity but it is social and cultural goods. Although water covers more than two-thirds of the Earth's surface and having a total volume of 1.332 billion cubic kilometers but most of the water on earth is in oceans and not useable for most purposes due to its salinity. Statistically, only 3% of the total water resources can be considered as freshwater and two-thirds of that freshwater is confined in natural ice-caps and glaciers. About 0.5% of the total freshwater which is available for human use comes from surface water, such as rivers, lakes/ponds, or groundwater drawn from dug wells, private bore-wells, and tube-wells. In context of consumption of water, there are three major stakeholders: (i). Agriculture (including irrigation, livestock, and aquaculture). (ii). Industry (including power generation), (iii). Households are accounting for 69%, 19%, and for 12% of annual water withdrawals, respectively (Aquastat). With pace of development and urbanization, Burek et al. (2016) estimated that water demand will increase @ 20%–30% by 2050 over the current level of water use.

Studies reveal that globally water demand has increased almost sixfold over past century and is, further, contributing an annual water demand of 1% on account of population thrust, economic development, and urbanization. Such increased water demand and shifting consumptive water pattern put a great threshold on freshwater resource and freshwater became a scarce commodity. Nevertheless, it is seen as a threat to generate the areas where water resources are presently, in abundance, into a water stress regions (The United Nations, 2020). Therefore, major concern is (i) whether the water crisis will deepen or intensify or (ii) whether key use trends can be changed toward sustainable water resources.

Sustainable Resource Management. https://doi.org/10.1016/B978-0-12-824342-8.00011-0

Globally, problems of water are nonhomogenous and inconsistent over time and space. It may vary significantly on regional, seasonal, and temporal basis. However, sustainable management of water also depends on several other factors:

(i) International, national, and regional attitudes and perceptions toward water management.
(ii) Social, economic, and climatic conditions of the concerned country.
(iii) Suitability and execution status of the legal/regulatory framework.
(iv) Governance approach together with practices, transparency, and political interference.
(v) Capacities and competence of the institutions that are managing water issues.
(vi) Efficacy and applicability of research/pilot projects conducted to invent new solutions for water issues in context of national/regional/local.
(vii) Applicability and adaptability of available technologies.

Apart from the above the water resources may affect and get affected by climatic changes which turn down the forecasting of water availability and its quality. Therefore, water can play a vital role in the global adherence toward a sustainable future.

2. Water resource management system

Naturally available freshwater is a limited resource and its ecosystems need to be protected (UN Economic and Social Council, 1997). Nevertheless, water is a complex system and may be divided into two halves; (1) ***Nature Managed System*** in which over a given basin and the intricacy of the ecosystem climate and common aptness by animals and plants to an altering environment, the natural water cycle optimizes and replenishes the nature need of a specific water. (2) ***Human Managed System*** in which human acts; (i) abstraction of freshwater surpassing its replenishment rate, (ii) water losses due to poor distribution system, (iii) adulterating various wastes into water and restraining its uses for other users including nature impacts the water circulatory system. From Fig. 6.1 it can be visualized that a complex system of water management depends on many interacting trends. Drastic and visionary challenges of environmental sustainability and application of membrane science in environmental engineering science that are of great relevance in industrial wastewater treatment and drinking water treatment are well addressed by Hussain and Mishra (2019). Moreover, real solutions require an integrated approach to water resource management (WRM). Most attempts in WRM are addressed at optimizing the uses of water and curtailing the impact on natural environment due to human activities. Nonetheless, only evaluating water demand of stakeholders and allocating water resources from natural system and optimizing them should not be considered as management of water as there are many human interventions which also affect freshwater quality. Commendable technological advance researches address management of environmental materials proposing eco-friendly solutions and designing policies that will sustain the environment for future generation (Hussain, 2019). However, in past few years, water security became a notable entity under international policy, highlighted

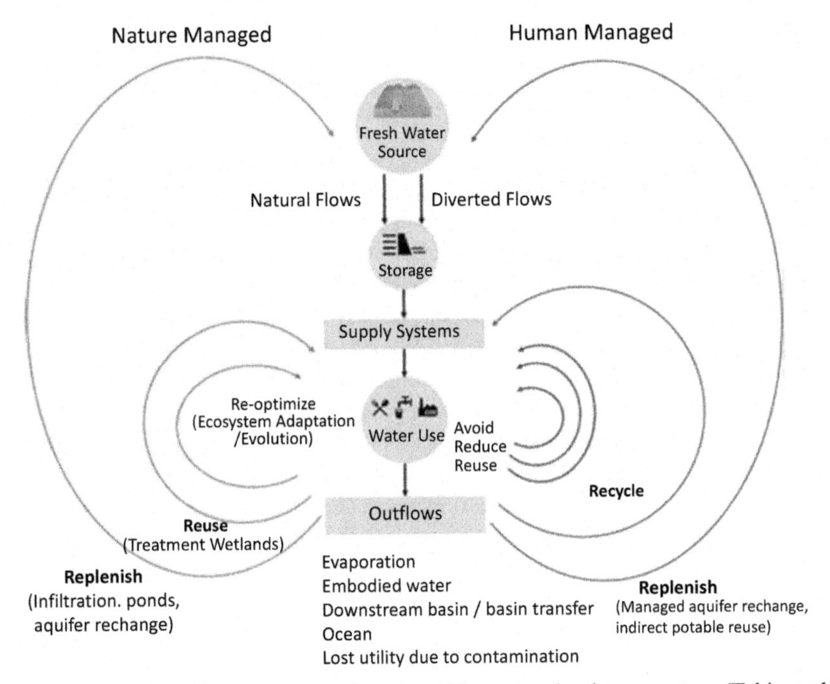

Figure 6.1 Natural and human managed cycle within water circulatory system (Tahir et al., 2018).

to social, economic, environmental, and political enforcement for availability of safe and adequate water, and qualifying sustainability factors of hydrology and climate (Zeitoun et al., 2016).

The more challenging issues that may be pivoted for a sustainable future is how technological and institutional innovation be stimulated to improve efficiency of water consumption—or preferably, enhance the capacity of water resources to reduce water stress and a sustainable WRM which aims for fulfilling the present water requirement needs of all stakeholders without hampering the need of future generation water demand.

With the above context, it is very clear that a common framework is needed which must cover all the aspects of sustainable water management. Several researches and actions are elaborated by theoretical and policy frameworks for understanding the change in quantity and quality of water at different scales. The major components of the frameworks are identified as (i). water demand, (ii). water supply, (iii). water users, (iv). rainfall runoff, (v). evapotranspiration and infiltration, and (vi). pollution. It is really difficult to fit all the components into a single framework. Hence, an integration approach is required to include every aspect of the water sector. Chapter 18 of Agenda 21 sets the basis for integration of water management which includes different aspects of water availability and consumption practices (GWP-TAC, 2000).

3. Water resource management: an integrated approach

3.1 Historical background

WRM for human use, a subset of management of natural water cycle, targets for optimal use of water resource through planning, development, and distribution. Major stakeholders of water consumption are agriculture in rural sector and domestic, industrial, and other institutional use in urban sector. Since water use for agriculture is discussed elsewhere in this book, we are more focused on WRM in urban sector for which European Union (also Dublin Statement 1992) sets the objective. Here, city was adopted as a project area and WRM for city development plan was basically based on social and economic aspect and technological option, service-level benchmark, and "willingness to pay" were selected as the key indicators. Further, various attempts were made by the water managers considered interrelation among the competing demands of resource from different stakeholders to evaluate the effective use of water resource. However, it was observed that ecology of the river basin which is affected by the human intervention remained unaddressed. Therefore, to address the water use for ecological aspect along with economic and social, the scope was extended to entire water resource basin. Researchers came forward with WRM framework considering quantity, quality of water resource, use pattern, wastewater generation, and its reuse, but this framework was limited to water supply and wastewater infrastructures, i.e., the physical characteristics of urban water cycle. But the interdependencies, complexities, and uncertainties of the water environments, further, continued due to the enhanced pollution of water resources from discharge of waste from domestic and industrial sources and climate change which draw the attention of researchers to develop a more integrated WRM.

3.2 Review of water resource management frameworks

A system-level design based on water balance modeling (Fig. 6.2) was developed as a comprehensive WRM framework which interrelates different components of social, economic, and ecology considering water basin as a fundamental spatial unit. Such a comprehensive WRM framework considered four dimensions: (i) **water resources** dimension which specifies the complete water cycle including water reserves and natural flows, (ii) **water users** which connect to consumption of water for humans including all stakeholders and economic interests, (iii) **spatial scale** which relates distribution and uses of water within the geographical area such as watershed, catchments, basin, etc., and (iv) **temporal scale** for water resources which indicates the variation in demand periodically.

Although comprehensive WRM framework opens the more holistic path for basin-level management, but there is no such generic model or any single method sufficient to address the complexities and uncertainties of urban water sub-basin. Rather, the mix of approaches reflects local sociocultural and economic conditions. Various

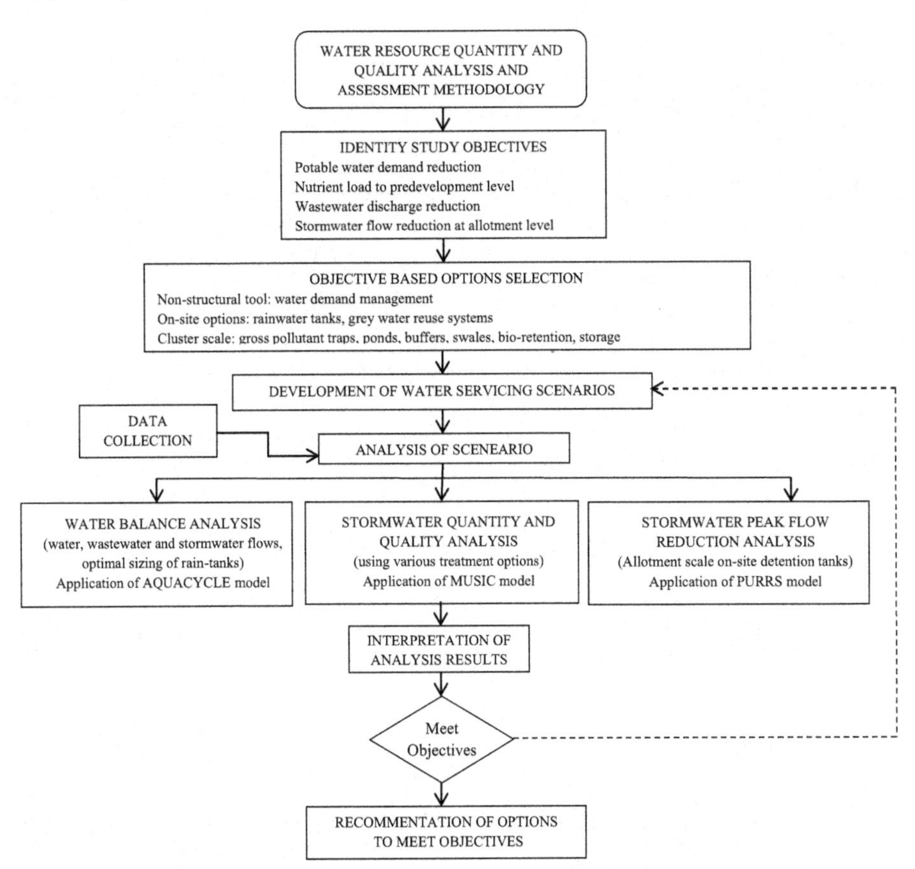

Figure 6.2 Assessment methodology for water resource management (Sharma et al., 2008).

researchers developed Integrated Water Management framework for specific urban geospatial region given herewith in Table 6.1.

It is obvious from Table 6.1 that growing technical and scientific knowledge attributed significant advances for development of various integrated water resource. Most of the frameworks are configured as unidirectional flow path in a treelike structure and covered various social, economic, and ecological aspects of urban water cycle over a sub-basin such as quality and quantity of water supply resource, wastewater generation, storm water, and nonconventional reuse as well as water efficiency. However, several aspects such as uncertainty, sensitivity analysis, economic and social aspects of nonuse water, pollution of water supply resource still need special attention for sustainable WRM.

3.3 *Key issues for sustainable water management*

Dynamics of change of urban water cycle, nonconventional water reuse, ecological response of water bodies over time, and evaluation of predictive sensitivity/uncertainty

Table 6.1 Frameworks developed for integrated water management for urban water use.

Sr. No.	Focus/objective	Modeling criteria/aspects	Limitations/ shortcomings	Geospatial region
1.	To analyze urban water systems with respect to sustainability including multidimensional criteria. (Hellstorm et al., 2000)	Water supply resource, water use, and wastewater.	Physical characteristics, water supply, and wastewater infrastructure in urban sub-basin.	Swedish Research Programme
2.	Urban water balance model (Aquacycle). (Mitchell et al., 2001)	Water demand, water supply, and reuse of wastewater, storm water.	Only quantitative. Qualitative aspects are missing.	Woden Valley Canberra, Australia
3.	Life Cycle Assessment of urban water based on Environmental Sustainability Index (ESI). (Lundin and Morrison, 2002)	Freshwater withdrawal use, wastewater, handling waste/sludge.	Storm water is not considered explicitly, ESI is area specific.	Goteborg, Sweden, King William's Town, South Africa
4.	Total Water Cycle Management (TWCM). (Chanan and Woods, 2006)	Reduction of water use, storm water, and wastewater in urban sub-basin.	Environmental issues and water quality not addressed.	Kogarah Council Initiative, Sydney
5.	Urban water management at household scale Using UWOT software for sustainable option selection. (Makropoulos et al., 2008)	Water demand, wastewater production, runoff, social, economic, and risk to human health and reliability.	Not applicable for large combination of communities.	Elvetham Heath Hampshire, England

Table 6.1 Frameworks developed for integrated water management for urban water use.—cont'd

Sr. No.	Focus/objective	Modeling criteria/aspects	Limitations/ shortcomings	Geospatial region
6.	Water balance analysis for efficient use of water resources over urban sub-basin. (Sharma et al., (2008)	Reduction in water demand, and wastewater discharge.	Data availability and explicit interdependence of the parameters used in different analysis.	Canberra city, Australia
7.	Wastewater use and recycling for reduction of freshwater demand in developing country setup. (Mekala et al., 2008)	Cost recovery, wastewater reuse and recycle.	Mainly focused wastewater use for agriculture and fisheries for cost recovery.	IWMI, India and Australia
8.	Integration of recycling of wastewater for urban water sustainability. (Mackay and Last, 2010)	Water use, wastewater, runoff, gray water, reclaimed water, and rainwater harvesting.	Measurement of sustainability is not well defined.	SWITCH Project, Birmingham, UK
9.	Framework for urban water supply service system through sustainability criteria. (i.e., (Okeola and Sule, 2012)	Environmental, economic, technical, institutional, and sociocultural aspect of water use.	Subjectivity of the indicators.	Kwara State, Nigeria
10.	"Dynamic Urban Water Simulation Model (DUWSiM)." (Willuweit and O'Sullivan, 2013)	Water demand, storm water and wastewater generation, water recycling potential.	Technical aspect has not been included in evaluation process of sustainability.	Dublin City, Ireland

Continued

Table 6.1 Frameworks developed for integrated water management for urban water use.—cont'd

Sr. No.	Focus/objective	Modeling criteria/aspects	Limitations/ shortcomings	Geospatial region
11.	To improve reliability of UWOT with increasing urban water demand due to expanding urban area. (Razos and Makropoulos, 2013; Makropolous et al., 2008)	Gray water, green water, and wastewater.	No sustainability assessment tool.	Athens City, Greece
12.	Conceptual framework for water sustainability which integrates important components of local and distant human–nature interactions that affect water dynamics. (Yang et al., 2016)	Population, urbanization, policy/ technology, and economic development.	Mostly focused on policy interventions.	Beijing, China

are some of the critical aspects that need to be integrated for a sustainable WRM specifically in urban sector. Such diversified aspect of urban water system pushed researchers to come up with more detailed and responsive integrated WRM. Probably, to address the complexity and diversification, we have to move toward new spatial temporal behavior and new attributes with significant change in approach and simulation modeling. There is broad range of assessment tools, approaches, and other methods that may effectively be integrated with pressure, state, and response framework to overcome the complexities and uncertainties of WRM. Fig. 6.3 represents the conceptual framework of sustainable water resource model.

The different criteria of sustainability with their respective measures made the objective fall into a MCDM problem. In this way, the water sustainability problem considering water cycle components and sustainability criteria deduces to a problem of sustainability index calculation. In context to different spatial–temporal scenario, various indices such as Water Stress Index, Water Resource Vulnerability Index,

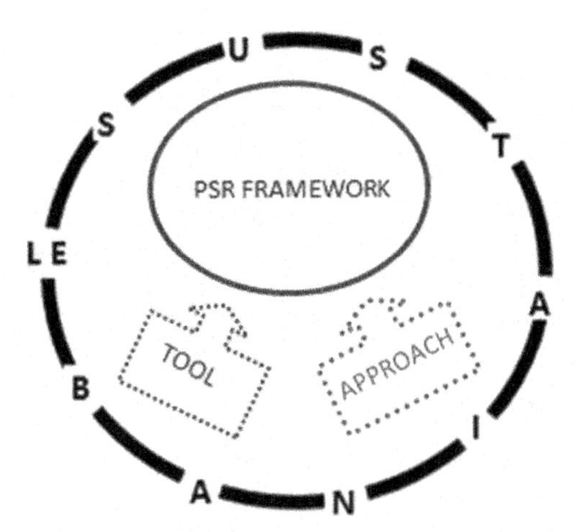

Figure 6.3 Conceptual framework of sustainable water resource model.

Canadian Water Sustainability Index, Sustainability Index for Urban Water Management, Water Scarcity Index, and Blue City Index, etc., were developed by various researchers. Though there are several indicators identified under different criteria for sustainability index measure. Nonetheless, there is no single measure which can evaluate the overall sustainability of a particular urban area/city. An attempt to develop Water for Development Planning Index (WDPI), a single measure of sustainability, was made by Maurya et al. (2020). WDPI is a single measure which signifies the availability of water for new developments in an urban area if urban boundary is enough sustainable in terms of water.

For the purpose, a framework of Water for Development Planning (WDP) is developed which is based on pressure-state response (PSR) framework under Urban Water Balance (UWB) constraints (Fig. 6.4). The water sustainability of an urban area could

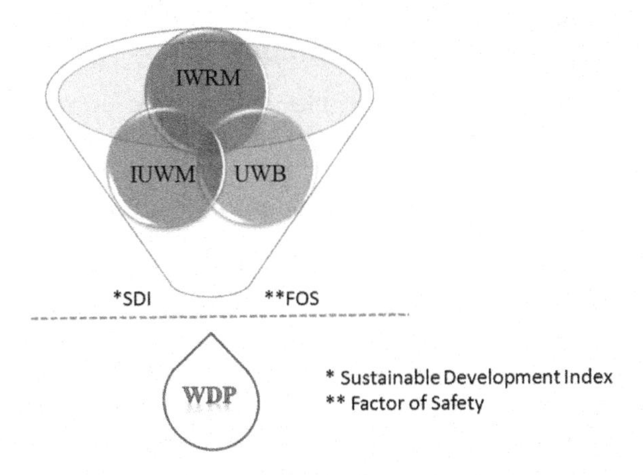

Figure 6.4 WDP framework derived from the concept of IWRM, IUWM, and UWB.

be evaluated using an index named WDPI which is derived from 22 measures categorized in eight indicators.

Furthermore, a spatial decision support system has been implemented with computer-aided programs to evaluate the sustainability of any urban area under WDP framework (Fig. 6.5). Though WDPI succeeded to translate different indices into a single measure but the values of measures or subindicators associated with different indicators as well as objectives are interdependent within their category and the weights associated it have some degree of subjectivity.

4. Interventions of modern computation techniques for sustainable water management

The quantitative as well as qualitative analyses of urban water cycle over a basin and socioeconomic uses and ecology of water body are relatively difficult to model as it has vivid measures with no common indicative units. Nevertheless, several methods applied to evaluate the WRM with various contexts but could not address the problem to its full extent. The dynamics of water with human activities and developments are in circle various dimensions for sustainable WRM over a basin. The uncertainties involve regarding inflow of water to basin, further complicates the problems. Since estimation or forecast of water inflow and water use pattern is nonlinear phenomena, so to address the sustainability of WRM, the probabilistic approach with defined risk of failure attracted researchers to set a policy framework for decision-makers. Nowadays, in addition to information technology, approaches like decision-making models, Agent-Based Modeling (ABM) and Machine Learning (ML) models are paving its path into other sectors. The successful use of genetic algorithms, fuzzy logic, neural networks, etc., used in hydrological modeling, pushed the scientist and technologist to consider artificial intelligence, also, applications as a useful tool for sustainable WRM.

4.1 Decision-making model approach

Decision-making model approach that is comprised of three phase (intelligence, design, and alternate choice) is the cognitive process for selecting a single or combination of course of actions among available alternatives based on the values, principles, preferences, and beliefs of the decision-maker (Fig. 6.5).

A problem may have multiple objectives while any solution represents a scenario which makes the choice process challenging. In the context of water, Multiple-Criteria Decision Analysis (MCDA), all the constraints/criteria are considered simultaneously in the interaction with the environment. It appears a better method because its results are based on quantitative as well as qualitative criteria. In recent trends, immense potential of nanomembranes in environmental sustainability has been pin pointed and need to be integrated with soft-computing–based prediction models have been implemented in the field of environmental engineering (Hussain, 2018).

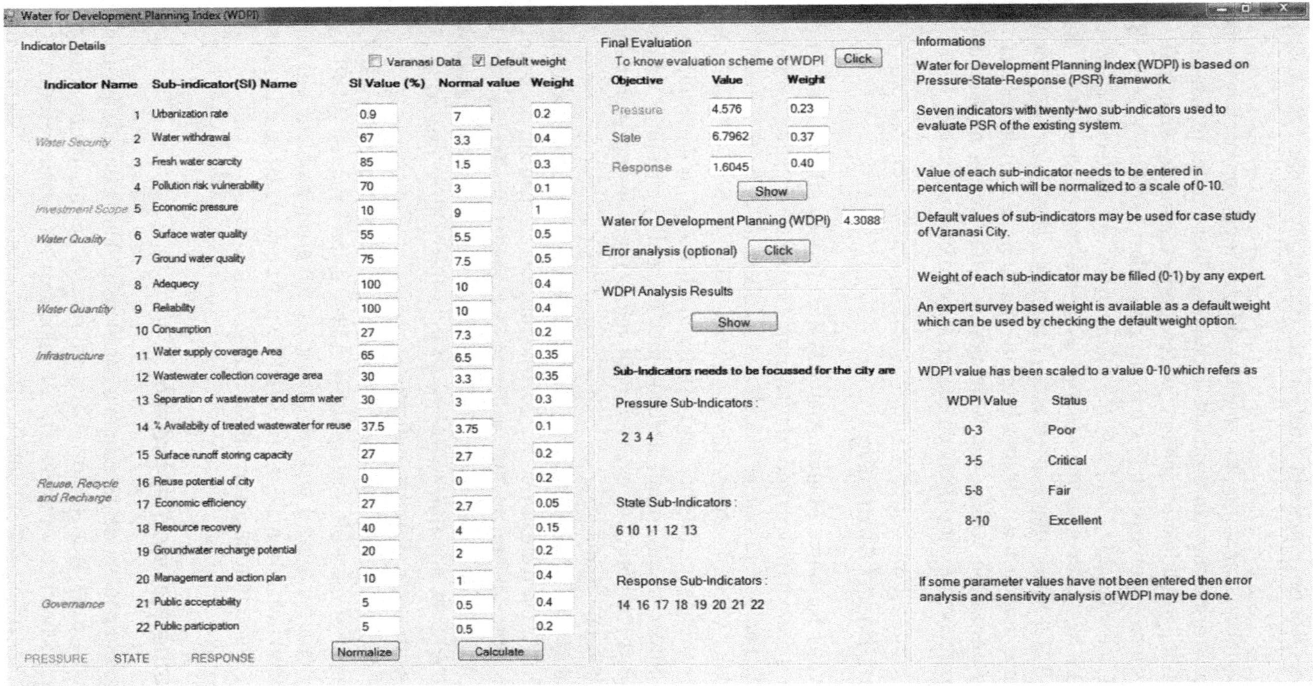

Water for Development Planning Index (WDPI)

Indicator Details

☐ Varanasi Data ☑ Default weight

Indicator Name		Sub-indicator(SI) Name	SI Value (%)	Normal value	Weight
Water Security	1	Urbanization rate	0.9	7	0.2
	2	Water withdrawal	67	3.3	0.4
	3	Fresh water scarcity	85	1.5	0.3
	4	Pollution risk vulnerability	70	3	0.1
Investment Scope	5	Economic pressure	10	9	1
Water Quality	6	Surface water quality	55	5.5	0.5
	7	Ground water quality	75	7.5	0.5
Water Quantity	8	Adequecy	100	10	0.4
	9	Reliability	100	10	0.4
	10	Consumption	27	7.3	0.2
Infrastructure	11	Water supply coverage Area	65	6.5	0.35
	12	Wastewater collection coverage area	30	3.3	0.35
	13	Separation of wastewater and storm water	30	3	0.3
	14	% Availability of treated wastewater for reuse	37.5	3.75	0.1
	15	Surface runoff storing capacity	27	2.7	0.2
Reuse, Recycle and Recharge	16	Reuse potential of city	0	0	0.2
	17	Economic efficiency	27	2.7	0.05
	18	Resource recovery	40	4	0.15
	19	Groundwater recharge potential	20	2	0.2
Governance	20	Management and action plan	10	1	0.4
	21	Public acceptability	5	0.5	0.4
	22	Public participation	5	0.5	0.2

Normalize Calculate

PRESSURE STATE RESPONSE

Final Evaluation

To know evaluation scheme of WDPI Click

Objective	Value	Weight
Pressure	4.576	0.23
State	6.7962	0.37
Response	1.6045	0.40

Show

Water for Development Planning (WDPI) 4.3088

Error analysis (optional) Click

WDPI Analysis Results

Show

Sub-Indicators needs to be focussed for the city are

Pressure Sub-Indicators :

2 3 4

State Sub-Indicators :

6 10 11 12 13

Response Sub-Indicators :

14 16 17 18 19 20 21 22

Informations

Water for Development Planning Index (WDPI) is based on Pressure-State-Response (PSR) framework.

Seven indicators with twenty-two sub-indicators used to evaluate PSR of the existing system.

Value of each sub-indicator needs to be entered in percentage which will be normalized to a scale of 0-10.

Default values of sub-indicators may be used for case study of Varanasi City.

Weight of each sub-indicator may be filled (0-1) by any expert.

An expert survey based weight is available as a default weight which can be used by checking the default weight option.

WDPI value has been scaled to a value 0-10 which refers as

WDPI Value	Status
0-3	Poor
3-5	Critical
5-8	Fair
8-10	Excellent

If some parameter values have not been entered then error analysis and sensitivity analysis of WDPI may be done.

Figure 6.5 Developed application to calculate sustainable water for development planning index (WDPI) and validated with data of Varanasi city.

Techniques such as Fuzzy Logic, Analytic Hierarchy Process (AHP), Technique for Order of Preference by Similarity to Ideal Solution (TOPSIS), Ordered Weighted Analysis, and Data Envelopment Analysis are commonly applied for managing water resources.

To address the sustainability, the rule-based fuzzy logic approach is vividly applied by the different researchers in different dimensions like aquifer management, rainfall runoff modeling, groundwater recharge, water supply alternates using AHP, and storm water management. Moreover, a few hybrid approaches like Fuzzy-TOPSIS for reservoir system management, ranking the alternatives of sustainable water supply, GRA-TOPSIS for water resource security with integration concept for Beijing city Fuzzy-AHP-TOPSIS for Indus reservoir system have, also, been used for WRM.

But in these approaches, identification of indicators and their respective weights with their prioritization within the evaluation scheme are set in accordance to the problem as perceived by the decision-maker as solidarity objective.

4.2 Agent-Based Modeling

Integrated Water Resource Management (IWRM) framework can be combined with ABM which is a modern dynamic computational systems modeling paradigm of interacting agents. The system can be modeled as a collection of basic entities "agent" for decision-making which is autonomous in nature. This idea is derived from the basic concepts of artificial intelligence and can be much useful in the complex problems of water resources management, decision-making, policy planning, and sustainable solutions. ABM has ability of asynchronous interactions among agents and between agents and their environments.

Many researchers attempted ABM for water resources planning and management (Berglund, 2015; Darbandsari et al., 2020), complex mechanism adoption for water conservation (Rasoulkhani et al., 2017). A conceptual framework for IWRM of ABM-based WDPI for urban water sustainability evaluation has been developed by authors (Fig. 6.6). In this framework the PSR factors will be treated as agents who will interact with seven different indicators.

The individual weight of indicators based on their importance is worked out within the environment of the PSR system using ABM. If the value of WDPI satisfies the sustainability criteria, several scenarios based on the given problem will be generated and presented to decision-makers else WDPI will be calculated again followed by the sustainability correction alternate options (Fig. 6.7).

4.3 Machine learning approach

ML is a computational model paradigm in latest trend. The system learning is based on algorithm that allows it to improve its performance in due course based on data. To solve multiobjective optimization of allocation of water resources, to assess groundwater potential, and to predict the water quality index various researchers

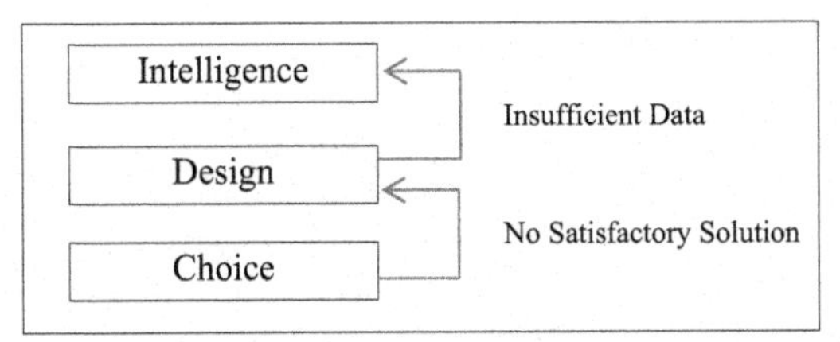

Figure 6.6 Three phases of decision-making process (Simon, 1965).

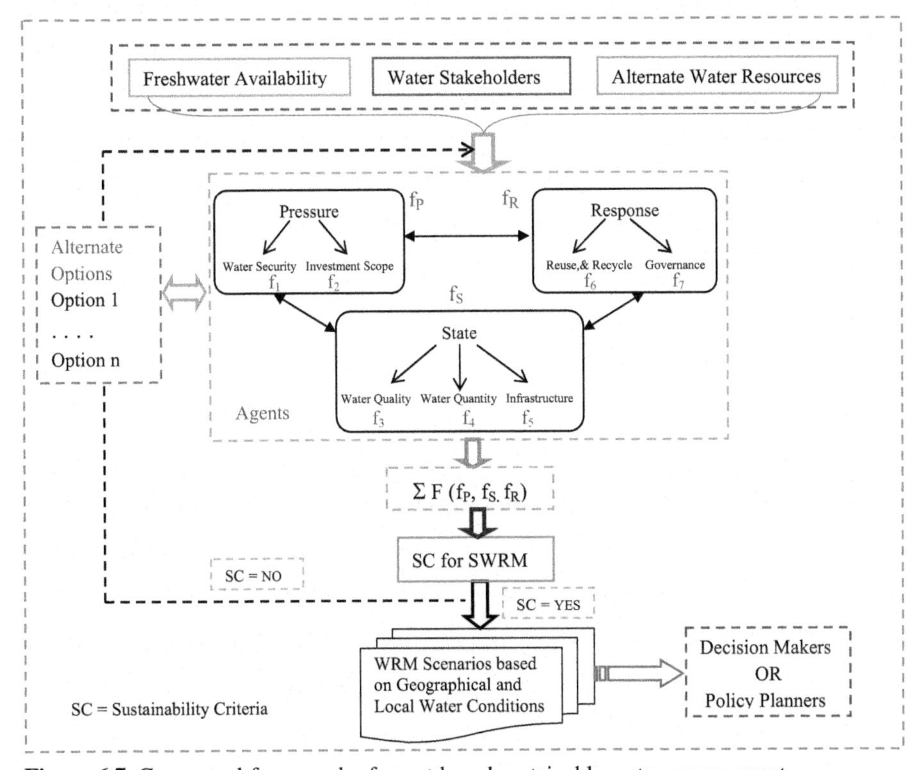

Figure 6.7 Conceptual framework of agent-based sustainable water management.

have used ML approach using multiobjective evolutionary algorithm (MOEAs), support vector machine (SVM), and genetic algorithm optimized random forest (RFGA). Such efforts indicate that these techniques have turned up as an effective option for forecasting and policy planning for water resources applications. Empirical equation–based models are computationally challenging as they require significant amount of data to assemble the model according to parameters.

5. Concluding remarks

Human interventions and urbanization have posed great threat on sustainability of freshwater resource. This chapter is focused on complexities of water system and WRM. Although comprehensive WRM frameworks developed for specific urban geospatial region open more holistic path for basin-level management, most of the approaches reflect local sociocultural and economic conditions. There is no such generic model or any single method sufficient to address the complexities and uncertainties of urban water sub-basin. As a single measure index, WDPI is seen a useful tool to address the sustainability of an urban area. Currently, modern computation techniques such as MCDA, ABM, and ML based on the probabilistic approach with defined risk of failure are paving its path to address the sustainability of WRM. To solve multiobjective optimization of allocation of water resources, to assess groundwater potential, and to predict the water quality index various researchers have used ML approach using MOEAs, SVM, and RFGA. Such efforts indicate that these techniques have turned up as an effective option for forecasting and policy planning for water resources applications. Such efforts indicate that modern computational techniques may prove to be a very effective tool for predictive model for sustainable WRM.

References

Aquastat. n.d. Aquastat website. Food and Agriculture Organization of the United Nations (FAO). www.fao.org/nr/water/aquastat/water_use/index.stm. [Accessed 14 June 2020].

Berglund, E.Z., 2015. Using agent-based modeling for water resources planning and management. J. Water Resour. Plann. Manag. https://doi.org/10.1061/(ASCE)WR.1943-5452.0000544.

Burek, P., Satoh, Y., Fischer, G., Kahil, M.T., Scherzer, A., Tramberend, S., Nava, L.F., Wada, Y., Eisner, S., Flörke, M., Hanasaki, N., Magnuszewski, P., Cosgrove, B., Wiberg, D., 2016. Water Futures and Solution: Fast Track Initiative (Final Report). International Institute for Applied Systems Analysis (IIASA), Laxenburg, Austria. IIASA Working Paper. pure.iiasa.ac.at/13008/.

Chanan, A., Woods, P., 2006. Introducing total water cycle management in Sydney: a Kogarah Council initiative. Desalination 187, 11−16.

Darbandsaria, P., Kerachianb, R., Malakpour-Estalakic, S., Khorasani, H., 2020. An agent-based conflict resolution model for urban water resources management. Sustain. Cities .Soci. 57, 102112.

GWP-TAC (Global Water Partnership Technical Advisory Committee), 2000. Integrated Water Resources Management. TAC Background Paper No. 4, Stockholm, Sweden.

Hellstrom, D., Jeppsson, U., Karrman, E., 2000. A framework for systems analysis of sustainable urban water management. Environ. Impact Assess. Rev. 20, 311−321.

Hussain, C.M., 2018. Handbook of Environmental Materials Management, first ed. Springer International Publishing.

Hussain, C.M., Mishra, A.K., 2019. Nanotechnology in Environmental Science, Vol. 2. John Wiley & Sons.

Lundin, M., Morrison, G.M., 2002. A life cycle assessment based procedure for development of environmental sustainability indicators for urban water systems. Urban Water 4, 145–152.
Mackay, R., Last, E., 2010. SWITCH city water balance: a scoping model for integrated urban water management. Rev. Environ. Sci. Biotechnol. 9, 291–296.
Makropolous, C.K., Natsis, K., Liu, S., Mittas, K., Butler, D., 2008. Decision support for sustainable option selection in integrated urban water management. Environ. Model. Software 23, 1448–1460.
Maurya, S.P., Singh, P.K., Ohri, A., Singh, R., 2020. Identification of indicators for sustainable urban water development planning. Ecol. Indicat. 108.
Mekala, G.D., Davidson, B., Samad, M., Boland, A.M., 2008. Wastewater Reuse and Recycling Systems: A Perspective into India and Australia. Working Paper No 129.
Mitchell, V.G., McMahon, T.A., Mein, R.G., 2001. Modelling the urban water cycle. Environ. Model. Software 16, 615–629.
Okeola, O.G., Sule, B.F., 2012. Evaluation of management alternatives for urban water supply system using multicriteria decision analysis. J. King Saud Univ. – Eng. Sci. 24, 19–24.
Rasoulkhani, K., Logasa, B., Reyes, M.R., Mostafavi, A., 2017. Agent-based modeling framework for simulation of complex adaptive mechanisms underlying household water conservation technology adoption. In: Chan, K.V., D'Ambrogio, A., Zacharewicz, G., Mustafee, N., Wainer, G., Page, E. (Eds.), Proceedings of the 2017 Winter Simulation Conference W. Las Vegas, NV, USA. https://doi.org/10.1109/WSC.2017.8247859.
Rozos, E., Makropoulos, C., 2013. Source to tap urban water cycle modelling. Environ. Model. Software 41, 139–150.
Sharma, A.K., Gray, S., Diaper, C., Liston, P., Howe, C., 2008. Assessing integrated water management options for urban developments – canberra case study. Urban Water J. 5 (2), 147–159.
Simon, H.A., 1965. The Shape of Automation for Men and Management, 1st. Harper and Row.
Tahir, S., Steele, K., Shouler, M., Steichen, T., Penning, P., Martin, N., 2018. Water and Circular Economy. White Paper (Draft 2-b). https://www.ellenmacarthurfoundation.org/assets/downloads/ce100/Water-and-CircularEconomy -White-paper-WIP-2018-04-13.pdf. (Accessed 24 June 2020).
The United Nations World Water Development Report, 2020. Water and Climate Change. United Nations Educational, Scientific and Cultural Organization.
UN Economic and Social Council, 1997. Comprehensive assessment of the freshwater resources of the world, A Report of the Secretary-General. https://www.un.org/esa/documents/ecosoc/cn17/1997/ecn171997-9.htm. (Accessed June 2020).
Willuweit, L., O'Sullivan, J.J., 2013. A decision support tool for sustainable planning of urban water systems: presenting the Dynamic Urban Water Simulation Model. Water Res. 47, 7206–7220.
Yang, W., Hyndman, D.W., Winkler, J.A., Viña, A., Deines, J., Lupi, F., Luo, L., Li, Y., Basso, B., Zheng, C., Ma, D., Li, S., Liu, X., Zheng, H., Cao, G., Meng, Q., Ouyang, Z., Liu, J., 2016. Urban water sustainability: framework and application. Ecol. Soc. 21 (4), 4. https://doi.org/10.5751/ES-08685-210404.
Zeitoun, M., Lankford, B., Krueger, T., Forsyth, T., Carter, R., Hoekstra, A.Y., Taylor, R., Varish, O., Cleaver, F., Boelens, R., 2016. Reductionist and integrative research approaches to complex water security policy challenges. Global Environ. Change 39, 143–154.

Further reading

Hussain, C.M., 2020. The Handbook of Environmental Remediation: Classic and Modern Techniques, first ed. Royal Society of Chemistry.

Important websites

https://www.unwater.org/publication_categories/world-water-development-report/.
http://www.indiawaterweek.in/.
https://www.giz.de/en/worldwide/20206.html.
https://ec.europa.eu/environment/index_en.htm.

Minerals and metal Industry in the global scenario and environmental sustainability

7

Sukanchan Palit[1], *Chaudhery Mustansar Hussain*[2]
[1]Department of Chemical Engineering, University of Petroleum and Energy Studies, Energy Acres, Dehradun, Uttarakhand, India; [2]Department of Chemistry and Environmental Sciences, New Jersey Institute of Technology, University Heights, Newark, NJ, United States

1. Introduction

Technological and engineering advancements in the field of industrial manufacturing, mineral, and metal industry are in the wide path of scientific forbearance and rejuvenation. In the global scenario, science and engineering of sustainable development today are in the middle of vast revamping. Industrial wastewater treatment techniques such as conventional and nonconventional environmental engineering tools need to be revamped and reorganized as mankind surges ahead. Technological prowess and scientific understanding in the field of environmental sustainability and mineral and metal industry are the immediate needs of the hour. Thus, the need of a comprehensive treatise in sustainable development in mineral and metal industry. The scientific march and the scientific triumph in the field of energy and environmental sustainability are immense as mankind progresses forward. Today manufacturing engineering and advanced manufacturing are in the path of newer danger because of industrial pollution. Air pollution control and industrial wastewater treatment are changing the face of science and engineering globally today. Mineral and metal industry are one of the worst polluters of environment. In this chapter, the authors deeply comprehend the needs of conventional and nonconventional environmental engineering techniques in the scientific research pursuit of humankind. A new visionary era in the field of industrial manufacturing, metallurgical industries, and environmental protection will emerge if nations around the world and the civil society takes proactive steps in mitigating environmental disasters.

2. The vision of this study

Scientific and technological challenges in the field of environmental protection are immense and versatile in the global scenario today. Future scientific thoughts and scientific endeavor should be targeted toward conventional and nonconventional environmental engineering tools. Minerals and metal industry are substantial polluters of

Sustainable Resource Management. https://doi.org/10.1016/B978-0-12-824342-8.00001-8

the global environment. Today the status of environmental engineering research globally is vastly devastating. Thus, the need of a proactive scientific initiative in mining engineering, metallurgical engineering, chemical process engineering, and environmental engineering science, thus, also the vision of this chapter. Zero-waste challenges are vital in today's research pursuit in environmental protection. The challenges, the vision, and the scientific and engineering targets need to be unfurled at the utmost. This chapter strongly addresses the scientific conscience and the scientific and technological ingenuity in the field of mining engineering and environmental engineering. A deep scientific thought in the field of mineral and metal industry and environmental protection are the immediate needs of the hour. Science and technology are today ebullient and strong as regards environmental protection. This chapter is an eye-opener to the scientific intricacies in the application of environmental engineering tools in mineral and mining industry. The authors deeply pronounce the success and ingenuity in advanced oxidation processes (AOPs) and novel separation processes such as membrane science. A new dawn in the field of membrane separation process will emerge and science and engineering will proliferate if scientists and engineers around the world surges ahead with vision and introspection. The application of AOPs in mitigating environmental disasters in mineral and metal industry are the other hallmarks of this study. The challenges and the barriers in application of environmental engineering tools are immense and are deliberated in details in this chapter.

3. The vast scientific doctrine of environmental sustainability

Environmental and energy sustainability are today in the middle of a deep scientific introspection and vision. Technology and engineering science today has no answers to the issues of air pollution and industrial wastewater treatment in manufacturing industries globally. Humankind is thus in a state of disaster. The vast scientific doctrine of environmental sustainability involves air pollution control, industrial wastewater treatment, and drinking water treatment. In developing nations and disadvantaged nations around the world, environmental sustainability is the need of the hour. The visionary words of Dr. Gro Harlem Brundtland, former Prime Minister of Norway, on the science of sustainability need to be reenvisioned and reorganized as humankind moves forward. Today the civilization is suffering enormously due rapid industrialization and rampant sustainable manufacturing. Air quality management, integrated water resource management, and wastewater treatment stands as major pillars of environmental engineering science. In developing nations around the world, arsenic and heavy metal groundwater and drinking water contamination are creating massive havoc. Thus, a deep scientific introspection and scientific comprehension are the necessities of the hour. Sustainable development in the global scenario involves social sustainability, economic sustainability, environmental and energy sustainability. Energy sustainability and renewable energy are the two opposite sides of the visionary coin. Renewable energy particularly biomass energy, solar energy, and wind energy are the visionary needs of human civilization and human scientific progress today. A profound scientific contemplation in the field of sustainable manufacturing, mineral, and metallurgical engineering will surely open new futuristic thoughts and futuristic vision in science and engineering globally.

4. Sustainable resource management, integrated water resource management, and the vast vision for the future

Scientific revelation and engineering vision are today the needs of sustainable resource management and sustainable manufacturing globally today. Integrated water resource management and industrial wastewater treatment are today changing the face of environmental engineering and chemical process engineering. Sustainable resource management and holistic sustainable development whether it is social, economic, energy, or environmental are the cornerstones of scientific progress today. Humankind needs sustainability whether it is energy or environmental in its scientific progress. The vast vision for the future in wastewater management and integrated water resource management needs to be revamped and reorganized as civilization trudges forward. Two different fields of scientific research such as human factor engineering and industrial systems engineering are today veritably integrated with water resource management and wastewater management. AOPs, hybrid AOPs, and membrane separation processes are the cornerstones of environmental engineering science globally today. Sustainable resource management today encompasses water treatment, integrated water resource management, and air quality management. The vision of industrial pollution control in developing and developed nations around the globe in the similar vein needs to be widely understood and reenvisioned as humankind gears forward. In this chapter, the authors deeply pronounce the human success of engineering science and environmental sustainability in the betterment of human society. A new day in the field of sustainable resource management will surely envision if the concepts of environmental sustainability, integrated water resource management, and air quality engineering and management are successfully implemented.

5. Today's mineral and metal industry and the needs of environmental sustainability

Today's mineral and metal industry is probably the highest polluter of humankind today. Environmental sustainability and environmental engineering science are the imperatives of science and engineering globally today. In developing nations around the world particularly South Asia, Bangladesh, and India, heavy metal and arsenic groundwater contamination are creating havoc in the scientific horizon. Thus, the immense scientific needs of environmental as well as energy sustainability.

Human suffering such as poverty is today changing the face of human civilization and humankind is in a state of immense disaster due to disregard to social and economic sustainability. Here comes the veritable need of holistic sustainable development and a deep scientific revelation. Mining engineering and metallurgical engineering are the hallmarks of a nation's economic progress and holistic development. Here comes the need of environmental sustainability and the science of environmental remediation. Environmental remediation is a revolutionary area of science and engineering globally today. There is a veritable connection of environmental remediation and manufacturing industry globally today. A new generation of environmental scientists and environmental engineers are today mesmerized at the global concerns of climate change, environmental degradation, and frequent environmental disasters. Technology, engineering, and science has practically no answers to the burning issues of climate change and environmental degradation. Mining industry and metallurgical industry in developing countries around the world are practically creating havoc and are a veritable cause of immense concern.

6. Recent scientific advancements in the field of environmental sustainability

Environmental sustainability and water resource management and wastewater management are today aligned with each other. A strong and deep introspection in the fields of engineering science and technology of environmental protection will open new windows of innovation in global research and development initiatives. In this section, the authors deeply portray the scientific ingenuity and the engineering vision in the field of application of sustainable development in human society. Climate change mitigation and the needs of environmental sustainability are the other cornerstones of this section.

Kuhlman and Farrington (2010) deeply discussed with vision, scientific grit, and perseverance the domain of sustainability. Sustainability as a highly project concept has its origin in Brundtland report of 1987. That paper was highly concerned with the aspirations of human mankind and the limitations imposed by nature on the other hand. In the course of time, the definition of sustainability changed as regards social, economic, and environmental sustainability (Kuhlman and Farrington, 2010). Human society, science, and engineering are thus in the process of major transformation (Kuhlman and Farrington, 2010). A serious scientific introspection of the science of sustainable development and the demarcation between weak and strong sustainability are the cornerstones of this treatise. The authors discussed in minute details history of the concept of sustainability, the definition of sustainability from the Brundtland commission, the definition of happiness, well-being, and welfare, and the resources and the future of sustainability and the definition of weak and strong sustainability. This paper had a lot of inspiration from the Brundtland commission and the visionary definition of "sustainability" by Dr. Gro Harlem Brundtland, former Prime Minister of Norway (Kuhlman and Farrington, 2010). When assessing sustainability, a clear distinction should be made for weak and strong sustainability (Kuhlman and Farrington, 2010). Human mankind's strong scientific and engineering vision and stance thus needs to be envisioned as civilization moves forward. The ideas advocated in this paper are not new and are deeply envisioned from the Brundtland report. A new era in the field of environmental sustainability will surely be envisioned if scientists and engineers takes a proactive effort in dealing with rampant environmental degradation (Kuhlman and Farrington, 2010).

United Nations Development Program (2011) discussed sustainability and inequality in human development. This paper deeply analyses the theoretical and empirical links between inequality in human development in one hand and sustainable development on the other hand. Inequality in many dimensions of human development is deeply analyzed with respect to weak and strong sustainability (United Nations Development Programme, 2011). Sustainable development and human progress are today interlinked with each other. Today a newer era of civilization is emerging as sustainability gains immense importance. Enabling everyone to be capable and free to do things are a newer avenue of social sustainability, economic sustainability, and sustainable development as a whole. Impact of inequality in human development on sustainability is a remarkable cornerstone of this report (United Nations Development Programme, 2011).

7. Recent scientific prowess and research endeavor in the field of environmental sustainability, wastewater treatment, and mineral and metal industry

Today the domain of mining and metallurgical engineering are moving at a rapid pace. The concerns of global environmental protection are rising. Increased industrialization and advances in manufacturing engineering are creating immense pressure on human habitat. In such a highly deplorable situation, a deep scientific comprehension is the need of the hour. In this section, the authors deeply elucidates the application of environmental sustainability in mineral and metal industry. Human civilization's deep scientific conscience and scientific prowess are in a state of peril. Nations around the world are gearing forward for implementation of United Nations Sustainable Development Goals.

Indian National Science Academy Report (2011) deeply elucidates the state of hazardous metals and minerals pollution in India and the domain of sources, toxicity, and management. Human civilization's immense scientific prowess and engineering vision in the field of hazardous materials management thus needs to be revisited with each step forward. In this report the authors deeply comprehend the genesis, sources of heavy metals and minerals, toxicity due to metals and minerals, and management of pollution from metals and minerals (Indian National Science Academy Report, 2011). Environmental degradation and deterioration has become a huge societal problem due large-scale anthropogenic activity besides natural causes. Entry of highly toxic heavy metals and minerals in human organism mainly through contaminated water and food leads to serious health issues (Indian National Science Academy Report, 2011). In such a situation, the world of science and technology needs to be envisioned and reframed as humankind trudges ahead. The major hazardous metals of concern in India in terms of their environmental health effects and other serious medical issues are lead, mercury, chromium, cadmium, copper, and aluminum (Indian National Science Academy Report, 2011). Their sources are mostly anthropogenic as well as natural. Natural causes like seepage from rocks, volcanic activity, and forest fires can highly contribute to this environmental disaster. Minerals like fluoride and arsenic salts are of highly natural origin but human activity can also deteriorate the situation. Today a major environmental disaster is taking place in the state of West Bengal as well as the neighboring country Bangladesh which is arsenic groundwater and drinking water contamination. Mankind, science and technology are thus in a deep stress and environmental trauma (Indian National Science Academy Report, 2011). This report discusses in details different heavy metals and its contaminations such as heavy metals and aluminum, fluoride, arsenic, lead, mercury, and cadmium. Management of pollution from metals and minerals and the remediation technologies are the other cornerstones of this well-researched treatise (Indian National Science Academy Report, 2011). A deep introspection and scientific perseverance, grit, and determination are the needs of the hour today. Any manufacturing industry are today in the state of immense disaster. The situation is immensely worse in developing nations around the world. Remediation technologies in developing countries like India involve release within safe limits and has three complimentary functions such as (1) technological, (2) management (implementation), and (3) regulatory (Indian National Science Academy Report, 2011). This report deeply portrays the scientific needs and the vast scientific ingenuity in the application of environmental sustainability and environmental engineering science in removing hazardous metals and compounds from industrial wastewater in a developing country like India (Indian National Science Academy Report, 2011).

OECD Report (2008) deeply elucidated with vision, girth, and scientific perseverance sustainable development linking economy, society, and development. Human race today stands in the middle of immense scientific contemplation (OECD Report, 2008). The OECD is a unique forum where the governments of 30 democracies work together to deeply address the economic, social, and environmental aspects of globalization. In the last 20 years, significant progress has been made in the field of implementation of sustainable development in human society. The OECD has been at the forefront of the massive effort to advance sustainable development (OECD Report, 2008). This report discussed in details the concept of sustainable development, the challenges of a global world, the future of mankind, production and consumption, how to measure sustainability, and the contribution of government and civil society. Growing awareness of the fragility of human habitat has caused mankind to look deeply into the issues of environmental sustainability (OECD Report, 2008). This well-researched treatise deeply explores the concept of sustainable development, the global dimension of sustainable development, the critical role played by producers and consumers in the application of sustainability, and how governments and civil society work together in the implementation of global sustainability. At the core of sustainable development, there are three pillars which are society, economy, and the environment. Thus, the urgent need of the practice of sustainability in different forms. Economic, social, environmental, and energy sustainability are the hallmarks of today's scientific pursuits and in progress of human civilization. The OECD report explores and deeply glances on the status of implementation of the concept of sustainable development in global standing. Scientific perseverance and scientific determination are the pillars of sustainable development today. Thus, the need of such a detailed treatise (OECD Report, 2008). Human life depends on a varied and complex set of interactions between people, environment, and the vastly prolific economic systems. The unprecedented growth seen during the 20th century has vastly affected these relationships in both positive and negative ways. High and unprecedented levels of industrial pollution have put immense stress on the human environment (OECD Report, 2008). Economic growth and advancements have created wealth in some areas of the globe but left others behind. Lack of food, shelter, drinking water, proper sanitation, human habitat, and sustainable development are degrading the entire scientific horizon. In such a difficult situation around the world, environmental and energy sustainability are the needs of the hour (OECD Report, 2008). Mineral and metal industry globally are standing in the middle of vision and deep scientific introspection. Every industrial sector and global industrial manufacturing are in need of a vast industrial and environmental overhaul. This brilliant report will surely be an eye-opener to the deep scientific issues of global sustainable development (OECD Report, 2008).

Klarin (2018) deeply explores with vision and vast scientific prowess the concept of sustainable development from its beginning to the contemporary issues. The concept of sustainable development has undergone various development phases since its introduction. The highly historical report and the development of the concept saw participation of various organizations and institutions, which nowadays work intensely on the implementation of its knowledge and its concepts (Klarin, 2018). Today science, engineering, and humankind are in the middle of a deep scientific and engineering contemplation as nations around the world grapple with global climate change and global environmental degradation. In such as a crucial juncture of human history, the concept of sustainability and sustainable development assumes immense importance (Klarin, 2018). There are certain goals in the implementation of sustainability globally today. Some goals have been updated and new goals are set (Klarin, 2018). These goals are united in the comprehensive framework of the United Nations Millennium Development Goals, 2015 (Klarin, 2018). These goals challenge the humanity in a way not only to achieve sustainable development but for the human survival on earth. Today United Nations Sustainable Development, 2015, explores different social and economic issues globally (Klarin, 2018). Throughout the world human suffering with respects to human habitat, water, food, shelter, and sustainability are increasing day by day (Klarin, 2018). Thus, the need of a massive reenvisioning and restructuring of United Nations Sustainable Development Goals. Overall development of humanity over the last 2 decades has led to immense and unfavorable climate changes and natural catastrophes but also wars and political and socioeconomic instability. The authors in this paper deeply discusses development, sustainability, history of the concept of sustainable development, contemporary challenges of the concept of sustainable development, chronological overview of the meaning of sustainable development in the years 1987–2015 (Klarin, 2018). This research endeavor also explores the difference between strong and weak sustainability. Sustainability and sustainable development should provide a veritable solution in terms of meeting basic human needs, absolutely with vision and prowess integrating environmental development and protection vastly achieving social determination, self-determination, and cultural diversity. United Nations Sustainable Development Goals focused deeply on a complex global situation, such pollution upheaval, environmental degradation, hunger, poverty, wars, and political instability. Now the serious question which arises as to how United Nations Sustainable Development Goals, 2015, are successfully implemented in developing and disadvantaged nations around the globe. These difficulties, trials, and tribulations are deeply discussed in this paper (Klarin, 2018).

Ghernaout and Elboughdiri (2019) deeply discussed with vision, lucidity, and scientific revelation water reuse, emerging contaminants elimination, and its progresses and trends. This paper concentrates on a previous comprehensive treatise on elimination of emerging contaminants using biological, chemical, and hybrid techniques in effluents from wastewater treatment plants (Ghernaout and Elboughdiri, 2019). Endocrine disruption chemicals are highly reduced by a membrane bioreactor, activated sludge, and aeration processes between biological processes and hybrid biological processes (Ghernaout and Elboughdiri, 2019). Microalgae treatment technologies may diminish nearly all sorts of endocrine disruption chemicals to a certain degree. Humankind's immense scientific judgment, prowess, and understanding need to be revamped as the science of environmental protection treads forward. Biodegradation techniques have vastly and conventionally been utilized in wastewater treatment devices for mitigation of endocrine disruptive chemicals. Today the engineering vision of environmental protection and environmental sustainability are in the middle of deep scientific ingenuity and scientific imagination (Ghernaout and Elboughdiri, 2019). Pharmaceutical organic compounds and personal care products contain analgesics, lipid regulators, antibiotics, diuretics, nonsteroid antiinflammatory drugs, stimulant drugs, stimulant drugs, antiseptics, analgesics, beta-blockers, antimicrobials, cosmetics, sunscreen agents, food supplements, fragrances, and their metabolites, and transformation products. Thus, a whole gamut of recalcitrant chemicals are the major industrial polluters (Ghernaout and Elboughdiri, 2019). Endocrine disrupter chemicals are exogenous substances or mixtures that vastly modify the roles of the endocrine systems and then induce negative influences in an intact organisms. A set of diverse physical, chemical, and biological treatment techniques have previously been utilized to eliminate or decompose the endocrine disrupter chemicals during the past few years (Ghernaout and Elboughdiri, 2019). Biological treatment techniques have been largely used for eliminating Endocrine Chemicals mostly via the veritable pathway of biodegradation. There are also chemical treatment techniques (Ghernaout and Elboughdiri, 2019). Human civilization and the progress of environmental protection and environmental sustainability are both progressing at a rapid rate. The status of global environment is extremely grave today (Ghernaout and Elboughdiri, 2019). This paper deeply targets the progresses and challenges in hybrid and nonhybrid systems.

The classical wastewater treatment methods are not appropriate for eliminating several endocrine chemicals. So the need of hybrid tools and much larger robust systems. Chemical oxidation tools have been observed to be the best techniques for reducing beta-blockers, pesticides, and pharmaceuticals (Ghernaout and Elboughdiri, 2019). Today hybrid and nonhybrid techniques globally are in the path of newer scientific divination. This research endeavor deeply targets the scientific success and the vast scientific subtleties in application of environmental engineering tools in removing hazardous chemicals from water and industrial wastewater (Ghernaout and Elboughdiri, 2019).

Kanchi (2014) discussed and deliberated with vision, scientific understanding, and prowess the application of nanotechnology for water treatment. Nanotechnology is one of the revolutionary areas of scientific endeavor globally. Humankind's deep scientific prowess, sagacity, and admiration are in the path of immense regeneration. The term nanotechnology describes a range of technologies performed on a nanometer scale with a wide area of applications (Kanchi, 2014). Today is a technology-driven society. The authors discussed in details significance of nanotechnology in wastewater treatment, instrument-based wastewater analysis with nanomaterials, and the other areas of membrane science and nanofiltration. Nanotechnology research and development is a highly promising technology and today it is highly path-breaking (Kanchi, 2014). Nanotechnology applications for sustainable water supplies include water filtration, water and industrial wastewater treatment, desalination science, and use of techniques such as sensors, nanoparticles, and catalysts. Desalination and water treatment are today in the path of newer scientific divination (Kanchi, 2014). Nanotechnology, nanomaterials, and engineered nanomaterials are the needs of environmental engineering science and chemical process engineering (Kanchi, 2014). A deep futuristic thought and a deep futuristic vision in the field of nanotechnology are the only needs of the hour. In this article, the author deeply ponders on sustainable environmental tools and sustainable nanotechnology techniques (Kanchi, 2014). Today novel routes in the development of nanomaterials and engineered nanomaterials are the most outstanding scientific technologies science and engineering can give to humankind. This paper reveals the scientific intricacies and the vast and wide scientific ingenuity in the applications of nanotechnology in water and wastewater treatment (Kanchi, 2014).

Geraci and Castranova (2010) elucidated with scientific vision challenges in assessing nanomaterial toxicology in a personal perspective. Civilization, science, and engineering today stands in the crossroads of vast challenges and difficulties (Geraci and Castranova, 2010). Nanotechnology exploits a deep fact that nanomaterials exhibit unique physicochemical properties which are highly distinct from fine-sized particles of the same composition (Geraci and Castranova, 2010). It is highly evident that nanoparticles exhibit distinct bioactivity and unique interaction with different biological systems. Today risk assessment and chemical process safety plays a major role in the application of nanotechnology and nanomaterials in different branches of science and engineering. A major challenge and hurdle in hazard assessment in nanotechnology is the large and rapidly growing number of possible nanoparticles to be tested for biological activity (Geraci and Castranova, 2010). Nanotechnology applications in science and engineering are highly daunting today. Humankind and scientific research pursuit are today in a deep gorge of scientific barriers, hurdles, and introspection. The authors deeply emphasize these important issues (Geraci and Castranova, 2010).

Varma (2012) discussed and deliberated with scientific grit and determination greener approach to nanomaterials and their sustainable applications. The alignment of green chemistry and green engineering principles into the rapidly evolving field of nanoscience is a necessity for the risk reduction. Green chemistry and green technology are the cornerstones of global scientific research endeavor. Several greener pathways have been developed to generate greener nanoparticles (Varma, 2012). The use of naturally occurring biodegradable materials such as vitamins, sugars, tea, or polyphenol-rich agricultural residues, which serve as reducing and capping agents, is demonstrated and may help assist in designing nanomaterials and engineered nanomaterials with reduced toxicity (Varma, 2012). The author discussed in details emerging areas of green chemistry, greener synthetic strategies, toxicity aspects, and environmental remediation (Varma, 2012). The benefits of the green-synthetic protocols for generation of nanoparticles over the conventionally used processes include (1) only naturally occurring nonhazardous materials, (2) no hazardous wastes are produced, (3) reduced processing effort is required, (4) the materials are more stable, and (5) the materials can be more readily produced. Green chemistry and green engineering are the zenith of scientific research pursuit globally. This paper is a veritable eye-opener toward the importance of green chemistry in proliferation of science and technology (Varma, 2012).

Sharma and Bhattacharya (2017) discussed drinking water contamination and treatment techniques. Water is of immense importance to humanity and human life. The vast synthesis and structure of cell constituents and the transport of nutrients into the cells as well as metabolism depends highly on water. The vicious contaminants present in water disturb the spontaneity of the chemical mechanism and result in long and short diseases (Sharma and Bhattacharya, 2017). This is a comprehensive treatise in water and wastewater treatment. Human scientific vision and vast scientific stance will surely lead a long way in finding the scientific truth in environmental remediation, water remediation, and environmental sustainability (Sharma and Bhattacharya, 2017). The review includes concepts and modalities of the technologies in comprehensive form (Sharma and Bhattacharya, 2017). It includes some meaningful hybrid technologies and promising awaited technologies in the coming years. Thus, human science and engineering vision will surely open new windows of innovation in water and wastewater treatment. The authors discussed in minute details inorganic contaminants, organic contaminants, biological contaminants, and radiological contaminants. Science and technology has come a long way since the first invention of membranes. Today membrane science and desalination science are linked to each other by an unsevered umbilical cord (Sharma and Bhattacharya, 2017). The authors stress in these areas also with vision and perseverance. The authors stressed on solving approaches, precipitation and coagulation, distillation, adsorption, application of activated carbon, application of activated alumina, silica gel, ion exchange, ion exchange resins, and membrane water treatment (Sharma and Bhattacharya, 2017). Other areas of research endeavor are electrodialysis membrane treatment, catalytic processes, hydrogenation of nitrate, photocatalytic method, electrocatalytic oxidation, bioremediation, magnetic separation, and the vast world of disinfection. Ultraviolet radiation disinfection plays an important role in water remediation and environmental remediation. Applications of ozone, chlorine, iodine, hydrogen peroxide, and other hybrid technologies are the other cornerstones of this well-researched treatise. The humankind is facing turbulent water future (Sharma and Bhattacharya, 2017). With the growing global economy and burgeoning population, the theme of all developing and developed nations around the world is to save water. Quantity and quality of water is the major issue today in many developed, disadvantaged, and poor nations around the world. A larger global awareness related to water conservation and safe drinking water is of immense importance and should be a good futuristic vision to the suffering human citizens around the world. This treatise deeply explores the science and engineering of water treatment, wastewater treatment, water quality, and wastewater management. The technological solutions depend on raw water characteristics, affordability of environmental engineering tools, and the level of application. A new dawn in the field of environmental protection and water remediation will surely emerge and usher in as civilization gears forward (Sharma and Bhattacharya, 2017).

Deng and Zhao (2015) discussed with deep vision and purpose AOPs in wastewater treatment. Today the technology of environmental remediation is moving at a rapid pace. Minerals and metal industry are the serious industrial polluter. So one needs to devise newer innovations in mitigating this pollution. AOPs were first proposed in the 1980s for drinking water treatment and industrial wastewater treatment (Deng and Zhao, 2015). During the AOP treatment of wastewater, hydroxyl radicals or sulfate radicals are generated in sufficient quantities to remove refractory organic matters (Deng and Zhao, 2015). Minerals and metal industry are the causes of immense concern globally today. In this entire chapter, the authors deeply stress on the success of different conventional and nonconventional environmental engineering tools to mitigate industrial pollution control. The authors discussed in minute details AOP, hydroxyl radical–based AOPs, ozone-based AOPs, ultraviolet-based AOPs, Fenton-related AOPs, and sulfate radical–based AOPs (Deng and Zhao, 2015). Today scientific regeneration in the field of AOPs are in the process of new restructuring (Deng and Zhao, 2015). Mankind, science, and engineering vision needs to be streamlined with the progress of water and industrial wastewater treatment. The other areas of intense research pursuit are AOPs for treatment of a high strength wastewater that is landfill leachate (Deng and Zhao, 2015). AOPs for treatment of effluent organic matters in biologically treated secondary effluents are the other cornerstones of this treatise. Traditional hydroxyl radical–based AOPs have been studied in minute details in the past (Deng and Zhao, 2015). The major purpose of HR AOP is to remove recalcitrant chemicals, traceable emerging contaminants in addition to inorganic pollutants (Deng and Zhao, 2015). This treatise elucidates these issues. It has also been demonstrated that AOPs are a viable proposition for leachate treatment and water reuse. Certainly a new dawn of environmental engineering science will evolve if researchers target conventional and nonconventional environmental tools and novel separation processes (Deng and Zhao, 2015).

Bina (2013) discussed and deliberated with vision, scientific determination, and grit the green economy and sustainable development also considering the uneasy balance. The United Nations Conference on Sustainable Development was conceived at a time when the global health system is in a precarious situation. In this scientific and technological atmosphere green economy was chosen as one of the two themes for the world-renowned conference (Bina, 2013). Technology and engineering science of environmental and energy sustainability needs to be reenvisioned as civilization treads forward. Mineral and metal industry are today one of the biggest industrial polluters (Bina, 2013). Water resource management, wastewater management, and industrial pollution control are today aligned with environmental sustainability. In the similar vein, industrial system engineering and human factor engineering are leading a long way in the true emancipation of sustainable development. In this chapter, the authors vehemently stress these vital environmental and sustainability issues. Through a systematic qualitative analysis of textual material, the three categories of discourse that can enshrine the meaning and the implementing of greening are identified: almost business as usual, greening, and change. The world today stands transfixed and mesmerized at the rapid changes in industrial manufacturing and sustainable development. These are the salient features of this paper. The authors in this paper discussed with immense vision and lucidity the green response to economies and environments in crisis (Bina, 2013). Human civilization's vast and varied scientific endeavor, the stance and success of science, and the scientific revelation will surely open new doors of innovation and scientific instinct in years to come. The other areas touched upon are approaches and methods, the roads to Rio+20 agenda and vision, green response, and the application of green and environmental sustainability to the progress of human society. Green chemistry and green engineering are the hallmarks of human civilization as well as human scientific pursuits. They are today highly related to environmental sustainability. The theme of a green economy in the context of sustainable development and poverty eradication was included as one of the emerging challenges during United Nations first preparatory meeting on 17–19 May 2010. The authors also discussed in minute details green economy and green responses in a detailed classification. A new beginning and a newer regeneration will occur in science and engineering as nations around the world start to think about sustainability in any forms (Bina, 2013).

Goni et al. (2015) discussed with scientific lucidity, grit, and determination environmental sustainability and its research growth and trends. The number of research in sustainability development is rapidly increasing along with the importance of sustainability. The primary aim and objective of this paper is to outline the research trends, define the literature characterization, and research focuses of environmental sustainability engineering research from the perspective of historical evaluation (Goni et al., 2015). This paper deeply pronounces the success, the ingenuity, and the vast scientific profundity in the application of environmental sustainability to human society. This paper discusses in details research trends in sustainability research, literature characterization of sustainability research, sustainability research focus, and the vast relevance of environment sustainability. Over the past couple of decades, the overall research in sustainability has progressed significantly (Goni et al., 2015).

Human scientific stance and the deep ingenuity of science will surely the forerunners toward a newer era of environmental sustainability as well as environmental protection (Goni et al., 2015).

8. Industrial wastewater treatment and mineral and metal industry

Technological advancements in the field of industrial wastewater treatment in mineral and metal industry stand in the middle of deep scientific revelation and vision. Arsenic and heavy metal groundwater contamination is a result of anthropogenic sources as well as geological sources. The science of industrial wastewater treatment needs to be reframed as civilization progresses forward. Challenges and difficulties are surmounting in today's scientific world. The status of environmental engineering research globally is in a state of disaster and deep comprehension. Research and development initiatives globally should concentrate on drinking water treatment and industrial wastewater treatment. Millions of people around the world are without pure drinking water and reasonable and proper sanitation. Poverty, deprivation, illiteracy, and lack of food and drinking water are destroying the social and economic fabric of nations around the world. Social, economic, energy, and environmental sustainability are today interlinked with each other. Thus, human civilization, scientific pursuit, and sustainable development are in the path of major disasters. In such a critical situation, industrial advancements in humankind need to be reenvisioned and revamped.

9. Heavy metal and arsenic groundwater remediation and the future of mineral and metal industry

Heavy metal poisoning in drinking water is a burning concern in many developing and developed nations around the world. Arsenic groundwater contamination in Bangladesh and the state of West Bengal, India, is in the midst of an unimaginable environmental engineering catastrophe. Mineral and metal industry are also the major polluters and the scientific and engineering vision of environmental engineering research globally needs to be changed. This is an inevitable burden to human civilization today. Environmental engineering concerns in minerals and metals industry need to be revitalized as humankind moves forward. South Asia and many developed countries around the world are in the pangs of world's largest disaster that is drinking water contamination. Sustainable resource management which includes water resource management and air quality management will surely pave the way toward newer futuristic vision in environmental remediation.

10. Future scientific recommendations and the future flow of scientific thoughts

Challenges, barriers, and hurdles are many in today's global scientific forays. Mining and metallurgical industry today stands in the deep abyss of scientific contemplation and provenance. Innovations in conventional and nonconventional environmental engineering techniques are developing with deep scientific and academic rigor day by day. Developing and developed nations around the world are taking concerted efforts in warlike footing to eradicate environmental degradation. Industrial wastewater treatment and drinking water treatment are the utmost need of the hour. Integrated water resource management and water quality management are today in the path of newer regeneration. In the developing countries around the world, technologies and inventions are in the incipient stage. More research forays need to be envisioned in the developing and disadvantaged countries around the globe. Human mankind is thus in a state of immense scientific and engineering distress as science and technology surges forward. The challenges of membrane science, AOPs, and novel separation processes need to move toward newer renovation. These are the future scientific recommendations and the future flow of scientific thoughts. India and many South East Asian countries are in the verge of an immense environmental disaster that is arsenic and heavy metal drinking water contamination. These are the targets of environmental engineering research in the distant future. Thus, minerals and metals industry also need to be reorganized as humankind moves forward.

11. Conclusion, summary, and scientific perspectives

Civilization is today moving fast toward a new scientific regeneration. In the similar vein, science and technology are today overcoming one barrier over another. Minerals and metal industry are today advancing very fast. New technologies and newer innovations are the cornerstones of advancements in mineral and metal industry today. Along with innovations, zero emissions concepts are the needs of human society and mankind today. Environmental sustainability and environmental protection are thus the needs of the hour. In this chapter, the authors deeply stress on the recent advances in mineral and metal industry and the relevant interfaces with environmental sustainability. Environmental remediation and water and wastewater treatment are the other cornerstones of this treatise. Due to environmental degradation, the civilization stands transfixed in the middle of scientific contemplation and vision. Pandemics and industrial disasters are part and parcel of mankind today. Scientific and environmental engineering perspectives need to be revamped if the civilization is to survive from global climate change, environmental degradation, and frequent epidemics. Rapid industrialization and rapid advances in the manufacturing industry are today veritably destroying the humankind's environment. The environmental degradation from mineral and metal industry in both developing and developed countries around the globe is a large and important concern for the humankind. In this chapter, the authors deeply delve into the science of environmental protection and the engineering science of environmental sustainability in present day human society. The authors deeply hope that there will be more innovations, more environmental mitigation tools, and a new prosperous beginning in the world of mining engineering, metallurgical engineering, and the science of environmental remediation. Future is extremely bright in the fields of mining engineering in the global scenario. Thus, scientific perspectives and engineering vision need to be revamped in the field of environmental protection in days to come. This is the larger hope and indeed a larger vision of mining engineering and metallurgical engineering today. A new dawn will surely emerge in the field of environmental engineering and the mining and metal industry if nations around the globe mitigate these immense environmental challenges.

References

Bina, O., 2013. The green economy and sustainable development: an uneasy balance? Environ. Plann. C Govern. Pol. 31, 1023–1047.

Deng, Y., Zhao, R., 2015. Advanced oxidation processes(AOPs)in wastewater treatment. Curr. Pollut. Rep. 1, 167–176.

Geraci, C.L., Castranova, V., 2010. Challenges in assessing nanomaterial toxicology: a personal perspective. In: WIREs Nanomedicine and Nanobiotechnology, vol. 2. John Wiley and Sons, Inc. November/December 2010.

Ghernaout, D., Elboughdiri, N., 2019. Water reuse : emerging contaminants elimination – progress and trends. Open Acc. Lib. J. 6, e5981, 2019.

Goni, F.A., Shukor, S.A., Mukhtar, M., Sahran, S., 2015. Environmental sustainability: research growth and trends. Am. Sci. Publ. Adv. Sci. Lett. 21, 192–195, 2015.

Indian National Science Academy Report, 2011. Hazardous Metals and Minerals Pollution in India: Sources, Toxicity and Management. A position paper, New Delhi. August,2011.

Kanchi, S., 2014. Nanotechnology for water treatment. J. Environ. Anal. Chem. 1 (2) https://doi.org/10.4172/jreac.1000e102.

Klarin, T., 2018. The concept of sustainable development: from its beginning to the contemporary issues. Zagreb Int. Rev. Econ. Bus. 21 (1), 67–94. Faculty of Economics and Business, University of Zagreb and Degruyter Open.

Kuhlman, T., Farrington, J., 2010. What is sustainability? Sustainability 2, 3436–3448. https://doi.org/10.3390/su2113436.

OECD Report, 2008. Sustainable Development- Linking Economy, Society and Environment, by Tracey Orange, Anne Bayley. Organization for Economic Co-Operation and Development.

Sharma, S., Bhattacharya, A., 2017. Drinking water contamination and treatment techniques. Appl. Water Sci. 7 (3), 1043–1067, 7.

United Nations Development Programme, 2011. Human development research paper, 2011/04, sustainability and inequality in human development. Eric Neumayer. 1, 1–27.

Varma, R.S., 2012. Greener approach to nanomaterials and their sustainable applications. Curr. Opin. Chem. Eng. 2012 (1), 123–128.

Sustainable land use and management

Juan F. Velasco-Muñoz, José A. Aznar-Sánchez, Belén López-Felices, Daniel García-Arca
Department of Economy and Business, Research Centre CIAIMBITAL, University of Almería, Almería, Spain

1. Introduction

Land plays a fundamental role in the survival of human beings (Yao et al., 2017; Xie et al., 2020a). Among the main contributions of the land we find the following (FAO and UNEP, 1999): it acts as a reservoir of wealth; it allows the production of food, fiber, and fuel; it provides biological habitats for plants, animals, and microorganisms; it allows the regulation of the flow of surface and groundwater; it stores minerals and raw materials; it filters chemical pollutants; it provides space for the settlement and development of different activities; and it regulates the movement of living beings between different areas.

Land use is defined as the direct link between land cover and the actions carried out by people in their environment, which lead to its transformation (FAO and UNEP, 1999; Martínez and Mollicone, 2012). The land is used for different purposes such as agricultural, residential, commercial, industrial, or recreational (Yao et al., 2017). Individuals determine the use they make of a specific area of land based on the expected benefit, which will vary depending on their individual objectives and limitations; the biophysical properties of the area; the institutional, cultural, and legal characteristics; as well as the cultural and socioeconomic environment (Cihlar and Jansen, 2001). About 75% of the world's ice-free land, including most of the most productive land, has some form of land use (IPCC, 2019). The earth's surface has been subject to a process of transformation as a result of changes in land use by humans throughout history, which has intensified in recent decades (Foley et al., 2005; Gauthier et al., 2015). Agricultural expansion and urbanization resulting from continued population growth have led to progressive land scarcity and significant negative impacts on the environment and quality of life (Yao et al., 2017; Seppelt et al., 2018). In addition, changes in land use have intensified the process of land degradation and the effects of climate change. This poses a major threat to biodiversity, food and energy security, desertification, and the development of sustainable human environments and socioeconomic systems (Xie et al., 2020a).

Sustainable Resource Management. https://doi.org/10.1016/B978-0-12-824342-8.00015-8

Sustainability has become an essential aspect of land use planning and management due to land depletion and deterioration (Yao et al., 2017; Dadashpoor et al., 2019; Xie et al., 2020b). The term sustainability first appeared in the Brundtland report by the World Commission on Environment and Development of the United Nations in 1987, which defined sustainable development as "development which meets the needs of the present, without compromising the ability of future generations to meet their own needs" (Brundtland et al., 1987). In this regard, optimal land use planning to ensure sustainable development aims to distribute the different activities that take place on the land in such a way that society's demands are met (Yao et al., 2017). Sustainable land management is based on the development, use, and protection of the resources it offers, considering the specific characteristics of each area and implementing appropriate forms of organization according to them (Xie et al., 2020b). Achieving sustainable land management will therefore require the development of policies that take into account the different objectives set at national and international level in relation to food security, climate, biodiversity, and forests, and the synergies between all these aspects (OECD, 2020). However, proper land planning and management is a complex task involving many factors such as the specific characteristics of each area and its socioeconomic context (Yao et al., 2017).

Given the importance of land as a resource for survival and human development, and the serious deterioration it is experiencing, this chapter focuses on sustainable land use management. The aim of this study is to provide an overview of the most relevant aspects of sustainable land management, so that the reader can get an overview of this study topic. Therefore, the rest of this chapter is structured based on the main aspects linked to sustainable land use and management. Thus, after this introduction some concepts are shown, as well as the evolution in the uses of the land and the changes produced in them (Section 2). The main impacts resulting from changes in land use are presented below, with special reference to climate change and general soil degradation (Section 3). Next, the concept of sustainable management applied to land is defined (Section 4), and it is exemplified with sustainable management practices of the different types of land use (Section 5). Section 6 includes the main barriers to the adoption of sustainable management practices. Finally, some future lines of research are presented (Section 7), before closing with the main conclusions (Section 8).

2. Land uses and changes

The terms land use and land cover are used indistinctly in many cases but have different meanings. Land cover is related to biophysical coverage of land such as bare soil, forests, or buildings (Cihlar and Jansen, 2001; Calvo-Buendia et al., 2019). Land uses are all those actions carried out by people to produce, transform, or conserve a certain type of land cover (Di Gregorio and Jansen, 1998; FAO and UNEP, 1999). Land uses therefore show the influence of humans on ecosystems (Turner et al., 2016). The relationship between the two terms can be better understood by linking land use to the socioeconomic factors that promote human actions (Martinez and Mollicone, 2012). Of the total land area, it is estimated that approximately 71% has habitability characteristics for humans. The uses within that area are distributed as follows (Ritchie and Roser, 2019): 50% agriculture, 37% forest, 11% shrubs and grasslands, 1% freshwater, and another 1% urban area (cities, towns, villages, roads, and other infrastructure).

Agriculture enables people to be fed and also plays a key role in generating biomass that can be used for biofuel production (Acevedo, 2011). Forests and their biodiversity fulfill essential functions for people by influencing carbon cycles, water, nutrients, and food (Aznar-Sánchez et al., 2018). Shrubs and grasslands perform several functions such as providing forage for animals, being a carbon sink, promoting soil and water cycles, performing biological pest control in agricultural environments, etc. (FAO, 2018). Water is a fundamental resource for human consumption and sanitation, as well as to produce food and other raw materials (WWAP, 2018). Finally, even though the urban area represents a small fraction, it includes many uses of great importance to society, such as homes, businesses, industries, institutions, and transport (Theobald, 2014). However, determining land use is a complex task since several uses can occur simultaneously on specific area of land (Potts et al., 2016; Turner et al., 2016).

Changes in land use generated by various factors such as population growth and urbanization, industrialization, and increased demand for resources and energy have caused the land surface to undergo a process of transformation (Ellis, 2011; Seppelt et al., 2018). Specifically, the use of land by humans throughout its history has meant the modification of more than 75% of the planet's land surface (Ellis and Ramankutty, 2008). We can find different levels in the processes of change of land uses in the global sphere that can be explained taking into account the different levels of economic development: agricultural, transport, and communication (Huston, 2005; Acevedo, 2011). Generally, in the initial phases of transformation, it is the demand for expansion of agricultural activities that initiates change in land use, transforming forests and other natural spaces. As economic growth takes place, the need for transport networks and infrastructures increases and, finally, for the urbanization of areas close to population centers (Acevedo, 2011). Therefore, the expected increase in the world's population, together with changes in their lifestyles and consumption habits, may further accentuate the pressure on land uses in the near future (FAO, 2018; WWAP, 2018). This, combined with the depletion of land because it is a limited resource, can lead to a great deal of rivalry between different land uses that can generate conflicts between different stakeholders (Yao et al., 2017).

3. Impacts of land use changes

Changes in land use have generated numerous negative impacts, both from a socioeconomic and environmental perspective. The recent process of migration from rural to urban areas, in search of better work options and living conditions, has led to the abandonment of farmland, as well as natural areas such as forests and grasslands

(Satterthwaite et al., 2010; Dadashpoor et al., 2019). In a context of a growing world population and increasing demand for food, rural exodus reduces the amount of land available to produce food and other raw materials. In addition, it is estimated that food production will need to increase by 70% by 2050, which will lead to increased demand for farmland, and greater competitiveness for available land (WWAP, 2018). As a result, speculation on the value of land is expected to increase, raising the level of inequality due to inaccessibility for groups with fewer resources (IPCC, 2019). On the other hand, the development of urbanization without adequate planning has meant on many occasions that large population centers do not have enough green spaces, impacting on air pollution and heat waves (Kruize et al., 2019). The "heat-island" effect implies the development of higher temperatures in the cities than in the surrounding countryside. This can have consequences on the habitability of cities, the health and well-being of the population by reducing air quality, favoring environmental degradation, as well as leading to a greater demand for energy used for cooling with an increase in the associated costs (Mohajerani et al., 2017). The accumulation of the population in large centers generates serious management problems such as the deposit of large quantities of solid waste, air pollution, and the generation of sewage. It is estimated that in 2050, urban residents will generate around 2.2 billion tons of municipal solid waste per year, with the consequent risk of proliferation and transmission of diseases through the air or water without adequate treatment (Hoornweg et al., 2009).

Changes in land use have also had serious environmental impacts. These changes have generated alterations in the hydrological cycles due to the excessive extraction of water from underground sources and the deviation of surface water to cover the growing demand of human activities (Pokhrel et al., 2017). Most of the changes in land use are due to the expansion of agriculture, which is the largest consumer of freshwater worldwide (Aznar-Sánchez et al., 2019). Agriculture is also a major source of water resource pollution due to excess nutrients, mainly nitrogen and phosphorus, which cause eutrophication (Adeyemi et al., 2017; Xia et al., 2020). In addition, the intensification in the use of groundwater, due to the development of irrigation systems, has caused in many cases the salinization of the land, impacting on its productivity (Daliakopoulos et al., 2016). Population movements and consequent urbanization have also had an impact on water resources through increased water demand in certain areas and pollution generated in cities (Bai et al., 2017). On the other hand, changes in land use, mainly agricultural expansion, have led to the loss of 420 million hectares of forest since 1990 (FAO and UNEP, 2020). These forest losses have a major impact on biodiversity, as forests host most of our planet's terrestrial biodiversity, providing habitat for 80% of amphibian species, 75% of bird species, and 68% of mammal species (MEA, 2005; Wu, 2008; Vié et al., 2009).

Climate change and land degradation are two particularly relevant processes that have a feedback relationship with land use transformations (UNCCD, 2015). Thus, changes in land use encourage the development of these processes which, in turn, negatively impact the state of the land and lead to further change to replace the deteriorated land. Terrestrial ecosystems play an essential role in the climate system because of their high carbon stocks and the flow of exchange with the atmosphere, which is altered as a result of changes in land use (Ciais et al., 2013; IPCC, 2019).

The main emitters of greenhouse gases are the industrial sector, and the energy sector, due to the use of fossil materials, with more than 25% of the total (Calvo-Buendia et al., 2019). Agriculture, forestry, and other land uses are estimated to emit between 21% and 24% of anthropogenic greenhouse gas emissions (Tubiello et al., 2015). On the other hand, vegetation and soil act as a CO_2 sink, so changes in land cover cause alterations in their absorption capacity, contributing to global warming (IPCC, 2019). In this way, the loss of forest cover has reduced the lands' capacity to neutralize greenhouse gases.

Climate change poses a threat to ecosystems by producing changes in their structures and functions, as well as in the goods and services they provide to society (Díaz et al., 2019; Weiskopf et al., 2020). These threats stem from increased variability in rainfall, the frequency of extreme weather events such as floods and droughts, and temperatures (UNCCD, 2015; Malhi et al., 2020). The resulting negative impacts include water scarcity, soil erosion, coastal degradation, vegetation loss, increased fire occurrence, melting of permafrost (naturally frozen layer of the earth), and food availability (IPCC, 2019). For example, rising temperatures influence the melting of ice, which increases sea levels and can lead to the loss of coastal areas; and they also increase the likelihood of fires, causing serious damage to vegetation (Malhi et al., 2020). All these factors influence changes in land use, in search of better adaptation and resilience to adverse phenomena (FAO, 2015).

The risks arising from changes caused by climate change are related to food security, human and ecosystem health, the commercial value of land, infrastructure, and communications (IPCC, 2019). Factors such as the scarcity of water resources or soil salinity as a result of climate change can produce instability in production and reduce food yields, endangering food security (Daliakopoulos et al., 2016). Human health can be threatened by factors such as extreme heat and cold, natural disasters or infectious diseases as climate conditions affect the transmission of these diseases through water or insects (WHO, 2014). On the other hand, the deterioration of livelihoods can accelerate migration processes to areas with available resources, leading to social conflicts over scarce resources (Carleton and Hsiang, 2016). Communications and infrastructure can also be affected by extreme weather events, for example, ice storms or high intensity winds can cause damage to electrical infrastructure (Abi-Samra and Malcolm, 2011; IPCC, 2019).

Land degradation is the long-term loss of the goods and services provided by terrestrial ecosystems (Mirzabaev et al., 2015). This degradation can have negative impacts on soil organic matter and nutrient content, deterioration of surface and groundwater quality, reduction of biodiversity, and decline in the flow of ecosystem services (Pacheco et al., 2018). As a result, there is a reduction in productivity, which negatively influences economic activity (Pacheco et al., 2018). It is estimated that the cost of land degradation can vary between US$ 18 billion and US$ 9.4 trillion annually and adversely affects 3.2 billion people (Nkonya et al., 2013; Mirzabaev et al., 2015; Díaz et al., 2019). In the case of agriculture, it has meant a decline in yields and in the income of people who depend on this sector (Farooq et al., 2019). This situation is particularly worrying in areas where agriculture is the main activity or where subsistence farming is practiced.

Land degradation processes are a global problem, especially related to situations of inequality because they can generate conflicts over scarce resources, increased migration, and higher food prices (Lal, 2004; von Braun et al., 2013). There are two types of causes of land degradation (Nkonya et al., 2013; Turner et al., 2016): proximate, those that directly affect the land ecosystem either through natural or anthropogenic causes; and underlying, those that have an indirect effect on proximate causes such as institutional, socioeconomic, or political factors (Fig. 8.1). Thus, the wide variety of factors influencing land degradation can have different effects depending on the characteristics of the area in which these factors occur, making it difficult to identify them (Mirzabaev et al., 2015).

4. Sustainable land management

Land plays a very important role in achieving sustainable development, as it is an essential natural resource on which other social and economic activities take place

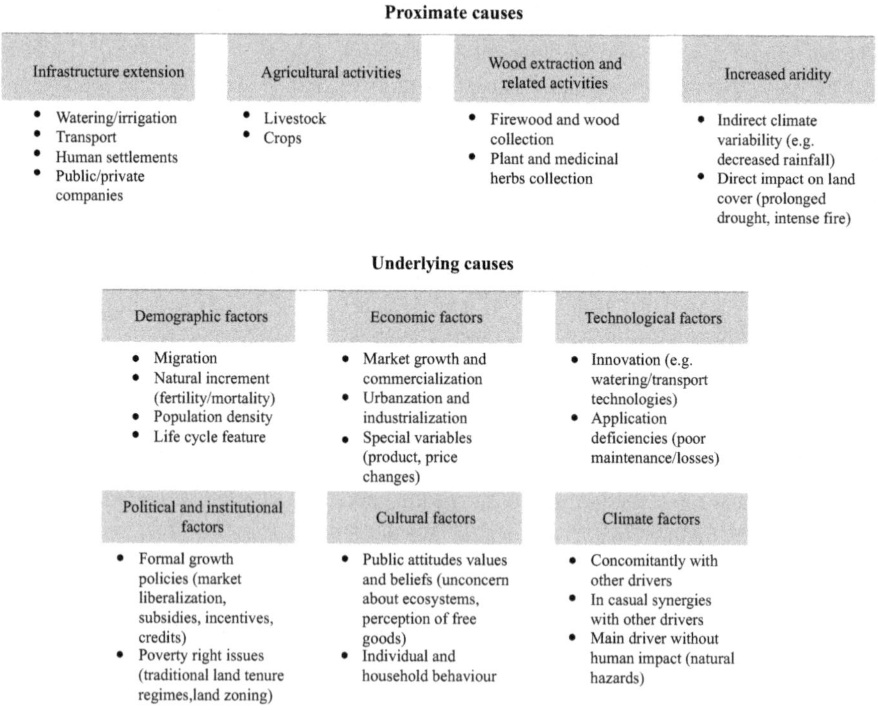

Proximate causes

Infrastructure extension
- Watering/irrigation
- Transport
- Human settlements
- Public/private companies

Agricultural activities
- Livestock
- Crops

Wood extraction and related activities
- Firewood and wood collection
- Plant and medicinal herbs collection

Increased aridity
- Indirect climate variability (e.g. decreased rainfall)
- Direct impact on land cover (prolonged drought, intense fire)

Underlying causes

Demographic factors
- Migration
- Natural increment (fertility/mortality)
- Population density
- Life cycle feature

Economic factors
- Market growth and commercialization
- Urbanzation and industrialization
- Special variables (product, price changes)

Technological factors
- Innovation (e.g. watering/transport technologies)
- Application deficiencies (poor maintenance/losses)

Political and institutional factors
- Formal growth policies (market liberalization, subsidies, incentives, credits)
- Poverty right issues (traditional land tenure regimes,land zoning)

Cultural factors
- Public attitudes values and beliefs (unconcern about ecosystems, perception of free goods)
- Individual and household behaviour

Climate factors
- Concomitantly with other drivers
- In casual synergies with other drivers
- Main driver without human impact (natural hazards)

Figure 8.1 Factors affecting land degradation.
Adapted from Turner, K.G., Anderson, S., Gonzales-Chang, M., Costanza, R., Courville, S., Dalgaard, T., Dominati, E., Kubiszewski, I., Ogilvy, S., Porfirio, L., 2016. A review of methods, data, and models to assess changes in the value of ecosystem services from land degradation and restoration. Ecol. Modell. 319, 190–207. https://doi.org/10.1016/j.ecolmodel.2015.07.017.

(Smith, 2018). Sustainable land management is defined as "knowledge-based procedure that helps integrate land, water, biodiversity, and environmental management (including input and output externalities) to meet rising food and fiber demands while sustaining ecosystem services and livelihoods" (Fernandes and Burcroff, 2006:14). Proper land management should aim to maximize the provision of ecosystem services in order to achieve the greatest social and environmental benefit (OECD, 2020). This will require appropriate planning of different land uses, considering all factors that can affect long-term sustainability and the constraints arising from the physical environment, the economy, and society (Ligmann-Zielinska et al., 2008). FAO (1993) defined land use planning as "the systematic assessment of land and water potential, alternatives for land use and economic and social conditions in order to select and adopt the best land-use options." Sustainable land planning is very useful for governing bodies at different scales as it allows them to manage land properly and avoid potential conflicts (Yao et al., 2017). However, planning and achieving sustainable land management is not easy because it will depend on the characteristics of the place where it is carried out (Yao et al., 2017). This has led to the development of extensive research in relation to sustainable land use, developing programs and action plans to control the factors that affect it and to achieve social, economic, and environmental objectives (Pacheco et al., 2018; Xie et al., 2020a). In this sense, sustainable land management is articulated as a holistic approach that seeks to achieve the highest productivity of ecosystems considering biophysical, sociocultural, and economic needs. This is particularly relevant in the case of less developed countries, which have fewer resources to plan and implement sustainable land management (Sivakumar and Stefanski, 2007; Dubovyk, 2017).

Land management is very important for the achievement of the objectives proposed in the United Nations Agenda 2030 for Sustainable Development (UN, 2015), as it is directly or indirectly related to the achievement of the 17 sustainable development goals (SDGs) included in the Agenda. Land management is fundamental to ensuring food supply and achieving the eradication of hunger (SDG 2). SDG 6 aims to guarantee the availability of water and its sustainable management and sanitation for all. In this sense, healthy soils can have a positive influence on this objective, since soils play a fundamental role in the supply of clean water by preventing contaminants from filtering into the water table, as well as minimizing evaporation and increasing the productivity and efficiency of water use in crops (FAO, 2015). Thus, proper urbanization planning can help make adequate sanitation and hygiene services available to all. Moreover, proper land management and use of bioenergy can provide affordable and clean energy (SDG 7). Efficient management of urban spaces can contribute to achieving sustainable cities (SDG 11). Certain practices, such as afforestation, can contribute to climate action (SDG 13) by reducing greenhouse gas emissions. Proper spatial planning and the incorporation of sustainable practices will face land degradation and desertification and reduce impacts on ecosystems and biodiversity (SDG 15).

All these objectives are interconnected, so the paths used to achieve some of them can have impacts on the others, which can be negative or positive depending on their effect. For example, intensification of agricultural land use is one of the most viable alternatives for increasing food and fiber production without expanding the area under cultivation. However, this could lead to a loss of biodiversity and generate more greenhouse gas emissions (Seppelt et al., 2018). The achievement of one objective may compromise others, leading to the development of compensation pathways at different scales, which can be moderated through proper land management (Pradhan et al., 2017; Seppelt et al., 2018). It should be noted that actions taken in a specific area may have an impact on the overall balance in the long term. In this regard, it has been shown that economically motivated expansion of agricultural land can lead to water degradation, jeopardizing the sustainability of the entire river basin under analysis (Hu et al., 2015; Xie et al., 2020b).

5. Sustainable land management practices

To ensure long-term sustainability, it is necessary to incorporate land management practices that help minimize the negative effects of land uses, seeking a balance between them and economic and social development. In the agricultural field, there are limitations to expanding the area dedicated to this activity because most of the available land is not suitable for agriculture and because of the high competition between sectors for its use (Bunning and De Pauw, 2017). Therefore, the most promising practice to increase production is the implementation of high performance and environmentally friendly intensive production systems. It is estimated that around 10% of the Earth's ice-free surface is managed intensively (Erb et al., 2016; IPCC, 2019). However, the intensification of production has been based on the use of large amounts of inputs, such as fertilizers, which have caused numerous environmental problems. Therefore, it is necessary to bet on a model of sustainable land intensification in which an optimal allocation of farmland is made according to the most suitable environmental conditions in order to achieve the highest productivity and be able to satisfy the growing demand (Garnett et al., 2013). Thus, a report prepared by the United Nations lists more than 100 technologies for achieving sustainable agricultural land management, especially those practices that address land degradation and the consequences of climate change (Sanz et al., 2017). In that report we found that the technologies most used in the case of croplands are related to the prevention of soil erosion (e.g., through the use of grass strips or riparian vegetation), soil degradation (integrated soil fertility management, minimizing soil disturbance per tillage), and the improvement of productivity and biodiversity (vegetation management, pest and disease control, sustainable irrigation systems, drainage and water harvesting systems). Regarding grazing lands, sustainable land management technologies are related to animal, plant, and fire management (e.g., rotational grazing, manure separation to better distribute organic matter, or grazing land rehabilitation with shrubs plantation).

In the case of forests, their sustainable management aims to maintain and increase the economic, social, and environmental value of these spaces (UN, 2008). Sustainable forest management practices will reduce the vulnerability of forests and maintain their productivity, as well as mitigate the effects of climate change (Sanz et al., 2017). These practices aim to reduce the impacts caused by forestry, through reforestation and reduction of deforestation. Some of the practices proposed in the United Nations report are afforestation with species mix at different scales, control anthropogenic disturbances such as fire and pest outbreaks, establishment of protected forest areas or selective logging. Thus, increasing the value of forests to local communities can help prevent their degradation, which will require the promotion of forest-based livelihoods through different sustainably generated forest services and products and the development of payments for carbon sequestration or other environmental services for companies (FAO and UNEP, 2020).

Cities are articulated as complex socioeconomic and natural systems in which human activities define land use patterns (He et al., 2017). The potential negative impacts of inadequate land management in cities are of particular concern in developing countries due to high rates of urbanization (Musakwa and Van Niekerk, 2013). It is therefore necessary to carry out actions to mitigate the negative effects on the well-being of the population and the natural environment resulting from increasing urbanization. The incorporation of green spaces in cities is presented as a promising alternative for achieving this objective (Chatzimentor et al., 2020). Urban green spaces can range from trees and shrubs located in the streets to different types of parks, nature reserves, forests, and botanical gardens (Shuvo et al., 2020). Such infrastructure can improve air and water quality, reduce noise pollution, and mitigate the effects of extreme weather events, as well as the "heat-island" effect (WHO, 2017). One of the main consequences of urbanization and population growth in cities is the increase in urban solid waste, which if it is not properly managed can cause diseases and pollution. The increase in waste and the scarcity of available and suitable sites for the installation of landfills makes waste management more complex (Pandey et al., 2012). Therefore, appropriate plans for recycling and waste-to-energy strategies are being implemented for sustainable management of resources and available land (Ikhlayel, 2018).

6. Barriers to sustainable management

Although work is underway to develop and improve possible strategies for sustainable land management, many of them are not easy to implement due to the limited amount of land available and the variety of uses to which it is put. In this sense, there are different technological, ecological, institutional, economic, and sociocultural factors that act as barriers when adopting sustainable land use management practices (Sanz et al., 2017; Ochoa et al., 2018; ELD, 2019; Tafazzoli et al., 2019). Lack of access to appropriate technologies and knowledge regarding sustainable land management practices are among the main constraints (Sanz et al., 2017). For example, in the case of agriculture, farmers' advisory services often focus mainly on achieving short-term benefits, to the detriment of sustainable land management, contributing to misinformation for farmers (Aznar-Sánchez et al., 2019). In the case of urbanization, many times the growth of population centers has been carried out without proper planning or foresight. This has produced management and supply problems due to a lack of infrastructure and action plans (De la Barrera et al., 2016).

On the other hand, despite the existence of an extensive literature, there is no common theoretical frame of reference at the international level regarding sustainable land management (Tafazzoli et al., 2019). This occurs mainly in urbanization, which is presented as a more recent phenomenon than agricultural expansion. It is estimated that most of the world's population will be concentrated in cities in 2050 (UN, 2019). The lack of frames of reference in this area gives rise to interpretations and errors when planning solutions to the challenges posed by this massification of the population. Therefore, the development of guidelines for action at a global level and the establishment of minimum requirements to be considered could represent a turning point in achieving better land management. However, this is complex because the biophysical and sociocultural characteristics of a particular area largely determine the best practices to be implemented.

It should be noted that strategies successfully implemented in one area do not have to work in the same way in a different area (Sanz et al., 2017). This fact is presented as a barrier since determining the best practices for a specific area involves a great effort derived from the need to collect a large amount of data, which can be very expensive both in terms of money and time. On the other hand, it must be taken into account that the implementation of a sustainable land management practice can have positive effects in an area, but influence areas at different levels (local, regional, or global) in a negative way (Ochoa et al., 2018). The complexity of studying these effects at a global level can also represent an obstacle when implementing land management strategies.

The implementation of land management strategies will require the establishment of a series of indicators to determine the results of land management (Tafazzoli et al., 2019). However, the construction of indicators is a complex task as it requires an in-depth understanding of the characteristics and functions of land in the particular area in which they are applied. In addition, the wide variety of existing indicators and the disparity of these between regions, as well as the difficulty of measuring some of them, can be a limitation when evaluating the achievement of the proposed objectives and, therefore, the implementation of sustainable land management strategies (De la Barrera et al., 2016).

At the institutional and government level, barriers can also arise from a variety of factors. In a large number of countries and regions, especially in less developed areas, there may be a lack of zeal in the management of resources by the competent authority. On the other hand, there is also often a lack of adequate regulation or a lack of concrete action plans to achieve long-term sustainability. Finally, the absence of coordination at different administrative levels can lead to contradictory measures at the regional and national levels, or to inaction due to a lack of definition of competencies (Sanz et al., 2017).

On the other hand, sustainable practices in relation to land, although they have environmental advantages, often do not bring immediate economic benefits, which limit their adoption by users (ELD, 2019). This is very common in the case of farmers, where low incomes and the lack of financing schemes or concrete support limit the adoption of land-friendly production systems. In the sociocultural field there can also be barriers to the implementation of different strategies due to the lack of awareness of users, habits, or social norms in a specific area (Liu et al., 2018).

7. Future lines of research

Firstly, it is of the utmost importance to analyze how to promote international cooperation in the field of sustainable land management (Xie et al., 2020b). Such cooperation involves global land management planning, equitable distribution of functions for the proper functioning of the planet, or adequate compensation in case of deviations. An example of deviation is the Amazon. This large forest mass plays a vital role for the whole planet, but its location imposes management responsibility on a group of countries. Although there are already international programs and initiatives working toward this goal, in many cases research is limited in space or time (Buizer et al., 2014; Adeyemi et al., 2017).

To facilitate this cooperation it will be necessary to unify the basic concepts in relation to sustainable land management, to specifically identify the key points for carrying out action plans, and to have appropriate metrics to measure the level of achievement of the objectives according to the characteristics of the area in which they are implemented. Factors such as climate change require large-scale analysis of time series that either do not exist or are not accessible. Therefore, unified and updated data collection systems are needed, for which information collection mechanisms based on remote sensing and Geographic Information System technologies will have to be perfected and developed (Dubovyk, 2017; Ranjan et al., 2016; Yao et al., 2017).

Although there is an extensive literature focusing on land use, land change, and sustainable regional development, most of it is focused on the environmental field (Wu, 2008; Carleton and Hsiang, 2016). There is a lack of studies in relation to the other areas of sustainability and, especially, studies that analyze the three areas together. Given the close relationship between these three areas and the influence of any measure in one of them on the other two, there is a need for a greater level of knowledge about synergy and trade-off. The information collected through these studies can be a turning point in ensuring strong sustainable development. For example, avoiding endangering the survival of economic activities in an area by proposing new activities based on more sustainable land management programs (Yao et al., 2017). In this sense, the development of studies that determine in a more concrete way the relationships, whether direct or indirect, between land use and management and the SDGs would be a great advance for development at a global level and especially in the poorest countries (Smith, 2018).

The correct planning and distribution of the land among the different uses that are made of it also requires a more in-depth investigation. In this sense, it is urgent to study specific problems and challenges such as food security, urban waste management, or landscape preservation. Land constraints mean that the increase in food production resulting from an increase in the area of land devoted to agriculture is unsustainable. Therefore, ensuring the provision of food to the current and future population will require the development of high-yield systems that are environmentally friendly and that minimize problems derived from agricultural activity such as water pollution from excessive use of fertilizers or high salinity of farmland that reduces its productivity (Foley et al., 2005; Aznar-Sánchez et al., 2019; Froná et al., 2019).

On the other hand, the search for mechanisms to promote sustainable development in cities must be expanded, considering economic development and social equity, and the links between urbanization and environmental impacts (Kauko et al., 2015; Bai et al., 2017). It is essential to study how changes in land use can affect landscape ecology, as changing land use can have significant effects on biodiversity. For example, the construction of structures that may act as barriers to the free movement of certain species may endanger their survival by preventing the genetic exchange necessary for their reproduction (Christensen et al., 2017; Gergel and Turner, 2017). It is therefore essential to develop studies in this area from a holistic point of view that will make it possible to determine the optimum distribution among the different land uses, taking into account the specific characteristics of the area and the impacts that may be generated at other levels.

To achieve long-term sustainable development, it will be necessary to take into account the objectives and preferences of different stakeholders when planning and conducting research, as this will allow for more comprehensive studies with a higher level of acceptance and practical application (UNCCD, 2015; Liu et al., 2018). Furthermore, the development of sustainable land management plans and the implementation of the practices covered by these plans will require the correct transmission of knowledge and information to the people involved (OECD, 2020). For example, if we want farmers to start using a sustainable land management practice, they will need to be informed about it and training programs adapted to their socioeconomic characteristics will have to be carried out (Branca et al., 2013; Buizer et al., 2014; IPCC, 2019).

8. Conclusions

As has been shown throughout this chapter, land is a fundamental resource for development, which is being compromised by various factors such as human action or the consequences of climate change. The multiple uses of this resource and the degradation process to which a large part of the available land is subjected are increasing competition for it. The combination of these factors threatens the long-term sustainability of society. Changes in land use, climate change, and land degradation are interrelated factors that can feed back into each other. Furthermore, their effects cause society to face a limitation in its capacity to generate livelihood resources. For example, deforestation processes to expand agricultural or urban areas result in a loss of carbon from the land and the vegetation on which it is stored, which in turn drives global warming.

Therefore, it is necessary to incorporate sustainable practices of land use and management that allow us to cover our current needs, as well as those of future generations. However, incorporating plans and actions to achieve sustainable land management is not a simple task because it requires a thorough knowledge of the current state of the land and the effects that these practices can generate, which will vary depending on the characteristics of each area. This highlights the need for further research in relation to sustainable land use and management, especially at a global level and considering the economic, social, and environmental fields, as well as the different uses made of the land. Such research will be of great value to governments at all levels of management in carrying out action plans in this area through which they can promote the implementation of practices that ensure the long-term sustainability of land.

Acknowledgments

This work was partially supported by the Spanish Ministry of Economy and Competitiveness and the European Regional Development Fund by means of the research project ECO2017-82347-P, and by the Research Plan of the University of Almería through a Gerty Cori Predoctoral Contract to Belén López Felices.

References

Abi-Samra, N.C., Malcolm, W.P., 2011. Extreme weather effects on power systems. In: Proceedings of the 2010 IEEE Power and Energy Society General Meeting. Detroit, MI, USA, pp. 24–29. https://doi.org/10.1109/PES.2011.6039594.

Acevedo, M.F., 2011. Interdisciplinary progress in food production, food security and environment research. Environ. Conserv. 38, 151–171. https://doi.org/10.1017/S0376892911000257.

Adeyemi, O., Grove, I., Peets, S., Norton, T., 2017. Advanced monitoring and management systems for improving sustainability in precision irrigation. Sustainability 9, 353. https://doi.org/10.3390/su9030353.

Aznar-Sánchez, J.A., Belmonte-Ureña, L.J., López-Serrano, M.J., Velasco-Muñoz, J.F., 2018. Forest ecosystem services: an analysis of worldwide research. Forests 9, 453. https://doi.org/10.3390/f9080453.

Aznar-Sánchez, J.A., Piquer-Rodríguez, M., Velasco-Muñoz, J.F., Manzano-Agugliaro, F., 2019. Worldwide research trends on sustainable land use in agriculture. Land Use Pol. 87, 104069. https://doi.org/10.1016/j.landusepol.2019.104069.

Bai, X., McPhearson, T., Cleugh, H., Nagendra, H., Tong, X., Zhu, T., Zhu, Y.G., 2017. Linking urbanization and the environment: conceptual and empirical advances. Annu. Rev. Environ. Resour. 42, 215–240. https://doi.org/10.1146/annurev-environ-102016-061128.

Branca, G., Lipper, L., McCarthy, N., Jolejole, M.C., 2013. Food security, climate change, and sustainable land management. A review. Agron. Sustain. Dev. 33, 635–650. https://doi.org/10.1007/s13593-013-0133-1.

Brundtland, G., Khalid, M., Agnelli, S., Al-Athel, S., Chidzero, B., Fadika, L., Hauff, V., Lang, I., Shijun, M., Okita, S., et al., 1987. Our Common Future ('Brundtland Report'). Oxford University Press, Oxford, UK, ISBN 019282080X.

Buizer, M., Humphreys, D., de Jong, W., 2014. Climate change and deforestation: the evolution of an intersecting policy domain. Environ. Sci. Pol. 35, 1–11. https://doi.org/10.1016/j.envsci.2013.06.001.

Bunning, S., De Pauw, E., 2017. Land Resource Planning for Sustainable Land Management. Eng No. 14; Land and Water Division Working Paper. FAO, Rome, Italy.

Calvo-Buendia, E., Tanabe, K., Kranjc, A., Baasansuren, J., Fukuda, M., Ngarize, S., Osako, A., Pyrozhenko, Y., Shermanau, P., Federici, S. (Eds.), 2019. Refinement to the 2006 IPCC Guidelines for National Greenhouse Gas Inventories. IPCC, Switzerland. Available online: http://www.ipcc-nggip.iges.or.jp/.

Carleton, T.A., Hsiang, S.M., 2016. Social and economic impacts of climate. Science 353, 6304. https://doi.org/10.1126/science.aad9837 aad9837.

Ciais, P., Sabine, C., Bala, G., Bopp, L., Brovkin, V., Canadell, J., Chhabra, A., DeFries, R., Galloway, J., Heimann, M., Jones, C., Le Quéré, C., Myneni, R.B., Piao, S., Thornton, P., 2013. Carbon and other biogeochemical cycles. In: Stocker, T.F., Qin, D., Plattner, G.-K., Tignor, M., Allen, S.K., Boschung, J., Nauels, A., Xia, Y., Bex, V., Midgley, P.M. (Eds.), Climate Change 2013: The Physical Science Basis. Contribution of Working Group I to the Fifth Assessment Report of the Intergovernmental Panel on Climate Change. Cambridge University Press, Cambridge, United Kingdom and New York, NY, USA, pp. 465–570.

Cihlar, J., Jansen, L.J.M., 2001. From land cover to land use: a methodology for efficient land use mapping over large areas. Prof. Geogr. 53, 275–289. https://doi.org/10.1111/0033-0124.00285.

Chatzimentor, A., Apostolopoulou, E., Mazaris, A., 2020. A review of green infrastructure research in Europe: challenges and opportunities. Landsc. Urban Plann. 198 https://doi.org/10.1016/j.landurbplan.2020.103775.

Christensen, A.A., Brandt, J., Svenningsen, S.R., 2017. Landscape Ecology. Int. Encycl. Geogr. People Earth Environ. & Technol. https://doi.org/10.1002/9781118786352.wbieg1168. Wiley.

Dadashpoor, H., Azizi, P., Moghadasi, M., 2019. Land use change, urbanization, and change in landscape pattern in a metropolitan area. Sci. Total Environ. 655, 707–719. https://doi.org/10.1016/j.scitotenv.2018.11.267.

Daliakopoulos, I.N., Tsanis, I.K., Koutroulis, A.G., Kourgialas, N., Varouchakis, E.A., Karatzas, G.P., Ritsema, C.J., 2016. The threat of soil salinity: a European scale review. Sci. Total Environ. 573, 727–739. https://doi.org/10.1016/j.scitotenv.2016.08.177.

De la Barrera, F., Reyes-Paecke, S., Banzhaf, E., 2016. Indicators for green spaces in contrasting urban settings. Ecol. Indicat. 62, 212–219. https://doi.org/10.1016/j.ecolind.2015.10.027.

Díaz, S., Settele, J., Brondízio, E., Ngo, H., Guèze, M., Agard, J., Arneth, A., Balvanera, P., Brauman, K., Butchart, S., et al., 2019. Summary for Policymakers of the Global Assessment Report on Biodiversity and Ecosystem Services of the Intergovernmental Science-Policy Platform on Biodiversity and Ecosystem Services. IPBES, Bonn, Germany.

Di Gregorio, A., Jansen, L.J.M., 1998. A new concept for a land cover classification system. Land 2, 55–65.

Dubovyk, O., 2017. The role of remote sensing in land degradation assessments: opportunities and challenges. Eur. J. Remote Sens. 50, 601–613. https://doi.org/10.1080/22797254.2017.1378926.

ELD Initiative, 2019. ELD Campus. Module: Land Degradation Versus Sustainable Land Management. 1st ed. 1. German Federal Ministry for Economic Cooperation and Development, pp. 1–32. Available from: http://www.eld-initiative.org.

Ellis, E.C., Ramankutty, N., 2008. Putting people in the map: anthropogenic biomes of the world. Front. Ecol. Environ. 6, 439–447. https://doi.org/10.1890/070062.

Ellis, E.C., 2011. Anthropogenic transformation of the terrestrial biosphere. Phil. Trans. Roy. Soc. Lond. 369, 1010–1035. https://doi.org/10.1098/rsta.2010.0331.

Erb, K.-H., Luyssaert, S., Meyfroidt, P., Pongratz, J., Don, A., Kloster, S., Kuemmerle, T., Fetzel, T., Fuchs, R., Herold, M., Haberl, H., Jones, C.D., Marín-Spiotta, E., McCallum, I., Robertson, E., Seufert, V., Fritz, S., Valade, A., Wiltshire, A., Dolman, A.J., 2016. Land management: data availability and process understanding for global change studies. Glob. Chang. Biol. 23, 512–533. https://doi.org/10.1111/gcb.13443.

Farooq, M., Rehman, A., Pisante, M., 2019. Sustainable agriculture and food security. In: Farooq, M., Pisante, M. (Eds.), Innovations in Sustainable Agriculture. Springer, Cham, Switzerland. https://doi.org/10.1007/978-3-030-23169-9_1.

Fernandes, E.C., Burcroff, R., 2006. Sustainable Land Management: Challenges, Opportunities, and Trade-Offs. World Bank, Washington, DC, USA, ISBN 978-0-8213-6597-7.

Foley, J.A., DeFries, R., Asner, G.P., Barford, C., Bonan, G., Carpenter, S.R., Chapin, F.S., Coe, M.T., Daily, G.C., Gibbs, H.K., et al., 2005. Global consequences of land use. Science 309, 570–574. https://doi.org/10.1126/science.1111772.

Food and Agriculture Organization (FAO), 1993. Guidelines for Land-Use Planning. Development Documents Series. Food and Agriculture Organization of the United Nations. http://www.fao.org/docrep/t0715e/t0715e00.htm.

Food and Agriculture Organization (FAO) and United Nations Environment Programme (UNEP), 1999. The Future of Our Land: Facing the Challenge. FAO and UNEP, Rome, Italy. http://www.fao.org/docrep/004/x3810e/x3810e00.htm.

Food and Agriculture Organization (FAO), 2015. Soils Store and Filter Water - Improving Food Security and Our Resilience to Floods and Droughts. Retrieved at: http://www.fao.org/documents/card/en/c/ab8d051e-af46-430c-9fa1-616e1347cba6/ (Accessed 3 July 2020).

Food and Agriculture Organization (FAO), 2018. The Future of Food and Agriculture – Alternative Pathways to 2050. Roma. http://www.fao.org/3/I8429EN/i8429en.pdf.

Food and Agriculture Organization (FAO), United Nations Environment Programme (UNEP), 2020. The State of the World's Forests 2020. Forests, Biodiversity and People. https://doi.org/10.4060/ca8642en. Rome.

Fróna, D., Szenderák, J., Harangi-Rákos, M., 2019. The challenge of feeding the world. Sustainability 11, 5816. https://doi.org/10.3390/su11205816.

Garnett, T., Appleby, M.C., Balmford, A., Bateman, I.J., Benton, T.G., Bloomer, P., Burlingame, B., Dawkins, M., Dolan, L., Fraser, D., Herrero, M., Hoffmann, I., Smith, P., Thornton, P.K., Toulmin, C., Vermeulen, S.J., Godfray, H.C.J., 2013. Sustainable intensification in agriculture: premises and policies. Sci. Magna 341, 33–34. https://doi.org/10.1126/science.1234485.

Gauthier, S., Bernier, P., Kuuluvainen, T., Shvidenko, A.Z., Schepaschenko, D.G., 2015. Boreal forest health and global change. Science 349, 819–822. https://doi.org/10.1126/science.aaa9092.

Gergel, S.E., Turner, M.G., 2017. Learning Landscape Ecology: A Practical Guide to Concepts and Techniques. Springer, New York, NY, USA.

He, C., Han, Q., de Vries, B., Wang, X., Guochao, Z., 2017. Evaluation of sustainable land management in urban area: a case study of Shanghai, China. Ecol. Indicat. 80, 106–113. https://doi.org/10.1016/j.ecolind.2017.05.008.

Hoornweg, D., Bhada-Tata, P., Joshi-Ghani, A., 2009. What a Waste: A Global Review of Solid Waste Management. Urban Development and Local Government Unit. World Bank, Washington, DC, USA.

Hu, X., Lu, L., Li, X., Wang, J., Guo, M., 2015. Land use/cover change in the middle reaches of the Heihe River Basin over 2000–2011 and its implications for sustainable water resource management. PLoS One 10. https://doi.org/10.1371/journal.pone.0128960.

Huston, M.A., 2005. The three phases of land-use change: implications for biodiversity. Ecol. Appl. 15 (6), 1864–1878. https://doi.org/10.1890/03-5281.

Ikhlayel, M., 2018. Development of management systems for sustainable municipal solid waste in developing countries: a systematic life cycle thinking approach. J. Clean. Prod. 180, 571–586. https://doi.org/10.1016/j.jclepro.2018.01.057.

IPCC, 2019. In: Shukla, P.R., Skea, J., Calvo Buendia, E., Masson-Delmotte, V., Pörtner, H.-O., Roberts, D.C., Zhai, P., Slade, R., Connors, S., van Diemen, R., Ferrat, M., Haughey, E., Luz, S., Neogi, S., Pathak, M., Petzold, J., Portugal Pereira, J., Vyas, P., Huntley, E., Kissick, K., Belkacemi, M., Malley, J. (Eds.), Climate Change and Land: An IPCC Special Report on Climate Change, Desertification, Land Degradation, Sustainable Land Management, Food Security, and Greenhouse Gas Fluxes in Terrestrial Ecosystems (in press).

Kauko, T., Siniak, N., Źróbek, S., 2015. Sustainable land development in an urban context. J. Tow. Nauk. Nieruchom 23, 110–119. https://doi.org/10.1515/remav-2015-0030.

Kruize, H., van der Vliet, N., Staatsen, B., Bell, R., Chiabai, A., Muiños, G., Higgins, S., Quiroga, S., Martinez-Juarez, P., Aberg Yngwe, M., Tsichlas, F., Karnaki, P., Lima, M.L., García de Jalón, S., Khan, M., Morris, G., Stegeman, I., 2019. Urban green space: creating a triple win for environmental sustainability, health, and health equity through behavior change. Int. J. Environ. Res. Publ. Health 16, 4403. https://doi.org/10.3390/ijerph16224403.

Lal, R., 2004. Soil carbon sequestration impacts on global climate change and food security. Science 304, 1623–1627. https://doi.org/10.1126/science.1097396.

Ligmann-Zielinska, A., Church Richard, L., Jankowski, P., 2008. Spatial optimization as a generative technique for sustainable multiobjective land-use allocation. Int. J. Geogr. Inf. Sci. 22, 601–622. https://doi.org/10.1080/13658810701587495.

Liu, T., Bruins, R.J., Heberling, M.T., 2018. Factors influencing farmers' adoption of best management practices: a review and synthesis. Sustainability 10, 432. https://doi.org/10.3390/su10020432.

Malhi, Y., Franklin, J., Seddon, N., Solan, M., Turner, M.G., Field, C.B., Knowlton, N., 2020. Climate change and ecosystems: threats, opportunities and solutions. Philos. Trans. R. Soc. B Biol. Sci. 375 https://doi.org/10.1098/rstb.2019.0104.

Martínez, S., Mollicone, D., 2012. From land cover to land use: a methodology to assess land use from remote sensing data. Rem. Sens. 4, 1024–1045. https://doi.org/10.3390/rs4041024.

Millennium Ecosystem Assessment (MEA), 2005. Ecosystems and Human Well-Being: Current State and Trends. Island Press, Washington, DC.

Mirzabaev, A., Nkonya, E., von Braun, J., 2015. Economics of sustainable land management. Curr. Opin. Environ. Sustain. 15, 9–19. https://doi.org/10.1016/j.cosust.2015.07.004.

Mohajerani, A., Bakaric, J., Jeffrey-Bailey, T., 2017. The urban heat island effect, its causes, and mitigation, with reference to the thermal properties of asphalt concrete. J. Environ. Manag. 197, 522–538. https://doi.org/10.1016/j.jenvman.2017.03.095.

Musakwa, W., Van Niekerk, A., 2013. Implications of land use change for the sustainability of urban areas: a case study of Stellenbosch, South Africa. Cities 32, 143–156. https://doi.org/10.1016/j.cities.2013.01.004.

Nkonya, E., von Braun, J., Mirzabaev, A., Le, Q.B., Kwon, H.Y., Kirui, O., 2013. Economics of Land Degradation Initiative: Methods and Approach for Global and National Assessments. ZEF - Discussion Papers on Development Policy No. 183, 1. SSRN. Available at: https://papers.ssrn.com/sol3/papers.cfm?abstract_id=2343636.

Ochoa, J.J., Tan, Y., Qian, Q.K., Shen, L., Moreno, E.L., 2018. Learning from best practices in sustainable urbanization. Habitat Int. 78, 83–95. https://doi.org/10.1016/j.habitatint.2018.05.013.

Organisation for Economic Co-operation and Development (OECD), 2020. Towards Sustainable Land Use: Aligning Biodiversity, Climate and Food Policies. OECD Publishing, Paris. https://doi.org/10.1787/3809b6a1-en.

Pacheco, F.A.L., Sanches Fernandes, L.F., Valle Junior, R.F., Valera, C.A., Pissarra, T.C.T., 2018. Land degradation: multiple environmental consequences and routes to neutrality. Curr. Opin. Environ. Sci. Health 5, 79–86. https://doi.org/10.1016/j.coesh.2018.07.002.

Pandey, P.C., Sharma, L.K., Nathawat, M.S., 2012. Geospatial strategy for sustainable management of municipal solid waste for growing urban environment. Environ. Monit. Assess. 184, 2419–2431. https://doi.org/10.1007/s10661-011-2127-2.

Pradhan, P., Costa, L., Rybski, D., Lucht, W., Kropp, J.P., 2017. A systematic study of sustainable development goal (SDG) interactions. Earth's Future 5, 1169–1179. https://doi.org/10.1002/2017EF000632.

Pokhrel, Y.N., Felfelani, F., Shin, S., Yamada, T.J., Satoh, Y., 2017. Modeling large-scale human alteration of land surface hydrology and climate. Geosci. Lett. 4, 10. https://doi.org/10.1186/s40562-017-0076-5.

Potts, S.G., Imperatriz-Fonseca, V., Ngo, H.T., Aizen, M.A., Biesmeijer, J.C., Breeze, T.D., Dicks, L.V., Garibaldi, L.A., Hill, R., Settele, J., et al., 2016. Safeguarding pollinators and their values to human well-being. Nature 540, 220–229. https://doi.org/10.1038/nature20588.

Ranjan, A.K., Vallisree, S., Singh, R.K., 2016. Role of geographic information system and remote sensing in monitoring and management of urban and watershed environment: overview. J. Remote Sens. GIS 7, 1–14.

Ritchie, H., Roser, M., 2019. Land Use. In Our World in Data. Available online: https://ourworldindata.org/land-use. (Accessed 28 June 2020).

Sanz, M.J., de Vente, J., Chotte, J.-L., Bernoux, M., Kust, G., Ruiz, I., Almagro, M., Alloza, J.-A., Vallejo, R., Castillo, V., Hebel, A., Akhtar-Schuster, M., 2017. Sustainable land management contribution to successful land-based climate change adaptation and mitigation. In: A Report of the Science-Policy Interface. United Nations Convention to Combat Desertification (UNCCD), Bonn, Germany.

Satterthwaite, D., McGranahan, G., Tacoli, C.M., 2010. Urbanization and its implication for food and farming. Philos. Trans. R. Soc. Lond. B Biol. Sci. 365, 2809–2820. https://doi.org/10.1098/rstb.2010.0136.

Seppelt, R., Verburg, P., Norström, A., Cramer, W., Vaclavik, T., 2018. Focus on cross-scale feedbacks in global sustainable land management. Environ. Res. Lett. 13, 9. https://doi.org/10.1088/1748-9326/aadc45.

Shuvo, F.K., Feng, X., Akaraci, S., Astell-Burt, T., 2020. Urban green space and health in low and middle-income countries: a critical review. Urban For. Urban Green. 52, 126662. https://doi.org/10.1016/j.ufug.2020.126662.

Sivakumar, M., Stefanski, R., 2007. Climate and Land Degradation. Springer, Berlin.

Smith, P., 2018. Managing the global land resource. Proc. R. Soc. B Biol. Sci. 285 https://doi.org/10.1098/rspb.2017.2798.

Tafazzoli, M., Ashkan, N., Ali, K., 2019. Investigating barriers to sustainable urbanization. In: Proceedings of the International Conference on Sustainable Infrastructure, Los Angeles, CA, USA, 7–9 November.

Theobald, D.M., 2014. Development and applications of a comprehensive land use classification and map for the US. PLoS One 9, e94628. https://doi.org/10.1371/journal.pone.0094628.
Tubiello, F.N., Salvatore, M., Ferrara, A.F., House, J., Federici, S., Rossi, S., Biancalani, R., Condor Golec, R.D., Jacobs, H., Flammini, A., Prosperi, P., Cardenas-Galindo, P., Schmidhuber, J., Sanz Sanchez, M.J., Srivastava, N., Smith, P., 2015. The contribution of agriculture, forestry and other land use activities to global warming, 1990–2012. Glob. Chang. Biol. 21, 2655–2660. https://doi.org/10.1111/gcb.12865.
Turner, K.G., Anderson, S., Gonzales-Chang, M., Costanza, R., Courville, S., Dalgaard, T., Dominati, E., Kubiszewski, I., Ogilvy, S., Porfirio, L., 2016. A review of methods, data, and models to assess changes in the value of ecosystem services from land degradation and restoration. Ecol. Model. 319, 190–207. https://doi.org/10.1016/j.ecolmodel.2015.07.017.
United Nations (UN), 2008. General Assembly 2007: Non-legally Binding Instrument on All Types of Forests (A/RES/62/98). United Nations, Bali, Indonesia.
United Nations (UN), 2015. General Assembly. Resolution Adopted by the General Assembly on 25 September 2015: 70/1. Transforming Our World: The 2030 Agenda for Sustainable Development. United Nations, New York, NY, USA.
United Nation (UN), 2019. Department of economic and social affairs, population division. In: World Urbanization Prospects: The 2018 Revision. United Nations, New York.
United Nations Convention to Combat Desertification (UNCCD), 2015. Climate change and land degradation: bridging knowledge and stakeholders. In: Outcomes from the UNCCD 3rd Scientific Conference, Cancún, Mexico. http://www.unccd.int/Lists/SiteDocumentLibrary/Publications/2015_Climate_LD_Outcomes_CST_Conf_ENG.pdf.
United Nations World Water Assessment Programme/UN-Water (WWAP), 2018. The United Nations World Water Development Report 2018: Nature-Based Solutions for Water. UNESCO, Paris, France. https://www.unwater.org/publications/world-water-development-report-2018/.
Vié, J.-C., Hilton-Taylor, C., Stuart, S.N., 2009. Wildlife in a Changing World: An Analysis of the 2008 IUCN Red List of Threatened Species. UICN, Gland, Suiza.
von Braun, J., Gerber, N., Mirzabaev, A., Nkonya, E., 2013. The Economics of Land Degradation. ZEF Working Paper 109. Center for Development Research, University of Bonn.
Weiskopf, S.R., Rubenstein, M.A., Crozier, L.G., Gaichas, S., Griffis, R., Halofsky, J.E., Hyde, K.J.W., Morelli, T.L., Morisette, J.T., Muñoz, R.C., Pershing, A.J., Peterson, D.L., Poudel, R., Staudinger, M.D., Sutton-Grier, A.E., Thompson, L., Vose, J., Weltzin, J.F., Whyte, K.P., 2020. Climate change effects on biodiversity, ecosystems, ecosystem services, and natural resource management in the United States. Sci. Total Environ. 733, 137782. https://doi.org/10.1016/j.scitotenv.2020.137782.
World Health Organization (WHO), 2014. Quantitative Risk Assessment of the Effects of Climate Change on Selected Causes of Death, 2030s and 2050s. Geneva, Switzerland. https://apps.who.int/iris/handle/10665/134014.
World Health Organization (WHO), 2017. Urban Green Spaces: A Brief for Action. Regional Office for Europe, Copenhagen, Denmark.
Wu, J., 2008. Land use changes: economic, social, and environmental impacts. Choice 23, 6–10.
Xia, Y., Zhang, M., Tsang, D.C.W., Geng, N., Lu, D., Zhu, L., Igalavithana, A.D., Dissanayake, P.D., Rinklebe, J., Yang, X., 2020. Recent advances in control technologies for non-point source pollution with nitrogen and phosphorous from agricultural runoff: current practices and future prospects. Appl. Biol. Chem. 63, 1–13. https://doi.org/10.1186/s13765-020-0493-6.

Xie, H., Zhang, Y., Wu, Z., Lv, T., 2020a. A bibliometric analysis on land degradation: current status, development, and future directions. Land 9, 28. https://doi.org/10.3390/land9010028.

Xie, H., Zhang, Y., Zeng, X., He, Y., 2020b. Sustainable land use and management research: a scientometric review. Landsc. Ecol. 35 (2) https://doi.org/10.1007/s10980-020-01002-y.

Yao, J., Zhang, X., Murray, A.T., 2017. Spatial optimization for land-use allocation: accounting for sustainability concerns. Int. Reg. Sci. Rev. 41, 569–600. https://doi.org/10.1177/0160017617728551.

Solid wastes: alternative materials for cementitious composites production

Shaswat Kumar Das [1,2,a], *Syed Mohammed Mustakim* [2,a], *Adeyemi Adesina* [3], *Subhabrata Mishra* [1], *Jyotirmoy Mishra* [4]

[1]Grøn Tek Concrete and Research, Bhubaneswar, Odisha, India; [2]E&S Department, CSIR-Institute of Minerals and Materials Technology, Bhubaneswar, Odisha, India; [3]Department of Civil and Environmental Engineering, University of Windsor, Windsor, ON, Canada; [4]Department of Civil Engineering, VSSUT, Burla, Odisha, India

1. Introduction

Cementitious composites are one of the fundamental materials required for the construction of our infrastructure. Compared to other types of construction materials, cementitious composites are versatile, possess outstanding durability and mechanical performance, and the materials used for their production are locally available all over the world. However, the increase in population alongside rapid urbanization has resulted in the alarming consumption of these composites. The high consumption of cementitious composites has yielded high consumption of natural resources coupled with a significant negative effect on the environment. For example, the Portland cement (PC) which is the primary binder used in cementitious composites is responsible for more than 5% of the world's human-induced carbon emission (Andrew, 2018; Flower and Sanjayan, 2007). Apart from the negative impact of the high consumption of cementitious composites, there is also a possible scarcity of raw materials for the production of cementitious composites in the nearest future if alternative materials are not sourced.

On the other hand, the world's population increase coupled with the expanding urbanization has also resulted in a high amount of various solid wastes generated by various processes (Palit and Hussain, 2018; Hussain and Keçili, 2020; Palit and Hussain, 2018; Palit and Hussain, 2020). The high quantities of these wastes have resulted in their improper disposal and they mostly end up as contaminants in the environment. Some of these wastes can be processed and used as a cementitious component to produce composites. The use of these wastes to produce cementitious composites would help to supplement the current reserve of raw materials used and would also create a way to effectively manage these wastes. Hence, this aim of this chapter is to explore solid wastes from various sources that can be incorporated into cementitious composites.

[a] Fly Ash Demonstration Area, IMMT, Sachivalaya Marg, RRL Campus, Acharya Vihar, Bhubaneswar, Odisha, India.

Sustainable Resource Management. https://doi.org/10.1016/B978-0-12-824342-8.00012-2

2. Wastes used in cementitious composites

Solid wastes can be incorporated into cementitious composites as the binder and/or aggregate component. The choice of the use of the waste material as components in cementitious composites is dependent on the physical and chemical properties of the wastes. Generally, in order to be able to use solid wastes as a binder component, it has to be processed to very fine size and possess pozzolanic properties that would contribute to the performance of the composites. Some of the wastes that can be used as a binder component in cementitious composites are fly ash (FA), slag, rice husk ash (RHA), etc.

Wastes with higher particle size can be used as either coarse or fine aggregate in cementitious composites where they act as fillers. Solid wastes such as plastics, ceramics, etc., can be recycled and used as aggregates in cementitious composites. Fig. 9.1 shows a summary of the sources of solid wastes and the components of cementitious composites they can be used for. The types of solid wastes used in cementitious composites are further discussed and classified based on the source.

2.1 Industrial solid wastes

2.1.1 Fly ash

The FA is the major waste generated from the thermal power plants. Relative to the volume of other waste generation from during the power generation from thermal power plants, FA is more than 80% of the total wastes generated. After the generation of FA in the thermal power plants, it is being stored in a silo and then transported for safe

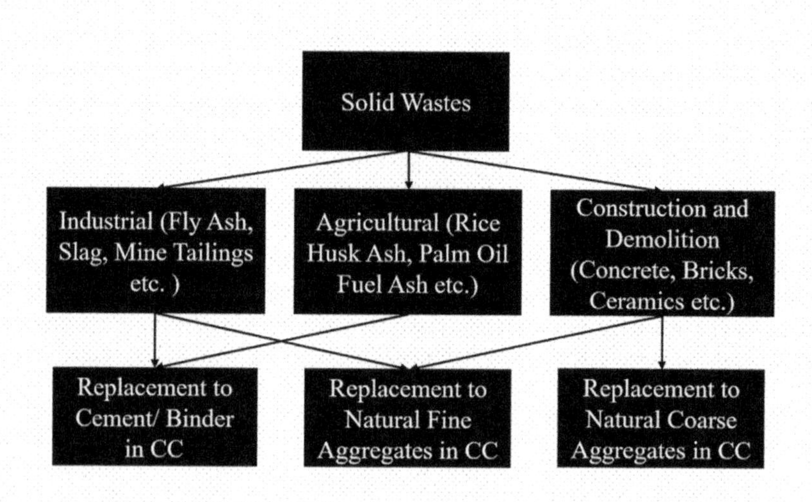

Figure 9.1 Source of solid wastes used in cementitious composites.

Figure 9.2 Transportation of FA from the thermal power plant to cement plants.
Picture was taken at National Thermal Power Plant (NTPC), Talcher, Odisha, India during a case study visit in 2019.

disposal or use in cement/construction/mining industries. A typical image of the FA transportation from a thermal power plant is represented in Fig. 9.2. FA is one of the major industrial wastes that has been used extensively in the concrete industry. The FA can be used either as partial replacement of PC (Guo and Pan, 2018; Li, 2004) or as a total replacement to produce alkali-activated materials (Mishra et al., 2020a,b; Das and Jyotirmoy Mishra, 2018; Mustakim et al., 2020; Das et al., 2020; Adesina, 2020).

The FA is one of the most used pozzolanic materials used in cementitious composites due to its availability and property enhancing traits. The suitable amount of both silica and alumina content of FA makes it appropriate for cement replacement. Several studies have been conducted using FA as a substitute to cement and also the replacement of clinker by FA in commercial cement production is widely accepted (Awoyera et al., 2019; Adesina and Awoyera, 2019; Poon et al., 2000; Dinakar et al., 2008; McCarthy and Dhir, 2005). As per the ASTM C 618 (ASTM C 618, 2010) classification, the FA is categorized into class-F and class-C based on the oxides of silicon, aluminum, and iron. FA containing a minimum 70% of the above oxides is considered as class-F, whereas the minimum value of 50% is the criteria for class-C. However, as per the ASTM requirement, the calcium content of class-F FA is below 10%, whereas class-C FA contains a higher value. The European standard EN 197-1 have also categorized the FA into two types, namely siliceous and calcareous (European Committee for Standartization, 2000). In contrast to ASTM C 618, this classification is primarily based on reactive calcium content. FA containing below 10% CaO falls under the siliceous category and above 10% is calcareous.

The similar chemistry of FA with PC and the reduction in heat of hydration resulting from the partial replacement of FA with PC has resulted in high interest by the engineers and scientists to utilize this waste as a binder component in various cementitious composites. Since the 1930s, the use of FA for major engineering application such as gravity dams has surged up, several laboratory research investigations were conducted to understand the material chemistry of FA and its effect in cement hydration. Davids et al. (1937) have given experimental evidence of the use of FA as a replacement of PC in cementitious composites. In the study, FA was incorporated up to 50% replacement of PC and several tests were conducted to evaluate the performance of the corresponding cementitious composites. Findings from this study showed that the optimum content of FA to replace PC in normal construction work is 30%. On the other hand, PC can be replaced by up to 50% with FA for heavy construction work (e.g., dams and harbors).

Earlier studies carried out by Smith (1967) also showed that the long-term performance of cementitious composites made with the optimum content of FA is similar to that of the conventional cementitious composite without FA. In general, several studies have shown that the optimum content of PC replaced with FA in cementitious composites is in the range of 30%−35% (Poon et al., 2000; Dinakar et al., 2008, 2013; McCarthy and Dhir, 2005). However, the use of FA as a replacement of higher content of PC in conventional cementitious composites would result in various detrimental effects. One of the major issues with high FA content in cementitious composites is the low early strength which is a result of the slow reaction mechanism. The use of high content of FA as a replacement of PC can also result in segregation issues during the placement of the fresh cementitious mixture. Hence, the commercial grade of FA-blended PC available in the market is manufactured by replacing 30%−33% of the clinker with FA.

Nevertheless, since the early 20th century, various research works have emerged where a high volume of FA (i.e., greater than 35%) is used as partial replacement of PC in cementitious composites (Bouzoubaâ et al., 2000, 2001). Cementitious composites made with a high content of FA as replacement of PC are generally referred to as high-FA cementitious composites. In high-FA cementitious composites, the aggregate composition is generally reduced to incorporate a large amount of FA into the composite thereby keeping the total volume of the composite. A typical model for the high-FA cementitious composite is provided in Fig. 9.3.

In recent years, there has been extensive use to produce geopolymer and alkali-activated composites. The high interest in these types of composites is due to the total elimination of PC as binders in the production of the composites. Palomo et al. (1999) used FA as an aluminosilicate precursor to produce composites with similar performance as that of the conventional composites. Several other studies have been carried out on the use of FA to produce these composites that totally eliminate PC as a binder (Mustakim et al., 2020; Prasanna et al., 2021; Ryu et al., 2013; Assaedi et al., 2020).

2.1.2 Slag

Slag is a by-product of metallurgical industries which is generated after the metal extraction process from the raw ore. The composition of slag varies significantly

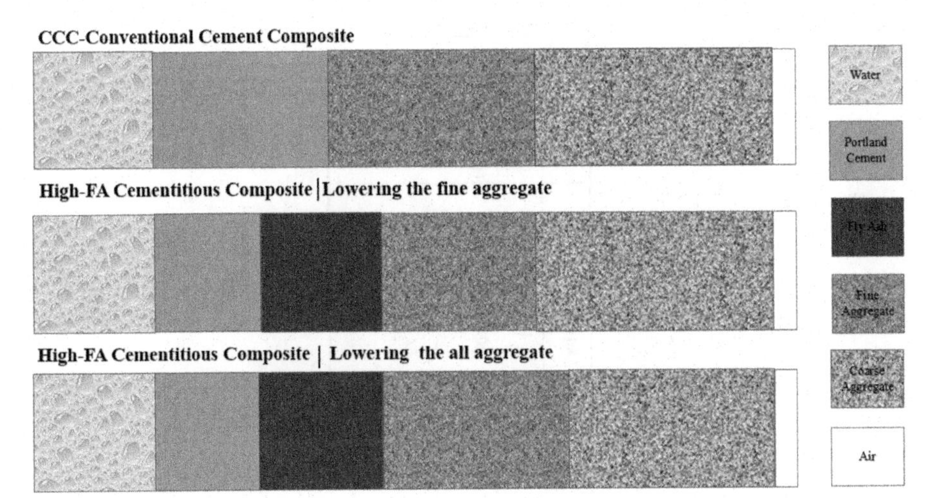

Figure 9.3 A model for high-FA cementitious composites.

compared to FA and it is dependent on the feed ore and process utilized for the separation operation. The major chemical composition of slag is oxides of silicon and other metal oxides. Depending on the type of metal and process involved, the slag can be classified. Types of slags include iron and steel slags, ferrochrome slag (FCS), ferromanganese slag (FMS), copper slag (CS), etc. The typical chemical compositions of the abovementioned slags are provided in Table 9.1.

The iron and steel industries alone contribute a major share in slag generation (Krishna et al., 2020), and the slag from these industries is classified into two categories, namely blast furnace slag (BFS) and steelmaking slag (SMS).

The BFS has been used extensively in the cement and construction industry as partial replacement of PC and there exist numerous studies on the performance evaluation of composites incorporating BFS. Whereas the SMS is not generally accepted for use in cementitious composites due to the volumetric expansion traits of SMS. The expansions in cementitious composites incorporating SMS is due to the hydration of free lime- or magnesia-forming phases like $Ca(OH)_2$ or $Mg(OH)_2$, or the phase change mechanism of dicalcium silicate, or a combination of both of the mechanisms (Al-Negheimish et al., 1997). Nonetheless, the SMS with desirable properties and limited expansion traits could be used as aggregates in cementitious composites.

The SMS has been utilized as aggregates to make asphalt concrete for pavement applications, and the results were very encouraging as far as the performance of the road pavement is concerned (Pasetto and Baldo, 2010). To promote the use of SMS in pavement construction, the ASTM International specified a standard code of practice (i.e., ASTM D 5106: Standard Specification for Steel Slag Aggregates for Bituminous Paving Mixtures (ASTM, 2015)). This standard propelled the wide acceptance of SMS as a construction material for pavements and hence opened a new way for sustainable disposal of large quantities of SMS.

Research incorporating SMS into cementitious composites is very limited. Findings from the study by Akinmusuru (1991) showed that SMS are preferably used as

Table 9.1 Typical chemical composition of different slags.

Slag	SiO_2	Al_2O_3	CaO	MgO	Cr_2O_3	MnO	Fe_2O_3
BFS	30%–40%	8%–15%	40%–50%	5%–10%	–	0.1%–1%	0.2%–3%
SMS	8%–30%	1%–8%	30%–50%	5%–15%	–	2%–8%	10%–30%
FCS	27%–30%	20%–30%	1%–10%	20%–30%	8%–15%	–	2%–5%
FMS	28%–30%	25%–30%	20%–30%	5%–6%	–	28%–30%	1%–2%
CS	25%–35%	0.2%–5%	0.2%–6%	–	–	–	50%–70%

aggregate in cementitious composites instead of the binder component. Martauz et al. (2017) investigated the possibility of use of SMS as aggregate in concrete, two types of cement has been taken in this, viz. PC and alkali-activated hybrid cement. The investigations were conducted using 100% SMS as aggregates in concrete, interestingly the mechanical performance of the concrete made from both of the cement types was identical but the durability performance (autoclave) for PC-SMS type concrete was comparatively better.

Several authors have discussed the advantages of using BFS as both major and minor precursors for geopolymer concrete (GPC). Rangan and Hardjito (2005) has discussed the usefulness of using BFS in achieving the strength of geopolymers at ambient temperature curing. The study showed that the calcium present in the BFS undergoes hydration and hence the heat generated from hydration helps in geopolymerization. Adesina and Das (2020); Awoyera and Adesina (2019); Ikponmwosa et al. (2020) extended their research in alkali-activated slag (AAS) using fibers and other additives, they suggested that AAS concrete is a viable replacement of PC as the binder in cementitious composites.

Das et al. (2020); Mishra et al. (2020a,b); Mustakim et al. (2020) also used BFS in a different type of GPC at it was suggested that the BFS could be used as a minor precursor in GPC taking FA or precursors. Bai et al. (2018) studied the effect of SMS on metakaolin-based GPC. The findings from the study showed that 10% of SMS in the GPC matrix was optimum as the final compressive strength was identical with the reference composite. It was also suggested that rather than the elevated or ambient curing, the moist curing should be employed in SMS-based GPC in order to facilitate the hydration of SMS resulting in good mechanical properties. However, even after such promising improvements in performances of cementitious composites by the slag of steel industries, a huge amount of slag is leftover and dumped in the open environment. In general, these dump yards consume a huge amount of land and hence causes great financial distress on the industrial economy, a typical image of a dump yard of a major steel industry (Steel Authority of India Limited (SAIL), Rourkela, Odisha, India) in India is presented in Fig. 9.4.

FCS is a major solid waste generated from electric arc furnaces during the production process of ferrochrome alloy (Panda et al., 2013). Generally for each tonne of ferrochrome alloy produced, about 1–1.2 tonnes of FCS were generated (Panda et al., 2013). FCS can be used in cementitious composites in place of both cement and aggregate. However, the current use of FCS as a binder component is very limited. Most of the current research utilized FCS as a coarse aggregate in cementitious composites such as concrete. The study by Zelić (2005) showed that the FCS has a very good mechanical strength similar to the basalt rock and could be used for high strength concrete applications. Panda et al. (2013) produced concrete with FCS as coarse and fine aggregate. The results from the study showed that concrete incorporating FCS showed excellent performance in terms of compressive strength and the viability to use the concrete to immobilize leachable chromium. Hence, such concrete is beneficial in terms of technical suitability and environmental compatibility. The study by Dash and Patro (2018) where water-cooled FCS was used as a replacement to fine aggregate up to 50% showed that the use of FCS as a 10% replacement of the natural aggregate is optimum.

Figure 9.4 Typical dump yard of a steel industry to dispose industrial slags.

Ongoing studies by the authors established that the FCS could be used as a replacement to PC in cementitious composites. The findings from the preliminary studies showed that the performance of the FCS-PC composite was better in terms of acid durability and water absorption compared to that of the conventional mixture. Similarly, the FCS has very limited research evidence as a precursor material in geopolymers. Karakoç et al. (2014) first reported that FCS could be used in geopolymers as the only binder. The FCS used in the Karakoç et al. (2014) study has a very low calcium content (typically FCS has low calcium content). Hence, ambient temperature curing of the resulting composites resulted in low compressive strength. A recent study by Nath (2018) proposed FCS-FA blends for geopolymeric binders. Similarly, Özcan and Karakoç (2019) recently reported that FCS incorporation in BFS-based geopolymer reduces the susceptibility of acid attack of the concrete when exposed to the acid environment, and also improves the residual compressive strength.

FMS is generally generated from the metallurgical process of making manganese series. For every tonne of alloys produced, an estimated 2.5 tonnes of FMS are generated (Rong-Jin et al., 2012). The use of FMS in cementitious composites as a replacement of PC has been reported by some studies. Rai et al. (2002) explored the possibility of the use of the two types of FMS in cementitious composites. The difference between the two types of FMS used is MnO content. The low MnO slag was found suitable as PC replacement up to 50% in accordance with the Indian standard of Portland slag cement IS 455:1989 (IS-455, 2015) requirements. It was recommended that the high MnO containing slag has deleterious effects on the performance of cementitious composites. Péra et al. (1999) also studied the influence of BFS with high MnO content which is similar to that of FMS on the performance of cementitious composites. The effect of FMS on durability characteristics has been studied by Liu et al. (Rong-Jin et al., 2012). The findings from the study showed that FMS majorly contains α-C_2S, C_3MS_2, $CaO \cdot MnO \cdot {}_2SiO_2$, and C_2AS and these compounds have the capability to perform hydraulic reactivity. Though the glassy nature of FMS caused a reduction in workability due to balling, the seawater resistance and freezing–thawing resistance of the FMS-modified concrete was improved. Results from the study also showed that high MnO content in the slag could result in volumetric instability. Hence, there is a need to pay critical attention to the percentage of MnO in FMS when used in cementitious composites.

CS is a solid by-product generated during the matte smelting and refining of copper from ore and the production of one tonne of copper generates about 2.2–3 tonnes of CS (Shi et al., 2008). There exist numerous research evidence that CS can be used as a partial replacement of PC in cementitious composites due to its pozzolanic properties of use of CS in CC (Ariño and Mobasher, 1999; Zain et al., 2004; Moura et al., 2007). Several other studies have also utilized CS as aggregate in cementitious composites (Al-Jabri et al., 2009; Khanzadi and Behnood, 2009; Murari et al., 2014). Ariño and Mobasher (1999) studied the effect of ground copper slag (GCS) up to 15% replacement of PC on the strength and fracture of cementitious composites. The results from the study showed that GCS enhanced the compressive strength of the cementitious composites but there was a corresponding reduction in the ductility. The fracture toughness results indicated that with increasing percentages of GCS the brittleness of the cementitious composite increases.

Similarly, Zain et al. (2004) investigated the possibility of using CS in cementitious composites as a replacement of the PC up to 10%. Results from the study showed that cementitious composites incorporating CS meet the limiting requirements for leaching as per the requirements in the Malaysian Environmental Quality Act. Also, the CS had a retarding effect on the initial and final setting times of the cementitious composites. However, the CS had no major impact on the primary hydration behavior of the cementitious composites and the optimum replacement level of CS was found to be 5% of the weight of the PC. A study conducted by Moura et al. (2007) showed that the addition of 20% CS (by wt. of the PC) enhanced both the mechanical and durability performances of the concrete.

Al-Jabri et al. (2009) conducted an experimental program to examine the effect of using CS in place of natural sand high-performance concrete (HPC). The natural sand was replaced by CS from 0% to 100% and the corresponding performance evaluated in terms of workability, density, compressive strength, tensile strength, flexural strength, and durability. The results from the study showed that when the CS percentage increases in the concrete mix, there is a slight increase (about 5%) in the high-performance concrete density. However, the workability improved rapidly with the increasing content of CS. Also, the addition of CS up to 50% yielded equivalent strength with that of the standard concrete mix and a further increase in the content of CS resulted in a reduction in the compressive strength. Mixtures with 80% and 100% CS substitute exhibited the lowest compressive strength value of about 80 MPa, which was almost 16% lower than the strength of the standard concrete mix. It was recommended to use CS up to replacement of 40% of the natural aggregate in order to prevent an increase in the permeability of the composites.

One of the major concerns that limits the use of CS in cementitious composites is the possibility of heavy metal leaching to the environment. Hence, the use of CS in cementitious composite should be keenly designed and the leaching activity evaluated and verified with standards provided by different regulatory boards. Though all the slags discussed above show several benefits including both the economic and environmental, however, more awareness is required among the policy-makers and practicing professionals for its wide practical implementations.

2.1.3 Mine tailings

The mine tailings (MTs) are generated from the ore beneficiation as a by-product. In general, there exist a large amount of gangue minerals that overcoat/attached with the ore/metal during the excavation process. Therefore, in order to reduce this unnecessary load on the ore, the mined ores go through beneficiation plants before the metal extraction process. The generated waste products during this initial ore beneficiation process are termed as MT. The MT has very fine particle size and is generally composed of large portion silt. It is worth mentioning that mining overburden (OB) is another mine waste, but the OB are the waste rocks and soil that covered the ore during the excavation.

The growing civilization has resulted in an increase in the demand for metals and related products. Hence, the production of MT and other mine wastes is increasing day-by-day. In a study by Jones and Boger (2012), it was disclosed that the world annual production of MT was around 14 billion tons in the year 2012. Such a large quantity of wastes can be recycled and utilized in the production of cementitious composites. Hence, both waste utilization and the need for sustainable alternatives for construction can be fulfilled. Numerous studies have been carried out to utilize MT in construction works and building material production. The MT can be used to prepare bricks (Ahmari and Zhang, 2012; Kim et al., 2019; Li et al., 2019), as filler in asphalt concrete and mortar (Oluwasola et al., 2015, 2016), as fine aggregates (Gao et al., 2020; Fisonga et al., 2019; Gallala et al., 2017; Zhang et al., 2020; Shettima et al., 2016), and as cement replacement (Dandautiya and Singh, 2019; Simonsen et al., 2020; Saedi et al., 2020; Kiventerä et al., 2019; Ince, 2019; Lyu et al., 2020) in cementitious composites. Kim et al. (2019) produced high porous bricks with red mud–MT blend and Li et al. (2019) produced bricks with high thermal stability using iron ore tailings. Oluwasola et al. (2015), (2016) extended the possibility of MT in the road construction sector by establishing its utilization in asphalt concrete preparation.

Gao et al. (2020) used molybdenum tailings as a replacement up to 100% of the natural sand in reinforced concrete (RC) members. The resulted RC members with 25%–50% MT had nearly the same properties as that of the conventional mixture. On the other hand, mixtures with 100% MT exhibited good enough performance to be used for practical applications. Fisonga et al. (2019) also replaced fine aggregates with MT in making highway traffic barriers. Good compressive strength of 31.51 MPa was achieved with the use of MT as a 70% replacement of the aggregates. It was also found in the study that the pozzolanic activity of the MT contributed to the good bonding observed between the binder and aggregates in the concrete.

Though much research has encouraged the use of MT as a filler material or as an aggregate in cementitious composites, it can also be utilized as supplementary cementitious material (SCM) for the PC. Dandautiya and Singh (2019) utilized MT alongside with FA to replace PC to produce concrete with a designed target strength of 30 MPa. Findings from the study showed that an increase of 8.27% in compressive strength was achieved when 5%MT and 10%FA was used as the replacement of PC. Simonsen et al. (2020) conducted a series of characterization evaluation on different types of MT from different origins to examine the potential of MT as SCM in CC. It is suggested that before the use of MT as an SCM in cementitious composites. Findings from the study showed that MT with high calcium or silica content can be used as SCM for their potential hydraulic or pozzolanic activity. On the other hand, less reactive MT can be used as filler or fine aggregate replacement in cementitious composites. Nonetheless, the low reactive MT can be activated in order to be used as SCM through several activation techniques such as thermal activation, chemical activation, physical activation, or compound activation (combination of two or more activation methods). The thermal activation includes the incineration of the MT at an elevated temperature, whereas in the physical activation method, the MT is grounded or crushed mechanically. The chemical activation methods include the activation of the MT with a reagent or chemical additives. Alkali activation method is one of the most preferred chemical activation strategies used to activate the MT in cementitious composites (Saedi et al., 2020).

2.2 Agricultural solid wastes

2.2.1 Rice husk ash

RHA is a waste generated from the process of biomass energy production using rice husk in boilers (Das et al., 2020). The RHA is an inorganic waste that contains a large amount of amorphous silica and could cause ecological issues if dumped in an open environment (Das et al., 2021). Since rice is cultivated as a primary crop in Asian countries like India, China, Bangladesh, Indonesia, etc., a huge amount of RHA is generated each year in these regions that creates severe problems for its safe disposal. A tentative picture of RHA disposal in an open environment is represented in Fig. 9.5. The image presented in Fig. 9.5 was taken during a case study of RHA production and disposal in Keonjhar District in the state of Odisha, India.

The high silica content of RHA makes suitable as an SCM for PC in cementitious composites (Padhi et al., 2018; Chindaprasirt et al., 2007; Chao-Lung et al., 2011; Adesina and Olutoge, 2019; Akinyemi and Adesina, 2020). The high silica content of RHA also widened its scope in synthesizing geopolymers. Several studies have been conducted using RHA for the production of alkali-activated composites (Das et al., 2020, 2021; Zhu et al., 2019a,b; Mehta and Siddique, 2018). The study by Chao-Lung et al. (2011) showed that the use of RHA as up to 20% replacement of PC is beneficial to both the strength and durability characteristics of the concrete. The strength gain process of RHA-blended concrete follows a similar trend like other SCMs used in concrete. However, due to the slow pozzolanic reaction mechanism, the early strength of the RHA concrete is lower than that of the control specimen

Figure 9.5 Disposal of RHA in open environment.

(Chindaprasirt et al., 2007). Nonetheless, the 90 days strength of the concrete with RHA percentage up to 20% has been found to be similar to that of the control concrete (Chao-Lung et al., 2011).

Adesina and Olutoge (2019) showed the usability of semiground RHA along with the addition of lime in cementitious composites up to 25% replacement of the PC. Though the strength performances of the RHA incorporated cementitious composite was not superior, it was sufficient to be used for building applications. Hence, it was concluded that RHA can be incorporated into cementitious composites in order to gain the economical and sustainability benefits. It is worth mentioning that ground RHA performs better than that of the raw RHA in cementitious composites because ground RHA possesses a large surface area coupled with high reactivity than that of partially or unground RHA (Chao-Lung et al., 2011; Adesina and Olutoge, 2019). Usually, the RHA particles are porous in nature with the presence of intrastructural pores (Das et al., 2020). Thus, the reduction in workability and high water requirements are major issues to deal with when RHA are used in cementitious composites. Hence, superplasticizers can be used to enhance the fresh properties of the RHA-blended concrete. If the mixture design is properly done, the RHA can be incorporated up to 40% replacement of PC (Chindaprasirt et al., 2007). The study by Makul (2019) showed promising results of the inclusion of RHA in developing high-performance self-compacting concrete. The produced concrete mixture with 10% RHA and 30% waste foundry sand exhibited the maximum compressive and splitting tensile strengths of all the samples tested compared to conventional concrete mixes. In addition, the self-compacting concrete requirements are met. The internal porous microstructure of RHA helps in storing water molecules, thus it helps in internal curing for a prolonged period promoting the continuous hydration reaction inside the cementitious matrix (Kang et al., 2019). Therefore, RHA can be used as a reactive filler in ultrahigh-performance concrete.

In geopolymers, the use of RHA generally improves the durability performance, particularly the sulfate resistance and permeability properties due to the densification of the geopolymer network through the honeycomb effect (Zhu et al., 2019a,b). Moreover, in geopolymers where the primary precursor contains less amount of silica, the RHA can be incorporated as a silica source to improve the geopolymerization. In red mud and laterite-based geopolymers, Sékou et al. (2017) and Bentchikou et al. (2012) used RHA to compensate for the scarcity of required amount of silica in primary precursors, respectively. They further added that the progressive addition of RHA in laterite-based geopolymers enhanced the phase evolution and microstructure of the product, thus improved its mechanical strength (Bentchikou et al., 2012).

2.2.2 *Wheat straw ash*

Wheat is the most cultivated cereal crop worldwide. As per the statistical reports, the European Union placed first in the global wheat production, producing about 151.6 million tons of wheat per annum (2017–18), followed by China and India (Qudoos et al., 2018). A huge amount of wheat straw is produced during the wheat harvesting process, about 1.3–1.4 kg of wheat straw is generated for obtaining each kilogram of wheat grains (Pan and Sano, 2005). These wastes can also be processed and used as an alternative binder component or as a filler in cementitious composites. In order to use wheat wastes as a binder component in cementitious composites, it needs to undergo combustion in order to produce wheat straw ash (WSA) with pozzolanic properties (Al-Akhras and Abu-Alfoul, 2002; Ataie and Riding, 2013).

The study by Biricik et al. (2000) showed that the use of WSA as a replacement of PC in cementitious composites resulted in lower compressive strength at 28 days compared to those made with only PC as the binder. However, the compressive strength of cementitious composites made with WSA is higher than 40 MPa indicating its suitability for construction applications. The lower compressive strength of cementitious composites with WSA was associated with the slow pozzolanic reaction of the WSA coupled with lower amorphous silica content. Hence, in the long term, the use of WSA as a replacement of PC is beneficial in enhancing the mechanical and durability properties of cementitious composites. Nevertheless, the effectiveness of WSA can be achieved by utilizing various pretreatment methods to enhance its reactivity (Ataie and Riding, 2013).

WSA can also be utilized in cementitious composites as a filler or aggregate. Hence, for WSA with low pozzolanic properties, it is beneficial to use it as a filler/aggregate rather than a replacement of the PC content in order to conserve the performance of the composites. Al-Akhras and Abu-Alfoul (2002) utilized WSA as partial replacement of fine aggregate up to 10.9% in mortar. The findings from the study showed that mortar made with WSA as a 10.9% replacement of the fine aggregate resulted in an enhancement in the compressive strength, tensile strength, and flexural strength by 87%, 67%, and 71%, respectively, compared to the control. Hence, it is recommended that when alternative solid wastes with higher pozzolanic properties exist as replacement of the PC content, WSA can be utilized as filler or aggregate.

3. Construction and demolition solid wastes

Construction and demolition (C&D) are some of the major solid wastes generated in large quantities all over the world. C&D wastes are generated during the construction of infrastructures, maintenance, and demolition. The C&D wastes generated during the construction process are mainly as a result of excess production of construction material for a particular project. Due to the need to rebuild infrastructures to meet current demand or as a result of other natural man-made and natural activities, a large amount of C&D wastes are also generated as a result of the demolition. These wastes are made up of the conventional building material and there is a need to effectively manage these wastes in order to prevent contamination of the environment. C&D wastes are majorly composed of brick/masonry, concrete, ceramics (tiles sanitary wares, etc.), steel, plastics, and wood.

With the increasing demand for raw materials for the production of cementitious composites, C&D wastes can be recycled and reused as components in new cementitious composites. The uses of some of these wastes in cementitious composites are further discussed.

3.1 Bricks

Bricks are common building materials used for the construction of walls, floors, arches, cornices, etc. Bricks with superior properties can also be used for the construction of elements in dams, buildings, bridges, etc. When such structures made with bricks are demolished as a result of either natural or man-made activities, they become a sustainability burden in the environment. However, these brick wastes can be recycled and used as aggregate in cementitious composites.

Corinaldesi (2012) conducted an experimental study to produce cement and hydraulic lime mortars by using a crushed red brick with different sizes as replacement of the aggregates. The results from the study showed that the use of coarsely crushed brick as aggregate resulted in a better performance compared to when fine crushed bricks were used. This observation was attributed to the angularity of the coarsely crushed bricks, both fine and coarse size to replace sand. Nevertheless, it was found out the mortar made with finer crushed bricks has better workability.

In a similar study by Silva et al. (2009), where ultrafine crushed red bricks were used as the replacement of the fine aggregate up to 10%, it was found out that mortars with the ultrafine crushed brick exhibited higher performance compared to the control. However, another study by the authors (Silva et al., 2010) showed that the use of the recycled brick wastes at higher content up to total replacement of the aggregate is detrimental to the performance of the composites. The study by Adamson et al. (2015) recommended that recycled bricks are more beneficial as aggregate in unreinforced cementitious composites compared to steel-reinforced ones.

Aliabdo et al. (2014) made an extensive study on the use of crushed clay bricks as fine and coarse aggregate in cementitious composites. The findings from the study revealed that the use of crushed clay bricks as replacement of fine aggregate is beneficial, while its use as coarse aggregate replacement is detrimental to the performance of the composites. The cementitious composite incorporating the crushed bricks as replacement of coarse aggregate exhibited lower compressive strength, while the compressive strength of those made with clay brick as fine aggregate is 9.9% higher than the control.

3.2 Ceramics

Ceramic products are utilized in various construction materials such as tiles, sanitary wares, chimney pipes, etc. These are the materials made from a mixture of silica sand, clay binder, and some impurities with water up to 30%. The wastes generated from the production and use of these materials can also be recycled and used as components in cementitious composites.

Senthamarai and Devadas Manoharan (2005) reported that about 30% of wastes are generated daily during the production of ceramic products. Due to the high resistance to biological, chemical, and physical degradation of these wastes, they can be utilized as components in cementitious composites. In a study by Medina et al. (2013), the rheology properties of cementitious material made by blending sanitary ware ceramics wastes at a dosage of 10% and 20% to replace PC was evaluated. The findings from the study showed that the addition of ceramic wastes resulted in a decrease in the shear yield stress and hydration reaction. Awoyera et al. (2018) used ceramic waste aggregate (CWA) to prepare concrete and evaluated its corresponding mechanical properties at different ages. The results showed that the CWA concrete exhibited better compressive strength and split tensile strength compared to that of the control concrete.

3.3 Concrete

The application of concrete in the construction field is widespread due to its promising mechanical and structural properties. Concrete possess longer service life with lower maintenance compared to other construction materials. Due to rapid urbanization, demolished materials (mostly contains concrete) are generated and often dumped on land which affects the fertility of soil adversely or poses various safety and health hazards. However, these concrete wastes can also be recycled and used in new construction works. Such recycled concrete used as aggregates are referred to as recycled concrete aggregate (RCA). The life cycle assessment of recycled concrete and conventional concrete carried out by Knoeri et al. (2013) revealed that the use of RCA in cementitious composites could reduce the environmental impacts up to 70% compared to that of the conventional concrete.

Ismail and Ramli (2013) conducted experiments to evaluate the engineering qualities of treated RCA for structural purposes. The acid treatment used on the RCA separated the mortar particles which were loosely attached to the surface and hence maximizes its mechanical and physical properties. Furthermore, the use of these treated RCA in concrete improved the surface contact between fresh cement paste and aggregates as compared to the untreated RCA. These improved surface contacts maximize the mechanical strength of the concrete. Similarly, McGinnis et al. (2017) studied the mechanical properties of concrete made with RCA as the replacement of natural aggregate up to 100%. The findings from the study reveal that the use of RCA as 50% and 100% replacement of the natural aggregates resulted in a decrease in strength by 16.6% and 26.4%, respectively. Also, the stiffness of the concrete reduces with a higher content of RCA in the concrete.

Bui et al. (2018) examined the mechanical properties of concrete made from 100% treated RCA. The coarse RCA sample was treated with the sodium silicate and then silica fume before it was used as aggregate in concrete. Results from the study showed that concrete made with treated RCA exhibited about 33%−50% improvement in the compressive strength compared to the control concrete without untreated RCA. The elastic modulus and tensile strength of concrete made with treated RCA were also higher than the concrete made with untreated RCA up to 42.5% and 41%, respectively. This observation corresponds with that of Kazemian et al. (2019) where the use of treated RCA as aggregates was found to yield densified microstructure, densified interfacial transition zone, and a corresponding enhancement in the crack resistance compared to when untreated RCA was used. However, findings from the study also reveal that the incorporation of RCA and treated RCA into concrete increases the brittle behavior.

Shaban et al. (2019) developed an economical and environmentally friendly method to strengthen RCA properties. Treatment of RCA with various pozzolan slurries, for example, FA, silica fume, etc., at various dosages and soaking time was carried out. Due to the surface treatments, the concrete incorporating the treated RCA exhibited improved performance. This was verified by scanning electron microscope analysis of the surface where it was observed that the microstructure showed surface homogeneity of the treated RCA and the Ca/Si ratio was decreased. The overall mechanical strength and durability of the concrete incorporating the treated RCA were improved due to the pozzolanic reaction around the surface of the RCA.

Another recent research done by Toghroli et al. (2020) evaluated the performance of RCA and pozzolanic materials compared to natural coarse aggregates (NCAs). The experimental results showed that 100% replacement of RCA with NCA resulted in a significant decrease in compressive and flexural strength. However, the addition of silica fume improves the surface properties of RCA, and hence compressive and flexural strength increased significantly. Similarly, the addition of steel fibers to the mixes increases the strength capacity of the composites. It was also concluded that pervious concrete for structural applications can be made by using RCA with specific amounts of pozzolanic additives and fibers.

4. Conclusions

This chapter explored some of the solid wastes products that can be incorporated into cementitious composites. Focus was placed on solid wastes that can be used to partially or totally replace PC in cementitious composites. Some discussions were also made on some types of solid wastes that can be used as both binder and aggregate in cementitious composites. The discussion made showed that solid wastes ranging from industrial wastes to agricultural wastes can be used to replace PC in cementitious composites.

Acknowledgments

All the authors are grateful to Grøn Tek Concrete and Research (GTCR), Bhubaneswar, India, for the necessary help and resources provided during this research. The first author is thankful to Mr. Saurabh Kumar Singh of Government College of Engineering, Keonjhar, India, for his valuable inputs and help during the case studies mentioned in the manuscript.

References

Adamson, M., Razmjoo, A., Poursaee, A., 2015. Durability of concrete incorporating crushed brick as coarse aggregate. Construct. Build. Mater. 94, 426–432.

Adesina, A., 2020. Performance and sustainability overview of alkali-activated self-compacting concrete. Waste Dispos Sust. Energy 1–11 [Internet] [cited 2020 Aug 17];1:3. Available from: http://link.springer.com/10.1007/s42768-020-00045-w.

Adesina, A., Awoyera, P., 2019. Overview of trends in the application of waste materials in self-compacting concrete production. SN Appl. Sci. 1 (9), 1–18.

Adesina, A., Das, S., 2020. Performance of green fibre-reinforced composite made with sodium-carbonate-activated slag as a binder. Innov. Infrastruct. Solut. 5 (2), 1–9.

Adesina, P.A., Olutoge, F.A., 2019. Structural properties of sustainable concrete developed using rice husk ash and hydrated lime. J. Build. Eng. 25, 100804.

Ahmari, S., Zhang, L., 2012. Production of eco-friendly bricks from copper mine tailings through geopolymerization. Construct. Build. Mater. 29, 323–331.

Akinmusuru, J.O., 1991. Potential beneficial uses of steel slag wastes for civil engineering purposes. Resour. Conserv. Recycl. 5 (1), 73–80.

Akinyemi, B.A., Adesina, A., 2020. Recent advancements in the use of biochar for cementitious applications: a review. J. Build. Eng. 101705 [Internet] [cited 2020 Sep 1];101705. Available from: https://linkinghub.elsevier.com/retrieve/pii/S2352710220318660.

Al-Akhras, N.M., Abu-Alfoul, B.A., 2002. Effect of wheat straw ash on mechanical properties of autoclaved mortar. Cement Concr. Res. 32 (6), 859–863.

Aliabdo, A.A., Abd-Elmoaty, A.E.M., Hassan, H.H., 2014. Utilization of crushed clay brick in concrete industry. Alexandria Eng. J. 53 (1), 151–168.

Al-Jabri, K.S., Hisada, M., Al-Oraimi, S.K., Al-Saidy, A.H., 2009. Copper slag as sand replacement for high performance concrete. Cement Concr. Compos. 31 (7), 483–488.

Al-Negheimish, A.I., Al-Sugair, F.H., Al-Zaid, R.Z., 1997. Utilization of local steelmaking slag in concrete. J. King Saud Univ. - Eng. Sci. 9 (1), 39–54.

Andrew, R.M., 2018. Global CO2 emissions from cement production. Earth Syst. Sci. Data 10 (1), 195–217.

Ariño, A.M., Mobasher, B., 1999. Effect of ground copper slag on strength and toughness of cementitious mixes. ACI Mater. J. 96 (1), 68–73.

Assaedi, H., Alomayri, T., Kaze, C.R., Jindal, B.B., Subaer, S., Shaikh, F., et al., 2020. Characterization and properties of geopolymer nanocomposites with different contents of nano-CaCO3. Construct. Build. Mater. 252, 119137.

ASTM C 618, 2010. Standard Specification for Coal Fly Ash and Raw or Calcined Natural Pozzolan for Use. Annu B ASTM Stand.

ASTM, 2015. Standard Specification for Steel Slag Aggregates for Bituminous Paving Mixtures. Astm.

Ataie, F.F., Riding, K.A., 2013. Thermochemical pretreatments for agricultural residue ash production for concrete. J. Mater. Civ. Eng. 25 (11), 1703–1711.

Awoyera, P., Adesina, A., 2019. Durability Properties of Alkali Activated Slag Composites: Short Overview. Silicon, pp. 987–996.

Awoyera, P.O., Ndambuki, J.M., Akinmusuru, J.O., Omole, D.O., 2018. Characterization of ceramic waste aggregate concrete. HBRC J. 14 (3), 282–287.

Awoyera, P.O., Adesina, A., Gobinath, R., July 3, 2019. Role of recycling fine materials as filler for improving performance of concrete - a review. Aust. J. Civ. Eng. 17 (2), 85–95 [Internet] Available from: https://www.tandfonline.com/doi/full/10.1080/14488353.2019.1626692.

Bai, T., Song, Z.G., Wu, Y.G., Hu, X Di, Bai, H., 2018. Influence of steel slag on the mechanical properties and curing time of metakaolin geopolymer. Ceram. Int. 44 (13), 15706–15713.

Bentchikou, M., Guidoum, A., Scrivener, K., Silhadi, K., Hanini, S., 2012. Effect of recycled cellulose fibres on the properties of lightweight cement composite matrix. Construct. Build. Mater. 34, 451–456.

Biricik, H., Aköz, F., Türker, F., Berktay, I., 2000. Resistance to magnesium sulfate and sodium sulfate attack of mortars containing wheat straw ash. Cement Concr. Res. 30 (8), 1189–1197.

Bouzoubaâ, N., Zhang, M.H., Malhotra, V.M., 2000. Laboratory-produced high-volume fly ash blended cements: compressive strength and resistance to the chloride-ion penetration of concrete. Cement Concr. Res. 30 (7), 1037–1046.

Bouzoubaâ, N., Zhang, M.H., Malhotra, V.M., 2001. Mechanical properties and durability of concrete made with high-volume fly ash blended cements using a coarse fly ash. Cement Concr. Res. 30 (10), 1393–1402.

Bui, N.K., Satomi, T., Takahashi, H., 2018. Mechanical properties of concrete containing 100% treated coarse recycled concrete aggregate. Construct. Build. Mater. 163, 496–507.

Chao-Lung, H., Anh-Tuan, B Le, Chun-Tsun, C., 2011. Effect of rice husk ash on the strength and durability characteristics of concrete. Construct. Build. Mater. 25 (9), 3768–3772.

Chindaprasirt, P., Kanchanda, P., Sathonsaowaphak, A., Cao, H.T., 2007. Sulfate resistance of blended cements containing fly ash and rice husk ash. Construct. Build. Mater. 21 (6), 1356–1361.

Corinaldesi, V., 2012. Environmentally-friendly bedding mortars for repair of historical buildings. Construct. Build. Mater. 35, 778–784.

Dandautiya, R., Singh, A.P., 2019. Utilization potential of fly ash and copper tailings in concrete as partial replacement of cement along with life cycle assessment. Waste Manag. 99, 90–101.

Das, S.K., Jyotirmoy Mishra, S.M.M., 2018. An overview of current research trends in geopolymer concrete. Int. Res. J. Eng. Technol. 5 (11), 376–381.

Das, S.K., Mishra, J., Mustakim, S.M, Adesina, A., Kaze, C.R, Das, D., 2021. Sustainable utilization of ultrafine rice husk ash in alkali activated concrete: characterization and performance evaluation. J. Sustain. Cem.-Based Mater. 1–19.

Das, S.K., Mishra, J., Singh, S.K., Mustakim, S.M., Patel, A., Das, S.K., et al., 2020. Characterization and utilization of rice husk ash (RHA) in fly ash – blast furnace slag based geopolymer concrete for sustainable future. Mater Today Proc. 33, 5162–5167.

Das, S.K., Singh, S.K., Mishra, J., Mustakim, S.M., 2020. Effect of rice husk ash and silica fume as strength-enhancing materials on properties of modern concrete—a comprehensive review. In: Babu, et al. (Eds.), Lecture Notes in Civil Engineering. Springer Nature, pp. 253–266.

Dash, M.K., Patro, S.K., 2018. Performance assessment of ferrochrome slag as partial replacement of fine aggregate in concrete. Eur. J. Environ. Civ. Eng. 1–20.

Davids, R., Carlso, R., Kelly, J., Davis, H., 1937. Properties of cements and concretes containing fly ash. ACI J. Proc. 33 (5), 577–612.

Dinakar, P., Babu, K.G., Santhanam, M., 2008. Durability properties of high volume fly ash self compacting concretes. Cement Concr. Compos. 30 (10), 880–886.

Dinakar, P., Kartik Reddy, M., Sharma, M., 2013. Behaviour of self compacting concrete using Portland pozzolana cement with different levels of fly ash. Mater. Des. 46, 609–616.

European Committee for Standartization, 2000. EN 197-1 Cement - Part 1: composition, specifications and conformity criteria for common cements. Eur. Stand.

Fisonga, M., Wang, F., Mutambo, V., 2019. Sustainable utilization of copper tailings and tyre-derived aggregates in highway concrete traffic barriers. Construct. Build. Mater. 216, 29–39.

Flower, D.J.M., Sanjayan, J.G., 2007. Green house gas emissions due to concrete manufacture. Int. J. Life Cycle Assess. 12 (5), 282–288.

Gallala, W., Hayouni, Y., Gaied, M.E., Fusco, M., Alsaied, J., Bailey, K., et al., 2017. Mechanical and radiation shielding properties of mortars with additive fine aggregate mine waste. Ann. Nucl. Energy 101, 600–606.

Gao, S., Cui, X., Kang, S., Ding, Y., 2020. Sustainable applications for utilizing molybdenum tailings in concrete. J. Clean. Prod. 226, 122020.

Guo, X., Pan, X., 2018. Mechanical properties and mechanisms of fiber reinforced fly ash–steel slag based geopolymer mortar. Construct. Build. Mater. 179, 633–641.

Hussain, C.M., Keçili, R., 2020. Environmental pollution and environmental analysis. In: Hussain, C.M., Kecili, R. (Eds.), Modern Environmental Analysis Techniques for Pollutants, 1st ed. Elsevier, pp. 1–36.

Ikponmwosa, E.E., Ehikhuenmen, S., Emeshie, J., Adesina, A., 2020. Performance of Coconut Shell Alkali-Activated Concrete: Experimental Investigation and Statistical Modelling. Silicon, pp. 1–6.

Ince, C., 2019. Reusing gold-mine tailings in cement mortars: mechanical properties and socio-economic developments for the Lefke-Xeros area of Cyprus. J. Clean. Prod. 238, 117871.

IS-455, 2015. Indian standard Portland slag cement — specification. Bur. Indian Stand.

Ismail, S., Ramli, M., 2013. Engineering properties of treated recycled concrete aggregate (RCA) for structural applications. Construct. Build. Mater. 44, 464–476.

Jones, H., Boger, D.V., 2012. Sustainability and waste management in the resource industries. Ind. Eng. Chem. Res. 51 (30), 10057–10065.

Kang, S.H., Hong, S.G., Moon, J., 2019. The use of rice husk ash as reactive filler in ultra-high performance concrete. Cement Concr. Res. 115, 389–400.

Karakoç, M.B., Türkmen, I., Maraş, M.M., Kantarci, F., Demirboȷa, R., Uȷur Toprak, M., 2014. Mechanical properties and setting time of ferrochrome slag based geopolymer paste and mortar. Construct. Build. Mater. 72, 283–292.

Kazemian, F., Rooholamini, H., Hassani, A., 2019. Mechanical and fracture properties of concrete containing treated and untreated recycled concrete aggregates. Construct. Build. Mater. 209, 690–700.

Khanzadi, M., Behnood, A., 2009. Mechanical properties of high-strength concrete incorporating copper slag as coarse aggregate. Construct. Build. Mater. 23 (6), 2183–2188.

Kim, Y., Lee, Y., Kim, M., Park, H., 2019. Preparation of high porosity bricks by utilizing red mud and mine tailing. J. Clean. Prod. 207, 490–497.

Kiventerä, J., Piekkari, K., Isteri, V., Ohenoja, K., Tanskanen, P., Illikainen, M., 2019. Solidification/stabilization of gold mine tailings using calcium sulfoaluminate-belite cement. J. Clean. Prod. 239, 118008.

Knoeri, C., Sanyé-Mengual, E., Althaus, H.J., 2013. Comparative LCA of recycled and conventional concrete for structural applications. Int. J. Life Cycle Assess. 18 (5), 909–918.

Krishna, R.S., Mishra, J., Meher, S., Das, S.K., Mustakim, S.M., Singh, S.K., 2020. Industrial solid waste management through sustainable green technology: case study insights from steel and mining industry in Keonjhar, India. Mater Today Proc. 33, 5243–5249.

Li, R., Zhou, Y., Li, C., Li, S., Huang, Z., 2019. Recycling of industrial waste iron tailings in porous bricks with low thermal conductivity. Construct. Build. Mater. 213, 43–50.

Li, G., 2004. Properties of high-volume fly ash concrete incorporating nano-SiO_2. Cement Concr. Res. 34 (6), 1043–1049.

Lyu, X., Yao, G., Wang, Z., Wang, Q., Li, L., 2020. Hydration kinetics and properties of cement blended with mechanically activated gold mine tailings. Thermochim. Acta 683, 178457.

Makul, N., 2019. Combined use of untreated-waste rice husk ash and foundry sand waste in high-performance self-consolidating concrete. Res. Mater. 1, 100014.

Martauz, P., Vaclavik, V., Cvopa, B., 2017. The use of steel slag in concrete. In: IOP Conference Series: Earth and Environmental Science, 92. IOP Publishing, 012041.

McCarthy, M.J., Dhir, R.K., 2005. Development of high volume fly ash cements for use in concrete construction. Fuel 84 (11), 1423–1432.

McGinnis, M.J., Davis, M., de la Rosa, A., Weldon, B.D., Kurama, Y.C., 2017. Strength and stiffness of concrete with recycled concrete aggregates. Construct. Build. Mater. 154, 258–269.

Medina, C., Banfill, P.F.G., Sánchez De Rojas, M.I., Frías, M., 2013. Rheological and calorimetric behaviour of cements blended with containing ceramic sanitary ware and construction/demolition waste. Construct. Build. Mater. 40, 822–831.

Mehta, A., Siddique, R., 2018. Sustainable geopolymer concrete using ground granulated blast furnace slag and rice husk ash: strength and permeability properties. J. Clean. Prod. 205, 49–57.

Mishra, J., Das, S.K., Krishna, R.S., Nanda, B., 2020. Utilization of ferrochrome ash as a source material for production of geopolymer concrete for a cleaner sustainable environment. Indian Concr. J. 94 (7), 40–49.

Mishra, J., Kumar Das, S., Krishna, R.S., Nanda, B., Kumar Patro, S., Mohammed Mustakim, S., 2020. Synthesis and characterization of a new class of geopolymer binder utilizing ferrochrome ash (FCA) for sustainable industrial waste management. Mater Today Proc.

Moura, W.A., Gonçalves, J.P., Lima, M.B.L., 2007. Copper slag waste as a supplementary cementing material to concrete. J. Mater. Sci. 42 (7), 2226–2230.

Murari, K., Siddique, R., Jain, K.K., 2014. Use of waste copper slag, a sustainable material. J. Mater. Cycles Waste Manag. 17 (1), 13–26.

Mustakim, S.M., Das, S.K., Mishra, J., Aftab, A., Alomayri, T.S., Assaedi, H.S., et al., 2020. Improvement in Fresh, Mechanical and Microstructural Properties of Fly Ash- Blast Furnace Slag Based Geopolymer Concrete by Addition of Nano and Micro Silica. Silicon, pp. 1–14.

Nath, S.K., 2018. Geopolymerization behavior of ferrochrome slag and fly ash blends. Construct. Build. Mater. 181, 487–494.

Oluwasola, E.A., Hainin, M.R., Aziz, M.M.A., 2015. Evaluation of asphalt mixtures incorporating electric arc furnace steel slag and copper mine tailings for road construction. Transp. Geotech. 2, 47–55.

Oluwasola, E.A., Hainin, M.R., Aziz, M.M.A., 2016. Comparative evaluation of dense-graded and gap-graded asphalt mix incorporating electric arc furnace steel slag and copper mine tailings. J. Clean. Prod. 122, 315–325.

Özcan, A., Karakoç, M.B., 2019. The resistance of blast furnace slag- and ferrochrome slag-based geopolymer concrete against acid attack. Int. J. Civ. Eng. 17 (10), 1571–1583.

Padhi, R.S., Patra, R.K., Mukharjee, B.B., Dey, T., 2018. Influence of incorporation of rice husk ash and coarse recycled concrete aggregates on properties of concrete. Construct. Build. Mater. 173, 289–297.

Palit, S., Hussain, C.M., 2018. Nanomaterials for environmental science: a recent and future perspective. Nanotechnol. Environ. Sci. 1–18.

Palit, S., Hussain, C.M., 2018. Environmental management and sustainable development: a vision for the future. In: Ahmed, Hussain (Eds.), Handbook of Environmental Materials Management, 2nd ed. Wiley.

Palit, S., Hussain, C.M., 2020. Chapter 1. Innovation in environmental remediation methods. In: Rahman, Hussain (Eds.), The Handbook of Environmental Remediation, 1. Elsevier.

Palomo, A., Grutzeck, M.W., Blanco, M.T., 1999. Alkali-activated fly ashes: a cement for the future. Cement Concr. Res. 29 (8), 1323–1329.

Pan, X., Sano, Y., 2005. Fractionation of wheat straw by atmospheric acetic acid process. Bioresour. Technol. 96 (11), 1256–1263.

Panda, C.R., Mishra, K.K., Panda, K.C., Nayak, B.D., Nayak, B.B., 2013. Environmental and technical assessment of ferrochrome slag as concrete aggregate material. Construct. Build. Mater. 49, 262–271.

Pasetto, M., Baldo, N., 2010. Experimental evaluation of high performance base course and road base asphalt concrete with electric arc furnace steel slags. J. Hazard Mater. 181 (1–3), 938–948.

Péra, J., Ambroise, J., Chabannet, M., 1999. Properties of blast-furnace slags containing high amounts of manganese. Cement Concr. Res. 29 (2), 171–177.

Poon, C.S., Lam, L., Wong, Y.L., 2000. Study on high strength concrete prepared with large volumes of low calcium fly ash. Cement Concr. Res. 30 (3), 447–455.

Prasanna, K.M., Tamboli, S., Das, B.B., 2021. Characterization of mechanical and micro-structural properties of FA and GGBS-based geopolymer mortar cured in ambient condition. In: Lecture Notes in Civil Engineering. Springer, pp. 751–768.

Qudoos, A., Kim, H.G., Atta-ur-Rehman, Ryou, J.S., 2018. Effect of mechanical processing on the pozzolanic efficiency and the microstructure development of wheat straw ash blended cement composites. Construct. Build. Mater. 193, 481–490.

Rai, A., Prabakar, J., Raju, C.B., Morchalle, R.K., 2002. Metallurgical slag as a component in blended cement. Construct. Build. Mater. 16 (8), 489–494.

Rangan, B., Hardjito, D., 2005. Studies on Fly Ash-Based Geopolymer Concrete, 28. Proc 4th World, pp. 133–137.

Rong-Jin, L., Qing-Jun, D., Ping, C., Guang-Yao, Y., 2012. Durability of Concrete Made with Manganese Slag as Supplementary Cementitious Materials. J Shanghai Jiaotong Univ, pp. 345–349.

Ryu, G.S., Lee, Y.B., Koh, K.T., Chung, Y.S., 2013. The mechanical properties of fly ash-based geopolymer concrete with alkaline activators. Construct. Build. Mater. 47, 409–418.

Saedi, A., Jamshidi-Zanjani, A., Darban, A.K., 2020. A review on different methods of activating tailings to improve their cementitious property as cemented paste and reusability. J. Environ. Manag. 270, 110881.

Sékou, T., Siné, D., Lanciné, T.D., Bakaridjan, C., 2017. Synthesis and characterization of a red mud and rice husk based geopolymer for engineering applications. Macromol. Symp. 373 (1), 1600090.

Senthamarai, R.M., Devadas Manoharan, P., 2005. Concrete with ceramic waste aggregate. Cement Concr. Compos. 27 (9–10), 910–913.

Shaban, W.M., Yang, J., Su, H., Liu, Q feng, Tsang, D.C.W., Wang, L., et al., 2019. Properties of recycled concrete aggregates strengthened by different types of pozzolan slurry. Construct. Build. Mater. 216, 632–647.

Shettima, A.U., Hussin, M.W., Ahmad, Y., Mirza, J., 2016. Evaluation of iron ore tailings as replacement for fine aggregate in concrete. Construct. Build. Mater. 120, 72–79.

Shi, C., Meyer, C., Behnood, A., 2008. Utilization of copper slag in cement and concrete. Resour. Conserv. Recycl. 52 (10), 1115–1120.

Silva, J., Brito, J de, Veiga, R., 2009. Incorporation of fine ceramics in mortars. Construct. Build. Mater. 23 (1), 556–564.

Silva, J., de Brito, J., Veiga, R., 2010. Recycled red-clay ceramic construction and demolition waste for mortars production. J. Mater. Civ. Eng. 22 (3), 236–244.

Simonsen, A.M.T., Solismaa, S., Hansen, H.K., Jensen, P.E., 2020. Evaluation of mine tailings' potential as supplementary cementitious materials based on chemical, mineralogical and physical characteristics. Waste Manag. 102, 710–721.

Smith, I.A., 1967. The design of fly-ash concretes. Proc. Inst. Civ. Eng. 36 (4), 769–790.

Toghroli, A., Mehrabi, P., Shariati, M., Trung, N.T., Jahandari, S., Rasekh, H., 2020. Evaluating the use of recycled concrete aggregate and pozzolanic additives in fiber-reinforced pervious concrete with industrial and recycled fibers. Construct. Build. Mater. 252, 118997.

Zain, M.F.M., Islam, M.N., Radin, S.S., Yap, S.G., 2004. Cement-based solidification for the safe disposal of blasted copper slag. Cement Concr. Compos. 26 (7), 845–851.

Zelić, J., 2005. Properties of concrete pavements prepared with ferrochromium slag as concrete aggregate. Cement Concr. Res. 35 (12), 2340–2349.

Zhang, W., Gu, X., Qiu, J., Liu, J., Zhao, Y., Li, X., 2020. Effects of iron ore tailings on the compressive strength and permeability of ultra-high performance concrete. Construct. Build. Mater. 260, 119917.

Zhu, H., Liang, G., Xu, J., Wu, Q., Zhai, M., 2019. Influence of rice husk ash on the waterproof properties of ultrafine fly ash based geopolymer. Construct. Build. Mater.

Zhu, H., Liang, G., Zhang, Z., Wu, Q., Du, J., 2019. Partial replacement of metakaolin with thermally treated rice husk ash in metakaolin-based geopolymer. Construct. Build. Mater.

Related websites

1. https://theconversation.com/plastics-could-help-build-a-sustainable-future-heres-how-133585.
2. https://canada.constructconnect.com/dcn/news/technology/2020/08/could-garbage-hold-the-key-to-carbon-friendly-concrete.
3. https://sflcn.com/how-your-building-firm-can-save-money-reusing-concrete/.
4. https://scientect.com/uncategorized/1262850/recycled-construction-aggregates-market-global-industry-analysis-trends-and-forecast-2018-to-2027/.

Comprehensive management of natural resources: a holistic vision

Milan Majerník[1], *Lucia Bednárová*[2], *Jana Naščáková*[3], *Peter Drábik*[1], *Marcela Malindžáková*[2]
[1]University of Economics in Bratislava, Research Institute of Trade and Sustainable Business, Bratislava, Slovak Republic; [2]Technical University of Košice, Faculty of Mining, Ecology, Process Control and Geotechnologies, Košice, Slovak Republic; [3]University of Economics in Bratislava, Faculty of Business Economy of the University of Economics in Bratislava with the Seat in Košice, Košice, Slovak Republic

1. Introduction

Only few people realize that large-scale environmental damage, whether on a regional, international, or global scale, generally has its roots in the failure of either the market or legal instruments. This fact has already been felt by industrial enterprises in Slovakia, which are currently going through a critical period.

Businesses will have to adapt to new market conditions as soon as possible, which means that they will have to introduce new management methods in order to be able to provide the necessary range of goods of the required quality, all within our standards and international environmental standards. These changes will have to be made with a view to reducing the negative environmental impacts of production and thus the costs of addressing them.

Knowledge of the basic principles of an environmentally oriented management system, which is one of the tools enabling companies to penetrate the foreign market faster, is becoming a prestigious matter for the top management of companies. At present, it is more than necessary to ensure that environmental protection is not only effectively financially secure. It is therefore important to take into account all economic aspects that can contribute to better financing of environmental protection, and in particular fees, taxes, duties, and other fiscal policy instruments.

The problem of the environment is one of the current problems of humanity, which have a global character. To ensure the well-being of man as a consumer with a modern way of life, science, technology, and, last but not least, the environment is very often used irrationally. The environmental and sustainability challenges that Europe faces today are rooted in global developments stretching back over decades. On the environmental degradation it has significantly contributed waste of natural resources and energy due to obsolete technology, and also advanced technology to globally enable more intensive consumption of natural resources, energy, and thus a significant deterioration of the environment. During this period, the "Great Acceleration" of social and economic activity has transformed humanity's relationship with the environment.

Sustainable Resource Management. https://doi.org/10.1016/B978-0-12-824342-8.00007-9

Since 1950, the global population has tripled to 7.5 billion; the number of people living in cities has quadrupled to more than four billion; economic output has expanded 12-fold, matched by a similar increase in the use of nitrogen, phosphate, and potassium fertilizers; and primary energy use has increased 5-fold. Looking ahead, these global developments look set to continue increasing pressures on the environment. The world's population is projected to grow by almost one-third to 10 billion by 2050. Globally, resource use could double by 2060, with water demand increasing 55% by 2050 and energy demand growing 30% by 2040. As a pioneer of industrialization, Europe has played a pivotal role in shaping these global changes. Today, it continues to consume more resources and contribute more to environmental degradation than many other world regions. To meet these high consumption levels, Europe depends on resources extracted or used in other parts of the world, such as water, land, biomass, and other materials. As a result, many of the environmental impacts associated with European production and consumption occur outside Europe. Collectively, these realities add up to a profound challenge for Europe and other world regions. The current trajectories of social and economic development are destroying the ecosystems that ultimately sustain humanity. Shifting onto sustainable pathways will require rapid and large-scale reductions in environmental pressures, going far beyond the current reductions. (The European environment—state and outlook 2020 Executive summary) Globalization, which is characterized by a reduction in barriers to the movement of goods, people, money, and technology across borders (O'Rourke, 2001), is the most important factor in determining economic development and the future of the planet (Intriligator, 2004). Opening borders for imported goods improves domestic institutions' financial development and quality, while the internationalization of domestic financial markets to other countries' financial systems promotes financial development in both the host country and the home country. In particular, developing countries can promote institutional development when they open their domestic markets to imported goods. Since the last few decades of the 20th century, natural resources have been included in sustainable development research. According to Boschini et al. (2007), better economic policies and high-quality political institutions can resolve the negative aspects of the resource curse. Development of the financial sector is one of the keys to enabling positive contributions from resource rents and accelerating economic growth, as financial development improves the use of natural resources, which leads to pressure for the development of financial institutions (Haider and Adil, 2019).

As the character and scale of global environmental and climate challenges has become clearer, policy frameworks have evolved. Europe's environmental policy framework—the environmental acquis—is increasingly shaped by ambitious long-term visions and targets. The overarching vision for Europe's environment and society is set out in the Seventh Environment Action Programme (seventh EAP). EU environmental policies are guided by three thematic policy priorities in the seventh EAP:

(1) to protect, conserve, and enhance the EU's natural capital;
(2) to turn the EU into a resource-efficient, green, and competitive low-carbon economy; and
(3) to safeguard the EU's citizens from environment-related pressures and risks to their health and well-being.

In recent years, the EU has also adopted a series of strategic framework policies that focus on transforming the EU economy and particular systems (e.g., energy, mobility) in ways that deliver prosperity and fairness, while also protecting ecosystems.

The United Nations (UN) Sustainable Development Goals complement these frameworks, providing a logic for transformative change that acknowledges the interdependence of social, economic, and environmental targets. It is clear that natural capital is not yet being protected, conserved, and enhanced in accordance with the ambitions of the seventh EAP. Small proportions of protected species (23%) and habitats (16%) are in favorable conservation status and Europe is not on track to meet its overall target of halting biodiversity loss by 2020. Europe has achieved its targets for designating terrestrial and marine protected areas and some species have recovered, but most other targets are likely to be missed (The European environment—state and outlook 2020 Executive summary).

2. Natural resources

This concept became more widely known in the 1970s. Natural resources are the basis of the functioning of the world economy, and in the past, the provision of natural resources was a fundamental factor in economic development. At present, where we have significant impacts of scientific and technological progress and the emergence of a global economy, the impact of natural resources has decreased. But despite progress, natural resources are still the backbone of any economy. Primary energy raw materials have a specific position, and also the overall raw material base of each state.

Natural resources are those parts of living or inanimate nature that man uses or can use to satisfy his needs. Natural resources are divided into two basic groups, namely:

1. renewable natural resources, which have the ability to be partially or completely regenerated during gradual consumption, either on their own or with human input (water, air, energy, forest, etc.),
2. nonrenewable natural resources disappear through consumption (various types of minerals, almost all mineral resources fall into this category of nonrenewable. The composition of minerals also includes fuel minerals, ores, and nonmetallic minerals.) (Fazekašova et al., 2018)

Renewable natural resources can be endlessly renewed with proper management and thus used appropriately. A characteristic feature of a renewable resource is its supply. It is not fixed, once it can fall, other times it can rise. If we want this resource to be renewed, we must keep a certain minimum supply when using it. This means that the amount of drawing of a given resource must be lower than its growth or production potential. If we meet this condition, we can resume drawing indefinitely, we call it sustainable development, or yield. In the case of nonrenewable resources, we cannot talk about sustainable development. If the extraction rate is positive, the resource is gradually depleted. It follows that in the case of renewable resources we speak of the optimal rate of use, but in the case of depletable resources we speak of the optimal rate of depletion (extraction). An exhaustible resource does not grow over time, its supply is limited, and its reproductive capacity is virtually zero.

Environmental resources are a broader concept than natural resources. These are all the resources and components of nature that have had and are important for the origin and maintenance of life on Earth. Natural resources cover only one of the main functions of the environment, the term environmental resources cover all four functions (natural resource storage, landscape, landfill, source of life). Environmental resources include ecosystems, which are the natural environment for the emergence and maintenance of various forms of life. An economic criterion is used to distinguish between natural and environmental resources. Natural resources are considered private goods, whose products are exchanged and valued in the market, environmental resources are public goods, they are not exchanged or valued in the market (Hronec et al., 2010).

Into the natural resources we also include minerals, forest, flora, and fauna. However, resource stocks are unevenly distributed and the result is that different countries and regions have different resources as well as different stocks of these resources. In a way, these resources are provided to the consumer for various purposes. If we wanted to define the provision of resources, it could be described as follows:

Provision of resources is basically the ratio between the extent of natural resources and the extent of their use. The availability of resources is expressed either in the number of years for which these resources should be sufficient or in the reserves of resources per capita. The resource supply indicator is influenced by the wealth or poverty of the area with natural resources, the rate of extraction, and the class of natural resources (IK-PTZ).

One of the ways to comprehensively manage natural resources is integrated landscape management (ILM), we can basically say that it is a tool for creating and promoting a model of land management and use, which is aimed at improving the overall quality of life, environmental protection and its components, respect for nature protection, stability and biodiversity of the area, protection as well as rational use of natural and cultural–historical resources. It is based on the understanding of the landscape as the integration of natural resources in a certain area. It is the space that represents the unifying framework in which all resources occur as intersecting layers (e.g., geological resources, water and soil resources, climatic conditions, biotic resources). The essence of ILM is to understand the space as an integration of these resources as well as to know the relationships between these resources.

The ILM model must be based on comprehensive country research in three basic dimensions:

- environmental,
- social,
- economic and to examine their interrelationships and connections.

In our conditions, the study of the relationship between the environmental and social dimensions is particularly sensitive, because, e.g., many companies with adverse environmental impacts have a high social and economic effect. In any case, we must not forget the five elements that must be incorporated in ILM:

1. Shared or agreed management objectives that encompass multiple benefits (the full range of goods and services needed) from the landscape.

2. Field, farm, and forest practices are designed to contribute to multiple objectives, including human well-being, food and fiber production, climate change mitigation, and conservation of biodiversity and ecosystem services.
3. Ecological, social, and economic interactions among different parts of the landscape are managed to realize positive synergies among interests and actors or to mitigate negative tradeoffs.
4. Collaborative, community-engaged processes for dialogue, planning, negotiating, and monitoring decisions are in place.
5. Markets and public policies are shaped to achieve the diverse set of landscape objectives and institutional requirements (Defining Integrated Landscape Management).

To achieve an effective and comprehensive application of ILM, the following are necessary:

Ensuring the landscape-ecologically optimal use of the territory is understood as a complex process of mutual harmonization of spatial requirements of economic and other human activities with the landscape-ecological conditions of the territory, which result from the structure of the landscape.

Implementation of technological measures—in particular the introduction of efficient technologies aimed at eliminating above-limit production of pollutants in order to minimize the load of individual components of the environment with foreign substances and other contaminants, as well as the application of technologies using alternative energy and renewable sources. It is also necessary to apply appropriate technologies in the field of agricultural and forestry management, etc. (Wikipedia).

Application of the regulation of landscape-ecologically optimal land use to all sectoral plans—setting limits on the use of individual resources by production and nonproduction entities so as not to prioritize the development of one sector at the expense of another and to prevent conflicts arising from conflicts of interest of individual sectors.

Promoting the principles of sustainable development in the consciousness of the population—creating an effective system of education in the field of ILM and sustainability. Not only a sufficiently educated public but also public administration, entrepreneurs, and other interest groups that are able to promote the principles and criteria of sustainability in real practice.

Enforcement of effective tools—ensuring the implementation of effective tools (e.g., legal, economic, conceptual, voluntary, etc.) supporting the rational use of natural resources, environmental protection, as well as the protection of public health (e.g., taxes, levies, fees, etc.) (Wikipedia).

2.1 Management of natural resources

Natural resources are components of the natural environment, they are not isolated in the biosphere, but, on the contrary, they are in certain mutual relations and ties to each other. At the same time, they are constantly changing. Changes in the relationship between natural resources can be qualitative or quantitative in nature. The dynamics of these relationships are very complex, so for the rational use and management of natural resources, human society must respect certain rules.

As early as 1958, the UN conference in Tokyo adopted the principles of rational use of natural resources:

- their use should be for the benefit of human society,
- the elimination of their damage and pollution in the process of production and consumption,
- maintaining the properties of harmonious relationships between resources and extraction,
- respecting the principles of conservation of resources for the future. (minzp)

Economical and efficient use of natural resources is one of the conditions for preserving the existence of life on our planet. Before we consider examples of rational and irrational environmental management, it is necessary to define this concept. There are two main interpretations:

The first definition considers nature management as a system of adequate consumption of resources, which allows to reduce the rate of processing, allows the restoration of nature. It is understood that a person does not violate the use of environmental resources, but improves his technology to make full use of every natural resource.

The second definition states that environmental management is a theoretical discipline that considers ways to improve the rational use of available resources. This science is looking for ways to optimize this problem.

Given current human activities, such as industrial production, agriculture, tourism, natural landscape change, it is sometimes difficult to say clearly which of these options is an example of rational nature management. Human activities ultimately affect our environment. Even a circular model in disposal with municipal waste can help us to make a more affective ratio in treated problem (Pavolova et al., 2019). Rational environmental management is called the most harmonious interaction of us with the world. This concept has several characteristics. The use of the gifts of nature is rational, if in the process of one's activity one applies new technologies as well as intensive approaches to production. For this purpose, methods for waste-free production of new products are introduced and all technological processes are automated. This approach to governance is characteristic of the developed countries of the world. They serve as an example for many other states.

2.2 Rational management of natural resources

The main objective of the rational management of nature is to ensure conditions for the existence of mankind and gain material benefits, maximize the use of all natural territorial complex, prevent or minimize possible harmful effects of manufacturing processes or other types of human activity, maintain and improve the productivity and attractiveness of nature, and to ensure and regulate economic development its resources.

An environmental management system can significantly reduce environmental pollution. Rational nature management is characteristic of an intensive economy, i.e., an economy that develops on the basis of scientific and technological progress and better work organization with high labor productivity. Examples of environmental management can be waste-free production or a waste-free production cycle in which waste is fully recovered, which leads to a reduction in the consumption of raw materials and minimizes environmental pollution (Bednárová and Witek, 2016) Production can use waste from its own production process as well as waste from other sectors. Thus, several enterprises in the same or different sectors may be included in a waste-free cycle.

Rational nature management is a nature management system in which:

1. Extracted natural resources shall be used to a sufficient extent and the amount of resources consumed shall be reduced accordingly;
2. The restoration of renewable natural resources is ensured;
3. Production waste is fully and reusable.

The components of rational nature management—the protection, development and transformation of nature—occur in different forms in relation to different types of natural resources. When using practically inexhaustible resources (solar and underground heat energy, tides, etc.), the rationality of environmental management is measured primarily through the lowest operating costs, the highest efficiency of the mining industry and plants. For resources obtained and at the same time nonrenewable (for example, mineral), the complexity and efficiency of production, waste reduction, etc., are important. The aim of protecting resources replenished during use is to maintain their productivity and resource turnover, and the operation should ensure their cost-effective integrated, waste-free production and be accompanied by measures to prevent damage to related types of resources.

Ensuring the sustainable use of natural resources has several basic aspects (Izakovičová and Hrnčiarová, 1999):

- organizational and spatial—focused on ecologically optimal use of natural resources,
- technological—focused on the greening of production technologies, protecting individual natural resources from the effects of stress factors,
- socioeconomic—aimed at economic stimulation of sustainable use of natural resources.

The socioeconomic causes of environmental damage can be derived from the functions that the environment performs in the economic system, as well as from the emergence of external effects that arise from its excessive use. The fact that the environment is for the most part a public good is characterized by:

- the possibility of simultaneous use by several individuals without competing with each other in its use, while respecting the principles of rational behavior and optimal use;
- the impossibility of excluding anyone from the use, for sociopolitical reasons, even if the exclusion would be technically possible.

In real life, this means that in the absence of environmental protection tools, no one can be excluded from using the environment, even if they do not contribute to its protection. However, the pursuit of profit maximization leads businesses to use environmental resources sparingly. This is especially the case when the incentive for its optimal use is not effective enough.

- political—focused on the institutional provision of the implementation of sustainable use of natural resources.

In this matter, it is important to constantly improve technology, approaches in the production of products. One of the main examples can be the creation of protected areas in which activities aimed at the protection and restoration of flora and fauna are actively carried out. Human activity deprives the habitats of many species of animals and plants. The changes are sometimes so strong that it is almost impossible to reverse them. Another example of rational nature management is the renewal of places for the development of natural resources, the creation of natural landscapes.

Given the rational use of natural resources, examples of which have been given above, real methods for improving them should be discussed. They are used successfully all over the world. In the first place, funding is provided for companies that carry out research into increasing the completeness of the development of natural resources (IK-PTZ). In the future, it will no longer be possible to use natural resources in the same way as they do today. One solution is the efficient use of natural resources, which will lead to the development of new products and services and the search for new ways to reduce inputs, minimize waste, improve resource management, change consumption patterns, optimize production processes and management and business methods, as well as to improve logistics and thus to transition from a linear to a circular economy.

2.3 Irrational management of natural resources

Irrational management of nature is reflected in the reduction of quality, waste and depletion of natural resources, the weakening of the regenerative forces of nature, environmental pollution, and the reduction of its health and esthetic benefits.

The irrational use of environmental resources is the other side of the same coin. If we want to define this situation in a simplified way, it is characterized as a disproportionately high consumption of resources. Consumers do not consider the consequences of their actions. The irrational approach also has its own characteristics. First of all, this includes a comprehensive approach to business. In this case, outdated technologies and production methods are used.

Such cycles are illogical and do not think to the end. The result is a high percentage of waste, and we often find that some of this waste damages the environment, human health, and even leads to the death of whole species of living organisms. If we would like to summarize the irrational management of nature, we can give the following examples:

- Burning of agricultural land, plowing of slopes on hills, leading to the formation of gorges, soil erosion, and destruction of the fertile layer of the soil (humus).
- Change of hydrological regime.
- Deforestation, destruction of protected areas, excessive grazing.
- Discharges of waste and sewage into rivers, lakes, and seas.
- Air pollution by chemical substances.
- Extermination of valuable species of plants, animals, and fishes.
- Open mining method.

Irrational management in the field of natural resources leads humanity into an abyss, an ecological crisis. This approach to management is characteristic of Latin

America, Asia, and Eastern Europe. There are several main activities that can be clearly attributed to one or the other group of environmental resource use. An example of environmental management is the use of technologies for production with waste elimination. For this purpose, enterprises with a closed or complete processing cycle are created.

Irrational nature management is a nature management system in which the most readily available natural resources are not used in large quantities and are usually not used in full, leading to a rapid depletion of resources. In this case, a large amount of waste is generated and the environment is highly polluted. Irrational nature management is typical of a large economy, that is, an economy developing through new construction, the development of new land, the use of natural resources, and increasing the number of employees. The large economy initially yields good results with a relatively low scientific and technical level of production, but quickly leads to the depletion of natural and labor resources. One of the many examples of unsustainable nature management may be agriculture, which is currently widespread in Southeast Asia. Soil burning leads to wood destruction, air pollution, poorly controlled fires, etc. Irrational nature management is often the result of narrow sectoral interests and the interests of multinational companies that have their own harmful industries in developing countries (Peskiadmin).

Irrational management of nature is reflected in the reduction of quality, waste and depletion of natural resources, the weakening of the regenerative forces of nature, environmental pollution, and the reduction of its health and esthetic benefits.

The changing ways in which resources are used in recent decades are an indication that progress in resource efficiency is making real progress. Over the last 20 years, recycling has become commonplace in businesses and households across the EU, with significant implications for industries such as paper, glass, and resource extraction. Legislation at EU level has also helped to reduce carbon emissions: EU greenhouse gas emissions have fallen by more than 10% since 1990, while European economies have grown by 40% over the same period. There are four golden rules for maximizing economic growth while easing pressure on the resource base:

- save: seize existing opportunities to save resources wherever possible—some EU economies are 16 times more efficient than others;
- recycle: increase the recycling of materials and reuse elements in products (the most recent example being mobile phones);
- replace: replace primary source inputs with alternatives that are more efficient and have lower environmental impacts throughout their life cycle (for example, by phasing out mercury);
- set value: policy makers need to find ways to take the right value of natural resources into account when making decisions, which will help to better manage our natural resource base. When we learn to value and value ecosystem services and natural resources, we will help alleviate the pressure on the environment. In the same way are also treated a natural raw recourse as we can see in Rybár et al. (2005).

In principle, it can be said that natural resources are becoming one of the main components of the circular economy. For a better understanding, I'll start with what it looks like at the moment. Today, we still operate on the principles of a linear economy in the vast majority. This means that we extract something bloodily, make some of it with expensive technologies and a lot of work, then use/consume something quickly, and then throw it away. Since a lot of scarce natural and human resources and energy are used for products throughout their life cycle, from design, production to consumption, it is a pity that we then just throw away the products. Often, the waste we produce is toxic and a great burden on our environment. At best, we recycle the product, but that is still not enough. Recycling is an energy-intensive process and it is not always possible to actually recycle a product (Cirkulárna ekonomika).

However, there is a model in which such waste of resources and nonenvironmental behavior does not occur. It is a model called circular (circular) economy. "The economy is a part of social life focused on the material maintenance and provision of human life. It is an area of social activity through which goods are produced, distributed and consumed. The economy deals with the basic economic problems, which are: What to produce? How to produce? For whom to produce?" (Recyklator).

The circular economy is therefore a strategy of sustainable development, in which man cooperates with nature, does not go against it. It is a model where waste as such does not actually exist. All raw materials, products, and packaging are sealed in long cycles (Fig. 10.1). The whole cycle is prepared to be sustainable. All raw materials are reusable, all products are repairable, treatable, recyclable, reusable, and processable.

The circular economy aims to keep products, equipment, and infrastructure in use for longer, thus improving the productivity of these resources. All "waste" should become "food" for another process: either a by-product or recovered resource for another industrial process or as regenerative resources for nature (e.g., compost). This regenerative approach is in contrast to the traditional linear economy, which has a "take, make, dispose" model of production. (Ellen MacArthur Foundation, 2012; accessed July 15, 2020).

Material flows are also very close to the circular economy, helping to optimize the necessary changes that must take place in order to achieve resource sustainability (Fig. 10.2).

2.4 Material flows and their use

The whole life cycle of material resources, from extraction, through use in the production and consumption of goods and services to the end of their life as waste, can lead to environmental impacts. Material Flow Accounting (MFA) throughout the economy is a tool for the systematic registration of all material input and output flows that cross the

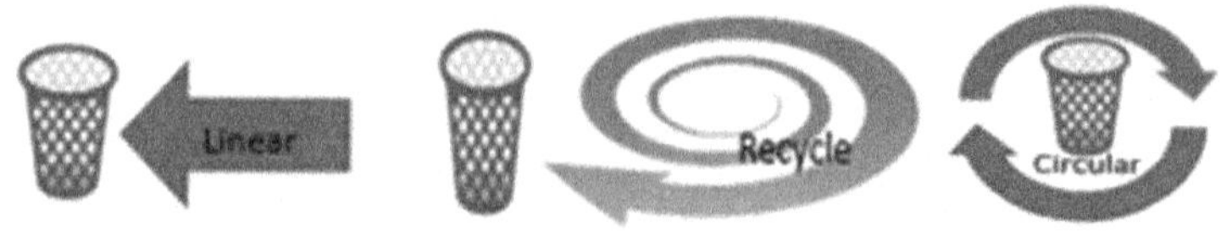

Figure 10.1 Difference between Linear, recycle and Circular economy.
Source: Owen processing.

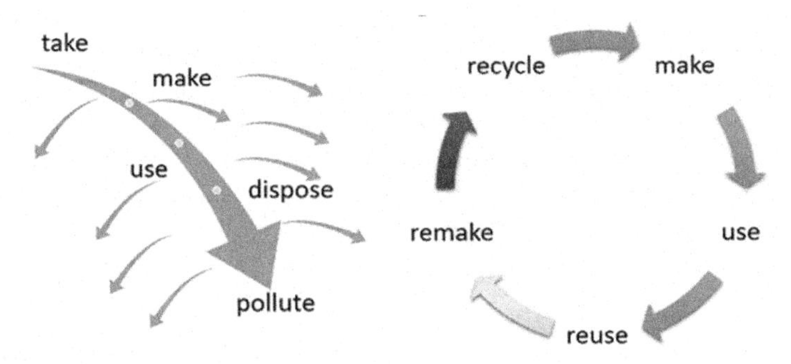

Figure 10.2 Supply chain revolution.
Source: https://www.rethinkglobal.info/ accessed July 15, 2020.

functional boundaries between the economy and the environment, including imports and exports (Bednárová and Liberko, 2007).

MFA provides economy-wide data on material use. The principle concept underlying MFA is a simple model of this interrelation between the economy and the environment, in which the economy is an embedded subsystem of the environment. Similar to living beings, this subsystem is dependent on a constant throughput of materials and energy. Raw materials, water, and air are extracted from the natural system as inputs, transformed into products, and finally retransferred to the natural system as outputs (waste and emissions). In order to highlight the similarity to natural metabolic processes, the terms "industrial" or "societal" metabolism have been introduced (IS4IE).

Environmental accounting is a system for providing accurate and quantitative measures of the costs and resulting effects, efficiency, and effectiveness of investments in environmental preservation activities. The benefit of undertaking a corporate environmental accounting initiative is that the identification and greater awareness of environment-related costs often provides the opportunity to find ways to reduce or avoid these costs, while also improving environmental performance. Adoption of environmental accounting techniques increase visibility of environmental costs and benefits, thus increasing cost manageability.

Environmental accounting gives the opportunity to:

- significantly reduce or eliminate environmental costs, and
- improve environmental performance.

Accountants need to respond to this challenge by finding ways to accommodate environmental costs within their accounting systems—environmental management accounting does this. The need for environmental management accounting was conceived in recognition of some of the limitations of conventional management accounting approaches for management activities and decisions involving significant environmental costs and significant environmental consequences impacts.

In MFA studies for a region or on a national level the flows of materials between the natural environment and the economy are analyzed and quantified on a physical level. The focus may be on individual substances (e.g., cadmium flows), specific materials, or bulk material flows (e.g., steel and steel scrap flows within an economy). Researchers in this field are organized in the Socio-Economic Metabolism (SEM) section OECD Glossary of the International Society for Industrial Ecology (ISIE) (OECD, a.b).

Statistics related to MFA are usually compiled by national statistical offices, using economic, agricultural, and trade statistics measuring the exchange of material between different products available in an economy (Fig. 10.3).

The data collected in MFA are used to calculate several different standardized indicators:

- Direct material input is a measure of the total material inputs into an economy and is calculated as the sum of domestic extraction (DE) and imports.
- The physical trade balance (PTB) is a measure of net imports and is calculated as the difference between imports and exports. Reflecting that material and money flow in opposite directions during trade, this is a contrast to the monetary trade balance which calculates net exports.
- Domestic material consumption is a measure of apparent consumption and calculated from domestic extraction plus imports minus exports (or DE plus PTB).

Economy-wide MFA is a satellite system to the system of national accounts and provides a rich empirical database for analytical studies. More information on how the statistics are collected, under what legal framework, and how they are defined is available in economy-wide material flow accounts.

In addition, the following indicators may be used in MFA:

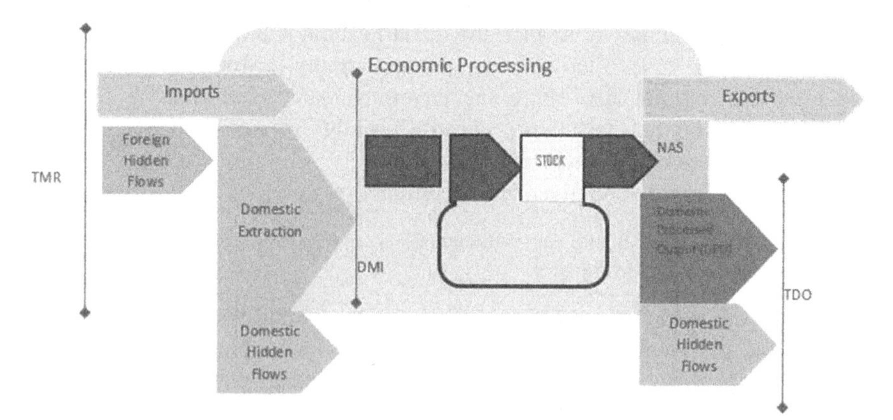

Figure 10.3 Schematic diagram of the material flows and principal indicators in Material Flow Accounting. This diagram is based on similar diagrams depicting societal or industrial metabolism by Bringezu (e.g., Material Flow Analysis for the European Union and beyond. Implications for science and policy. Presentation held at the inaugural ISIE conference on Nov 13, 2001 at Nordwijkerhood.
Source: http://is4ie.org/images/Bringezu.pdf accessed July 15, 2020.

- Total material requirement includes the domestic extraction of resources (minerals, fossil fuels, biomass), the indirect flows caused by and associated with the domestic extraction (called "hidden flows") and the imports.
- Hidden flows are materials that are extracted or moved, but do not enter the economy. According to OECD, the "displacement of environmental assets without absorption into the economic sphere" such as overburden from mining operations.
- Domestic processed output (DPO) is defined by the OECD as "the total mass of materials which have been used in the national economy, before flowing into the environment. These flows occur at the processing, manufacturing, use, and final disposal stages of the economic production-consumption chain." (OECD, a,b; Projectdefinition)
- Total domestic output includes the DPO plus the hidden flows associated with the domestic production.
- Raw material consumption (RMC) includes the raw materials that are embodied in traded products (e.g., metal ores from which metals have been extracted). RMC does not include hidden flows. RMC is most commonly calculated via environmentally extended input–output analysis.

As the results of Europe's green industries suggest, more efficient use of resources and pollution control can be the main drivers of economic growth. The sector has grown by around 8% a year in recent years, with an annual turnover of € 319 billion, representing around 2.5% of Europe's GDP. A large share of recent growth has been concentrated in the management of resources incorporating new technologies such as solar and wind energy. The environmental market is a global opportunity for European companies: the global market for green industries, which is currently worth around € 1000 billion a year, is expected to triple in 2030. The EU has about a third of the world market and is a net exporter, with many European producers benefiting from being the first in the market. Strong export markets, including China and other developing countries, are pursuing environmentally friendly growth. The world market is growing by about 5% per year.

2.5 Management of natural resources in the context of EU environmental policy

Since the 1970s, a wide range of environmental legislation has been put into practice in the European Union, which currently forms the most complete set of modern legal norms in the world. The main priorities and goals of international environmental policy are again based on the concept of sustainable development—in addition to the natural emphasis on the environmental aspect, there is a social dimension and economic dimension in each (UN, 2015; UNEP, 2016; EU, 2013; Delereux and Happaerts, 2016, etc.). However, the emphasis on individual priorities and the formulation of objectives and tools for policy implementation are different. It is important to add in accordance with several authors (e.g., Mezřický et al., 2012, Miklós in Izakovičová et al., 1998) that the "fourth" cross-sectional dimension also includes institutional security (development of the legal system, functional public administration in the field of the environment) as well as support for raising environmental public awareness. It is also generally acknowledged that in addition to quality environmental legislation and the system of its application, the so-called horizontal policy integration (OECD, 2012), in particular the integration of environmental objectives into economic and sectoral policies, as well as the context of other global challenges such as food and energy security, poverty eradication, and the like.

The invocation of public interventions, whether international, state, or self-governing, into the economy or the environment is, in essence, caused by the need to harmonize their desired state with that which is actually taking place. The causes of this discrepancy have their roots primarily in human behavior, on the side of production or consumption. In terms of motivating factors influencing environmental behavior, we can quite clearly distinguish on the one hand directive, centralist, normatively oriented instruments of environmental policy and on the other hand motivational, decentralist, economically oriented tools relying on economically rational decision-making of free entities. Typical representatives in the first group of instruments include various regulations, orders, standards, and limits, while the second group includes, in particular, the application of liability and compensation in accordance with the rule of law, property rights, and moral principles. Carefully defined criteria for its evaluation should play an important role in assessing the appropriateness of using individual environmental policy instruments, taking into account the specific conditions of each country (Cohen, 2014; Field, 2006; Tietenberg and Lewis, 2009; Vig and Kraft, 2012).

Sustainable access to and use of raw materials are key elements of EU sustainability policy. They are the basis for the current and future competitiveness of EU industries (Bednárová et al., 2013) The supply chains for raw materials—both primary and secondary—are real economic sectors that create jobs and wealth in Europe. Recycling is an economic activity that contributes significantly to EU GDP. The collection of used materials and products involves citizens, municipalities, and public authorities that have invested in efficient systems to meet the growing demand for long-term sustainability (Chovancova and Tej, 2020).

The body of EU environmental law, known as the environmental acquis, consists of around 500 regulations, rules, and decisions. Environmental policies have contributed to some progress toward a sustainable green economy, and economy in which policies and innovation enable society to use resources efficiently, thereby increasing human well-being in an inclusive way while preserving natural systems (Bednárová, 2008). These policies have also stimulated innovation and investment in environmental goods and services, creating jobs and export opportunities (EU, 2013).

In addition, the integration of environmental objectives into sectoral policies, such as those governing agriculture, transport, or energy, has provided financial incentives for environmental protection. The optimal use of natural resources must also be strongly supported by environmental policy instruments, both direct and indirect, which at least partially—although not completely—internalize the negative externalities arising from excessive use of the environment (EUR-Lex).

In the past, specific environmental issues, often with local impacts, have been addressed through targeted single-issue policies and tools. These included waste management issues or the protection of biodiversity. Recognition of diffuse pressures from various sources has led since the 1990s to place greater emphasis on integrating environmental aspects into sectoral policies, e.g., into transport or agriculture, leading to different results. In response to the financial crisis, many European countries adopted recovery policies with a focus on the green economy in 2008 and 2009. Although the attention of policy makers has subsequently shifted to fiscal consolidation and the sovereign debt crisis, a recent survey of European citizens' attitudes to the environment shows that interest in environmental issues has not diminished. European citizens believe that more needs to be done to protect the environment at all levels and that progress at national level should be measured on the basis of environmental, social, and economic criteria (EU, 2014).

A very important starting point for the implementation of the EP are functioning instruments—mostly a distinction is made between economic instruments (market—fees, taxes, subsidies, tradable permits, payments ...), normative instruments (regulation by legislation and legal system—laws, regulations, planning ...), and voluntary instruments (participation, dissemination of information, auditing, certification, labeling). None of these types can be described as working best on their own. Čech (2015), e.g., states that while normative instruments are fast, effective and easy to control, they do not provide incentives to reduce pollution. Economic instruments, in turn, require "tuning" in order to ensure an efficient level of management and protection of natural resources (increases in taxes and charges must not exceed a certain economically viable level). Finally, he adds that the most ethically advanced tools of the EP could be voluntary and free instruments with a bottom-up initiative. Most authors consider various economic instruments to be promising—although they are based in part on voluntariness, they can gradually be incorporated into the economic accounts of countries. We agree that the most appropriate comprehensive and long-term functioning instrument could be an ecological tax reform based on a change in the entire system of valuation of natural resources, as well as inputs and outputs of the entire system of the economy (OECD, 2008). Although it does not yet work comprehensively in any country, the SEEA (UN, 2014) and European environmental accounting systems are being developed. We consider information tools, education, and training to be extremely important tools for implementing the EP.

According to the EU, the UN, and the OECD, the green economy represents a strategic approach to systemic challenges in the areas of global environmental degradation, natural resource security, employment, and competitiveness. Policy initiatives to support the green economy can be found in a number of key EU strategies, including the strategy Europe 2020, the seventh Environmental Action Program, the EU Framework Program for Sector Policy Research and Innovation, such as transport and energy.

The green economy approach emphasizes resource-efficient economic development, which takes place within environmental constraints and fairly throughout society. This requires the simultaneous pursuit of economic, environmental, and social objectives (Chovancová and Vavrek, 2020). The prevailing policy practice remains largely fragmented and shaped by established governance structures, so that the opportunities offered by the green economy perspective to address systemic challenges and exploit synergies have not yet been fully realized.

In the field of natural resources, there is no policy to address this issue at international level, as is the case with climate change, when the United Nations Conference on Environment and Development in cooperation with the IPCC (Intergovernmental Panel on Climate Change) led in 1988–92 to form the UN Framework Convention on Climate Change (UNFCCC). Bringezu and Bleischwitz (2009) point out the importance of such a policy and express the view that a policy addressing natural resources in a global context will be formulated in the near future. The main reasons for adopting an international agreement on the sustainable management of natural resources include:

- existing initiatives and documents are legally nonbinding; are based on voluntary participation and uncertain continuity,
- competitive advantages through material efficiency are outweighed by the destructive use of natural resources, environmental dumping, and the uneconomical disposal of waste,
- growing pressure from problems and thus the growing potential for conflict.

One of the important steps in this area was the establishment of the International Panel on Sustainable Resource Management (IPSRM). This panel was established at the end of 2007 at the initiative of the European Commission under the United Nations Environment Program. Its task is to bring together relevant scientific knowledge from around the world and formulate proposals for:

- the environmental burden of extracting and exploiting natural resources from a life cycle perspective,
- strategies and approaches to decouple the negative environmental impacts of production and consumption from economic growth,
- support for capacity building in developing and new industrialized countries.

The IPSMR can become a pioneer in the development of international policies and the preparation of an international consensus based on a scientific approach and the evaluation of strategies and measures in the field of sustainable management of natural resources (Mederly, 2017).

3. Summary

Both the economy and the ecology significantly affect the economic efficiency of each company. If we want our society to continue to develop, we must adhere to the principles of sustainable development and environmental policy. The benefit for manufacturing companies should be not only the economic value of their investments but also ecological. Economic reproduction in manufacturing companies depends on natural resources, which form the material and energy base for the implementation of the production process, and thus allow to meet the needs of consumers—customers. In the conditions of the company, it is necessary to realize that natural raw materials are available only in limited quantities and constant economic growth and development reduces their quantity for future generations.

The importance of environmental policy in today's world is extremely great, which is related to the need to respond to unprecedented and accelerating change in the planetary environment, the most visible (though by no means the only) manifestation of which is global climate change. This fact is primarily linked to the process of globalization of the economy and politics, where on the one hand increasing wealth and power are concentrated in the hands of "elites" (who are often anonymous) and on the other hand deepening social and environmental problems, which grow to serious political meetings (Dobson, 2016; Harris et al., 2016). As a result, the polarization of the interests of the various groups representing the "North" on the one hand and the "South" states, and in particular their people on the other, is currently intensifying.

The current trend in the state of the environment in the world is not favorable. In view of the above, the most important drivers of global development are not, in fact, states or international organizations, but rather strong global economic players and the strongest countries, whose decisions and activities are often not favorable in terms of environmental impact. However, for the actual implementation of a sustainable policy, the "weights" of individual global actors would have to change in favor of the competences of supranational institutions, the strengthening of control and sanction mechanisms, and the detriment of the economic interests of different groups or large countries. Likewise, the real priorities of most countries would have to shift from economic–social to environmental–social. However, the real potential for change is represented by the people themselves—Dobson (2016) sees it in the lived experiences of the large masses of the population, which will lead to their involvement or political organization, to make concrete demands—even to change a system that behaves unsustainably and creates imbalances.

This chapter is part of the solution of scientific project KEGA 026EU-4/2018 and it was issued with ist support.

References

Bednárová, L., 2008. Ekonomická efektívnosť environmentálneho manažérstva. Environmentálne účtovníctvo, Grafotlač Prešov, ISBN 978-80-8068-733-5.

Bednárová, L., Liberko, I., Rovňák, M., 2013. Environmental benchmarking in small and medium sized enterprises and there impact on environment, International Multidisciplinary Scientific GeoConference-SGEM, Albena. Bulgaria 141–146. ISSN 1314-2704.

Bednárová, L., Liberko, I., 2007. Monitoring account system for environmental costs by ekological tools. In: 10. medzinárodná vedecká konferencia Trendy v systémoch riadenia podnikov, Vysoké tatry –Štrbské pleso. október, pp. 15–17. ISBN 978-80-8073-885-5.

Bednárová, L., Witek, L., et al., 2016. Assessment methods of the influence on environment in the context of eco-design process. In: Majernik, M., et al. (Eds.), Production Management and Engineering Sciences. CRC Press-Taylor & Francis Group, pp. 15–16.

Boschini, A.D., Pettersson, J., Roine, J., 2007. Resource curse or not: a question of appropriability. Scand. J. Econ. 109, 593–617.

Brigenzu, S., Bleischwitz, R., 2009. Sustainable Resource Management: Global Trends, Visions and Policies, vol. 338. Greenleaf Publishing, Abingdon, ISBN 9781906093266.

Cirkulárna ekonomika, n.d. https://www.sumne.sk/l/co-je-to-cirkularna-ekonomika/.
Čech, J., 2015. Nástroje súčasnej environmentálnej politiky. In: Životné prostredie, vol. 49, pp. 111−115. ISSN 0044-4863 (2).
Chovancová, J., Tej, J., 2020. Decoupling economic growth from greenhouse gas emissions: the case of the energy sector in V4 countries. Equilibrium. Q. J. Econ. & Econ. Pol. 15 (2), 235−251.
Cohen, S., 2014. Understanding Environmental Policy. Columbia University Press, New York, p. 232.
Delreux, T., Happaerts, S., 2016. Environmental Policy and Politics in the European Union. Palgrave MacMillan, U.K., ISBN 978-0-23-024426-9, p. 320
Dobson, A., 2016. Environmental Politics: A Very Short Introduction. Oxford University Press, Oxford, ISBN 978-0-19-966557-0, p. 135.
EU, 2013. Decision No 1386/2013/EU of the European Parliament and of the Council (Accessed 15 August 2020).
EU, 2014. Directive 2014/52/EU of the European Parliament and of the Council of 16 April 2014 amending Directive 2011/92/EU on the assessment of the effects of certain public and private projects on the environment.
Fazekašova, D., et al., 2018. Prírodné zdroje a ich využitie v podmienkach udržateľného rozvoja, 190, ISBN 978-80-8165-305-6. Grafotlač Prešov, s.r.o.
Field, B.C., 2006. Environmental Policy: An Introduction. Waveland Press Inc., Long Grove, Illinois, p. 438.
Haider, S., Adil, M.H., 2019. Does financial development and trade openness enhance industrial energy consumption? A sustainable developmental perspective. Manag. Environ. Qual. 30 (6), 1297−1313. https://doi.org/10.1108/MEQ-03-2019-0060.
Harris, C., Kharecha, P., Goble, P., Goble, R., 2016. The Climate is A-Changin:' Teaching civic competence for a sustainable climate. Soc. Stud. Young Learner 28 (3), 17−20.
Hronec, O., Boltižiar, M., Bujnovský, R., Michaeli, E., Minďáš, J., Schwarczová, H., Tomeková, B., 2010. Manažment a oceňovanie prírodných zdrojov. Stredoeurópska vysoká škola v Skalici, Skalica, ISBN 978-80-89391-19-6, p. 202.
Defining Integrated Landscape Management, n.d. https://www.researchgate.net/publication/262996374_Defining_Integrated_Landscape_Management_for_Policy_Makers. (Accessed 12 August 2020).
IK-PTZ n.d. https://ik-ptz.ru/sk/diktanty-po-russkomu-yazyku–2-klass/racionalnoe-ispolzovanie-resursov-primery-racionalnoe.html.
n.d. Integrated landscape management. http://www.fao.org/land-water/overview/integrated-landscape-management/en/.
Intriligator, M.D., 2004. Globalisation of the world economy: potential benefits and costs and a net assessment. J. Pol. Model. 24, 485−498.
IS4IE, n.d.b http://www.is4ie.org/.
Izakovičová, Z., Hrnčiarová, T., 1999. Trvalo udržateľné využívanie prírodných zdrojov. Životné prostredie 33 (5), 250−254.
Izakovičová, Z., Kozová, M., Pauitšová, E. (Eds.), 1998. Implementácia trvalo udržateľného rozvoja, 357. ÚKE SAV, Bratislava, ISBN 80-968120-0-9.
lowwastewellness, n.d. https://lowwastewellness.com.
Mederly, P., 2017. Habilitačná prednáška, Východiská, súčasnosť a perspektívy environmentálnej politiky vo svete a na Slovensku, p. 46.
Materialflows.net, n.d. Definition taken from: http://www.materialflows.net.
Mezřický, V., et al., 2012. Environmentální politika a udržitelný rozvoj. Portál, Praha, ISBN 978-80-262-0249-3, p. 207.

OECD Glossary, n.d.a. http://stats.oecd.org/glossary/detail.asp?ID=6403.
OECD Glossary, n.d.b. http://stats.oecd.org/glossary/detail.asp?ID=6472.
OECD, 2008. OECD Environmental Outlook to 2030. OECD Publishing, Paris, ISBN 978-92-64-04048-9, p. 517.
OECD, 2012. OECD Environmental Outlook to 2050. OECD Publishing, Paris, ISBN 978-92-64-12216-1, p. 353.
O'Rourke, H.,K., 2001. In: Globalization and Inequality: Historical Trends Annual World Bank Conference on Development Economics (2001/2) (No. NBER Working Paper No. 8339).
Pavolová, H., Lacko, R., Hajduová, Z., Šimková, Z., Rovňák, M., 2019. The circular model in disposal with municip waste. A case study of Slovakia tors of raw material policy in Slovakia. In: The First Interregional Conference "Sustainable Development of Eurasian Mining Regions. - London (Veľká Británia): Édition Diffusion Presse Sciences s, pp. 1–12.
Peskiadmin, n.d. https://peskiadmin.ru/en/racionalnoe-i-neracionalnoe-prirodopolzovanie-primery-tablica.html.
Projectdefinition, n.d. https://www.theprojectdefinition.com/domestic-processed-output-dpo/.
Recyklator, n.d. https://recyklator.org/co-je-to-cirkularna-ekonomika-a-preco-ju-potrebujeme/.
Rethinkglobal, n.d. https://www.rethinkglobal.info/a-supply-chain-revolution-how-the-circular-economy-unlocks-new-value/.
Rybár, P., Cehlár, M., Engel, J., Mihok, J., 2005. Mineral deposit. Evaluation of Mineral Deposits. Edičné stredisko/AMS, Košice, pp. 20–35.
n.d. The European Environment — State and Outlook 2020 Executive Summary.
Tietenberg, T., Lewis, L., 2009. Environmental Economics and Policy. Prentice Hall, Upper Saddle River, New Persey, p. 560.
Towards the Circular Economy: An Economic and Business Rationale for an Accelerated Transition, 2012. Ellen MacArthur Foundation, p. 24. Archived from the original on 2013-01-10. Retrieved 2012-01-30.
UN, 2014. System of Environmental-Economic Accounting 2012 – Experimental Ecosystem Accounting. United Nations, Statist. Division, New York, ISBN 978-92-1-161575-3, p. 177.
UN, 2015. Transforming Our World: The 2030 Agenda for Sustainable Development, vol. 2015. United Nations, New York, p. 40 (Accessed 13 August 2020). https://sustainabledevelopment.un.org/content/documents/21252030%20Agenda%20for%20Sustainable%20Development%20web.pdf.
UNEP, 2016. Annual Report 2016 (Accessed 15 August 2020) Source: http://web.unep.org/annualreport/2016/index.php?page=1&lang=en.
Vig, N.J., Kraft, M.E., 2012. Environmental Policy: New Directions for the Twenty-First Century. SAGE Publications, Inc., Thousand Oaks, Kalifornia, p. 451.
Wikipedia, n.d. https://sk.wikipedia.org/wiki/Integrovan%C3%BD_mana%C5%BEment_krajiny.

Further reading

Bringezu, S., Bleischwitz, R., 2009. Sustainable Resource management: Global Trends, Visions and Policies. Greenleaf Publishing Limited, Sheffield.
Chovancová, J., Vavrek, R., 2020. (De)coupling analysis with focus on energy consumption in EU countries and its spatial evaluation. Pol. J. Environ. Stud. 29 (3).
n.d. https://d2ouvy59p0dg6k.cloudfront.net/downloads/soer_2020_executive_summary_embargoed_4_december.pdf.

Environment Resource Efficiency, n.d. https://ec.europa.eu/environment/resource_efficiency/documents/factsheet_sk.pdf.

EUR-Lex, n.d.b https://eur-lex.europa.eu/LexUriServ/LexUriServ.do?uri=OJ:C:2011:107:0001:0006:SK:PDF.

Freeman, O.E., Duguma, L.A., Minang, P.A., 2015. Operationalizing the integrated landscape approach in practice. Eco. & Soc. 20 (1), 24. https://doi.org/10.5751/ES-07175-200124.

minzp, n.d. https://www.minzp.sk/strategicke-dokumenty/agenda-21.html.

Mishkin, F.S., 2009. Globalization and financial development. J. Dev. Econ. 89, 164−169. https://doi.org/10.1016/j.jdeveco.2007.11.004.

Vidová, J., 2014. Európa efektívne využívajúca zdroje. FOR FIN. odborný mesačník pre financie a investovanie 1 (7−8). ISSN 1339-5416.

Zaidi, S.A.H., Wei, Z., Gedikli, A., Zafarf, M.W., Hou, F., Iftikha, Y., December 2019. The impact of globalization, natural resources abundance, and human capital on financial development: evidence from thirty-one OECD countries. Resour. Pol. 64, 101476 (Accessed 29 August 2020). https://www.sciencedirect.com/science/article/pii/S0301420719304611?via%3Dihub.

Shades of green: HOPF for standardized environmental performance indicators

Helen S.Y. Chen, T.C.E. Cheng
Faculty of Business, The Hong Kong Polytechnic University, Hong Kong SAR, PR China

1. Introduction

1.1 Background and motivation

Sustainability reporting has become the mainstream in the past decade (G&A Institute, 2020). Growing equally rapidly is the size of such reports. With more than 120,000 sustainability reports currently available publicly (Corporate Register, 2020) and ranging from 50 to 100 pages each (WBCSD, 2018), it is humanly impossible to sift through the millions of pages of reports to extract insights in corporate sustainability performance. The increasingly prevalent use of environmental, social, and governance (ESG) ratings that compile composite metrics does not help with clarity, as they themselves have come under fire due to questionable credibility (Financial Times, 2018). The myriad of metrics currently in use have further exacerbated the chaos (WBCSD, 2020) and the field of sustainability reporting is desperately in search of standardization (Deutsche Bank, 2019).

The inception of sustainability reporting can be traced back to the 1970s and 1980s, propelled by the then emerging demand for nonfinancial performance information to pursue market-based social activism led by religious leaders and early-day social investors (Brown et al., 2009). The momentum continued to grow in the 1990s and voluntary reporting started to arise, as multinational corporations adopted these ideas in order to get a better share of voice in setting the agenda and preempt activists' call for international norms of corporate behaviors (Brown et al., 2009). As a descendant of the social movement and an institutionalized torch-carrier, the Global Reporting Initiative (GRI) was envisioned by its founders to "*improve corporate accountability by ensuring that all stakeholders … have access to standardized, comparable, and consistent environmental information akin to corporate financial reporting … to empower NGOs around the globe with the information they need to hold corporations accountable*" (CERES, 1997). So, if today's sustainability reporting is to stay true to its founding values, it should be about making transparent corporations' environmental and social performance in order to hold them accountable accordingly. Transparency, as the means to the end of corporate accountability, is key to sustainability reporting.

In this chapter, we follow Bourne et al. (2000) and conceptualize performance measurement as a decision support information system comprising three phases: indicator

Sustainable Resource Management. https://doi.org/10.1016/B978-0-12-824342-8.00017-1

design, data collection and analysis, and performance information reporting. Traditional performance measurement is concerned only with the economic aspect; sustainability performance measurement (SPM) broadens the scope by incorporating the environmental and social dimensions. Therefore, the empirical process of SPM is multiphased (design, implementation, reporting) and multidimensional (environmental, social, economic). A sustainability report is part of the output from the third phase, reporting, which covers performance information on the environmental and social dimensions as well as economic externalities that are not traditionally included in financial reports. GRI defines it as a report published by a company or an organization about the economic, environmental, and social impacts caused by its everyday activities (GRI, 2016).

Against this backdrop, this project is focused on designing environmental performance indicators (EPIs) within the broader system of SPM. The rationale is that in order to ensure transparency, sustainability performance information needs to be grounded on the solid foundation of physical flows along the environmental dimension. Physical flows provide a concrete structure for SPM to build upon as they are not just "material carriers" (environmental) but also "activity carriers" (social) and "cost/revenue carriers" (economic) that move through an organization's value chain. The intangible social and economic performance indicators can be developed and organized systematically on the foundation of the tangible physical-based EPIs, and they together can construct an architecture of indicators for the implementation and reporting of the multiphased multidimensional SPM.

1.2 Definitions and terminology

Sustainability or sustainable development refers to "*development that meets the needs of the present without compromising the ability of future generations to meet their own needs*" (World Commission on Environment and Development, 1987, p. 37). Sustainability performance is similar to GRI's definition of "*impact,*" referring to "*the effect an organization has on the economy, the environment, and/or society, which in turn can indicate its contribution (positive or negative) to sustainable development*" (GRI, 2016).

Environmental aspect (EA) is the "*element of an organization's activities or products or services that can interact with the environment*" (ISO 14001, 2015). Environmental impact (EI) refers to "*any change to the environment, whether adverse or beneficial, wholly or partially resulting from an organization's environmental aspects*" (ISO 14001, 2015). Environmental performance (EP) is the "*measurable results of an organization's management of its environmental aspects*" (ISO 14031, 2013). An EPI is "*measurable representation (that) provides information about an organization's environmental performance*" (ISO 14031, 2013). EPIs can be classified as operational performance indicators (OPIs) that provide information about the EP of an organization's operational processes; management performance indicators (MPIs) that assess the management activities to influence an organization's EP; and environmental condition indicators (ECIs) that provide information about the local, regional, national, or global condition of the environment (ISO 14031, 2013).

OPIs are the primary indicators of an organization's EP since they are based directly on the physical input/output flows in the operational processes (ISO 14031, 2013), including examples such as energy consumption and emission generation in total amounts as well as in relation to production volumes. MPIs are relevant but secondary to EP as they supply information on management activities to control EAs, so an excessive use of MPIs do not reflect the actual results of EP and may even camouflage them (UNDSD, 2001). Examples of MPIs include staff training, participation in environmental programs, and supplier auditing. ECIs are primarily measured and monitored by government agencies unless the effects can be attributed to a single or a few companies, given that many factors can affect the environmental conditions such as the quality of air, water, and soil (ISO 14031, 2013). ECIs are usually referred to as "*environmental indicators*" (UNDSD, 2001) and are out of the scope of this study.

The terms "*industry*" and "*sector*" are used interchangeably in this study. The term "*material*" may refer to (1) "*physical objects rather than the mind or spirit*", or (2) "*important, essential, relevant*" (Oxford Dictionary), depending on the context.

2. Review

2.1 The field of sustainability reporting

The field of sustainability reporting can be broadly illustrated as a three-layer pyramid. As shown in Fig. 11.1, on the top of the pyramid is ESG Rating—this layer collects and compiles data related to ESG issues into composite metrics with the aim to support investment and procurement decisions. The middle layer is Sustainability Reporting—organizations disclose their environmental and social performance in their annual reports or standalone sustainability reports. At the based of the pyramid is Methodological Frameworks—the "plumbing" layer that sets guidelines and standards for SPM. This foundational layer is the focus of the project and will be closely studied in Section 4. The first two layers are reviewed as follows.

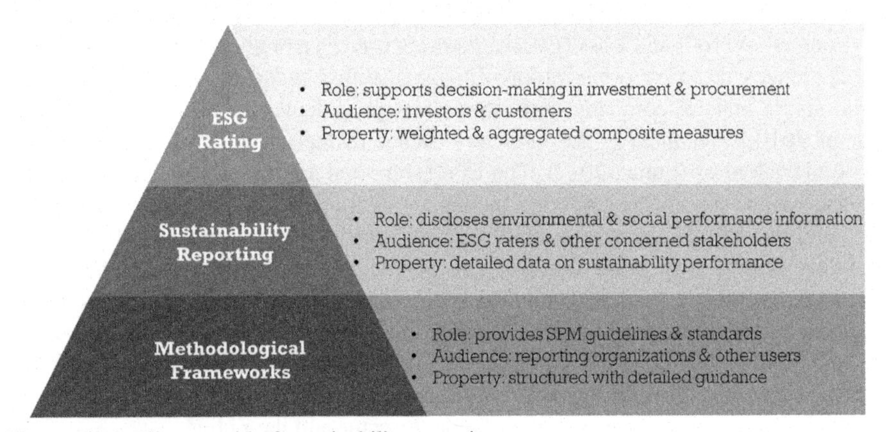

Figure 11.1 The pyramid of sustainability reporting.

2.1.1 ESG ratings

ESG ratings have been criticized for their rampant inconsistencies. Such examples abound—e.g., Tesla was rated last by FTSE for global auto ESG, best by MSCI, and average by Sustainalytics (Financial Times, 2018); and a study by the Japanese Government Pension Investment Fund found that the correlation between FTSE and MSCI rankings for the population of 430 Japanese companies is very low (GPIF, 2017). As the President and CEO of WBCSD points out, with more than 600 ESG rankings and 4500 performance metrics, the field is "a *bit of a zoo*" and in urgent need of standardization (WBCSD, 2020).

Underlying the inconsistencies are methodological issues. Despite the lack of full transparency, the general procedure to construct the ESG scores is to identify sector-specific key issues, collect and analyze data, assign evaluative scores, and aggregate data into a composite metric for rating and ranking. It is unsurprising that the results would come out differently as any minor variations in these steps could affect or even distort the final results—e.g., a sensitivity analysis found that different weighting approaches affect the financial significance of the compiled ESG scores (MSCI, 2020a). Another issue is the subjectivity involved in the evaluation process—e.g., MSCI's key issue score for carbon emissions does not measure companies' carbon emissions, but instead assess how they manage their emissions relative to their exposure to potential regulatory risks (MSCI, 2020b). Furthermore, there are risks of oversimplification and loss of information by weighting and aggregating a set of diverse indicators with different properties (such as measurement boundary and temporal orientation) into one single composite indicator. Such issues render uncertain the validity and reliability of the ESG rating methodologies. The ESG rating scores hence might not faithfully reflect companies' performance in these areas. As a leading expert in this field points out, they are mostly just "*quantified prejudice.*"

2.1.2 Sustainability reporting

The quality of insights derived from ESG ratings can only be as good as the quality of their data sources, provided that the methodologies are solid. In this section, Puma's Environmental Profit and Loss (EP&L) report is used as a mini case study to illustrate the challenges faced by organizations in sustainability reporting. Puma, a German-based sports apparel company, published an unprecedented EP&L in 2012 for the year of 2010 by estimating the monetary values of the negative EIs caused by its economic activities (Puma, 2012). The EP&L reported approximately EUR 8 million of unaccounted environmental costs in Puma's own operations, and additional EUR 137 million in its upstream supply chain from raw material production to manufacturing to the point of sale (environmental costs cover greenhouse gas (GHG) emissions, water use, land use, air pollutants, and wastes—the details are extracted and summarized in Appendix 1). Puma also published its EP&L methodology and invited other companies to join force in moving toward more transparency in, and accountability for, businesses' environmental responsibilities. This initiative received wide acclaim from both the corporate sector and the public.

However, this pioneering practice did not sustain itself—eight years on, this is the only EP&L Puma has published and made public (it was mentioned in Puma's, 2018 Annual Report that another EP&L was published in 2017 (Puma, 2018), yet the document is not publicly accessible). Why so? The resource intensity of producing such a report might be a plausible explanation. The methodology for the EP&L was codeveloped with PwC, a consulting firm, and TruCost, an ESG advisory specialized in monetizing environmental externalities. The majority of data used in the calculation (representing 88% of the total EIs in monetary value) were not directly available but had to be modeled using TruCost's environmentally extended economic input–output model (Puma, 2012). The model applied government census data, environmental resource use and pollution information, Puma's expenditure data for first tier suppliers, and these suppliers' own revenue information to estimate their, and their supply chains', shares of environmental resources and pollution costs (Puma, 2012). The modeling approach is data-heavy and relies on several key assumptions such as that Puma's suppliers are typical within their sectors and that the valuation results from previous studies can be transferred to new contexts, making the efforts resource-intensive and the results afflicted with uncertainty. Plus, TruCost's monetization model is proprietary and might not be available to Puma for it to continue producing such EP&L year after year independently.

Another plausible explanation for discontinuing the EP&L initiative might be a deliberate choice to avoid inviting public scrutiny and pressure to demonstrate performance improvement. A thorough examination of the environmental sustainability sections in Puma's five Annual Reports for the years from 2014 to 2018 finds that the EPIs used are inconsistent and the key contextual information is ambiguous (e.g., indicating the coverage of reported data is for "*core suppliers*" without specifying the percentage of the supply base, or reporting pollution per piece/pair without providing absolute figures[1]). Such reporting practices, intended or not, can effectively make attempts to construct time series data for performance comparison over time very difficult, if not impossible.

The learnings from this mini case study of Puma's EP&L are twofold. First, there needs to be an ecosystem of sustainability reporting that is inducive to such pioneering EP&L initiatives, such as providing technical support to lighten the burden on the reporting organizations and having other companies join force to sustain the momentum. Second, even for a supposedly environmentally enlightened company like Puma, it is still hard to get transparency in their EP from the five reports prepared "*in accordance with the GRI standards*". It seems that transparency is not assured by the image of the corporation or the number of EPIs in the sustainability reports, but how these indicators are selected and constructed. Therefore, it is necessary to standardize and systematize EPIs in order to address the challenges in sustainability reporting and concerns on greenwashing.

[1] A brief summary of the findings is available upon request.

2.2 Environmental performance indicators in research

A systematic literature review[2] was conducted on academic research in the environmental–economic intersection of corporate sustainability (referred to as green competitiveness, GC). After screening for relevance and quality, the review retained and analyzed 171 articles on this topic published between 1975 and 2019. The analysis found that the operationalization of EP in research is highly fragmented, spreading across ten categories with a myriad of different EPIs. As shown in Table 11.1, the top three EPI categories are MPIs that cannot be classified elsewhere (80 studies, 47% in the dataset), pollution output indicators (32 studies, 19%), and indicators of green supply chain management (15 studies, 9%).

Table 11.2 shows the type of data used for the EPIs in GC research. The top three categories are perceptual data collected from surveys which may raise validity concerns due to the method's susceptibility to social bias (84 studies, 49% in the dataset), evaluative data (33 studies, 19%), and physical data (29 studies, 17%).

Table 11.1 EPIs in GC Research.

EPI	Description	No.	%
Not elsewhere classified	Environmental strategies/policies/capabilities, staff training/education/pay-links, participation in environmental programs, and other composite measures not elsewhere classified	80	47%
Pollution output	Pollution output after pollution prevention/control	32	19%
Green SCM	Green procurement, logistics, customer collaboration	15	9%
Certification	EMS 14001 and other environmental certifications	12	7%
Green innovation	R&D and innovation to reduce process/product EIs	11	6%
Rating, ranking, and awards	Third-party environmental rating/ranking/awards	10	6%
Investments	Investment and expenditure on environmental management	5	3%
Pollution reduction practices	Preventive/corrective practices for pollution reduction	3	2%
Incidents and penalties	Environmental incidents/penalties/lawsuits	2	1%
Product design	Designing of product features that reduce EIs	1	1%

[2] This subsection is based on the first author's PhD thesis, *On Sustainability,* available upon request.

Table 11.2 EPI data types of GC research.

Data type	Description	No.	%
Perceptual	Data collected from surveys	84	49%
Evaluative	Evaluative data based on third-party ratings/rankings	33	19%
Physical	Physical measures such as GHG emissions in metric tonnes	29	17%
Communicative	Communicative data from corporate announcements or media	18	11%
Financial	Financial data from corporate annual reports or financial databases	5	3%
Anecdotal	Anecdotal evidence such as "best practices" or "success stories"	2	1%

The review finds that the EPIs used in GC research are highly inconsistent with heavy reliance on perceptual data. The findings imply that in order to improve research validity and facilitate knowledge accumulation, it is necessary to develop a holistic conceptual framework to enable reliable EP operationalization.

2.3 Gaps in sustainability performance measurement research

Research in SPM has been growing rapidly, exploring different aspects of the domain. Mura et al. (2018) conducted the first bibliometric review of SPM studies and identified eight research areas, from sustainability performance disclosure to the diffusion of reporting standards, conducted by scholars from multiple disciplines with different theoretical/a-theoretical perspectives. For the area of critical environmental accounting, the most relevant to this essay, they summarized the main arguments—which we intend to refute through this study—as follows: "*(a) fully reliable set of indicators that measure sustainability at corporate level can never be developed. Problems relate to: (1) unit of analysis, (2) scope of analysis, (3) the impossibility to adopt a really systemic view*" (Mura et al., 2018, p. 674). Overall, their findings suggest that the lack of a comprehensive view of SPM has led to the fragmentation of the field, resulting in duplicated efforts, incomplete problem framing, and proposal of partial solutions (Mura et al., 2018).

The need for a holistic SPM framework has been called for by both academics and practitioners. From a theoretical perspective, Bititci et al. (2018) advised that there is an urgent need to develop a conceptual framework as a skeletal structure "*to lay the theoretical foundations for the field and … progress us towards developing a theoretical framework for performance measurement within which all this knowledge may be integrated*" (p. 656). Schaltegger and Burritt (2010) pointed out the value of pragmatism in SPM research, calling for the development of meaningful indicators and a sustainability accounting system that is "*reliable and transparent, and, thereby, provides a credible basis for decision making and accountability*" (p. 383). The European Environmental Agency (EEA) underscores the need for a holistic conceptual framework, stating that it can provide a rationale for indicator selection in a transparent and logical manner, identifying their interrelations, and exposing gaps where indicators are not available to guide the use of best available proxy measures (EEA, 2014).

3. Methodology

3.1 Research objectives

The above review highlights the urgency to standardize EPIs and develop a conceptual SPM framework that can integrate fragmented studies and bring closer research and practice. This project intends to answer the call, with the research objective to *develop a system of generic EPIs in a holistic organizational performance framework (HOPF) as the foundation for EPM standardization.*

In view of the review findings, the generic EPIs and HOPF shall possess three characteristics: (1) relevant—not too disconnected from operations or too complex that users lose understanding of how an indicator may be influenced; (2) transparent—the sources and transformation of data as well as derivation of composite indicators can be verified and replicated; (3) scalable—able to incorporate social and economic performance indicators into an integrated SPM framework. These ideal characteristics are taken as guiding principles in the design of the solution.

Broadly speaking, the natural environment interacts with all industries in the economy in a similar way: providing valuable resource inputs and bearing undesirable residual outputs. However, although the nature of such environmental–economic interactions to a large extent is similar across industries, the degree of the exchanges varies—e.g., although both the beverage industry and the transporation sector use water and emit greenhouse gases (GHGs), the former uses much more water than the latter, while the latter emits much more GHGs than the former. We therefore propose that *EPIs for all industries can be considered "same, but different"—i.e., all industries may use the "same" set of EPIs with "different" weights.* The standardization of EPM can then be divided broadly into two phases: (1) Identifying the "*same*" set of EPIs; (2) Developing the "*different*" context-contingent weights of the EPIs. This essay is focused on the first phase and will discuss how a weighting scheme may be built upon the "plumbing" work done through the project.

The next section introduces the methodology. Section 4 presents the data and Section 5 the results. Section 6 discusses the implications. Section 7 concludes the study and explores future research.

3.2 Research model

The research model of this project is shown in Fig. 11.2. From the perspective of EPM, an organization is in a physical exchange relation with the natural environment and other organizations in the economy via physical material flows. Measuring these physical flows (inputs or "antecedents") across an organization's boundary can therefore provide instructive information on the organization's management of its EAs

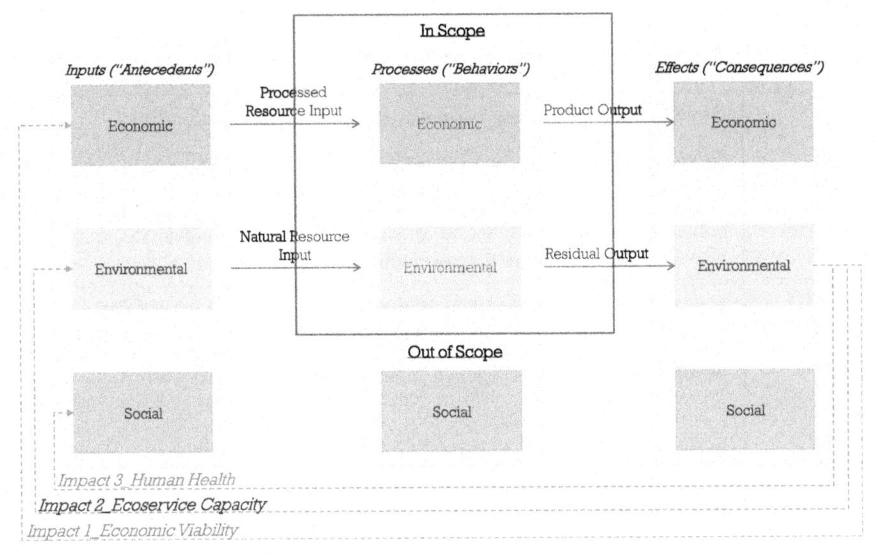

Figure 11.2 Research model.

(processes or "behaviors") and the resultant effects ("consequences"). On the environmental dimension, these effects may be immediate (e.g., resource use and residual release), midterm (e.g., natural capital depletion and degradation), or long-term (e.g., impaired ecosystem service capacities). This project is focused only, as a first step, on the immediate environmental effects of the physical material flows. Such physical flows of resource use and pollution release in economic activities can also affect the economy and society in the short-, mid-, and long-term (e.g., rising prices of certain materials due to resource depletion and scarcity, reduced leisure and cultural activities, or impaired human health). These are out of the scope of this study.

3.3 Research design

Guided by design thinking (Beckman and Barry, 2007), this project includes four stages: (1) Observation, (2) Framing, (3) Imperatives, and (4) Solution. The Observation stage takes place in Section 2, grounding the project in deep understanding of the practical and academic contexts of the problem and the prospective use of the solution. Based on insights generated from the observation, the Framing stage identifies what is missing (practical challenges and research gaps) and formulates the problem that needs solving (research objectives). Imperatives are a set of design principles distilled from the insights that will guide the design of the solution (the Imperatives are the three ideal characteristics of HOPF—relevant, transparent, and scalable). Finally, Solution is to be developed with guidance from the Imperatives—the research model in Fig. 11.2 can be considered a prototype of HOPF that will go through design details formulation and improvement in Section 4 before the solution is presented as results in Section 5.

Table 11.3 Methodological frameworks as data sources.

Framework	Description	Relevance
GRI	Global Reporting Initiative Standards	Indicators
SASB	Sustainability Accounting Standards Board Standards	Indicators
UNEMA	United Nations Environmental Management Accounting Procedures and Principles	Indicators
SEEA CF	United Nations System of Environmental-Economic Accounting 2012 Central Framework	Nomenclature
EEA indicators	European Environment Agency Indicators	Indicator framework
EW-MFA	European Commission's methodological guide for economy-wide material flow accounts	Material accounting structure and unit
YCW index	Environmental Performance Index developed by Yale, Colombia, and the World Economic Forum	Composite index structure

As shown in Table 11.3, seven methodological frameworks relevant to SPM are selected as data sources for the formulation of design details. The first three frameworks are primary sources and chosen for their representative nature, practical relevance, and pragmatic validity. They complement each other and together provide a holistic perspective for SPM at the corporate level: (1) GRI as external sustainability reporting standards covering a broad range of externality impacts of corporate economic activities; (2) SASB is forward-looking and focused on financially relevant sustainability risks/returns with investors as its target audience; (3) UNEMA is backward-looking and aims to support internal management decision-making with environmental management accounting (EMA). The other four macro level frameworks are chosen for their technical rigor and scientific validity, providing guidelines on nomenclature, principles, and considerations on potential methodological alignment and data aggregation at different measurement levels.

The seven frameworks are examined in detail with regard to their purposes, characteristics, and implications for EPI and SPM standardization. The indicators that fit the purpose of the study are first selected, refined, and harmonized according to the design principles; and then synthesized, classified, and organized in HOPF in a systematic manner that is exhaustive, mutually exclusive, and logically homogeneous.

4. Data

4.1 GRI

Founded by the Coalition for Environmentally Responsible Economies (CERES) in 1997, GRI is the most widely adopted sustainability reporting standards globally

(GRI, 2020). GRI has published four generations of guidelines since 2000 and transitioned from providing guidelines to setting global standards in 2016, publishing the current GRI Sustainability Reporting Standards 2016 (the "Standards") (GRI, 2020). The Standards include a set of Universal standards applicable to all organizations (GRI 101, 102, 103) and 33 topic-specific standards organized into Economic (GRI 200), Environmental (GRI 300), and Social (GRI 400) series.

A close examination finds that despite the title, the GRI Standards are actually quite flexible with regard to reporting scope, topic boundary, and reporting timespan. First, the reporting scope according to the Standards shall cover "*topics that reflect a reporting organization's significant economic, environmental and social impacts; or that substantively influence the assessments and decisions of stakeholders*" (GRI, 2016, p. 10). A topic's inclusion in the report depends on its "*materiality*", which is determined by the reporting organization's considerations of various factors, such as its "*overall mission and competitive strategy, the concerns expressed directly by stakeholders ... broader societal expectations, and by the organization's influence on upstream entities ... or downstream entities*" (GRI, 2016, p. 10). Given the myriad of factors involved in determining a topic's materiality, the inclusion of topics can be subjective and organization-specific rather than standardized. Second, topic boundaries that describe "*where the impacts occur for a material topic, and the organization's involvement with those impacts ... (can) vary based on the topics reported*" (GRI, 2016, p. 28). Third, the reporting timespan is decided by the reporting organization, which can be one year or two years. The flexibility in these areas can obstruct data aggregation at the corporate level for overall EP assessment and may also create loopholes for selective reporting and greenwashing.

GRI has been a powerful driving force in mainstreaming sustainability reporting, due in no small part to its broad multistakeholder engagement—since its inception, the initiative has taken an inclusive and participative approach in setting the guidelines ("by *the users and for the users*") with the intention that businesses would not see it as an adversial move by social activists but embrace it as a win-win friendly effort that delivers efficiency gains (Brown et al., 2009). Underneath its pragmatic multistakeholder participative approach though, the vision that CERES set for GRI was for the initiative "*(to) improve corporate accountability by ensuring that all stakeholders ... have access to standardized, comparable, and consistent environmental information akin to corporate financial reporting ... to empower NGOs around the globe with the information they need to hold corporations accountable*" (CERES, 1997). Yet, although GRI may have succeeded in normalizing the practice of sustainability reporting, it seems to have fallen short of empowering NGOs and other stakeholders with "*standardized, comparable, and consistent environmental information.*" Some critics have questioned "*the excessive influence of the business sector on the guidelines ... (indicating) we felt that GRI was a movement but then it evolved into a service organization*" (Brown et al., 2009, p. 27). The abovementioned flexibility within the Standards' fundamental structure might be the result of corporations' preemptive move against activists' demand for transparency and accountability, or the inevitable consequences of the complexities involved in setting globally applicable, sector-neutral reporting standards covering the multiplexes of sustainability.

The primary implications of GRI on EPM standardization are to precisely define the scope and focus on measuring material flows using physical indicators. Hence, the Standards' Environmental disclosure topics that meet such criteria are adopted as source indicators to be further adapted as EPIs. They are highlighted in gray in Appendix 2 which includes a full list of the disclosures in GRI's Environmental series and a brief analysis of their key characteristics.

4.2 SASB

SASB is a nonprofit organization established in 2011 with the mission "*to develop and disseminate sustainability accounting standards that help public corporations disclose material, decision-useful information to investors*" (SASB, 2017, p. 1). SASB refers to sustainability as "*corporate activities that maintain or enhance the ability of the company to create value over the long term*" and defines "*sustainability accounting (as) the measurement, management, and reporting of such corporate activities*" (SASB, 2017, p. 3). The SASB standards were officially published in 2018, including a materiality map to help identify relevant and significant ESG issues for 77 industries and a set of industry-specific sustainability accounting standards covering five dimensions: Environment, Social Capital, Human Capital, Business Model and Innovation, and Leadership and Governance. The accounting boundary and timespan are designated as the same as financial reporting, and the material topics and accounting metrics are specified by industry (SASB, 2017). Following the SASB standards, it is feasible to aggregate data at the corporate level for within-industry performance comparison. Judging from these characteristics, SASB seems to be one step ahead of GRI with regard to enabling "*standardized, comparable, and consistent environmental information akin to corporate financial reporting.*"

However, since the SASB standards are only concerned with ESG issues that are financially relevant to a reporting organization, their coverage does not include sustainability externalities which are the focus of GRI. In other words, sustainability issues are off the radar for SASB unless they mean business to business—with high probability of materialization and significant financial impacts. Essentially, they are merely improved financial reporting standards adapted to a new context where sustainability concerns in society represent relevant and significant financial risks and opportunities. This new approach of reporting expands the traditional backward-looking financial accounting reports by adding forward-looking evaluation of risks and opportunities associated with environmental and social issues in order to help financial capital providers better assess the reporting organizations' future operational performance and financial positions. As such, GRI is stakeholder-oriented with broad coverage of sustainability issues, while the SASB standards are shareholder-centered involving a narrow set of reporting topics. The Puma EP&L provides an example to illustrate such differences. In the two areas with significant negative EIs, Puma reported (1) 110,100 tons CO_2e of GHG emissions (~EUR 7 million) from its own operations and 607,400 tons CO_2e (~EUR 40 million) in its upstream supply chains, and (2) 77.4 million CBM (~EUR 46 million) of water use in its upstream supply chains (Puma, 2012). Yet, both topics are not covered in the SASB standards—in fact, according to SASB's "financial materiality" criterion for inclusion, there are no environmental topics that need to be disclosed by the Apparel, Accessories, and Footwear industry (SASB, 2020) as the environmental externalities are not currently, nor perceived soon to be, internalized.

Yet, such significant differences between GRI and SASB are not immediately obvious to most and both are often seen as peer frameworks complementing each other. Why so? The one critical factor that has muddied the water might be the different definitions of "sustainability". As opposed to the common definition of sustainability in GRI following the Brundtland Report, "*(f)or the purposes of SASB standards, sustainability refers to corporate activities that maintain or enhance the ability of the company to create value over the long term*" (SASB, 2017, p. 2). Both use the term to mean "*the ability to exist*", yet GRI refers to which of humankind, while SASB refers to which a company—the latter's use of languate is perfectly legitimate, it unfortunately just might be perfectly misleading. Whether it is intended or innocent is beyond this study, but it seems that to name things ambidextrously can indeed add to the misfortune of the world, if we may paraphrase the French philosopher Albert Camus here.

So, if ESG ratings are based on data in corporate sustainability reports that are based on the SASB standards which are based on a financialized, individualized, self-serving definition of "sustainability", are they still fit for the purpose of impact investing, which by definition is about generating "*positive, measurable social and environmental impact alongside a financial return*"(GIIN, 2020)? Probably not. It seems that a healthy dose of skepticism would not hurt here and much of the sustainability talks nowadays would better be taken with a generous grain of salt.

For the purpose of EPM standardization, the SASB standards' primary strength of industry-specificity can be adapted to compensate for the generality of GRI disclosure topics. Appendix 3 lists SASB's seven environmental issue categories relevant to this project, with six of them adapted from the Environment dimension and one from Business Model & Innovation. The key differences between GRI and SASB are summarized in Appendix 4.

4.3 UNEMA

UNEMA is a methodological framework that provides guidelines and techniques for quantifying environmental expenditures and costs for EMA. EMA is the identification, collection, estimation, analysis, internal reporting, and use of physical flow information, environmental cost information, and other monetary information for decision-making within an organization (UNDSD, 2001). EMA can be carried out with either monetary or physical units: (1) monetary EMA provides information for calculating potential savings, budgeting, product pricing, or disclosure of environmental expenditures and investments; (2) physical EMA provides information for environmental management systems and internal benchmarking or external sustainability reporting. EMA is not full accounting of environmental costs since it focuses only on those actual, materialized financial costs borne by organizations.

UNEMA provides detailed techniques to account for material flows and indicates that most of the data required for physical EMA can be found and extracted from material management systems and other internal systems. It also offers guidelines for using product output as a denominator to track changes in environmental efficiency for internal benchmarking. The framework offers an insider perspective on how data in environmental sustainability reports are sourced and processed, so may be used as a reference for sustainability report auditing. Most of the indicators in UNEMA overlap with either GRI or SASB, although their scopes of measurement differ.

4.4 SEEA CF

SEEA CF is a methodological framework adopted by the United Nations Statistical Commission in 2012 as the international statistical standard to guide the collection of comparable, coherent, and reliable data for holistic policymaking and research related to the environmental–economic intersection (UN, 2014). It is used as a background reference framework for this project, since (1) it contains the internationally agreed concepts, definitions, and classifications (the "nomenclature") for compiling environmental data that are globally comparable, (2) its accounting nomenclature is consistent with the System of National Accounts (SNA) which can facilitate the integration of environmental and economic statistics, and (3) it has been developed with the view of application at industry, enterprise, plant, or product level for data aggregation across different levels of accounting (UN, 2020).

Four categories of environmental indicators have been developed as an illustration of the application and extension of SEEA CF (UN, 2017). The most relevant to this project is the category of resource management indicators, grouped broadly as resource use and environmental intensity. Resource use indicators characterize resource intensity by relating environmental input variables with economic variables (e.g., joules of energy use per unit of GDP in local currency). Environmental intensity indicators characterize pollution intensity by relating environmental output variables with economic variables (e.g., tons of air pollutant emission per unit of GDP in local currency). Such categorization can be related to the input/output delineation of HOPF in Fig. 11.2, and the EPIs developed in next section shall follow a similar structure.

4.5 EEA indicators

EEA has developed a set of environmental indicators to support environmental policy making. There are currently 122 indicators maintained, covering 13 environmental topics: air pollution, biodiversity/ecosystems, climate change adaptation, climate change mitigation, energy, environment and health, industry, land use, resource efficiency and waste, soil, sustainability transitions, transport, and water and marine environment (EEA, 2020). EEA uses a DPSIR (Driving force, Pressure, State, Impact, and Response) framework to structure thinking about the interactions between socioeconomic activities and the natural environment (EEA, 2014). According to the DPSIR framework, socioeconomic developments *Drive* alternations that exert *Pressure* on, and change the *State* of, the environment, leading to *Impacts* on ecosystem functioning, human health, as well as the economy stipulating societal and political *Responses* to remedy the situation (EEA, 2014). The 122 EEA indicators cover all the aspects of the DPSIR framework, providing a holistic decision support tool for environmental policy making.

Using the DPSIR typology, OPIs—the focus of this project—are Pressure indicators that characterize "*the release of substances, physical and biological agents, (and) the use of resources and land*"; and MPIs are Responses indicators that describe "*responses ... that attempt to prevent, compensate, ameliorate, or adapt to changes in the state of the environment*" (EEA, 2014). The other three types of indicators (Driving force, State, and Impact) are not currently measured and managed at the corporate level. From the perspective of sustainable development, developing indicators across the full logic chain of DPSIR might be a key first step to identify and quantify environmental externalities for internalization.

4.6 EW-MFA

The European Union's EW-MFA is a statistical accounting framework describing the physical interaction of the national economy with the natural environment and the other economies in the world in terms of material flows (EU, 2018). EW-MFA is compatible with the principles and definitions set by SEEA CF, although it has a narrower scope focusing only on material flows in and out of the economy without covering intraeconomy flows or environmental asset accounts (EU, 2018). At a different measurement level, this project's defined scope of EPM for an organization is conceptually consistent with EW-MFA's scope of material accounting for a nation, in that the Processed Resource Input/Product Output flows depicted in the research model in Fig. 11.2 can be analogized to the import/export flows in EW-MFA.

One limitation with the EW-MFA indicators is that they add together materials in different stages of production (raw materials, work-in-progress, and final products) (EU, 2018). This asymmetry can be overcome by accounting for materials that are expressed in terms of their "material footprints", i.e., the amount of raw materials necessary to produce them (raw material equivalents, RME) (EU, 2020). MFA-RME material indicators are more granular metrics of material intensity, so the concept of RME and its potential extensions are adapted to envision the future of EPM standardization with HOPF.

4.7 YEPI

YEPI is a composite measure developed at Yale to quantify and rank environmental states of nations around the world (Wendling et al., 2020). As an information tool to support policy efforts toward sustainable development, YEPI has been published biannually since 2006 with the latest 2020 report covering 180 countries. It includes 32 subindicators across 11 categories that are classified into two policy objectives: environmental health (four categories with seven indicators) and ecosystem vitality (seven categories with 25 indicators). Source data for YEPI are collected from national statistics, normalized, weighted, and aggregated into the composite indices (Wendling et al., 2020). The key strength of YEPI is its simplicity and clarity in communicating national environmental conditions to policymakers and the general public unfamiliar with environmental issues. Another of its strengths is the transparency in the methodology and data sources, making the published results verifiable and credible.

The main weakness of YEPI is its lack of explicit consideration of the interplays between the economy and the environment, which may conceal the driving forces underlying the environmental state changes (such as offshoring instead of greening domestic manufacturing). This is further exacerbated by the inconsistency of the temporal boundaries for its comprising subindicators (e.g., in the 2020 report, the CO_2 growth rate is calculated as the average annual change in raw emissions over the years 2008–2017, SO_2 growth rate over 2005–2014, and GHG per capita is based on data in 2017) (Wendling et al., 2020). Another issue with YEPI is that it encapsulates logically inconsistent proxy measures into one supposedly State indicator, including Pressure (e.g., pollution emissions), State (e.g., fish stock status), and Impact (e.g., Ozone exposure proxied by the number of age-standardized disability-adjusted life-years) indicators across the DPSIR chain. Such simplification does not enable the identification of the areas where policy interventions are needed and may even mislead by fostering a false sense of confidence and clarity. YEPI provides an example of how a standardized environmental index at the macro level is constructed. Keeping its strengths and weaknesses in mind, this project aims to build a solid analytic foundation with HOPF as the basis for such an EP index at the corporate level.

5. Results

5.1 HOPF as a structural framework

A critical first step toward EPM standardization is to identify the measurement scope of the indicators. As shown in Fig. 11.3, HOPF proposes that an organization is

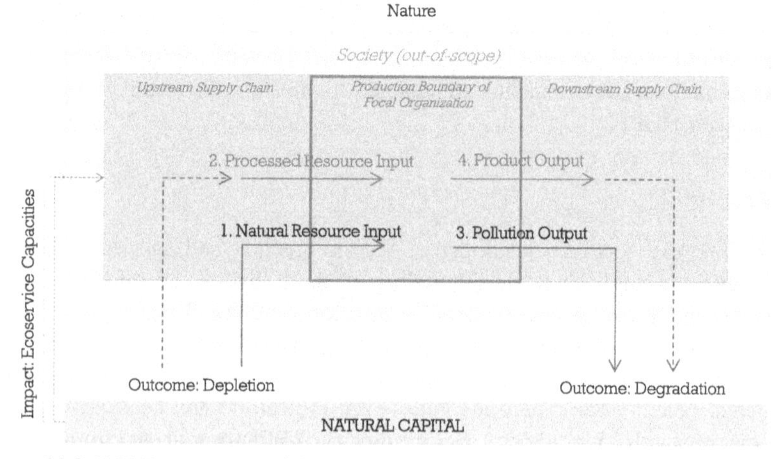

Figure 11.3 HOPF as a structural framework.

embedded in its supply chains embedded in the economy embedded in the broader social–biophysical ecosystems. HOPF is built upon the first author's PhD thesis and the External Rate of Return framework that measures an investment's purpose and wider impact on society, the economy, and the environment (Time Partners, 2020). The measurement scope of EPIs is delineated in HOPF as the focal organization's production boundary, defined as "*comprising a specific set of economic activities carried out under the control and responsibility of institutional units in which inputs of labour, capital, and goods and services are used to produce outputs of goods and services (products)*" (UN, 2014, p. 39).

The physical flows crossing an organization's production boundary can be categorized into four groups: (1) Flow 1—Natural Resource Input from the environment directly to the organization (including biomass resources, metals, nonmetallic minerals, energy carriers, and water), (2) Flow 2—Processed Resource Input from upstream supply chain to the organization (including processed raw materials and semiprocessed components, carrying in them "embodied" Natural Resource Inputs that have been processed and transformed by upstream production processes to different degrees), (3) Flow 3—Pollution Output from the organization to the environment (including flows of gaseous, liquid, and solid residuals released and discarded, instead of recycled, by the organization), and (4) Flow 4—Product Output from the organization to downstream supply chain (either for intermediate consumption as Flow 1 for institutional customers or for final consumption by ultimate customers—both, together with the products stocked in the organization, will eventually become residuals in Flow 3 within or beyond the production boundary spanning across different accounting periods).

The above scope specified in HOPF is similar to which of UNEMA (with different EPIs) and EW-MFA (at different measurement levels), but broader than GRI and SASB since it covers Processed Resource Input (Flow 2) and Product Output (Flow 4). Such a broader scope can be justified from both practical and theoretical perspectives. First, from a pragmatic standpoint, including Processed Resource Input (Flow 2) in EPM can enable the identification of the "shared" environmental responsibilities among the producers and the customers of goods in the supply chain. With such an approach, the suppliers of goods are responsible for reducing EIs in their production processes, and the customers (including the focal organization in this case) are responsible for reducing the volume of their consumption. Applying the same logic consistently along the supply chain can help ensure that responsibilities for the environmental pressures throughout the lifecycle of physical goods are clearly understood and properly managed rather than simply displaced somewhere else (DEFRA, 2011).

Second, including data on Product Output (Flow 4) can provide key information to contextualize the other EPIs for meaningful interpretation. For manufacturing organizations, product output is the primary driving force of environmental pressures pulling in resource inputs (Flow 1 and 2) and pushing out pollution outputs (Flow 3). Therefore, the absolute figures of these EPIs need to be related to product output in order to assess a manufacturer's EP in terms of eco-intensity or eco-efficiency. With similar reasoning, SASB stipulates the reporting of industry-specific "*activity metrics*" that quantify the scale of a company's business in order to enable data normalization for within-industry performance comparison (SASB, 2017).

Third, from a theoretical perspective, covering Flow 2 in EPM reconciliates the competing "either production- or consumption-based" perspective of the "polluter pays principle" into a compatible "production- and consumption-based" perspective. Indicators with the production-based view account for the environmental flows directly "used" (extraction of natural resources) and/or "produced" (residual flows) by producers as their "direct" environmental responsibilities, while all the other flows are seen as "indirect"—e.g., the Scope 1 of "direct" GHG emissions and Scope 2 and 3 of "indirect" emissions as defined by the GHG Protocol (2004). On the other hand, consumption-based indicators draw on the premise that the consumer of a product is ultimately responsible for EIs "embodied" in the product (e.g., calculation of carbon footprint or product lifecycle analysis). HOPF adopts a holistic perspective and proposes that as an economic enterprise, an organization is both a producer (of Flow 3 and Flow 4) and a consumer (of Flow 1 and Flow 2). By accounting for Flow 2 of Processed Resource Input and Flow 4 of Product Output in EPM, it makes transparent the company's true and full environmental responsibilities. EW-MFA includes imported, in addition to domestically extracted, materials in its accounts by applying the same logic (EU, 2018). Furthermore, such a broadened scope does not result in "double counting" but instead enables "data coupling" which may address the current challenge that Scope 3 data are virtually unavailable.

In sum, HOPF's proposed scope of EPM is sufficiently, although not absolutely, exhaustive to gauge an organization's EIs in full and avoid partial coverage that may mislead and misinform. Furthermore, HOPF classifies physical flows into homogenous categories and organizes them with a supply chain structure, making it operationally relatable, straightforward, and decomposable at a more granular level within a coherent framework. These characteristics of HOPF prove that the solution design follows the imperatives of relevant, transparent, and scalable.

5.2 Generic EPIs as the building blocks

Table 11.4 presents a system of generic EPIs applicable to different industries. These are the refined and synthesized indicators adapted from the three primary data sources: GRI, SASB, and UNEMA. They are constructed in a mutually exclusive and logically coherent structure with HOPF. The three-level hierarchy based on YEPI can enable data transformation and aggregation into a composite index in future research. A full list of the indicators with detailed descriptions and their primary EIs is available in Appendix 5.

The system of generic EPIs is organized in three levels covering two EAs, nine categories, and 15 indicators. The first level of classification is based on the direction of physical flows, including environment-organization inflows on the environmental input aspect (EIA) and organization–environment outflows on the environmental output aspect (EOA). The second level of classification is based on material physical properties and environmental mediums (i.e., sources for EIA and destinations for EOA). EIA includes five categories of resource input (Flow 1 and Flow 2): Processed

Table 11.4 A system of genenc EPIs.

EA	Category	Indicator	Measure	GRI	SASB	UNEMA
EIA	Processed materials	Raw processed materials (% recycled)	Tons (%)	301-1,2	410a	UNEMA
		Packaging materials (% recycled/ renewable)	Tons (%)	301-1	n/a	UNEMA
	Natural materials	Raw natural materials	Tons	301-1	n/a	UNEMA
	Energy	Energy purchased (% renewable)	Joules (%)	302-1	110a	UNEMA
		Energy self-generated (% renewable)	Joules (%)	302-1	110a	UNEMA
	Water	Water purchased	Tons	–	n/a	UNEMA
		Water self-withdrawn	Tons	303-3	140a	UNEMA
	Land use	Land use biosensitive areas	M2	304-1	160a	n/a
EOA	Air emissions	GHG emissions in CO_2e (scope 1 direct)	Tons	305-1	110a	UNEMA
		GHG emissions in CO_2e (scope 2 indirect)	Tons	305-2	n/a	n/a
		Air pollutants	Tons	305-7	120a	n/a
		Ozone-depleting (ODS) substances	Tons	305-6	n/a	n/a
	Water effluents	Water effluents	Tons	303-4	n/a	UNEMA
	Solid wastes	Waste generated (% hazardous, % recycled)	Tons	306-3,4	150a	UNEMA
	Product output	Products produced	Tons/ Units	–	000.A	UNEMA

Materials, Natural Materials, Energy, Water, and Land Use. EOA includes three categories of Pollution Output (Flow 3): Air Emissions,[3] Water Effluents, Solid Wastes; and the category of Product Output (Flow 4).

The indicators are based on physical flows measured in absolute amount with physical units, serving as a solid and unequivocal basis for further data transformation, aggregation, comparison, and evaluation. For example, by relating the indicators with one another, an overall picture of an organization's EP can be broadly depicted: (1) EIA and Pollution Output indicators on EOA represent the overall environmental pressures exerted by the organization's economic activities; (2) EIA over Product Output indicates its material intensity (with its inverse indicating material productivity); (3) Pollution Output over Product Output indicates its pollution intensity; and (4) Material intensity and pollution intensity combined represent the organization's eco-intensity.

HOPF and the system of generic EPIs together are the project's deliverable solution. As a proof of concept, the next section discusses how they can form a firm foundation for EPM standardization.

5.3 Toward EPM standardization

For the purpose of EPM standardization, this project sets out with the supposition that *EPIs for all industries can be considered "same, but different"*—i.e., *all industries may use the "same" set of EPIs with "different" weights.* Now that the "*same*" set of generic EPIs have been developed, the next step in this endeavor is to develop "*different*" context-contingent weighting schemes for data transformation and aggregation. A key challenge facing this task is finding a common measurement unit for the myriad of material types in the physical flows. Money has been used as a common currency (as in Puma EP&L and TruCost's monetization model), since it speaks the same language as financial indicators that everybody understands. However, valuation of EIs with monetary units is a complex and controversial issue involving assumptions and value judgments (Sadowska and Lulek, 2016).

Using physical units for normalization would be less equivocal since they are based on intrinsic and tangible physical properties of materials. As mentioned previously, RME can be used to convert different materials into a common unit for data aggregation. A similar concept is the Global Warming Potential factors that convert the other six GHGs into CO_2 equivalent (CO_2e) based on their respective radiative forcing impacts over a specified time horizon (GHG Protocol, 2004). Despite the current limitation that MFA-RME indicators consider only the volume of material mass (which reflects only one aspect of the materials' overall EIs), it is one step forward toward more granular material accounting involving a multitude of physical flows. In order to serve as a common converter for EPM standardization, RME needs to build on a material-based analytic structure—as has been laid out by HOPF and the system

[3] The Indirect Scope 2 GHG emissions are not covered by HOPF but included in the system of EPIs in order to stay consistent with existing practices.

Table 11.5 Envisioning EPM standardization with RME and context-specific weighting.

EA	Category	WT-2	Indicator	Measure RME WT-3
EIA (WT-1)	Processed materials		Raw processed materials (% recycled)	Tons (%)
			Packaging materials (% recycled/ renewable)	Tons (%)
	Natural materials		Raw natural materials	Tons
	Energy		Energy purchased (% renewable)	Joules (%)
			Energy self-generated (% renewable)	Joules (%)
	Water		Water purchased	Tons
			Water self-withdrawn	Tons
	Land use		Land use biosensitive areas	M2

of generic EPIs. Using EIA as an example, Table 11.5 envisions how RME, together with a context-specific weighting scheme that is to be developed in future research, fits into the broader picture of EPM standardization with HOPF.

Regarding the unit of measurement for EPM standardization, product or product groups with similar material contents would be the logical choice since both the EPIs and RME are material-based. Furthermore, as most EIs are localized, the products' production locations are also necessary to contextualize the EPIs for meaningful interpretation (e.g., evaluating the impacts of water withdrawal from water stressed areas). Therefore, the ideal unit of measurement for HOPF-based EPIs would be location-specific product(s). EPIs data collected at this level of granularity can then be normalized and weighted within the specific contexts, and transformed and aggregated into a corporate level EP index that is standardized and comparable across industrial sectors and geographical regions.

6. Implications

Practitioners may use the designed solution to improve consistency and transparency in sustainability reporting and ESG ratings. Here are two examples of using the design as a reference framework in archival analysis for this purpose. One application is to use HOPF and the EPIs to guide the collection and compilation of longitudinal quantitative data in corporate sustainability reports and assess the potential data gaps and inconsistencies. If identified, text analysis can then be performed to seek justifications and explanations from the supplementary contextual information (e.g., outsourcing of manufacturing or merge and acquisitions). Inconsistencies in reported data not justified or explained in the contexts are potential red flags of greenwashing, and recommendations can be made accordingly to rectify such practices. Another application of the design is for cross-sectional check of consistencies among different data sources. The quantitative data collected from corporate sustainability reports can be processed, rather crudely at this stage, into an organization level score to compare with ESG ratings by different advisories. There are likely to be significant discrepancies—which do not necessarily discredit any of the ESG ratings, but the results can identify and expose the areas of differences for reconciliation and improvement.

Researchers on the environmental–economic interface can refer to HOPF for the operationalization EP measures. The generic set of EPIs can be used to guide data collection directly from archival documents instead of from secondary databases—which might or might not be compiled with verifiable data sources through transparent methodologies. Together, they can enhance the validity of GC research by grounding EP measurement on material flows with physical units instead of using proxy measures and perceptual data collected from surveys.

7. Conclusions and future research

This project designs a system of generic EPIs and HOPF as the foundation for EPM standardization. The results address the urgent need to standardize EPIs, integrate fragmented research areas, and bring closer SPM research and practice. HOPF as a structural framework for EPM possesses three ideal characteristics: relevant—grounded in material flows along the supply chains; transparent—verifiable and replicable; and scalable—capable of decomposing into more granular levels and incorporating social and economic indicators into a multidimensional SPM framework. Practitioners in the field may use HOPF to improve transparency by identifying greenwashing behaviors and improving reporting and rating consistencies. Academics may use it as a reference to operationalize EP measures and collect data directly from primary archival sources.

The key limitation of the generic EPIs is that they do not cover industry-specific indicators. However, for the purpose of EPM standardization and given the diversity and complexity of material flows, a generic list of EPIs applicable to all industries should first be designed as a shared foundation for breadth and alignment before industry-specific indicators can be developed for depth and precision. Therefore, the project's trade-off of indicator specificity for generality is necessary and justifiable at the early stage of EPM standardization.

Future research may go in two directions. First, expanding HOPF by incorporating the social dimension of SPM. The task is likely to be more intricate than this project due to the intangible and fluid nature of social sustainability. It is necessary to first build a solid conceptual foundation synthesizing divergent perspectives from different disciplines before attempting to quantify and standardize social performance indicators. Second, developing RMEs and weighting schemes in the pursuit of EPM standardization. With a system of EPIs calibrated to contexts and structured with HOPF, it will be one step closer to sifting through the millions of pages of sustainability reports and generating credible EP indices in a transparent way using artificial intelligence, assuming that it will also possess the critical characteristic of authentic integrity to wade through the muddied water of sustainability.

Appendix 1: Environmental costs estimated in Puma EP&L

Item	Cost (EUR)	Range (EUR)
GHG emissions (Ton of CO_2e)		
GHG	66	
Air pollutant (Ton)		
Particulates	14,983	1,285–191,743
Ammonia	1,673	1,133–5,670
Sulfur dioxide	2,077	783–6,422
Nitrogen oxides	1,186	664–3,179
Volatile organic compounds	836	425–1,1002
Water use (CBM)		
Water	0.81	0.03–18.45
Historic Ecoregion (Hectare)	Avg. 347	
Grassland	229	
Temperate forest	283	
Subtropical forest and woodland	251	
Tropical forest	1,352	
Inland wetland	5,792	
Costal wetland	18,653	
Waste (Ton)		
Landfilled	73	36–87
Incinerated	51	35–63

Appendix 2: GRI environmental disclosures (with adapted disclosures highlighted in gray)

GRI disclosures	Type	Measurement	Scope	Application
301-1 Materials used by weight or volume	OPI	Quan (absolute)	Within	Generic
301-2 Recycled input materials used	OPI	Quan (relative, %)	Within	Generic
301-3 Reclaimed products and their packaging materials	MPI	Quan (relative, %)	Within	Specific
302-1 Energy consumption within the organization	OPI	Quan (absolute)	Within	Generic
302-2 Energy consumption outside of the organization	OPI	Quan (absolute)	Beyond	Specific
302-3 Energy intensity	OPI	Quan (relative, vs. self-chosen denominator)	Within	Generic
302-4 Reduction of energy consumption	MPI	Quan (absolute)	Within	Generic
302-5 Reduction in energy requirements of products and services	OPI	Quan (absolute)	Beyond	Generic
303-1 Interactions with water as a shared resource	MPI	Qual (descriptive)	Within	Specific
303-2 Management of water discharge-related impacts	MPI	Qual (descriptive)	Within	Specific
303-3 Water withdrawal	OPI	Quan (absolute)	Within	Specific
303-4 Water discharge	OPI	Quan (absolute)	Within	Generic
303-5 Water consumption	OPI	Quan (absolute)	Within	Generic
304-1 Operational sites owned, leased, managed in, or adjacent to, protected areas and areas of high biodiversity value outside protected areas	OPI	Various (type of operation, size)	Within	Specific
304-2 Significant impacts of activities, products, and services on biodiversity	OPI	Qual (descriptive)	Within	Specific

GRI disclosures	Type	Measurement	Scope	Application
304-3 Habitats protected or restored	ECI	Various (location, size, status)	Within	Specific
304-4 IUCN Red List species and national conservation list species with habitats in areas affected by operations	ECI	Quan (absolute)	Within	Specific
305-1 Direct (Scope 1) GHG emissions	OPI	Quan (absolute)	Within	Generic
305-2 Energy indirect (Scope 2) GHG emissions	OPI	Quan (absolute)	Within	Generic
305-3 Other indirect (Scope 3) GHG emissions	OPI	Quan (absolute)	Beyond	Specific
305-4 GHG emissions intensity	OPI	Quan (relative, vs. self-chosen denominator)	Within	Generic
305-5 Reduction of GHG emissions	MPI	Quan (absolute)	Within	Specific
305-6 Emissions of ozone-depleting substances (ODS)	OPI	Quan (absolute)	Within	Specific
305-7 Nitrogen oxides (NOX), sulfur oxides (SOX), and other significant air emissions	OPI	Quan (absolute)	Within	Specific
306-1 Waste generation and significant waste-related impacts	MPI	Qual (descriptive)	Within	Specific
306-2 Management of significant waste-related impacts	MPI	Qual (descriptive)	Within	Specific
306-3 Waste generated	OPI	Quan (absolute)	Within	Generic
306-4 Waste diverted from disposal	OPI	Quan (absolute)	Within	Specific
306-5 Waste directed to disposal	OPI	Quan (absolute)	Within	Generic
307-1 Noncompliance with environmental laws and regulations	MPI	Various (e.g., fines, no. of sanctions)	Within	Specific
308-1 New suppliers that were screened using environmental criteria	MPI	Quan (relative, %)	Within	Specific

Continued

GRI disclosures	Type	Measurement	Scope	Application
308-2 Negative environmental impacts in the supply chain and actions taken	MPI	Various (e.g., significant impacts identified)	Beyond	Specific

Appendix 3: SASB environmental issue categories

Issure category	Accounting metric	Unit
Greenhouse gas emissions	110a. 1_Gross global scope 1 emissions	Metric tons (t) CO_2e
	110a.3_Fuel consumed by (1) road transport, percentage (a) natural gas and (b) renewable, and (2) air transport, percentage (a) alternative and (b) sustainable	Gigajoules (GJ), %
Air quality	120a. 1_Air emissions of the following pollutants: (1) NOx (excluding N_2O), (2) SOx, and (3) particulate matter (PM10)	Metric tons (t)
Energy management	130a. 1_(1) Total energy consumed, (2) percentage grid electricity, (3) percentage renewable	Gigajoules (GJ), %
Water and wastewater management	140a. 1_(1) Total water withdrawn, (2) total water consumed, percentage of each in regions with high or extremely high baseline water stress;	Thousand cubic meters (m^3), %
Waste and hazardous materials management	150a. 1_(1) Total amount of waste from manufacturing, (2) percentage hazardous, (3) percentage recycled	Metric tons (t), %
Ecological impacts	0101-11_(Operations) in or near sites with protected conservation status or endangered species habitat	(Varies)
Product design and lifecycle management	410a. 1_Weight of end-of-life products recovered, percentage recycled	Metric tons (t), percentage (%)

Appendix 4: GRI versus SASB

	GRI	SASB
Mission	Enabling transparency and open dialogue about impacts by providing global sustainability disclosure standards and catalyzing the change for a sustainable future	Enabling businesses around the world to identify, manage, and communicate financially material sustainability information to their investors
"Sustainability"	An organization's contribution to SDG: *"sustainability (refers to) development that meets the needs of the present without compromising the ability of future generations to meet their own needs"*	An organization's ability to sustain: *"for the purposes of SASB standards, sustainability refers to corporate activities that maintain or enhance the ability of the company to create value over the long term"*
"Impact"	Externalities: *"the effect an organization has on the economy, the environment, and/or society, which in turn can indicate its contribution (positive or negative) to sustainable development"*	Internalities: sustainability-related risks and opportunities that are relevant and may have significant impact an organization's operational performance and financial position
Reporting scope	Topics with material sustainability impact flexible—not specified and may vary by topic	Topics with material financial impact specific—same as financial reporting
Measurement scope	Flexible - not specified and may vary by topic	Specific—same as financial reporting
Reporting timespan	Flexible	Specific—same as financial reporting
Data consolidation	Data consolidation at corporate level—n/a	Data consolidation at corporate level—yes
Industry-specific	No, industry-neutral	Yes, standards are specific to 77 industries

Appendix 5: Generic environmental performance indicators

EA	Category	Indicator	Measure	Description	GRI	SASB	UNEMA	Primary EI
EIA	Processed materials	Raw processed materials (% recycled)	Tons (%)	Weight of raw materials and components other than natural material inputs that are part of the final products; by type (% recycled, incl. from recovered products)	301-1,2	410a	UNEMA	Resource depletion, waste generation
		Packaging materials (% recycled/ renewable)	Tons (%)	Materials for packaging purposes; by type	301-1	n/a	UNEMA	
	Natural materials	Raw natural materials	Tons	Weight of raw natural materials used for production, incl. minerals, metals, and biomass resources	301-1	n/a	UNEMA	Resource depletion, waste generation
	Energy	Energy purchased (% renewable)	Joules (%)	Purchased electricity/heating/cooling/steam for consumption (% renewable)	302-1	110a	UNEMA	Resource depletion, air emissions
		Energy self-generated (% renewable)	Joules (%)	Self-generated electricity/heating/cooling/steam for own consumption (% renewable)	302-1	110a	UNEMA	
	Water	Water purchased	Tons	Municipal water supplies or other public or private water utilities	–	n/a	UNEMA	Resource depletion, water effluents
		Water self-withdrawn	Tons	Water withdrawal by source	303-3	140a	UNEMA	
	Land/Space use	Land/space use in biosensitive areas	M2	Land/space use in or near sites with protected conservation status or endangered species habitat	304-1	160a	n/a	Biodiversity loss

EOA	Air emissions	GHG emissions in CO2e (Scope 1_Direct)	Tons	Incl. but not limited to fuel consumption, generation of electricity/heating/cooling/steam, physical/chemical processing, and transportation of materials/people	305-1	110a	UNEMA	Climate change, health impact
		GHG emissions in CO_2e (Scope 2_Indirect)	Tons	Incl. emissions from purchased and acquired energy as per 302-1	305-2	n/a	n/a	
		Air pollutants	Tons	Air emissions of the following pollutants: NOx (excluding N_2O), SOx, and particulate matter	305-7	120a	n/a	
		Ozone-depleting (ODS) substances	Tons	Production, imports, and exports of ODS, specifying substances included	305-6	n/a	n/a	
	Water effluents	Water effluents	Tons	Total water discharge, by destination	303-4	n/a	UNEMA	Biodiversity loss, health impact
	Solid wastes	Waste generated (% hazardous, % recycled)	Tons	Total weight of waste generated (% hazardous, % recycled)	306-3,4	150a	UNEMA	Emissions, disamenity, health impact
	Product output	Products produced	Tons/ Units	Total weight/units of physical products produced, by type	–	000.A	UNEMA	Indirect residual generation

References

Beckman, S.L., Barry, M., 2007. Innovation as a learning process: embedding design thinking. Calif. Manag. Rev. 50, 25–56.

Bititci, U.S., Bourne, M., Cross, J.A.F., Nudurupati, S.S., Sang, K., 2018. Towards a Theoretical Foundation for Performance Measurement and Management.

Bourne, M., Mills, J., Wilcox, M., Neely, A., Platts, K., 2000. Designing, implementing and updating performance measurement systems. Int. J. Oper. Prod. Manag. 20 (7), 754–771.

Brown, H.S., De Jong, M., Lessidrenska, T., 2009. The rise of the Global Reporting Initiative: a case of institutional entrepreneurship. Environ. Polit. 18, 182–200.

CERES, 1997. CERES Global Reporting Initiative Concept Paper. Coalition of Environmentally Responsible Economies, CERES, Boston, MA.

Corporate Register, 2020. Search 127,642 CR Reports Across 2Ü,4Ü5 Organizations[Online]. Available: https://www.corporateregister.com/. (Accessed 9 August 2020).

DEFRA, 2011. Environmental Key Performance Indicators - Reporting Guidelines for UK Business. Department for Environment, Food and Rural Affairs, UK.

Deutsche Bank, 2019. Why ESG Investingis Desperately in Search of Standards [Online]. Available: https://deutschewealth.com/en/our-capabilities/why-responsible-investing-is-desperately-in-search-of-standards.html#:~:text=Investors are now showing much,on society and the enviromrient&text=In the worst cases, companies,d11bbed 'green-washing [Accessed 5 2029].

EEA, 2014. Digest of EEA Indicators 2014.

EEA, 2020. EEA Indicators [Online]. Available: https://www.eea.europa.eu/data-and-maps/indicators/about. (Accessed 19 August 2020).

EU, 2018. Economy-wide Material Flow Accounts Handbook. European Union.

EU, 2020. Handbook for Estimating Raw Material Equivalents of Imports and Exports and RME-Based Indicators on the Country Level - Based on Eurostat's EU RME Model.

Financial Times, 2018. Lies, Damned Lies and ESG Rating Methodologies [Online]. Available: httpsV/ftalphavilleftcom/2018/12/06/1544076001000/Lies–damned-lies-and-ESG-rating-methodologies/. (Accessed 11 August 2020).

G&A Institute, 2020. 90% of S&P500 Index Companies Publish Sustainability/Responsibility Reports in 2019 [Online]. Available: https://www.ga-instiftite.com/research-reports/flash-reports/2020-sp-500-flash-report.html. (Accessed 23 July 2020).

GHG Protocol, 2004. The Greenhouse Gas Protocol: A Corporate Accounting and Reporting Standard (Revised Edition). WRI and WBCSD.

GIIN, 2020. What Is Impact Investing? [Online]. Available: https://thegiin.org/impact-investing/need-to-know/#:~:text=What is impact investing?,-impact investments im&text=NOUN: Impact investments are investments,impact alongside a financial return. (Accessed 2 July 2020).

GPIF, 2017. Results of ESG Index Selection. The Japanese Government Pension Investment Fund.

GRI, 2016. GRI 101: Foundation.

GRI, 2020. Our Mission and History [Online]. Available: https://www.globalreporting.org/about-gri/mission-history/. (Accessed 17 June 2020).

ISO 14001, 2015. ISO 14001:2015 Environmental Management Systems - Requirements with Guidance for Use. ISO, Geneva.

ISO 14031, 2013. ISO 14031:2013 Environmental Management - Environmental Performance Evaluation - Guidelines. ISO, Geneva.

MSCI, 2020a. Combining E, S, and G Scores: An Exploration of Alternative Weighting Schemes.
MSCI, 2020b. Deconstructing ESG Ratings Performance: Risk and Return for E, S and G by Time Horizon, Sector and Weighting.
Mura, M., Longo, M., Micheli, P., Bolzani, D., 2018. The evolution of sustainability measurement research. Int. J. Manag. Rev. 20, 661−695.
Puma, 2012. Puma's Environmental Profit and Loss Account for the Year Ended 31 December 2010.
Puma, 2018. Puma Annual Report 2018.
SASB, 2017. SASB Conceptual Framework. Sustainability Accounting Standards Board (SASB).
SASB, 2020. SASB Materiality Map [Online]. Available: https://materiality.sasb.org/.
Sadowska, B., Lulek, A., 2016. Measuring and Valuation in Accounting-Theoretical Basis and Contemporary Dilemmas. World Scientific News, pp. 247−256.
Schaltegger, S., Burritt, R.L., 2010. Sustainability accounting for companies: catchphrase or decision support for business leaders? J. World Bus. 45, 375−384.
Time Partners, 2020. The External Rate of Return Framework [Online]. Available: httpsV/time-partners.com/advisory-services/. (Accessed 2 July 2020).
UN, 2014. System of Environmental Economic Accounting 2012 - Central Framework. United Nations.
UN, 2017. SEEA Applications and Extensions. United Nations.
UN, 2020. System of Environmental Economic Accounting: What Is SEEA? [Online]. Available: https://seea.un.org/. (Accessed 23 July 2020).
UNDSD, 2001. Environmental Management Accounting: Policies and Linkages. United Nations Division for Sustainable Development, New York.
WBCSD, 2018. Report Matters - WBCSD 2018 Report.
WBCSD, 2020. ESGratings… It's a Bit Ofa Zoo [Online]. Available: https://www.wbcsd.org/Overview/News-Insights/General/News/ESG-ratings-it-s-a-bit-of-a-zoo. (Accessed 20 June 2020).
Wendling, Z.A., Emerson, J.W., De Sherbinin, A., Esty, D.C., et al., 2020. 2020 Environmental Performance Index Technical Appendix. Yale Center for Environmental Law & Policy, New Haven, CT.
World Commission on Environment and Development, 1987. Our Common Future ("The Brundtland Report"). Oxford University Press, Oxford.

Social indicators of sustainable resource management

Almudena Guarnido-Rueda, Ignacio Amate-Fortes
Department of Economics and Business, University of Almería, La Cañada de San Urbano, Almería, Spain

1. Introduction

The issue of sustainable development is very broad, encompassing sustainability over time of economic, social, and environmental growth, at a time when human dominance over the earth system has led to a number of problems such as biodiversity loss, global climate change, overexploitation of natural resources, degradation of environmental quality, and socioeconomic inequality and instability that need to be measured and addressed. Therefore, it can be said that achieving sustainable development is the most pressing issue of our time.

When referring to sustainable development, it is necessary to start from the definition presented in the **Brundtland report** (WCED, 1987) prepared by the World Commission on Environment and Development under the title "Our Common Future." This report focuses on development models and their impact on the functioning of ecological systems, stressing that the problems of the environment, and therefore of the possibilities of a "sustainable development model" are directly related to the problems of poverty, of satisfying basic food, health, and housing needs, of an energy matrix that favors renewable sources, and of the process of technological innovation. Within this framework of ideas, the Brundtland Report proposes the search for "sustainable development" as an alternative, leaving a definition that operates in relation to a destiny, which implies intergenerational responsibility, by proposing this as "a new path of progress that allows the needs and aspirations of the present to be met without compromising the ability of future generations to satisfy their own needs."

Since this report, the United Nations Environment Programme has promoted documents, conferences, commitments, and resources from countries to protect and care for our planet from the negative effects of growth. The ***1992 Rio Declaration*** is important in this regard. It places human beings at the center of concerns related to sustainable development: "They are entitled to a healthy and productive life in harmony with nature" (Principle 1). It is the power of states to guarantee the protection of the environment and natural resources for present and future generations. The environment is an integrating element of development and all states must cooperate with the essential task of eradicating poverty as an indispensable requirement of

Sustainable Resource Management. https://doi.org/10.1016/B978-0-12-824342-8.00020-1

sustainable development. In this sense, it establishes the new concept of **Sustainable Human Development**, since Human Development cannot be understood without sustainability, nor sustainability without Human Development. That is why the director of the United Nations Development Programme defines it as follows: "We must unite sustainable development and human development and unite them not only in word but in deeds, every day, on the ground, all over the world. Sustainable human development not only generates growth, but also distributes its benefits equitably; regenerates the environment rather than destroying it; empowers people rather than marginalizing them; expands people's choices and opportunities; and enables them to participate in the decisions that affect their lives. Sustainable human development is pro-poor, pro-nature, pro-employment and pro-women. It emphasizes growth, but growth with jobs, growth with environmental protection, growth that empowers people, growth with equity." (UNDP, 1994).

Agenda 21 (within the 1992 Rio Declaration on Environment and Development) called on countries to "develop sustainable development indicators" in ways that "contribute to self-regulating sustainability," an action plan endorsed by more than 170 national governments.

The challenge in measuring sustainable human development lies in developing new indicators, and there is a need to combine them through accounting frameworks, global approaches, and the creation of composite indices.

The main goal of this chapter is to analyze how the creation of these indicators has evolved and improved, especially the social indicators that are becoming more and more important (Fig. 12.1).

International systems (OECD, UN, EU) have emerged as an important reference for the elaboration of national and regional systems, especially in developed countries. They have provided new frameworks and guidelines to adapt them to other

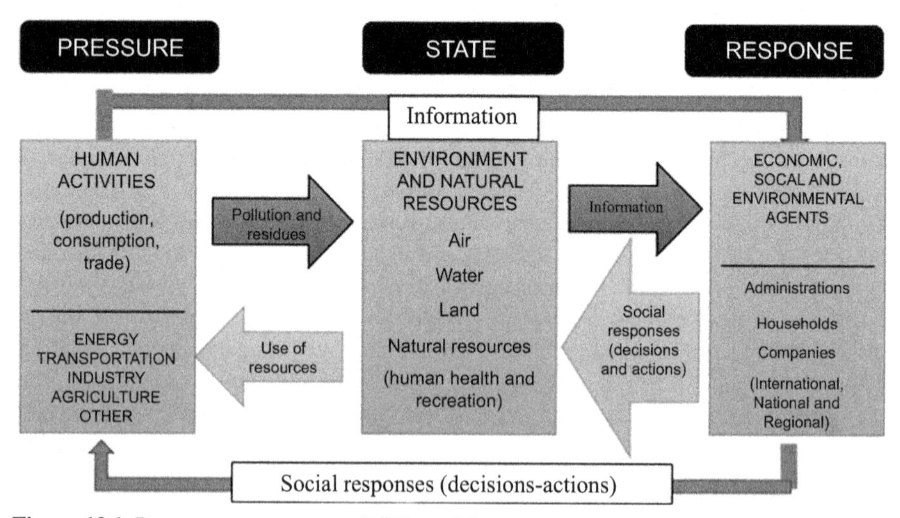

Figure 12.1 Pressure-state-response (PSR) model.
Own elaboration based on OECD (Organization for Economic Co-Operation and Development), 1993; Sotelo and Lastra, 2011.

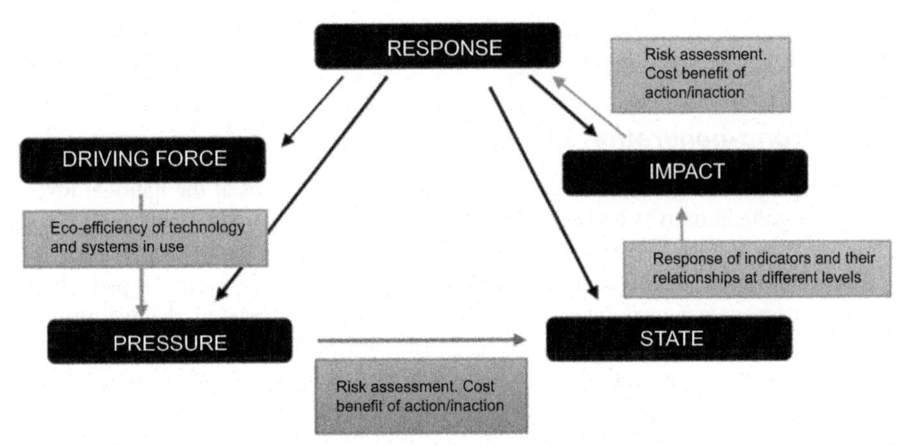

Figure 12.2 Driving Force-pressure-state-impact-response (DPSIR) model.
Own elaboration based on OECD (Organization for Economic Co-Operation and Development), 1993; Sotelo and Lastra, 2011.

geographical scales, as well as a large number of benchmarks and technical procedures for the calculation of indicators. Generally, they take as objective values those that have been drawn up in the different treaties, conventions, or objectives at a global level (Kyoto Protocol, Millennium Goals, Agenda 21, Gothenburg Strategy, even the current Sustainable Development Goals) and they seek that the countries use their indicators as a guide for the implementation of Sustainable Development Programs and Plans (Fig. 12.2).

2. Measuring sustainability

The Sustainability Indicator Systems have been undergoing 30 years of continuous change and progress toward interdimensional models that allow an easy and reliable evaluation of development processes. The first ***Sustainability Indicator Systems*** **(*SIS*)** models, mainly of an environmental type and of a national or supranational scope, have evolved toward recent systems that address the triple economic, environmental, and social dimension in an integrated manner. Thus, we can identify an evolution of the indices that we could classify them in.

2.1 First-generation systems

Their origin dates back to the 1980s, from the publications collected by the OECD (1993), and they are characterized by being very theoretical and exclusively environmental. The main "frameworks" used were

- Pressure-State-Response (PSR)
- Driving Force-State-Response (DSR)
- Driving Force-Pressure-State-Response (DPSR)
- Driving Force-Pressure-State-Impact-Response (DPSIR)

Of these, the most widely used were the PSR, by the OECD,[1] and the DPSIR, by the European Environment Agency (EEA)[2]

2.2 Second-generation systems

Its use began in the 1990s, through the development of systems at the national level, highlighting the initiatives undertaken by Mexico, Chile, the United States, the United Kingdom, Spain (OSS, 2005 and 2006), etc.

The multidimensional (economic, environmental, and social) approach of sustainable development is incorporated. In recent years, a fourth dimension, the institutional one, has gained strength due to the relevance and influence of the policies dictated by the control bodies (local and national governments, international organizations, etc.). The development of these systems has been led by the United Nations Commission on Sustainable Development, with indicators that are included in each of the dimensions of development, but without being linked to each other.

The **United Nations Commission on Sustainable Development's System of Indicators** was launched in order to develop Chapter 40 of Agenda 21 through the "Work Programme on Indicators of Sustainable Development," and produced a list of Sustainability Indicators, developed by methodological sheets, which were published in 1998 in the book *Indicators of Sustainable Development. Framework and Methodologies* or also known as the "Blue Book of Indicators" (Quiroga, 2001). It initially used the FSR framework until 2001 when it developed a new framework based on dimensions, topics, and subtopics (United Nations, 2001; CSD, 2006a; CSD, 2007; Quiroga, 2001).

In 2006, the Expert Group on Sustainability Indicators met to review the list published in 2001, resulting in 14 issues, which group the 96 indicators developed (United Nations, 2001; CSD, 2006b; CSD, 2007). The indicators, applied in a national geographic scope, have been classified into basic (relevant for most countries) and noncore (or complementary that provide additional information or refer to problems that are relevant in some countries). The 14 agreed topics are Poverty, Governance, Health, Education, Demography, Natural Hazards, Atmosphere, Land, Oceans, Seas

[1] **The OECD System of Indicators**, and its PSR framework, is composed of indicators that are considered to be partial, and which account for the complex phenomenon from a productive sector or from the singularity or a reduced number of dimensions. Its importance lies in obtaining indicators with a level of rigor and quality similar to the economic and social indicators developed previously, and its application is limited at the international level, especially in developed countries. It is one of the pioneers in the development of indicators at a global level, and its work provides an interesting vision that links environmental problems and opportunities to economic processes (Quiroga, 2001).

[2] The **European Environment Agency's (EEA) Indicator System** was introduced in 2004, with the aim of establishing a manageable and stable system of indicator-based reporting of priority data from the European Environmental Observation and Information Network (EIONET), and streamlining the input of the EEA and EIONET to other European and global indicator initiatives (Ministry of Environment, 2006). It uses the DPSIR framework to provide the basis for the analysis of the interrelated factors affecting the environment. It includes basic indicators covering six environmental themes (air pollution and ozone depletion, climate change, waste, water, biodiversity, and land environment) and four sectors (agriculture, energy, transport, and fisheries) without developing social or health indicators.

and Coasts; Drinking Water, Biodiversity, Economic Development, World Economic Cooperation and Consumption and Production Patterns (CSD, 2006b).

Other systems that highlighted in this second generation are as follows:

- **Capital frameworks** attempt to calculate a nation's or region's wealth based on different types of capital (UN, 2007). The capital approach borrows the concept of capital from the economy and expands it to include several types of capital: manufactured or built capital (all the produced assets that make up the human economy in a traditional sense), natural capital (the environment and natural resources), human capital (people's capabilities to work, including knowledge, skills, and health), and social capital (the stock of social networks, trust, and institutional arrangements). An example of a capital-based framework is the *Daly Triangle* system advocated by the **Balaton Group**, an international network of researchers and practitioners in the field of sustainable development, which identifies natural, built, human, and social capital (Meadows, 1998).

  The different forms of capital are usually expressed in the same currency terms, so they can be aggregated. Sustainable development, in this context, can be interpreted differently depending on whether a strong or weak sustainability perspective is adopted. Can natural capital be replaced by other types of capital? Which natural resources and ecosystem services are substitutable? What are the limits of such substitutions? These are some of the critical questions that need to be addressed when using capital frameworks. Other challenges of using a capital framework include problems of monetization of different forms of capital, disputes over substitutability, and issues of intragenerational equity (UN, 2007).
- **Integrated accounting frameworks refer** to synthesized economic and environmental accounting systems based on national accounting methodologies. The most prominent example is the System of Environmental Economic Accounting (SEEA), developed jointly by the United Nations, the European Commission, the International Monetary Fund, the OECD, and the World Bank in 2003 (Hecht, 2006; UN, 2007). The SEEA integrates environmental accounting with the standard System of National Accounts (SNA). It includes four categories of accounts: (1) physical data on material and energy flows, (2) data on environmental management and environment-related transactions, (3) environmental asset accounts, and (4) transaction and adjustment accounts related to the impact of the economy on the environment. Therefore, rather than just producing a set of indicators, SEEA provides comprehensive accounts of environmental and economic capital and flows. It has also been used to develop indicators and policy analysis.

  However, integrated accounting frameworks such as SEEA have not taken into account the social and institutional dimensions of sustainable development, although efforts are being made to incorporate human and social capital (UN, 2007). Nevertheless, the implementation of SEEA not only serves its own purposes but also benefits the application of other indicator frameworks, particularly those based on capital. Thematic frameworks can also benefit from SEEA, as a coherent database facilitates the development and disaggregation of thematic indicators that can support analysis and policy design. Therefore, the latest version of the UNCSD indicator framework has strengthened its relationship with SEEA (UN, 2007).
- **The Bossel orientation framework** is a theoretical framework of systems used to develop sustainable development indicators (Bossel, 1977, 1999, 2001). This approach aims to provide a holistic and comprehensive conceptual structure to guide the development of indicators. The framework has a nested hierarchical appearance that somewhat resembles the Daly Triangle. The orientor theory assumes that any ecological and socioeconomic system can be characterized by six fundamental environmental properties: normal environmental status, resource scarcity, variety, variability, change, and other systems. These properties "restrict development possibilities and limit management opportunities at all spatial scales," and make "orientors," which are broad categories of our key concerns, values, or interests that "guide most of

our decisions" (Bossel, 1999, 2001). In the context of sustainable development, Bossel (2001) described the following seven basic orientators: existence, effectiveness, freedom of action, security, adaptability, coexistence, and psychological needs.

2.3 Third-generation indicators

In recent years, the need to link the dimensions of development and its indicators with each other generated systems of composite indicators, sustainability indices.

There are a large number of these indices, but we focus on a small set of the most commonly used ones.

- **Green GDP** was developed to take into account the effects of natural resources consumption and pollution on human well-being (Wu and Wu, 2010). Green GDP aims to correct these biases by extending the coverage of accounting in order to include many, but certainly not all, of the values that people derive from nature.
- **The Human Development Index (HDI),** developed since 1990, quantifies the key capabilities it aims to achieve: (1) enjoy a long and healthy life as measured by life expectancy at birth, (2) acquire knowledge as measured by the adult literacy rate (with a two-thirds weighting) and the combined gross enrollment rate in primary, secondary, and tertiary schools (with a one-third weighting), and (3) have access to the resources needed to achieve a decent standard of living as measured by GDP per capita in terms of purchasing power parity:

$$\boldsymbol{HDI} = \frac{1}{3}[\boldsymbol{I}(\boldsymbol{H})+\boldsymbol{I}(\boldsymbol{E})+\boldsymbol{I}(\boldsymbol{Y})] \quad (12.1)$$

where $I(H)$ is the health indicator, $I(E)$ education index, and $I(Y)$ the income index.

Since 2010, the World Human Development Report has introduced a new methodology for calculating the HDI[3] that refines the education and income dimensions and adjusts the international benchmarks of all its variables and the way they are

[3] Before calculating the HDI, it is necessary to create an index for each of these dimensions. To calculate these indices (life expectancy, education, and GDP), minimum and maximum values (thresholds) are chosen for each of the basic indicators. The construction of the health index, I(H), is immediate. The maximum value is 83.2 years (corresponding to the effective value of Japan in 2010) and the minimum value is 20 years. To calculate the education index, I(E), each of the two components of this partial index is normalized and then the geometric mean of both indexes (the square root of their product) is taken. The maximum value for average years of education is 13.2 years (corresponding to the United States in 2000) and the maximum value for expected years of schooling is 20.6 (corresponding to Australia in 2002). In both cases, zero is taken as the minimum value. The normalization is again applied to the resulting value by taking zero as the minimum value and the maximum of the observed values of this geometric mean as the maximum value (which is 0.951 for New Zealand in 2010). In order to normalize the income index, I(Y), the use of logarithms is maintained with the idea of approximating the consumption capacities generated by income. Gross National Income per capita is expressed in US dollars in terms of purchasing power parity. The maximum and minimum values are $108,221 (United Arab Emirates in 1980) and $163 (Zimbague in 2008), respectively. Life expectancy data are from UNDESA. The data for average years of study come from the work of Barro and Lee (2010). Data for expected years of study are taken from UNESCO (2010). Data on Gross National Income per capita are obtained from the World Bank (2010) and the International Monetary Fund (2010).

synthesized. The changes in the education dimension seek to better capture the differences, so the Education Index is now obtained by expected schooling and average years of schooling for people under and over 25, respectively. On the other hand, changes in the income dimension seek to reflect, more accurately, the country's internal resources by using the Gross National Income (GNI) per capita, instead of the GDP. Thus, the conjunction of the three aspects with a geometric mean introduces the notion of complementarity between the dimensions and, above all, gives rise to inequality between them.

$$\boldsymbol{HDI}(2010) = \sqrt[3]{\boldsymbol{I}(\boldsymbol{H}) \times \boldsymbol{I}(\boldsymbol{E}) \times \boldsymbol{I}(\boldsymbol{Y})} \tag{12.2}$$

This new aggregation criterion produces important changes in the index values. The geometric mean produces lower values for the indicator, so the more unequal the achievements of each country in the different dimensions, the lower the indicator. The impact of this new formula on the global ranking is moderate.

- **Inclusive wealth (IW) and genuine savings (GS)**. Unlike GDP and Green GDP, which are "flow" measures, IW and GS are based on stocks. IW and GS derive data on national income and resource use from the UN-designed SEEA. It is an amendment to the more standard SNA used in more traditional economic performance calculations. Economic patterns of production and consumption necessarily depend on the availability and configuration of the disposable resource base, or capital. Thus, intertemporal (e.g., intergenerational) transfers of economic opportunities are best represented by the value of the capital stock. The terms "inclusive" and "genuine" in the nomenclature are largely derived from the incorporation of natural resources into economic accounting.

  An intuitive criterion of sustainability emerges from this formulation: a country or region is "sustainable" for a given period if its per capita IW or GS does not decrease during that time. A country can grow in both GDP and HDI while at the same time decreasing its per capita wealth (Arrow et al., 2004). When compared to this new standard of capital-oriented development, and taking into account the role of natural resources, optimism about welfare expansion is greatly tempered by much lower (often negative) values. Despite the potential for a more accurate approach to sustainability, it is not common practice to maintain stock-based measures of economic development, and GDP and HDI continue to prevail as the most common development indicators at the national and regional levels.
- **Genuine Progress Indicator (GPI) and Index of Sustainable Economic Welfare (ISEW)** are essentially equivalent measures, the former being more widely recognized and implemented (Lawn, 2003). Both take Green GDP as reference, modifying the standard metric based on the flow of economic results in order to consider the role of environmental welfare. However, unlike Green GDP, which is primarily a project to augment the standard national accounting framework, GPI or ISEW divides economic transactions into those that contribute positively to human well-being and those that contribute negatively. In addition, the GPI and ISEW also include imputed values of nontraded goods and services (both social and ecological), and are adjusted for income distribution effects. Thus, the GPI and ISEW seek to separate "goods" from "evils" and provide a more holistic and sensible assessment of economic activities.

- **Material Flow Accounting (MFA).** The material flow through an economy can provide an insightful indication of the sustainability of the system in relation to its resource base. Unlike Green GDP, which is also a "flow-based" measure, MFA attempts to quantify the physical value in weight, not the currency value (Matthews et al., 2000). It tracks the weight of a number of different material flows, including production inputs and outputs, the material that moves in the environment to access resources (such as excavated soil), and the waste material from the production process (Hecht, 2006). This focus on the "direct ingredients" of production and consumption eliminates the "middleman" of monetary valuation. MFA produces a single measure of system performance by aggregating different material flows into the total summarized indicator material in order to give a picture of the entire physical metabolism of the economic system. Although monetary accounting is still more widespread, MFA has been expanding and has been carried out in several countries and regions (Ness et al., 2007). The Statistical Office of the European Communities (Eurostat), a research institution responsible, inter alia, for the collection and calculation of comparative data related to the performance of European countries, developed an economy-wide MFA which is the most standardized assessment tool (Ness et al., 2007). This guideline divides the material flows into three categories: input, output, and consumption. In addition, within each of these categories there are levels that indicate whether the flows involve domestic, foreign, and/or hidden materials, which are materials not included in economic accounting, such as soil erosion (Matthews et al., 2000).
- **Ecological Footprint (EF)** is primarily a measure of human appropriation of natural resources, and is defined as the area of land (and water) that would be required to sustain indefinitely a human population defined in terms of providing all energy/material resources consumed and absorbing all waste discharged (Wackernagel and Rees, 1996). The basic unit of measurement is the "global hectare," a standardized unit that captures the average biocapacity of all hectares of all biologically productive land in the world. Consumption patterns of natural resources, from energy to biomass, can be converted to this common metric. This simple but comprehensive measure allows us to compare our demands on the planet's ecosystems with the capacity of those ecosystems to regenerate. In this way, we can create a direct correspondence between our current standard of consumption and the capacity of the biosphere to support that standard. Since its creation in the early 1990s, the EF has become a very influential indicator of the impact on the human environment, which has given rise to a large volume of literature and analysis.
- **Happy Planet Index (HPI)** was developed by the New Economics Foundation as an alternative to measures such as the HDI or GDP. While still advocating human-oriented measures, the HPI directly combines human well-being with human consumption of natural resources. Unlike the standard welfare account, which is largely defined as a simple consumption function, the HPI defines human welfare in "happy life years," a combination of life expectancy and life expectancy. The indicator is then calculated as the ratio of happy life years to environmental impact, which is measured by the EF. The HPI is intended to measure "the ecological efficiency with which human well-being is provided worldwide" (New Economics Foundation, 2009).
- **Environmental Sustainability Index (ESI) and Environmental Performance Index (EPI).** Published between 1999 and 2005, the ESI is, like the MFA and the EF, primarily a biophysical indicator of humans' use of natural resources. The ESI was created by the Yale Center for Environmental Law and Policy at Yale University and the Center for International Earth Science Information Network (CIESIN) at Columbia University, in collaboration with

the World Economic Forum and the European Commission's Joint Research Center. The computational methodology behind the ESI involves combining 76 variables into 21 metrics, which are then averaged to obtain a single index. The ESI was succeeded by the EPI, which was developed by the same institutions and has been published in 2006, 2008, and 2010.

Environmental performance, with a focus on assessing current environmental conditions, differs from ESI's original issue of measuring the long-term environmental trend. While the ESI had five assessment topics (i.e., the states of environmental systems, reduction of environmental stress, reduction of human vulnerability, institutional capacity, and global governance), the EPI reduces its purposes to two objectives: environmental health and ecosystem vitality. The shift in focus is illustrated by the way the two indices differ in the analysis of forest management. The 2005 ESI focused on the annual change in forest cover and the percentage of total forest area certified for sustainable management. On the other hand, the 2008 EPI simply used the change in stock growth as a proxy for management performance. The EPI is supposed to provide a report of "more immediate value to policymakers" and, since the 2010 publication, 163 countries are included in the analysis.

3. How can the social dimension be measured?

3.1 Interpretation of social sustainability criteria

According to the Merriam-Webster dictionary, the term social refers to "human society, the interaction of the individual and the group, or the welfare of human beings as members of society."

Social sustainability was originally introduced as part of the concept of social sustainability in the Brundtland report (WCED, 1987). The main definition of Social Sustainability, that is, "development that meets the needs of present without including the ability of future generations to satisfy their own needs," has a clear social imperative. The Brundtland Report focused on issues such as health and the income gap between rich and poor with the aim of reducing poverty worldwide.

The 1992 Rio conference introduced social sustainability as the right to live a decent life; intergenerational, intragenerational, and international social justice; and local participation in sustainable development processes. This was further elaborated by including issues such as well-being, security, and a healthy environment, access to education, learning opportunities, identity, a sense of place and public participation. The concept of social sustainability continues to develop (Table 12.1).

Thin (2002) describes social justice, solidarity, participation, and security as social values. Social values can be characterized as conditions associated with quality of life in the earth's landscape, including such aspects as equity, participation in democratic life, security, and health (Rosenstrom et al., 2006). Recent additions include concepts such as human well-being, happiness, and quality of life (Colantonio, 2007).

While the measurement of social values presents methodological challenges (Scazzosi, 2004; Tress et al., 2006; Naveh 2007), a set of concepts and corresponding

Table 12.1 Defined social criteria.

Traditional
• Welfare, housing, and environmental health
• Education and job skills
• Equity
• Human rights and gender
• Poverty
• social justice
Emerging
• Demographic change (aging, migration, and mobility)
• Integration and social cohesion
• Identity, sense of place, and access
• Health and safety
• Social capital
• Welfare, happiness, and quality of life

Own elaboration based on UNESCO (1972, 2003), Saastamoinen (2005), Colantonio (2007) and the Rio+20 process (Culture 21, 2011).

measurements can be discerned that are being used by scholars and policymakers to identify social values. Following the extensive review by Magis and Shinn (2009), we interpret social values as four groups of indicators:

- **Democratic civil society**, including participation in the development process at the local level. The process of transition from government to governance is an important part of this indicator and a prerequisite for further democratic development in many societies.
- **Living environment**, which includes the well-being and security of human beings in relation to natural disasters and social unrest, the need to understand esthetic values, health preferences, and the impact of the population on health with respect to the environment. Considerable research has been conducted on human perceptions of different landscapes and landscape features using a variety of methods, including surveys, photo-based studies, and in-depth interview studies (Herzog, 1987; Gyllin and Grahn, 2005; Grahn and Stigsdotter, 2010). In addition, some studies have demonstrated direct links between landscape and human health (De Jong et al., 2012; Tzoulas et al., 2007).
- **Human development** related to health, education, income, and potentially other parameters. There are several indicator frameworks and indices designed to provide information on quality of life, with statistical measures, at international, national, and local levels (UN, 1996, 2007; Bartelmus, 1997; Bell and Morse 2008; Carraro et al., 2009).
- **Equity** as equal rights, opportunities, education, income, and health (Uslaner, 1999; Rothstein and Uslane, 2005).

Nevertheless, and as mentioned above, the fertile literature on the concept of sustainable development has historically dealt mainly with the environmental dimension of the concept, followed by the economic, generally relegating the social perspective to the last level, as its weakest pillar (Lehtonen, 2004; Murphy, 2012), and in any case, when the social dimension has been recognized, it has been integrated into the category of social welfare and as an addition, but with no link to environmental protection.

It has been with the recent Sustainable Development Goals (SDG)[4] that the social dimension of sustainable development has been widely incorporated.

Recognizing this need, the UN Statistical Commission in January 2016 published the proposed indicators Report of the Inter-Agency and Expert Group on Sustainable Development Goal Indicators, prepared by a working group of experts from the UN and other institutions. The list of indicators is included in Annex III of the proposal. For some targets, several indicators are proposed, and for some, they have not yet been developed. It is an ongoing work. According to the report: "*Based on the level of methodological development and the availability of information, the indicators contained in this proposal will be grouped into three levels: (a) a first level for indicators for which there is an established methodology and information is available (Level I); (b) a second level for indicators for which there is an established methodology but information is not readily available (Level II); (c) a third level for indicators for which there is no internationally agreed methodology (Level III).*"

Developing indicators for 169 targets, many of which have language that is not conducive to implementation, is not easy and in some cases, indicators should be proxy measures of achievement. This does not mean that the indicators are not useful, but it does mean that, for the current time, they are incomplete and, in some cases, not relevant.

4. Iuuses and problems inherent in social indicators

As we have already verified, the social dimension of sustainable development has proved particularly difficult to measure.

There are major differences between OECD countries in the way they interpret social sustainability: from concerns about poverty in the developing world, to the health consequences of environmental change, to issues of ethnic minorities and gender balance, to broader considerations of quality of life and social relations (poverty, crime, education, etc.). The diversity of concerns and the lack of a common approach is one of the obstacles to identifying appropriate measures for the social side.

It has been difficult to obtain indicators of "social capital," comparable to economic and environmental capital that can be measured in terms of aggregate stocks.

[4] The Sustainable Development Goals are as follows: (1) No Poverty; (2) Zero Hunger; (3) Good Health and Well-being; (4) Quality Education; (5) Gender Equality; (6) Clean Water and Sanitation; (7) Affordable and Clean Energy; (8) Decent Work and Economic Growth; (9) Industry, Innovation, and Infrastructure; (10) Reduced Inequalities; (11) Sustainable Cities and Communities; (12) Responsible Consumption and Production; (13) Climate Action; (14) Life Below Water; (15) Life on Land; (16) Peace and Justice Strong Institutions; (17) Partnership to Achieve the Goals.

Many social concerns relate to access to opportunities, for example, whether some individuals or groups lack opportunities for education and training, adequate health care, or affordable housing. To develop statistical measures of the social dimension of sustainable development, data are needed on how economic, environmental, and social resources are distributed in society.

Social capital also reflects the shared norms and values in society that bring benefits to individuals and groups by facilitating cooperation and reducing opportunistic behavior. This has led to attempts to develop indicators of social cohesion in countries, regions, and cities, but these tend to be highly subjective.

The design of appropriate measures for the social dimension of sustainable development faces a number of practical and conceptual obstacles. As in the economic and environmental fields, the selection of indicators relating to the social dimension is a political act. Through this selection, Governments convey a sense of their priorities, commit to action, and indicate that they are prepared to respond to their voters for failures to progress.

In the Sustainable Development Goals (UN, 2015), despite the fact that they have culminated in the inclusion of the social perspective, where 10 of the 17 goals would constitute the social pillar of sustainable development, the problem is their more or less integrated articulation, particularly with the environmental pillar, as well as the possible contradictions (Cereceda et al., 2016) among the 162 specific goals established for the achievement of these objectives.

In explaining the historical scarcity of attention received in the literature to the social aspects of sustainability, the following ideas can be put into practice:

I. As argued by Bebbington and Dillard (2009: 157), the social dimension appears to present different and more complicated challenges in terms of specification, understanding, and communication than those presented by environmental sustainability, since there is no commonly accepted scientific basis for analysis or a common unit of measurement, as is the case with the economic dimension of sustainability.

II. Global indicator criteria and frameworks have been developed and there is a need to collect data on social indicators at various scales or appropriate policy units (e.g., district, municipality, regional, etc.).

III. A third issue is the range of types of social indicators on a spectrum from the objective to the subjective. In the past, social indicators have often been objective measures such as demographic characteristics (e.g., gender and ethnic composition) taken from statistical records on population, income, health, and employment. Increasingly, however, sociologists have employed more subjective indices such as community cohesion, social capital, and social alienation. These can be more difficult to define and measure (Nadeau et al., 1999), unless indicated and proxy measures are used (e.g., divorce rates, youth migration, and crime statistics). In either case, quantitative indicators are not necessarily ideal. In particular, the economic valuation of quality of life can distort and trivialize those things that people value such as clean air, traditional practices, or freedom.

IV. Finally, there is much heterogeneity within a community, people have different opinions, preferences, and needs that are often expressed as competing views for the future of the community. In addition, perceptions can change over time with shifting events and conditions.

5. Conclusions

The concept of sustainable development is founded on the union formed by the economy, social justice, and environmental protection. However, this concept has been based since its origins mainly on its environmental and economic position, to the detriment of its social dimension.

The recent proposal of 17 United Nations Sustainable Development Goals opens a great opportunity for the implementation of the social perspective of sustainable development.

Efforts have been made to obtain sustainable development indicators over time but the measurement approaches have been based on purely economic concepts of well-being. These tend to neglect the environmental and social aspects of sustainability. In addition, the social dimension of sustainable development has proved particularly difficult to measure.

With the challenge of the Sustainable Development Goals, countries face the realization of policies with the achievement of objectives that are sometimes not very concrete, precise, and very difficult to interpret and measure.

The work of creating and developing indicators and eventually collecting them should be a lesson for the preparation of the next goals to be proposed for the period 20312045, as it is obvious that many of these 17 goals will remain unfulfilled (poverty will not have been eradicated, equity.......etc., sustainable development will not have been achieved).

If we really want sustainable development, we will need the participation of governments with conductive policies, relevant regulations and institutions, and resources adequate to the challenges and the application of new technologies (Hussain and Mishra, 2019; Hussain and Keçili, 2019; Hussain, 2019, 2020). In addition, very concrete assessment criteria and indicators will be necessary for all the dimensions or creation of indicators where there is real equity in the weighting of the economic, social, institutional, and environmental dimensions. But in parallel, the collaboration of companies and all citizens will be required, which we must comply with and demand compliance with.

References

Arrow, K., Dasgupta, P., Goulder, L., Daily, G., Ehrlich, P., Heal, G., Levin, S., Mäler, K.G., Schneider, S., Starrett, D., Walker, B., 2004. Are we consuming too much? J. Econ. Perspect. 18, 147–172.

Bartelmus, P., 1997. Measuring sustainability: data linkage and integration. In: Moldan, B., Billharz, S., Matravers, R. (Eds.), Sustainability Indicators: A Report on Indicators of Sustainable Development. Wiley, Chichester, pp. 116–118.

Barro, R.J., Lee, J.W., 2010. A new data set of educational attainment in the world, 1950-2010. In: NBER Working Paper 15902. National Bureau of Economic Research, Cambridge, MA.

Bebbington, J., Dillard, J., 2009. Social sustainability: an organizational-level analysis. In: Dillard, J., Du- jon, V., King, M.C. (Eds.), Understanding the Social Dimension of Sustainability. Routledge, London.

Bell, S., Morse, S., 2008. Sustainability Indicators: Measuring the Immeasurable? Earthscan, London.

Bossel, H., 1977. Orientors of nonroutine behaviour. In: Bossel, H. (Ed.), *Concepts and Tools of Computer-Assisted Policy Analysis*. Birkhäuser, Basel, pp. 227–265.

Bossel, H., 1999. Indicators for Sustainable Development: Theory, Method, Applications (A Report to the Balaton Group). International Institute for Sustainable Development, Winnipeg.

Bossel, H., 2001. Assessing viability and sustainability: a systems-based approach for deriving comprehensive indicator sets. Ecol. Soc. 5 (2), 12.

Carraro, C., Cruciani, C., Ciampalini, F., Giove, S., Lanzi, E., 2009. Aggregated projection of sustainability indicators: a new approach. In: Research Paper Presented at the 3rd OECD World Forum on "Statistics, Knowledge and Policy". Charting Progress, Building Visions, Improving Life, 27–30 October 2009, Busan, Korea.

Cereceda, R.C., Hernández, S.R., Rivera, E.D.O., 2016. Los Objetivos de Desarrollo Sostenible y los Retos para su Implementación. Revista Pluralidad y Consenso 5 (26), 55–84.

Colantonio, A., 2007. Social Sustainability: An Exploratory Analysis of its Definition, Assessment Methods Metrics and Tools. EIBURS Working Paper Series, Oxford.

CSD, 2006a. Global Trends and Status of Indicators of Sustainable Developments. UN Department of Economic and Social Affairs, New York. Available from: http://www.un.org/esa/sustdev/csd/csd14/documents/bp2_2006.pdf.

CSD, 2006b. Expert Group Meeting on Indicators of Sustainable Development. Report. UN Department of Economic and Social Affairs, New York.

CSD, 2007. CSD Indicators of Sustainable Development, third ed. Fact Sheet. Available from: http://www.un.org/esa/sustdev/natlinfo/indicators/factsheet.pdf.

Culture 21, 2011. Lobbying for Culture as the 4th Pillar of Sustainable Development in the Process of the Rio?20 Summit. Agenda 21 for Culture. United Cities and Local Governments—Committee on Culture.

De Jong, K., Albin, M., Skärbäck, E., Grahn, P., Björk, J., 2012. Perceived green qualities were associated with neighborhood satisfaction, physical activity, and general health: results from a cross-sectional study in suburban and rural scania, southern Sweden. Health Place 18, 1374–1380.

Grahn, P., Stigsdotter, U.K., 2010. The relation between perceived sensory dimensions of urban green space and stress restoration. Landsc. Urban Plann. 94, 264–275.

Gyllin, M., Grahn, P., 2005. A semantic model for assessing the experience of urban biodiversity. Urban Greening 3, 149–161. Urban Forest.

Hecht, J.E., 2006. Can Indicators and Accounts Really Measure Sustainability? Considerations for the U.S. Environmental Protection Agency. US EPA Workshop on Sustainability. Available from: http://www.epa.gov/sustainability/other resources.htm.

Herzog, T.R., 1987. A cognitive analysis of preference for natural environments: mountains, canyons, deserts. Landsc. Res. 6, 140–152.

Hussain, C.M., Mishra, A.K., 2019. Nanotechnology in Environmental Science, vol. 2. John Wiley & Sons.

Hussain, C.M., Keçili, R., 2019. Modern Environmental Analysis Techniques for Pollutants, first ed. Elsevier.

Hussain, C.M., 2019. Handbook of Environmental Materials Management, first ed. Springer International Publishing.

Hussain, C.M., 2020. The Handbook of Environmental Remediation: Classic and Modern Techniques, first ed. Royal Society of Chemistry.

Lawn, P.A., 2003. A theoretical foundation to support the index of sustainable economic welfare (ISEW), genuine progress indicator (GPI) and other related indexes. Ecol. Econ. 44, 105–118.

Lehtonen, M., 2004. The environmental-social interface of sustainable development: capabilities, social capital, institutions. Ecol. Econ. 49, 199–214.

Matthews, E., Amann, C., Bringezu, S., Fischer-Kowalski, M., Huuttler, W., Kleijn, R., Moriguchi, Y., Ottke, C., Rodenburg, E., Rogich, D., Schandl, H., Schuu tz, H., van der Voet, E., Weisz, H., 2000. The Weight of Nations: Material Outflows from Industrial Economies. World Resources Institute, Washington, D. C.

Meadows, D., 1998. Indicators and Information Systems for Sustainable Development. Sustainability Institute, Hartland/VT.

Magis, K., Shinn, C., 2009. Emergent principles of social sustainability. In: Dillard, J., Dujon, V., King, M.C. (Eds.), Understanding the Social Dimension of Sustainability. Routledge, New York and London, pp. 15–44.

Ministry of Environment, 2006. Conjunto Básico de Indicadores de la AEMA. Ministerio de Medio Ambiente, Madrid.

Murphy, K., 2012. The social pillar of sustainable development: a literature review and framework for policy analysis. Sustain. Sci. Pract. Pol. 8 (1), 1529.

Nadeau, S., Shindler, B., Kakoyannis, C., 1999. Forest Communities: new frameworks for assessing sustainability. For. Chron. 75 (5), 747–754.

Naveh, Z., 2007. Transdisciplinary Challenges in Landscape Ecology and Restoration Ecology. An Anthology. Springer, Dordrecht.

Ness, B., Urbel-Piirsalu, E., Anderberg, S., Olsson, L., 2007. Categorising tools for sustainability assessment. Ecol. Econ. 60, 498–508.

New Economics Foundation, 2009. The (Un)Happy Planet Index 2.0. Available from: http://www.happyplanetindex.org/public-data/files/happy-planet-index-2-0.pdf.

OECD (Organization for Economic Co-Operation and Development), 1993. OECD Core Set of Indicators for Environmental Performance Reviews. OECD, Paris.

OECD, 2001. Sustainable Development: Critical Issues. OECD, Paris.

OSS (Observatory of Sustainability in Spain), 2005. Sostenibilidad en España 2005. Ministerio de Medio Ambiente-Fundación Biodiversidad-Fundación Universidad de Alcalá-Mundi-Prensa Libros, Madrid.

OSS, 2006. Sostenibilidad en España 2006. Ministerio de Medio Ambiente- Fundación Biodiversidad-Fundación Universidad de Alcalá-Mundi-Prensa Libros, Madrid.

Quiroga, R., 2001. Indicadores de Sostenibilidad Ambiental y de Desarrollo Sostenible: Estado del Arte y Perspectivas. Santiago de Chile, vol. 16. CEPAL-ECLAC, Serie Manuales.

Rosenström, U., Mickwitz, P., Melanen, M., 2006. Participation and empowerment-based development of socio-cultural indicators supporting regional decision-making for eco-efficiency. Local Environ. Int. J. Justice Sustain. 11, 183–200.

Rothstein, B., Uslaner, E.M., 2005. All for all: equality, corruption, and social trust. World Polit. 58, 41–72.

Saastamoinen, O., 2005. Multiple ethics for multidimensional sustainability on forestry? Silva Carelica 49, 37–53.

Scazzosi, L., 2004. Reading and assessing the landscape as cultural and historical heritage. Landsc. Res. 29, 335–355.

Sotelo, T., Lastra, 2011. Indicadores por y para el Desarrollo Sostenible, un Estudio de Caso. Estud. Geográficos 72, 611–654.

Thin, N., 2002. Social Progress and Sustainable Development. Kumarian Press, Bloomfield.

Tress, B., Tress, G., Fry, G., 2006. Defining concepts and the process of knowledge production in integrative research. In: Tress, B., et al. (Eds.), From Landscape Research to Landscape Planning. Springer, Dordrecht, pp. 13–26.
Tzoulas, K., Korpela, K., Venn, S., Ylipelkonen, V., Kazmierczak, A., Niemela, J., James, P., 2007. Promoting ecosystem and human health in urban areas using green infrastructure: a literature review. Landsc. Urban Plann. 81, 167–178.
UN (United Nations), 1993. Informe de la Conferencia de Naciones Unidas sobre Medio ambiente y Desarrollo. UN, New York.
UN, 1996. Indicators of Sustainable Development, Frameworks and Methodologies. United Nations, New York.
UN, 2001. Indicators of Sustainable Development: Guidelines and Methodologies. UN, New York. Report.
UN, 2007. Indicators of Sustainable Development: Guidelines and Methodologies. United Nations, New York.
UN, 2015. Objetivos de Desarrollo Sostenible (ODS). Nueva York: Naciones Unidas. Available from: http://www.undp.org/content/undp/es/home/mdgoverview/post-2015- developmentagenda.
UNDP, 2007. Human Development Report 2007/2008. UN, New York. Technical Note 1.
UNDP (United Nations Development Programme), 1994. Human Development Report 1994. American Writting Corporation, Washington, D. C., Estados Unidos de América.
UNESCO, 1972. Convention Concerning the Protection of the World Cultural and Natural Heritage. Adopted by the General Conference at its Seventeenth Session. 16 November 1972 Paris, France. UNESCO, Paris.
UNESCO, 2003. Cultural landscapes: the challenges of conservation. In: Proceedings of the Workshop, 11–12 November 2002, Ferrara, Italy. UNESCO World Heritage Papers, No. 7. UNESCO, Paris.
UNESCO, 2010. The Power of Culture for Development. UNESCO, Paris.
Uslaner, E., 1999. Democracy and social capital. In: Warren, M.E. (Ed.), Democracy and Trust. Cambridge University Press, Cambridge, pp. 121–150.
Wackernagel, M., Rees, W.E., 1996. Our Ecological Footprint: Reducing Human Impact on the Earth. New Society Publishers, Gabriola Island, BC.
WCED, 1987. Our Common Future. Oxford University Press, Oxford and New York.
Wu, J., Wu, T., 2010. Green GDP. In: Christensen, D., et al. (Eds.), Berkshire Encyclopedia of Sustainability, Vol. II. The Business of Sustainability. Berkshire Publishing, Great Barrington, pp. 248–250.

Economic indicators of sustainable resource management

Diego Martínez-Navarro, Francisco J. Oliver-Márquez
Department of Economics and Business (Applied Economy), University of Almería, Almería, Spain

1. Introduction

Overexploitation of certain renewable resources (such as fishing grounds and forest land) and, above all, the threat of fossil combustibles depletion were some of the facts that, by the 1970s, highlighted the unsustainable nature of the unprecedented industrialization process that the world had begun in the mid-20th century. Likewise, Geoergescu-Roegen (1971) warned of the exhaustion of available resources, as well as of the limits of technological efficiency, relying on the laws of thermodynamics and advocating the use of those resources that dissipate less energy, such as solar radiation. Meanwhile, Meadows et al. (1972) pointed to the limits of economic growth, proposing restrictions on population growth, as well as on the consumption of goods and, consequently, on the use of raw materials for their manufacture.

However, this type of approach did not begin to take on sufficient importance until the outbreak of the 1973 and 1979 energy crises. But precisely this climate also led to the discrediting of Keynesian and Socialist policies and, consequently, to the rise of a series of Neoliberal measures that were framed in the Washington Consensus (Williamson, 1990). These types of measures are more compatible with the approach of Meadows et al. (1972) which, in the end, proposes to put conditions on the market system to make a more sustainable use of available resources, unlike the more disruptive vision of Georgescu-Roegen (1971) which, moreover, implies a shift in the modes of production and life.

Since then there have been many initiatives around the sustainable management of resources, including the Montreal (1987) and Kyoto (1997) Protocols, the Rio de Janeiro Earth Summit (1992) and the Earth Summit + 5 (1997), the Johannesburg Summit (2002), Rio + 20 (2012) or the United Nations Summit on Sustainable Development (2015), which established the 2030 Agenda for Sustainable Development, currently in force. Especially important was the creation in 1983 of the World Commission on Environment and Development, which led to the Brundtland Report

Sustainable Resource Management. https://doi.org/10.1016/B978-0-12-824342-8.00002-X

(WCED, 1987). The latter contains the definition of sustainable development as "that meets the needs of the present without compromising the ability of future generations to meet their own needs" (WCED, 1987, p. 43).

To ensure that sustainable resource management did not become a diffuse objective and that all these initiatives were not in vain, the need arose to define and create different indicators of such sustainable management. This chapter deals with the most important of these.

2. Economic indicators of sustainable resource management

Sustainable Development definition (WCED, 1987, p. 43) implies that the present generations have to manage the resources currently available to meet their needs in such a way that these resources remain available for future generations and that they can also meet their needs with them. However, it is not really possible to know the preferences of future generations, so that the present generations end up arbitrating the value of the resources in the future. To achieve a truly sustainable use of resources, it is necessary that the preference rate of present generations for current consumption be less than or equal to the resources reproduction rate.

Now then, nonrenewable resources record a reproduction rate equal to zero, so the income obtained through them must be reinvested in sustainable resource management patterns, in accordance with the rule of Hartwick (1977). This leads to a distinction between weak and strong sustainability. Weak sustainability means that there is substitutability between different types of resources, the important thing being that the total stock of resources is maintained or increased over time and that, therefore, the present generation leaves to the future at least a stock equal to that enjoyed by the first. Strong sustainability means that there is a stock of nonsubstitutable resources that must not decrease, while the stock of substitutable resources must be kept constant or increase. Therefore, it implies that there are certain resources that must be conserved in order to continue satisfying needs.

In this sense, next indicators could be classified as weak or strong economic indicators of sustainable management. Exactly, all indicators we address in the following pages are weak sustainability indicators, except for ecological footprint and biocapacity (Section 2.1), which are strong sustainability indicators.

2.1 Resource demand and supply: ecological footprint and biocapacity

Ecosystem provides resources that people demand to meet their needs. The pressure that people put on available resources (*biocapacity*) is known as *ecological footprint* (Wackernagel and Ress, 1996). This indicator represents the biologically productive area needed by people to sustain their consumption and waste generation patterns.

The ecological footprint of a given territory depends on the amount and type of consumption carried out by the people living there, or net consumption. According to the Global Footprint Network (2019), net consumption is calculated by adding imports and subtracting exports from total production.

Each product's net consumption is divided by the overall average productivity per hectare of each product. Carbon dioxide emissions are divided by the assimilation capacity of forests, subtracting the absorption capacity of the oceans. Bio-productive areas considered are crop land, grazing land, forest land, fishing grounds, carbon dioxide absorption areas, and constructed area. Therefore, the area required to cover food needs (crop land, grazing land, and fishing grounds), wood demand (forest land), the damage from energy generation and consumption (carbon dioxide absorption areas), and the direct occupation of land with infrastructure (constructed area) are estimated.

But all these areas are not homogenous in terms of their productivity. For example, 1 hectare of crop land is not as productive as 1 hectare of grazing land. Therefore, each standard hectare must be transformed into a global hectare, which is identified with the average biological productivity of the planet. In other words, to provide a single aggregate figure for resource demand or supply (ecological footprint or biocapacity), each area must be given a weighting based on its productivity. These weights are called *equivalence factors*. The value of the equivalence factor of a given resource will be

- Equal to 1 when its bio-productivity is equal to the global average bio-productivity.
- Less than 1 when its bio-productivity is less than the overall average bio-productivity.
- Greater than 1 when its bio-productivity is greater than the overall average bio-productivity.

According to the Global Footprint Network (2019), grazing land has a bio-productivity equal to 0.46. This is because the productivity of this type of land is approximately half of the average bio-productivity of the Earth's surface. Table 13.1 provides the current value of equivalency factors for each of the bio-productive areas. Since a global hectare is equal to the average bio-productivity of the planet and this may vary slightly from year to year, the equivalence factors are also not immutable.

Table 13.1 Current equivalence factors for each of the bio-productive areas.

Bio-productive areas	Equivalence factors
Crop land	2.52
Grazing land	0.46
Forest land	1.29
Fishing grounds	0.37
Carbon dioxide absorption areas	1.29
Constructed area (infrastructure)	2.52

Extracted from the Global Footprint Network (2019, p. 55).

If people demand resources to meet their needs, then biocapacity is a key aspect of achieving economic and social well-being. The Millennium Ecosystem Assessment (2005) already warned of this association between biocapacity and human well-being. This well-being is threatened when the demand for resources exceeds their supply. That is, when the ecological footprint exceeds the biocapacity. If this happens in a certain territory, it means that the people who live there are consuming more resources than they have. In other words, they are meeting their needs at the expense of the resources of other territories, or at the expense of the resources belonging to future generations, or even both.

In view of the above, another important aspect when analyzing the ecological footprint and biocapacity is to know how many people live in each territory, which allows comparisons to be made. Thus, the following table shows the data for these two indicators, measured in global hectares per capita, for the main regions of the planet in 2016 (the last year for which data exist).

Data reveal that, worldwide, the ecological footprint is currently higher than the biocapacity. This phenomenon is replicated in almost all the regions we analyze. Oceania, Latin America and the Caribbean, Middle Africa, and Eastern Europe are the regions that still have resource reserves. In other words, in these regions' resources demand does not yet exceed the supply. This results in the regions with the highest deficits covering their demand with the supply from the regions in surplus, thus cutting off the welfare of current generations living in those regions, as well as future generations in all regions. To avoid this situation, it would be necessary to achieve greater convergence between the ecological footprint and biocapacity in the most deficit regions. In other words, people living in these regions should adopt consumption habits that are more in line with the resources available in their territory.

Table 13.2 is completed by the next map (Map 13.1). It shows in red color (dark gray color in printed version) those countries whose ecological footprint is greater than their biocapacity (resource deficit). Meanwhile, those countries whose biocapacity exceeds their ecological footprint are shown in green (gray in print version) (resource reserve).

Despite the information capacity on sustainable resource management provided by the ecological footprint and biocapacity, these indicators have many limitations. Thus, it identifies air pollution with carbon dioxide emissions, omitting other important substances that pollute the atmosphere such as carbon monoxide, sulfur dioxide, nitrogen dioxide, asbestos, or zinc, among others. Furthermore, it does not consider the phenomenon of erosion, which also implies considering that the productivity of soils is immutable. It also excludes water pollution. However, these shortcomings do not invalidate the ecological footprint and biocapacity, although they do highlight the need for improvement.

2.2 Green net national product

Gross National Product (GNP) is a traditional indicator of economic growth. It measures the monetary value of the final goods and services produced within an economy's borders over a given period (usually 1 year) by national production factors, regardless of whether they are located within or outside those borders. However, GNP does not consider the fact that capital is depreciating. However, this fact is considered by Net National Product (NNP), which is the GNP minus depreciation.

Table 13.2 Demand and supply of resources: ecological footprint and biocapacity: deficit and reserve.

Region	Ecological footprint (gha. per person)	Biocapacity (gha. per person)	Deficit or reserve
North America	8.07	4.83	Deficit
Latin America and the Caribbean	2.59	5.06	Reserve
Western Europe	4.86	1.78	Deficit
Eastern Europe	4.40	4.73	Reserve
Northern Europe	5.10	3.52	Deficit
Southern Europe	4.11	1.31	Deficit
Northern Africa	1.84	0.70	Deficit
Southern Africa	3.03	1.23	Deficit
Middle Africa	0.99	2.85	Reserve
Western Africa	1.27	1.05	Deficit
Eastern Africa	1.02	0.89	Deficit
Central Asia	2.99	1.68	Deficit
South-Eastern Asia	1.92	1.21	Deficit
Western Asia	3.41	0.75	Deficit
Oceania	5.22	9.42	Reserve
World	2.75	1.63	Deficit

Source: Own elaboration using data from the Global Footprint Network, 2020. Advancing the science of sustainability. Country Trends. Available from: http://data.footprintnetwork.org/?_ga=2.11617591.26475017.1594741980-2030800832.1594741980#/countryTrends?cn=5001&type=BCpc,EFCpc. (Accessed July 2020).

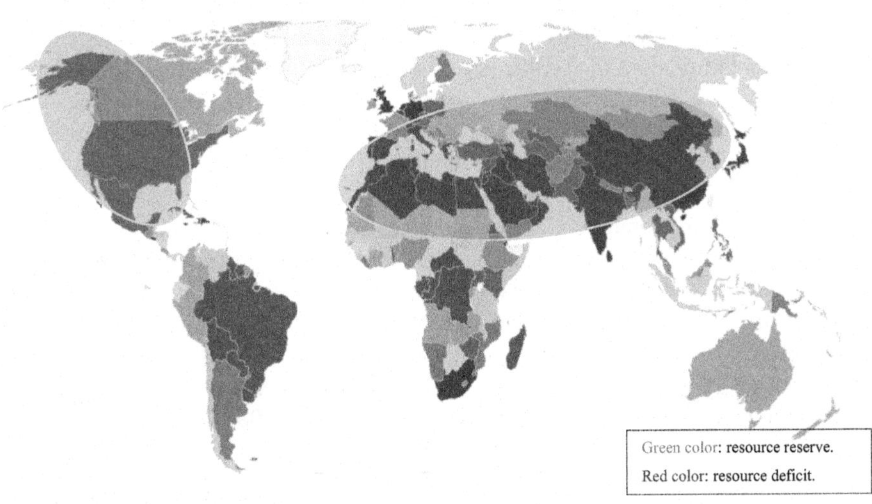

Map 13.1 Demand and supply of resources around the world. Extracted from the Global Footprint Network, 2020.

Therefore, NNP is more accurate than GNP. But neither of these two indicators includes environmental resources, which are a form of capital. Thus, they do not consider environmental degradation and overexploitation of resources associated with economic activities. This fact pushed Hamilton (1994) to make green adjustments, giving rise to the Environmentally Adjusted Net National Product, better known as Green Net National Product (gNNP). Hamilton (1994) linked a series of specific environmental variables (renewable resources, exhaustible resources, pollution flows and discoveries) to a neoclassical model of growth with a constant discount rate. According to the methodology of Hamilton and Clemens (1999), the gNNP can be expressed as follows:

$$gNNP = C + K - \sum_{i}^{n}(p_i - C_i)(R_i - g_i) - \sum_{j}^{m} b_j(E_j - d_j) \tag{13.1}$$

Where

- $gNNP$: Green National Net Product.
- C: Total Consumption.
- K: Net Investment.
- p_i: market price of the resource i.
- C_i: marginal cost of production or extraction of the resource i.
- R_i: production or extraction of the resource i.
- g_i: growth of the resource i, which is 0 if i is a nonliving resource.
- b_j: marginal cost of abatement of the pollutant j.
- E_j: emission of the pollutant j.
- d_j: natural dissipation of the pollutant j, which is 0 if j is a pure cumulative pollutant.

Thus, the economic value of the depreciation of nonrenewable resources (such as oil, coal, gas, etc.) is equivalent to the income from the resources extracted. The income per unit extracted is equal to the difference between the unit market price and the marginal cost of extraction. If an economy has $i = 1, \ldots, n$ exhaustible resources, depreciation is calculated as the product of income per unit and the difference between the amount extracted from the resource and the new deposits discovered in period $m_i - n_i$. In the case of renewable resources, the biological reproductivity of the resource is considered. Likewise, the gNNP includes the depreciation of the assimilative quality, which is equivalent to the damage caused by the polluting emissions.

Regarding its interpretation, gNNP indicates the maximum consumption that we can make today without reducing the possibilities of consumption in the future. Therefore, increases in an economy's gNNP imply increases in its sustainability. Therefore, to analyze the advances or setbacks of an economy in terms of sustainability in the use of its resources, it is more useful to use the growth rate of the gNNP than the gNNP in absolute terms. We do so in the Graph 13.1, which represents the growth rate of the NNP in the main regions of the world during the period 2008–2018[1].

[1] Except for Middle East and North Africa, for which the latest available data refer to 2017.

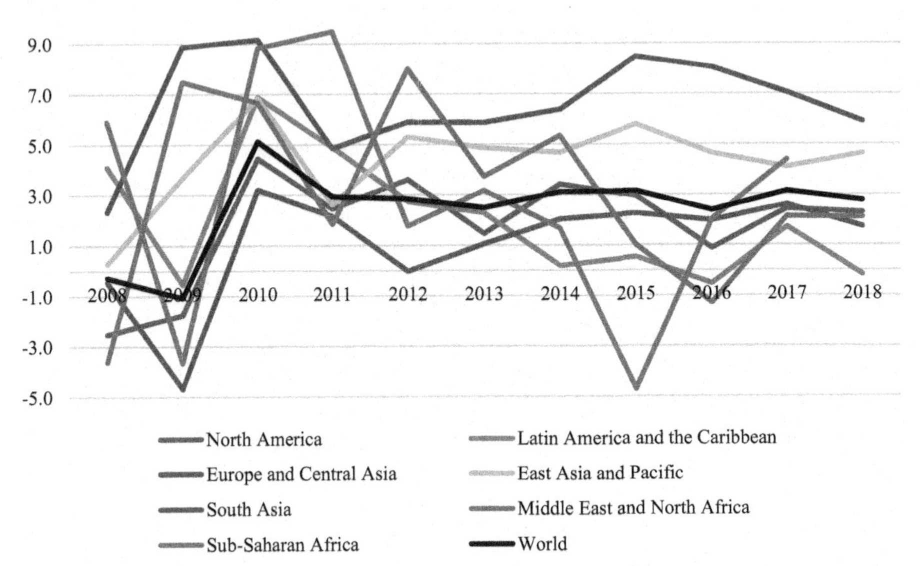

Graph 13.1 Green net national product (annual % growth) in different regions in the world during 2008–18.
Source: World Bank Database.

This graph reveals that there are strong advances and setbacks in the gNNP annual growth rate. In other words, the different regions of the planet are unable to maintain a constant trend in the sustainable use of their resources. However, during the last years of the analysis period, some regions describe a less unstable behavior. This is the case of Europe and Central Asia, North America, East Asia, and Pacific and South Asia. Meanwhile, both Middle East and North Africa are currently registering progress after several years of significant setbacks. A similar behavior is registered by Sub-Saharan Africa, although with less abrupt turns. In contrast, Latin America and the Caribbean reflects a downward trend since 2010 and although it registered a slight upturn in 2017, it is currently continuing to decline. Finally, the fact that the World is maintaining an almost constant trend over the last 7 years suggests that there are regions that are meeting their own needs by using the resources of others.

Although the gNNP is the weak indicator of sustainability par excellence, it has several limitations. Thus, it is not always easy to obtain the value of the marginal costs of extraction or capture of resources. For this reason, it is common to resort to average costs instead of marginal costs (Nourry, 2008). In fact, estimating environmental – damage in monetary units is not a simple task. Nevertheless, gNNP provides important information about the origin of the problem of unsustainable resource use and therefore constitutes a starting point for achieving sustainable resource use and, with it, sustainable development.

2.3 Adjusted net savings or genuine savings

In the System of National Accounts (UNSD, 2008) only fixed capital formation counts as investment and therefore as an increase in the value of the available assets in a society. Similarly, the standard national accounts only include the depreciation of the value of the capital produced as a decrease in the value of a company's assets when calculating the net savings rate. Indeed, Adjusted Net Savings (ANS), also known as Genuine Savings, goes further by considering that the development of economies also relies on other forms of capital, such as human and natural capital (Pearce and Atkinson, 1993; Hamilton, 2000a; World Bank, 1995, 2018).

Why consider these two forms of capital? On the one hand, when a renewable resource is overexploited (or a nonrenewable resource is depleted), the value of that resource decreases, leading to disinvestment and a consequent decline in future productivity and development. On the other hand, educating people and creating a skilled labor force implies increases in future productivity and development. Therefore, an economy with a positive standard net saving may have a negative ANS because it overexploits its renewable resources (or depletes its nonrenewable resources) and because it has a poorly educated and skilled labor force. This logic also shows that ANS is a weak indicator of sustainability, since it assumes that any kind of capital is perfectly replaceable by natural capital.

Therefore, ANS can be defined as a measure of the net savings obtained by adding spending on education to gross domestic savings and deducting from it, in addition to depreciation of material capital (or consumption of fixed capital), depletion or overexploitation of natural resources, and damage caused by emissions of pollutants. Exactly, the World Bank (2018) calculates the ANS as follows:

$$ANS = GNS - CFC + EDU - NRD - GHG - POL \tag{13.2}$$

Where

- *GNS*: Gross National Saving, that is, Gross National Income (GNI) minus public and private consumption. These all are standard national accounting items.
- *CFC*: Consumption of Fixed Capital, which is also a standard national accounting item, and consists of the replacement value of the capital used in the production process.
- *EDU*: Public Spending on Education. In standard national accounts, savings measures only consider as investment that part of education expenditure that is allocated to fixed capital (grossly, school infrastructure), while the rest is considered as consumption. However, the ANS definition conceives human capital as an important asset. Therefore, in its calculation it considers current spending on education, including wages and salaries and excluding capital investments in buildings and equipment.
- *NRD*: Natural Resource Depletion. It consists of the sum of the net depletion of forests, the depletion of metals and minerals (bauxite, copper, tin, iron, nickel, gold, silver, lead, phosphate rock, and zinc) and the depletion of fossil energy resources (coal, natural gas, and crude oil).
- *GHG*: Damage caused by carbon dioxide emissions from the use of fossil fuels and the manufacture of cement These damages are estimated at $30 per ton of carbon dioxide emitted.
- *POL*: Damages from exposing people to air pollution. It includes ambient concentrations of particles 2.5 microns in diameter, indoor concentrations of air pollution in homes cooking

with solid fuels, and finally, ambient ozone pollution. These damages are calculated as the loss of labor output because of premature death due to exposure to pollution.

ANS is a flow variable that is measured as a percentage of GNP. As regards its interpretation, it is desirable that savings should exceed the depreciation of assets, which include resources. Thus, if the ANS is negative, it means that economic welfare at some point in the future will be lower than at present (Hamilton and Clemens, 1999; Dasgupta and Mäler, 2000). Also, a consistently negative ANS over time is reporting unsustainable resource use. Graph 13.2 reports on the evolution of ANS in different regions of the world during the period 2008–2018[2].

This graph reveals that globally the ANS trend has been upward over almost the entire analyzed period. In fact, in this interval almost all regions considered have positive ANS. Sub-Saharan Africa is the exception. Resources are therefore being used unsustainably in this region. Similarly, in Middle East and North Africa, despite positive ANS, resource use has been declining throughout most of the period and has been increasing since 2016. Similarly, while South Asia has had the highest values for most of the period, its ANS has been gradually declining. Latin America and the Caribbean have performed similarly, albeit from considerably lower levels. The opposite has happened in East Asia and Pacific which has been increasing its ANS to become the region with the highest value in 2018. Similarly, North America's ANS has been rising steadily until 2014, when it began to decline, although it is currently at levels well above the initial ones. The trend in Europe and Central Asia has remained almost constant.

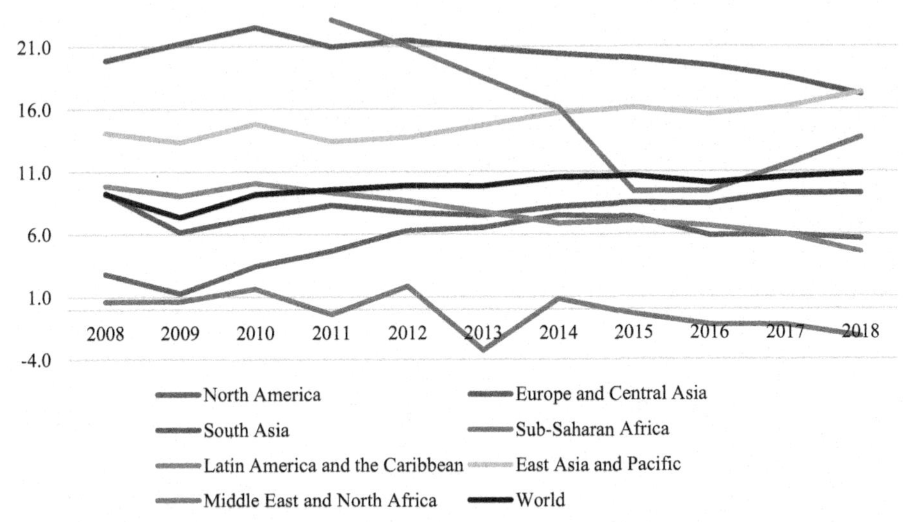

Graph 13.2 ANS (% of GNI) in different regions in the World during 2008–18.
Source: World Bank Database.

[2] Missing values for Middle East and North Africa in 2009 and 2010.

ANS is an advance compared to standard net savings, as it considers forms of capital (human and natural) that actually contribute to the productivity of economies and hence to their prosperity, which is why they cannot be excluded. ANS even reveals more information than gNNP, not only because it considers human capital but also because it has a simpler calculation methodology that also uses variables whose data are normally available for a large majority of countries while being easy to interpret. Nevertheless, estimating the depreciation of human and natural capital is not an easy task (Hamilton, 2004). Moreover, while ANS allows us to know whether an economy's resources have increased or decreased, it does not reveal information about whether that economy would be sustainable with a growing population (Hamilton, 2000b). In other words, it could be that ANS records positive values and that per capita total wealth decreases, which justifies the existence of the following indicator.

2.4 *Total wealth per capita*

Nations economic prosperity has traditionally been measured through income indicators (such as GDP or GNP, among others), with wealth taking a back seat. However, when a company sells some of its assets, although it obtains immediate income from the sale, its ability to generate future income is also hampered. The same applies to economies. Indeed, the sustained rapid income growth experienced by rich economies in the 1980s was largely due to the liquidation of their natural capital, which called into question their capacity to grow in the future (Repetto et al., 1989; Lange et al., 2018).

Therefore, income and wealth are complementary indicators. In other words, it is not only important to know whether an economy's income is increasing or decreasing (a fact reported by income indicators) but also whether that income and its growth can be sustained over the long term (which is what wealth indicators are for). Similarly, it may be that the total wealth of a given economy increases, but that the population increases more. This would hinder the sustainability of that economy, since the larger the population, the greater the needs to be covered. Therefore, it is important to express total wealth in per capita terms. It is also important if comparisons are to be made between different economies, since not all have the same population.

Officially, Total Wealth is defined as "the sum of produced capital, natural capital, human capital, and net foreign assets" (Lange et al., 2018, p. 44). The result of this sum is divided by the total population of the economy in question to obtain the Total Wealth per capita. Analytically

$$TW = PC + NC + HC + NFA \tag{13.3}$$

$$TW_{pc} = \frac{TW}{p} \tag{13.4}$$

Where

- TW: Total Wealth.
- TW_{pc}: Total Wealth per capita.

- *p*: Population or number of residents in the economy concerned.
- *PC*: Produced Capital. It includes machinery, buildings, equipment, and residential and nonresidential urban land at market prices.
- *NC*: Natural Capital. It consists of energy valuation of fossil fuels (coal, crude oil, and gas), minerals (bauxite, copper, tin, phosphate, iron ore, nickel, gold, silver, lead, and zinc), agricultural land (cropland and pasture), forests (wood and some nonwood forest products), and protected areas.
- *HC*: Human Capital. Like ANS, Total Wealth per capita considers human capital. This is calculated by updating the future income of the working population over their lifetime. It considers gender gaps as well as the different types of employment that exist.
- *NFA*: Net Foreign Assets, that is, the sum of a country's external assets and liabilities, such as foreign direct investment and reserve assets.

Table 13.3 shows the trend of Total Wealth per capita in the main regions of the World during the period 1995–2014, for which data are available. These are measured in constant 2014 US dollars. The last column shows the average growth rate, expressed as a percentage. North America has a much higher total wealth per capita than other regions over the entire period of analysis. In fact, in 2014, it is almost six times the value recorded by World. However, it is the region with the second lowest average growth rate. South Asia shows exactly the opposite behavior. In other words, it is

Table 13.3 Total Wealth per capita in different regions in the World (constant 2014 US $, using a country-specific GDP deflator). 1995–2014.

Region/Year	1995	2000	2005	2010	2014	Average growth rate (%)
North America	782.370	901.889	962.329	945.004	986.621	4.7
South Asia	9.251	10.523	12.511	15.710	18.400	14.7
Latin America and Caribbean	108.351	109.692	117.115	130.960	138.294	5.0
Middle East and North Africa	91.203	95.076	113.731	143.965	158.892	11.7
Europe and Central Asia	279.651	300.506	327.765	355.495	368.233	5.7
Sub-Saharan Africa	26.403	21.964	22.669	25.362	25.562	−0.6
East Asia and Pacific	76.102	83.618	89.773	117.983	140.042	13.0
World	128.929	138.064	145.891	158.363	168.580	5.5

Source: World Bank Database and Lange, C.M., Wodon, Q., Carey, K., 2018. The Changing Wealth of Nations 2018: Building a Sustainable Future. World Bank, Washington, D.C.

the region with the lowest total wealth per capita during the whole period, but the one with the highest average growth rate. It is followed by East Asia and Pacific and Middle East and North Africa, with rather modest levels of total wealth per capita.

Latin America and the Caribbean has similar levels of total wealth per capita as the latter two regions, but a lower average growth rate than World. Europe and Central Asia, with an average growth rate like that of World, is the second region with the highest total wealth per capita. Even so, it is not even half as rich as North America. The region in the worst position is Sub-Saharan Africa. First because it has low levels of total wealth per capita (although it exceeds South Asia). Second, because it is the only region with a negative average growth rate, which suggests that there are likely to be regions that are meeting their needs using the resources of this region.

Total Wealth per capita is an advance compared to both gNNP and ANS. It has also undergone several methodological improvements, achieving a more accurate measurement of sustainability in the use of resources (Lange et al., 2018). However, this indicator still has some limitations. Thus, natural capital is underrepresented, due to the difficulty of measuring some of the services provided by ecosystems. So much so that the market prices of natural assets may not really reflect the costs and benefits they bring. Similarly, no account is taken of forest degradation and the uncertainty associated with potential natural disasters, which could lead to irreversible damage. Finally, it should not be forgotten that Total Wealth per capita is a weak indicator of sustainability, i.e., it assumes substitutability between types of capital. But the truth is that there are natural capital assets that are critical, that is, that perform transcendental functions while being irreplaceable.

2.5 *Index of sustainable economic welfare*

The Index of Sustainable Economic Welfare (ISEW) is an economic indicator that tries to measure the social welfare of a country. Initially, the ISEW was designed to replace the Gross Domestic Product (GDP) since it has several shortcomings when measuring social welfare, this index is based on the concept of "Economic Welfare" published by Nordhaus and Tobin (1972) in their work entitled "*Is growth obsolete?* Likewise, from 1972 to the present, ISEW continued to evolve, being finally coined by Daly and Cobb (1989), although some costs would later be added to the definition of ISEW. The work of Daly and Cobb would also lead to a second macroeconomic indicator that would also try to measure sustainability, this is, Genuine Progress Indicator (GPI), this is an extension of the ISEW that focuses on the genuine and actual progress of a country for the purpose of measuring the ecological sustainability of an economy. Both macromagnitudes synthesize economic well-being into a single value according to the same logic by which GDP summarizes economic performance into a single figure. In addition to economic issues, social and environmental issues are included in monetary terms.

For this reason, the ISEW could be considered an extension of GDP, since it takes into account the goods and services produced in the economy, such as GDP, and also consumer spending, the utility provided by domestic work (which does not count as

GDP) and to these factors it deducts the cost of externalities associated with pollution and natural resource consumption. Thus, we can write the ISEW formula as follows:

$$ISEW = (C_P + C_C - C_{PD} - C_{DL}) + CF + S_{DL} + (\delta_{ED} + \delta_{NC})$$

where C_P is Personal Consumption; C_C is Public nondefensive expenditures; C_{PD} and C_{DL} are private and public defensive expenditures; CF is capital formation; S_{DL} is services from domestic labor; δ_{ED} is costs of environmental degradation; δ_{NC} is depreciation of natural capital.

One could compare the GDP of a closed economy with the ISEW as follows:

$$GDP = (C_{PT} + C_{CT}) + CF = ISEW + 2(C_{PD} + C_{DL}) - S_{DL} + \delta_{ED} + \delta_{NC}$$

where C_{PT} is personal consumption (including defensive expenditures); C_{CT} is public expenditure (including defensive expenditures).

The first review of the ISEW comes from Munday and Roberts (2006). These authors point out that it is a weak indicator of sustainability, since it is based on this paradigm, since it takes into account the environmental assets that are used to increase consumption. For this reason, the ISEW does not take into account the integrity, irreversibility, uncertainty, or existence of natural components that make unique contributions to human welfare, typical of the strong sustainability paradigm.

On the other hand, Ziegler (2007) points out that the ISEW is focused on consumption, then Daly and Cobb do not really achieve the "community approach" they describe in their 1989 work "For the Common Good" Therefore, the criticism that Ziegler makes on the ISEW is that it is a macromagnitude with an individualistic vision of collective welfare, this is, that this index cannot capture the collective conditions of work or the collective processes that underlie the processes of consumption or production.

Other authors such as Eisner (1994), Jackson et al. (1997), Neumayer (2000), Lawn (2005), and Bleys (2008) criticize the ISEW arguing that the methods to measure externalities at the monetary level are imprecise, then it is not possible to calculate precisely the environmental costs or the depreciation of natural capital, therefore the ISEW is also imprecise, as well as the poor foundation of some economic values.

In the face of all this criticism, Posner and Costanza (2011) defend "it is better to be approximately right tan precisely wrong" (pag. 1973).

The ISEW has been calculated for several countries, in the following graphs we can see the GDP per capita and the ISEW per capita calculated for two areas of Italy, Marche and Tuscany (Graph 13.3)

This is also the case for Romania, calculated by Butnairu and Luca (2019). In this case we have the GDP per capita in blue (gray in print version), the ISEW per capita in red (dark gray in print version), and the ISEW-K which is the basic ISEW but making some adjustments according to the capital (K) in green (Graph 13.4).

Finally, we will also show the case of Ecuador in Sánchez et al. (2020), in this case we find the GDP in the blue (gray in print version) line and the ISEW in the green (dark gray in print version) line (Graph 13.5).

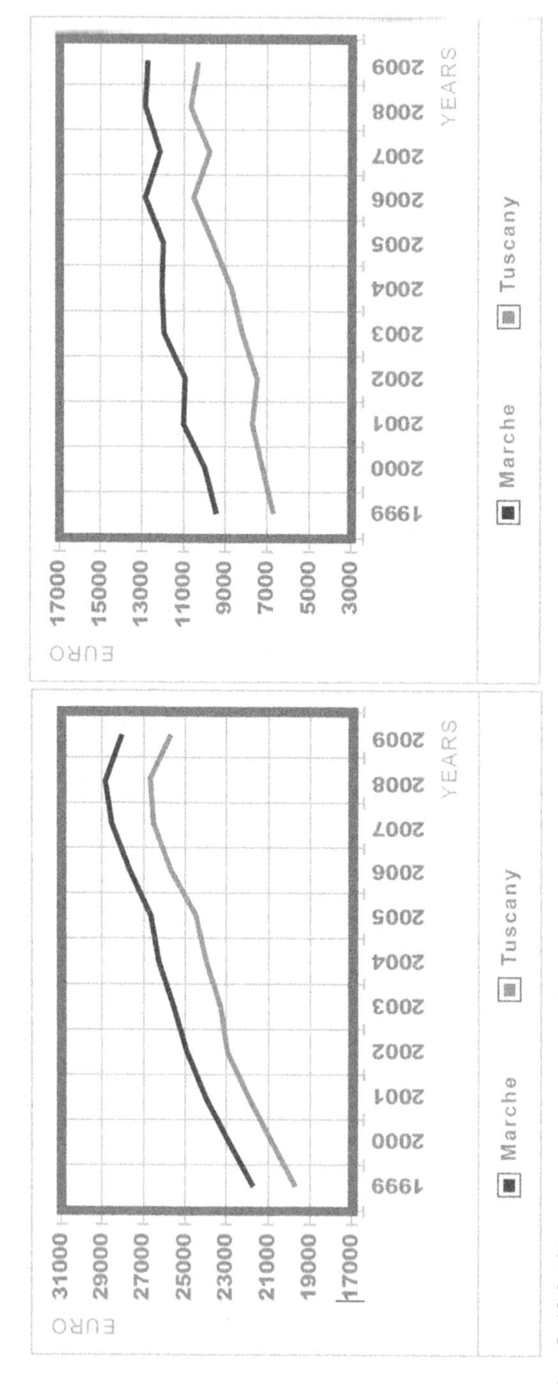

Graph 13.3 ISEW calculated for two Italian regions.
Source: Own elaboration based on Chelli et al., 2013.

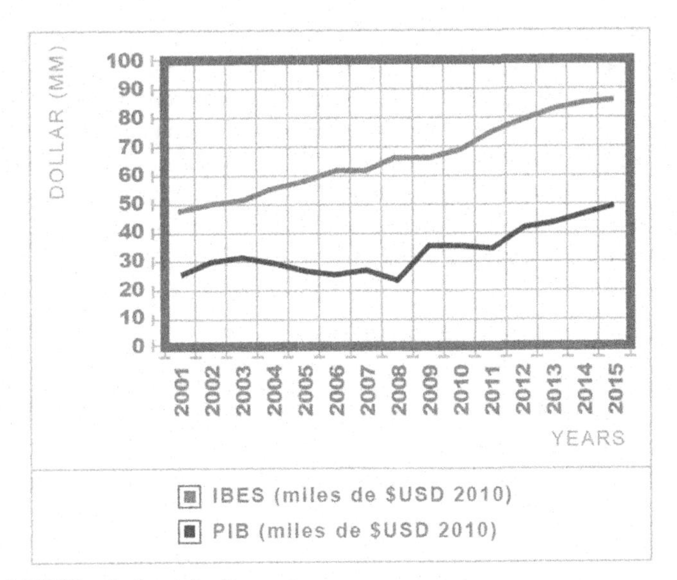

Graph 13.4 ISEW calculated for Romania.
Source: Own elaboration based on Butnariu and Luca, 2019.

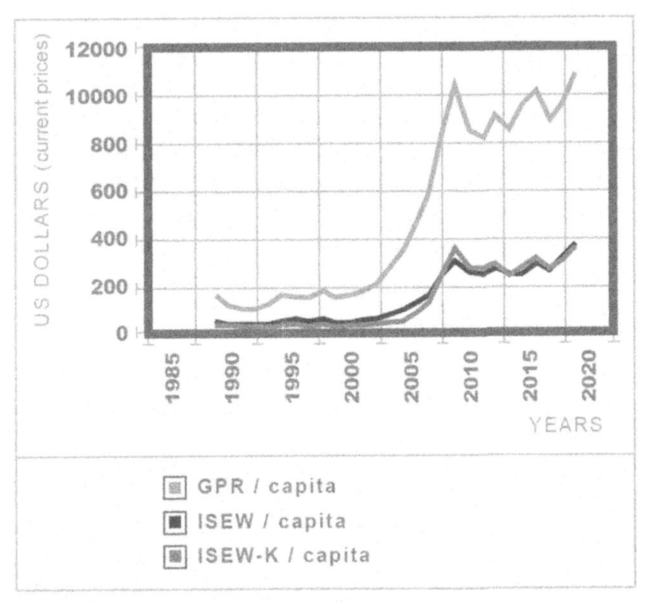

Graph 13.5 ISEW calculated for Ecuador.
Source: Own elaboration based on Sánchez et al., 2020.

In these graphs we can see how GDP and ISEW follow a similar pattern of growth. With the difference that the ISEW is notably smaller in value than GDP because in GDP the pollution adds up (if it is due to production), while in ISEW it subtracts, the difference between GDP and ISEW being greater, the higher the level of production as can be seen especially in the case of Romania of Butnairu and Luca (2019). It is also observed how the Great Recession of 2008 made the GDP fall and the ISEW increase a little due to the decrease in production, which caused less CO_2 emissions and pollution, although it subsequently fell in 2009.

2.6 Genuine progress indicator

The GPI, like the previous index, has its origin in Daly. This index tries to measure welfare as well as social progress, such as GDP, with the difference that it takes into account factors that are not exclusively monetary.

The aim of this index is to reflect activities that are not remunerated, and are therefore excluded from the calculation of GDP, as well as to reflect the environmental impact that production generates on the ecosystem. In this way, the GPI counts up the factors related to work (including unpaid work such as domestic or voluntary work), and counts down the loss of environmental resources, income inequality, and crime.

The calculation of the GPI is made through 26 indicators grouped in three different areas. These indicators can be seen in the following table (Table 13.4).

The GPI, like other alternative measures of wealth, is criticized as being based on variables that are difficult to quantify, such as the cost of crime or lost leisure time. However, this index has emphasized that it is more vulnerable to political manipulation since the data on which these measures are based are published by the state.

In conclusion, the GPI is an index that considers economic growth taking into account social and environmental factors, and therefore shows what is the economic trajectory of a country by penalizing it if it does not have a firm commitment to the environment or by raising the indicator if it does show such commitment.

2.7 Natural heritage accounts

The input–output matrix is an ordered representation of the supply and use balances of goods and services from different sectors of an economy. This matrix summarizes the situation of the economy in a given year.

In this matrix, Sejenovich (1996) proposed in his *Manual of Natural Heritage Accounts* the inclusion of a new economic sector, the preprimary sector. This sector would be analogous to the manufacturing sector, but taking into account the reproduction of nature through the capture of energy, that is, it would be the "factory of nature."

In the input–output matrix of this preprimary sector, the sectors of nature would be presented: flora, fauna, soil, water, air, biodiversity, etc., which would offer and consume the natural capital among themselves as well as with the other economic sectors. Once this has been done, the final columns would include the intermediate and final products, as well as the objective to be pursued in order to reach the condition

Table 13.4 Indicators and variables of the genuine progress indicator (GPI).

Component	Impact on IPG	Variable
GPI economic well-being indicators	Positive	**1.** Personal Consumption (PC)
	Split	**2.** Income Distribution Index (IDI)
	(Var. 1/Var. 2) * 100	**3.** Weighted personal Consumption
	Indeterminate (Pos/Neg)	**4.** Net Capital investment
	Negative	**5.** Cost of underemployment
	Negative	**6.** Cost of consumer durables
	Positive	**7.** Services of consumer durables
GPI environmental well-being indicators	Negative	**8.** Cost of water pollution
	Negative	**9.** Cost of air pollution
	Negative	**10.** Cost of noise pollution
	Negative	**11.** Loss of wetlands
	Negative	**12.** Loss of farmland
	Negative	**13.** Loss of primary Forests and damage from logging roads
	Negative	**14.** Carbon dioxide Emissions damage
	Negative	**15.** Cost of Ozone depletion
	Negative	**16.** Depletion of nonrenewable Energy resources
GPI social-human well-being indicators	Positive	**17.** Housework and the value of education for children
	Negative	**18.** Value of household work and parenting
	Negative	**19.** Cost of crime
	Negative	**20.** Cost of household pollution abatement
	Positive	**21.** Value of volunteer work
	Negative	**22.** Loss of leisure time
	Positive	**23.** Value of higher education
	Positive	**24.** Services of highways and streets
	Negative	**25.** Cost of commuting
	Negative	**26.** Cost of automobile accidents

Source: Table made by the authors based on Talberth, J., Cobb, C., Slattery, N., 2007. The Genuine Progress Indicator 2006. Redefining Progress, Oakland, C.A.

of sustainability. Once the matrix is established, it is necessary to define an "ecozone" whose function is to systematize the functioning of the inputs and outputs of this pre-primary sector, allowing this ecozone to understand the interrelations between the rows and the columns in a qualitative way. Likewise, an ecozone needs to be proposed for each ecosystem to be included in the matrix (forest, steppe, bush, etc.). The next step is to define the management costs in each cell or the cost of research needed to qualitatively define the cell. It should be noted that depending on the ecozone defined, the situation of each case will depend, as well as the definition of the columns.

If we take the case of a forest, the fact of cutting down a tree to sell it as a raw material of this preprimary sector generates diverse purchase and sale interactions within the matrix, that is, it provokes diverse ecosystemic interrelations with the other sectors of nature as well as with economic sectors. In this way, in order to be able to cut down the tree, nature will "buy" other resources necessary for the tree to be cut down as well as it will have to "pay" what corresponds to it in wages and salaries, inputs of "machinery" and "infrastructure" that maintain in each box the relationship described by being used that tree. That is, the sale of the tree for its cutting does not only imply the extraction of the tree from nature, but also the remaining ecosystem resources, such as the flora to which it provides shade and substances necessary for its development (proteins, carbohydrates, fiber, tannins, etc.), the fauna that serves as its habitat, the soil that collects it and protects it from rainfall and other types of erosion, the water that regulates it with its roots, etc.

The disadvantages of Natural Heritage Accounts analysis are that there are alterations in the environmental system that are irreversible, or that are uncertain because researchers cannot anticipate the future consequences of current changes. Likewise, at present we have no knowledge of the planet's real natural resource reserves, or the technology that will be used in the future, so we do not know what the real limit of natural capital production is. And finally, current and future monetary valuations can be arbitrary, generating even more error if we admit that it is not possible to reduce the diversity of units of the environmental system to a common unit.

2.8 Environmental sustainability index

The Environmental Sustainability Index (ESI) was first proposed by the World Economic Forum in 2000, under the name of the Pilot Index of Environmental Sustainability (IPSA). Since then, it has been modified by various authors to better capture environmental sustainability and was finally named the Environmental Sustainability Index. The ESI, like other indices, is between 0 and 100, with 0 being the result of a country with no capacity to achieve sustainable environmental development and 100 being a country with fully sustainable development.

The ESI is composed of five subscripts:

- Environmental systems. An economy is sustainable from an environmental perspective as long as its environmental systems (air, water, land, and biodiversity) are maintained at healthy levels, and these tend to improve.

- Reducing environmental stresses. This subindex considers those factors that can erode the environment, such as the evolution of the population, the generation of waste and emissions, deforestation, etc.
- Reducing human vulnerability. This subindex considers the health status and ability to meet the basic food needs of the population in the environmental system.
- Social and institutional capacity. A subindex that tries to describe how appropriate the environmental institutions are, the participation and debate on environmental issues, the resources that are allocated to environmental research, etc.
- Global stewardship. It takes into account the country's dedication to global pollution problems as well as participation in international treaties that seek to address such problems.

These five subindices of the ESI are in turn divided into 21 components, which are measured through 76 different variables, as can be seen in the following table taken from Schmiedeknecht (2013) (Table 13.5).

The results of the 2020 ESI can be seen in the following ESI's framework and result snapshot. It shows the countries of the world according to the ranking obtained following the ESI, being the countries highlighted in darker color those with a higher ESI, and from this the lighter countries have results of lower ESI (Graph 13.6).

The map shows that, according to the report published in 2020: the more income a country has, the more it can invest in policies and programs that lead to a desirable point of environmental exploitation, such as reduction of air pollution, control of hazardous substances and waste, etc.

The 2020 ESI also stresses that it is not incompatible that countries need to sacrifice the environment in order to make economic progress, as it shows how more developed countries can mobilize their communities to protect natural resources and human well-being, this being the result of good and effective governance. This phenomenon is particularly evident in the top-ranked countries. However, it is also pointed out that all countries still have issues to improve, as none of them can claim with certainty to be on a fully sustainable path.

It is also stressed that the countries that are lagging behind must multiply their efforts to achieve long-term sustainability, as some important countries such as India, Nigeria, Nepal, or Afghanistan are near the bottom, which highlights the lack of attention to sustainability policies such as air quality, water, biodiversity, or climate change. 2020 EPI speculates that such outcomes may be caused by weak governance.

3. Criticism of monetary sustainability indicators

In general, the criticisms made of these indicators, as we have seen throughout the chapter, are as follows:

- Some damage to the environmental system is irreversible, and therefore difficult to quantify.
- Damages to the environmental system are cumulative, so trying to measure new damages over the previous ones may cause difficulties in terms of their repercussion.
- Some alterations to the system are unpredictable and we cannot know their results.

Table 13.5 Indicators and variables of the environmental sustainability index.

Component	Indicator	Variable
Environmental system	**I.** Air quality	**1.** Urban population weighted NO_2 concentration
		2. Urban population weighted SO_2 concentration
		3. Urban population weighted SO_2 concentration
		4. Indoor air pollution from solid fuel use
	II. Biodiversity	**5.** Percentage of country's territory in threatened ecoregions
		6. Threatened bird species as percentage of known breeding bird species in each country
		7. Threatened mammal species as percentage of known mammal species in each country
		8. Threatened amphibian species as percentage of known amphibian species in each country
		9. National Biodiversity Index
	III. Land	**10.** Percentage of total land area (including inland waters) having very low anthropogenic impact
		11. Percentage of total land area (including inland waters) having very high anthropogenic impact
	IV. Water quality	**12.** Dissolved oxygen concentration
		13. Electrical conductivity
		14. Phosphorus concentration
		15. Suspended solids
	V. Water quantity	**16.** Freshwater availability per capita
		17. Internal groundwater availability per capita

Table 13.5 Indicators and variables of the environmental sustainability index.—cont'd

Component	Indicator	Variable
Reducing environmental stresses	**VI.** Reducing air pollution	**18.** Coal consumption per populated land area
		19. Anthropogenic NO_x emissions per populated land area
		20. Anthropogenic SO_2 emissions per populated land area
		21. Anthropogenic VOC emissions per populated land area
		22. Vehicles in use per populated land area
	VII. Reducing ecosystem stresses	**23.** Annual average forest cover change rate from 1990 to 2000
		24. Acidification exceedance from anthropogenic sulfur deposition
	VIII. Reducing population pressure	**25.** Percentage change in projected population 2004–50
		26. Total fertility rate
	IX. Reducing waste and consumption pressures	**27.** ecological Footprint per capita
		28. waste recycling rates
		29. Generation of hazardous waste
	X. Reducing water stress	**30.** Industrial organic water pollutant (BOD) emissions per available freshwater
		31. Fertilizer consumption per hectare of arable land
		32. Pesticide consumption per hectare of arable land
		33. Percentage of country under severe water stress
	XI. Natural resource management	**34.** Productivity overfishing
		35. Percentage of total forest area that is certified for sustainable management
		36. World Economic Forum Survey on subsidies
		37. Salinized area due to irrigation as percentage of total arable land
		38. Agricultural subsidies

Continued

Table 13.5 Indicators and variables of the environmental sustainability index.—cont'd

Component	Indicator	Variable
Reducing human vulnerability	**XII.** Environmental health	**39.** Death rate from intestinal infectious diseases
		40. Child death rate from respiratory diseases
		41. Children under five mortality rate per 1000 live births
	XIII. Basic human sustenance	**42.** Percentage of undernourished in total population
		43. Percentage of population with access to improved drinking water source
	XIV. Reducing environment-related natural disaster vulnerability	**44.** Average number of deaths per million inhabitants from floods, tropical cyclones, and droughts
		45. Environmental Hazard Exposure Index
Social and institutional capacity	**XV.** Environmental governance	**46.** Ratio of gasoline price to world average
		47. Corruption measure
		48. Government effectiveness
		49. Percentage of total land area under protected status
		50. World Economic Forum Survey on environmental governance
		51. Rule of law
		52. Local Agenda 21 initiatives per million people
		53. Civil and political liberties
		54. Percentage of variables missing from the CGSDI "Rio to Joburg Dashboard"
		55. IUCN member organizations per million population
		56. Knowledge creation in environmental science, technology, and policy
		57. Democracy measure

Table 13.5 Indicators and variables of the environmental sustainability index.—cont'd

Component	Indicator	Variable
	XVI. Eco-efficiency	**58** energy efficiency
		59. Hydropower and renewable energy production as a percentage of total energy consumption
	XVII. Private sector responsiveness	**60.** Dow Jones Sustainability Group Index (DJSGI)
		61. Average Innovest EcoValue rating of firms headquartered in a country
		62. number of ISO14001 certified companies per billion dollars GDP (PPP)
		63. World Economic Forum Survey on private sector environmental innovation
		64. Participation in the Responsible Care Program of the Chemical Manufacturer's Association
	XVIII. Science and technology	**65.** Innovation Index
		66. Digital Access Index
		67. Female primary education completion rate
		68. Gross tertiary enrollment rate
		69. number of researchers per million inhabitants
Global stewardship	**XIX.** Participation in international collaborative efforts	**70.** number of memberships in environmental intergovernmental organizations
		71. contribution to international and bilateral funding of environmental projects and development aid
		72. Participation in international environmental agreements
	XX. Greenhouse gas emissions	**73.** carbon emissions per million US dollars GDP
		74. Carbon emissions per capita
	XXI. Reducing transboundary environmental pressures	**75.** SO_2 exports
		76. Import of polluting goods and raw materials as percentage of total imports of goods and services

Source: Schmiedeknecht, M. H., 2013. Environmental sustainability index. Encycl. Corporate Soc. Responsibility, 1017–1024, pp. 3–4.

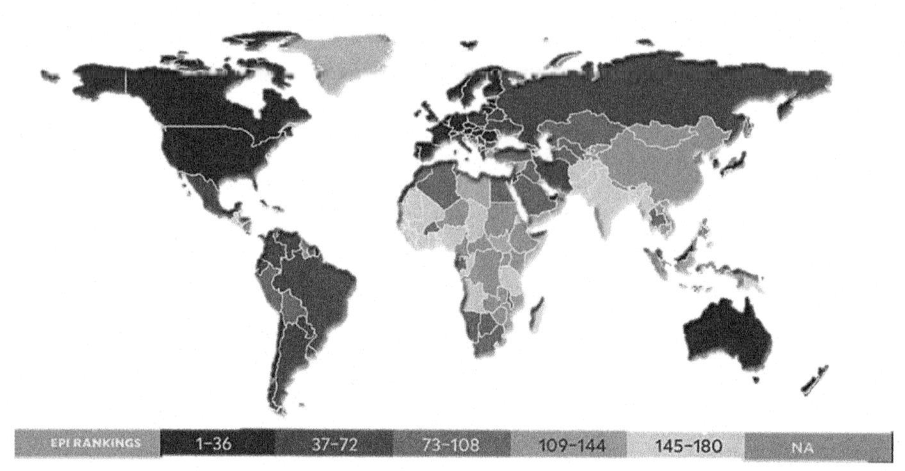

Graph 13.6 ESI results worldwide in 2020.
Source: Own elaboration based on Environmental performance index (epi.yale.edu).

- Knowledge of resource reserves is uncertain, as we do not know exactly what the planet's endowment is of each resource.
- Nothing (or very little) is known about the future technologies that will be available to us. This limitation can be seen in Malthus' theory that the population could not grow indefinitely because it would end up dying of hunger (as population growth was higher than food growth), in this case, Malthus did not have the technology that would allow him to increase food production.
- It is not possible to reduce the diversity of units in the environmental system to a common unit, which is a necessary requirement for accounting for the value of the natural assets used for the calculation of these indices.
- Current or future monetary valuations are arbitrary, as each author uses a different system.

4. Conclusions

The first step in solving a problem is certainly to accept it. Today, unlike decades ago, we care about the sustainability of our natural resources and want future generations to be able to enjoy these same natural resources at least as much as we do. This growing concern for the environment can be seen in the excellent work of Hussain and Mishra (2019), Hussain and Keçili (2019) as well as Hussain (2019, 2020). As Rachel Carson, a pioneer in the global environmental movement, said: "The more clearly we can focus our attention on the wonders and realities of the universe about us, the less taste we shall have for destruction."

It is a goal that is recognized as difficult, starting with knowing the state of those natural resources today and how sustainable our economies are. Humanity still has a long way to go; however, there is already environmental awareness and efforts are being made to preserve natural resources. As we have seen throughout this chapter, there are many economies in the world that are productive and respectful of the environment, that are carrying out policies that promote the defense of the environment and the results of these policies can be seen. The aim of this century is to make these economies more friendly to the environment and to export this awareness and behavior to the rest of the world, because only in this way will future generations be able to live on a natural planet.

References

Bleys, B., 2008. Proposed changes to the index of sustainable economic welfare: an application to Belgium. Ecol. Econ. 64, 741–751.

Butnariu, A., Luca, F., 2019. An index of sustainable economic welfare for Romania. New Trends Issues Proc. Human. Soc. Sci. 6 (8), 1–10.

Chelli, F.M., Ciommi, M., Gigliarano, C., 2013. The index of sustainable economic welfare: a comparison of two Italian regions. Proc. Soc. Behav. Sci. 81, 443–448.

Daly, H., Cobb, J., 1989. For the Common Good. Beacon Press, Boston.

Dasgupta, P.S., Mäler, K.G., 2000. Net national product, wealth, and social well-being. Environ. Dev. Econ. 5, 69–94.

Eisner, R., 1994. The index of sustainable welfare: comment. In: Cobb, C.J. (Ed.), The Green National Product: A Proposed Index of Sustainable Economic Welfare, Cobb. University Press of America, Lanham, pp. 97–110.

Georgescu-Roegen, N., 1971. The Entropy Law and the Economic Process. Harvard University Press, Cambridge.

Global Footprint Network, 2019. Working Guidebook to the National Footprint and Biocapacity Accounts. Global Footprint Network, Oakland, CA.

Global Footprint Network, 2020. Global footprint Network. Advancing the science of sustainability. Country Trends. http://data.footprintnetwork.org/?_ga=2.11617591.26475017.1594741980-2030800832.1594741980#/countryTrends?cn=5001&type=BCpc,EFCpc. (Accessed July 2020).

Hamilton, K., 2000a. Genuine Savings as a Sustainability Indicator. World Bank, Washington D.C.

Hamilton, K., 2000b. Sustaining Economic Welfare: Estimating Changes in Wealth per capita. World Bank, Washington, D.C.

Hamilton, K., 2004. Accounting for Sustainability. Measuring Sustainable Development: Integrated Economic Environmental and Social Frameworks. OECD, Paris, France.

Hamilton, K., Clemens, M., 1999. Genuine savings rates in developing countries. World Bank Econ. Rev. 13, 333–356.

Hamilton, K., 1994. Green adjustments to GDP. Resour. Pol. 20 (3), 155–168.

Hartwick, J.M., 1977. Intergenerational equity and the investing of rents from exhaustive resources. Am. Econ. Rev. 67 (5), 972–974.

Hussain, C.M., 2020. The Handbook of Environmental Remediation: Classic and Modern Techniques, 1st ed. Royal Society of Chemistry.

Hussain, C.M., Keçili, R., 2019. Modern Environmental Analysis Techniques for Pollutants, first ed. Elsevier.

Hussain, C.M., Mishra, A.K., 2019. Nanotechnology in Environmental Science, vol. 2. John Wiley & Sons.

Hussain, C.M., 2019. Handbook of Environmental Materials Management. Springer International Publishing.

Jackson, T., Laing, F., MacGillivray, A., Marks, N., Ralls, J., Stymne, S., 1997. An Index of Sustainable Economic Welfare for the UK 1950–1996. University of Surrey, Centre for Environmental Strategy, Guildford.

Lange, C.M., Wodon, Q., Carey, K., 2018. The Changing Wealth of Nations 2018: Building a Sustainable Future. World Bank, Washington, D.C.

Lawn, P.A., 2005. An assessment of the valuation methods used to calculate the index of sustainable economic welfare (ISEW), genuine progress indicator (GPI), and sustainable net benefit index (SNBI). Environ. Dev. Sustain. 7 (2), 185–208.

Meadows, D.H., Meadows, D.L., Randers, J., Behrens, W.W., 1972. The Limits to Growth: A Report for the Club of Rome's Project on the Predicament of Mankind. Universe Books, New York.

Millennium Ecosystem Assestment, 2005. Ecosystems and Human Well-Being: Synthesis. Island Press, Washington D.C.

Munday, M., Roberts, A., 2006. Developing approaches to measuring and monitoring sustainable development in wales: a review. Reg. Stud. 40, 535–554.

Neumayer, E., 2000. On the methodology of ISEW, GPI and related measures: some constructive suggestions and some doubt on the "threshold" hypothesis. Ecol. Econ. 34, 347–361.

Nordhaus, W., Tobin, J., 1972. Is Growth Obsolete? Columbia University Press, New York.

Nourry, M., 2008. Measuring sustainable development: some empirical evidence for France from eight alternative indicators. Ecol. Econ. 67, 441–456.

Pearce, D., Atkinson, G., 1993. Capital theory and the measurement of sustainable development. Ecol. Econ. 8, 103–108.

Posner, S.M., Constanza, R., 2011. A summary of ISEW and GPI studies at multiple scales and new estimates for Baltimore City, Baltimore County, and the State of Maryland. Ecol. Econ. 70 (11), 1972–1980.

Repetto, R., McGrath, W., Welss, M., Beer, C., Rossini, F., 1989. Wasting Assests: Natural Resources in the National Income and Product Accounts. World Resources Institute, Washington, D.C.

Sánchez, M., Toledo, E., Ordóñez, J., 2020. The relevance of index of sustainable economic wellbeing. Case study of Ecuador. Environ. Sustain. Ind. 100037.

Schmiedeknecht, M.H., 2013. Environmental sustainability index. Encycl. Corporate Soc. Responsibility 1017–1024.

Sejenovich, H., 1996. Manual de Cuentas Patrimoniales. Argentina: Programa de las Naciones Unidas para el Medio Ambiente. Fundación Bariloche e Instituto de Economía Energética.

Talberth, J., Cobb, C., Slattery, N., 2007. The Genuine Progress Indicator 2006. Redefining Progress, Oakland, C.A.

UNSD, 2008. System of National Accounts. European Commmunities, IMF, OECD, UN, and the World Bank, New York.

Wackernagel, M., Ress, W., 1996. Our Ecological Footprint: Reducing Human Impact on Earth. New Society Publishers, Gabriola Island.

WCED, 1987. Our Common Future. Oxford University Press, Oxford.

Williamson, J., 1990. What Washington means by policy reform. In: Williamson, J. (Ed.), Latin American Adjustment: How Much has happened? Institute for International Economics, Washington, pp. 7–20.

World Bank, 1995. Monitoring Environmental Progress. World Bank, Washington D.C.

World Bank, 2018. Building the World Bank's Wealth Accounts. World Bank, Washington D.C.

Ziegler, R., 2007. Political perception and ensemble of macro objectives and measures: the paradox of the index for sustainable economic welfare. Environ. Val. 16, 43–60.

Interesting web links

Bleys, B., Van der Slycken, J., 2019. De Index voor Duurzame Economische Welvaart (ISEW) voor Vlaanderen, 1990–2017. Universiteit Gent, Gent.

Galland, P., Lisitzin, K., Oudaille-Diethardt, A., Young, C., 2016. World Heritage in Europe Today. UNESCO Publishing, Paris.

— Woring Guidebook, 2019. To the National Frootprint and Biocapacity Accounts. Global Footprint Network, Oakland.

— Sustaining Economic Welfare, 2000b. Estimating Changes in Wealth Per Capita. World Bank, Washington, D.C.

— The Handbook of Environmental Remediation: Classic and Modern Techniques, first ed., 2020. Royal Society of Chemistry.

Schepelmann, P., Goossens, Y., Makipaa, A., 2010. Towards Sustainable Development: Alternatives to GDP for Measuring Progress (No. 42). Döppersberg (Wuppertal). Wuppertal Institut für Klima, Umwelt, Energie GmbH.

— Monitoring Environmental Progress, 1995. World Bank, Washington D.C.

Center for International Hearth Science Information Network. Socioeconomics Data and Applications Center (SEDAC). https://sedac.ciesin.columbia.edu/.

EEA. European Environment Agency. https://www.eea.europa.eu/.

European Commission. European Commission. https://ec.europa.eu/.

European Commission. Eurostat. https://ec.europa.eu/eurostat/data/database.

GFN. Global Footprint Network: Advancing the Science of Sustainability. https://www.footprintnetwork.org/about-us/our-history/.

Labandeira, X. Personal Web Page of Xavier Labandeira. https://labandeira.eu/index.php.

MA. Millennium Ecosystem Assessment. http://www.millenniumassessment.org/en/Index-2.html.

Pettinger, T. Economics Help. Obtenido de https://www.economicshelp.org/.

Sustainability Project. SustainWellbeing. http://www.sustainwellbeing.net/index.html.

The World Bank. DataBank | Wealth Accounts. https://databank.worldbank.org/source/wealth-accounts/Type/TABLE/preview/on.

The World Bank. DataBank | World Development Indicators. https://databank.worldbank.org/reports.aspx?source=2&series=NY.ADJ.SVNG.GN.ZS&country=.

UNEP. United Nations Environment Programme. Environmental Data Explorer: http://geodata.grid.unep.ch/.

United Nations. UNdata. http://data.un.org/Explorer.aspx?d=SDGs&f=series%3aSI_COV_SOCAST.

WWF. World Wide Fund for Nature. https://www.worldwildlife.org/?type1=%2F.

Yale Center for Environmental Law and Policy. Environmental Performance Index (EPI). https://epi.yale.edu/.

Indicators of sustainability

14

Somayeh Jangchi Kashani [1], *Mohammadhadi Hajian* [2]
[1]Department of Agriculture Development, Islamic Azad University, Sciences and Researches Branch, Tehran, Iran; [2]Department of Economics, Tarbiat Modares University, Tehran, Iran

1. Introduction

Because of its quantitative concept, the definition of sustainability has always been under discussion by experts. However, many scientists tried to define this concept in a way that can be measured easily. In order to achieve this target, specialists from different fields of science worked on defining indicators for sustainability that could be simply calculated.

Although discussed by different points of view—such as social, environmental and economic viewpoints—socio environmental degradation has been largely associated with the dominant development model concentrated on economic growth. For example, Meadows et al. (1972) stressed the restrictions of ecological systems in carrying out functions such as absorption and recycle of waste of anthropogenic activities, with concomitant problems in improving social, educational, health, and employment aspects. Sustainable development play a part as a pattern expected at endorsing balance between social, economic, and environmental aspects. The *Brundtland Report* (WCED, 1987) defined sustainable development as "development that meets the needs of the present without compromising the ability of future generations to meet their own needs." Such an idea is focused on the concept of intergenerational parity, so that future generations access at least the same level of welfare as does the present one. For Waas et al. (2014) sustainable development is a social paradigm with decision-making instructions based on values linked to current and future development, which must be understood in a dynamic environmental background. As suggested by Biasi et al. (2019), satisfying the needs of present and future generations depends on appropriate circumstances of the human, natural, and economic capitals to provide for human welfare. Apprehending and defining sustainable development dimensions and aspects arises from initial works, such as the UN's 1992 Conference on the Environment and Development—which produced the Rio Declaration on the Environment and *Agenda 21*—and continued progress, e.g., in the Rio20 Conference leading to additional documents and debates and the preparation of current sustainable development aims. Biasi et al. (2019) suggest that measuring sustainable development necessitates observing and assessment of capital changes. From such a standpoint, one of the main challenges is to evaluate different actions and processes related to sustainability. Theoretical and operational problems endure in relation to promoting sustainable development, which can only be surpassed with the progression of assessment models and methods (Fernandes et al., 2012).

Sustainable Resource Management. https://doi.org/10.1016/B978-0-12-824342-8.00009-2

Sustainable development monitoring approaches using indicator systems are normally organized in terms of its key dimensions (e.g., Sartori et al., 2014; Raynaut et al., 2018), which tend to cover physical–natural and socioeconomic aspects related to the pursuit of the UN's 17 sustainable development goals and 169 associated subgoals (Pinsky et al., 2015; UN, 2019). Such goals focus on and are planned to impact public and private policy toward alleviating life-threatening poverty while encouraging social inclusion, environmental sustainability, and governance for security and peace. In such an environment, the definition of indicators should be united to a sustainable development strategy to cover key developments and outcomes, as well as to serve as social mobilization tools through its capacity to inform on socioenvironmental conditions and assist decision-making practices with communal political contribution (Meadows, 1998; OECD, 2003; United Nations Economic and Social Council, 2017). Indicators must provide information access on applicable dimensions and features in an integrated manner (Meadows, 1998; Guimaraes and Feichas, 2009; Pissourios, 2013; Wang et al., 2013; Waas et al., 2014).

Wang et al. (2019) argued that the challenge of the 193 countries that committed to fulfilling Sustainable Development Goals Agenda of the United Nations is to certify that their governments meet such an Agenda, which necessitates overcoming resistance related to the level of autonomy of administrative authorities, demanding political capital on the part of who coordinates the Agenda in such countries. Among the coordinating strategies is the requirement for ensuring inclusion of sustainability commitments in governmental plans to pursue scientific contribution for negotiation among ministries on sustainability goals (Aud, 2013).

Therefore, using indicators to monitor sustainable development can be an effective way to show to governments the social, economic, and environmental outcomes acquired, thanks to development policies. The intricacy of sustainable development indicator (SDI) systems is obvious in terms of their significant number of indicators. For example, the UN's SDI approach, applied in Brazil, includes 63 indicators organized in terms of four dimensions including environmental, social, economic, and institutional dimension (da Silva et al., 2020). Also, UNESCO's Dashboard of Sustainability involves 60 indicators similarly separated in such four dimensions (UNESCO, 2006), while OECD (2003) uses 50 indicators to evaluate how the environmental dimension interacts and is measured in sectorial socioeconomic policy-making (OECD, 2003). Indicators are known as key communication tools on how countries are coping with internal and external aims and conditions in relation to their environmental commitments. Switzerland is a pioneer country in SDI system (MONET—Monitoring Sustainable Development) organized in terms of three target dimensions: social unity, economic efficiency, and environmental sustainability, which were divided into 26 topics leading to 163 indicators. Choosing and applying indicators in the framework of a sustainable development assessment needs a well-determined analytical background and criteria, or the analysis may be weakened by the large availability of indicators from several sectors. SDI systems have ever been under criticism for their extensive range and low ability to inform on cause-and-effect relationships. Conditional to the choice made, an extra demonstration of special features may take place, while other main components may not be adequately involved. Highlights that choosing and computing sustainability indicators are among the key issues demanding additional inquiry in socioenvironmental researches. For Venturelli and Galli (2006), obtaining indicators of the interactions among ecology, economy, and society is a major problem, while pointed out the existence of a large number of sustainability measurement systems in the literature, but no clear suggestion of which are the most proper.

Meadows (1998) suggested a measurement viewpoint of interrelating assets of sustainable development, which comprises the natural, human, social, and physical capitals. The natural capital embodies low entropy energy and matter that create natural resources and services with considerable market value, which must be well preserved and involved in the cost of goods and services. The physical capital is linked to the production substructure and is described as a basic element to endorse development and reduce environmental destruction triggered by such development. The social capital includes values, community behavior, institutions, human relations, and instructions that regulate the quality and quantity of social communications, and can be regarded together with the human capital as characteristics of a society, such as health, education, culture, and security levels that enable it to self-organize and get results.

2. Concept of sustainable development

Sustainable Development is a concept that at its center is groundbreaking, yet extremely hard to be practically defined. The history behind sustainable development is not that long. Strains that may exist in the concept of sustainable development are abundant, stretching from its unclear and imprecise meaning to the failure of reaching a worldwide practical and operational framework (Mariann, 2018). The great challenge that encounters with sustainable development is not only the necessity to teach it to the people but to first describe it in a way people will comprehend it. The concept of sustainable development is one that disputably is multidisciplinary, multifaceted, and systematic; yet defining the concept is undoubtedly a great task. Sustainable development was a word first invented in 1980, when the meaning of the concept was simply basic. It was in the World Conservation Strategy, an association between three paramount environmental nongovernmental organizations IUCN, WWF, and UNEP, where sustainable development was defined as "conserving the earth's natural resources" (World Conservation Strategy, 2011). What the World Conservation Strategy had recognized is that economic growth in the world has obtained with the price of degradation of natural resources. The innovative, and unique, aim of the World Conservation Strategy was to bring countries together to prevent the destruction of natural resources, which in turn was negatively affecting the environment. Sustainable development was consequently purely seen on fairly a basic level, at the time of its invention. Not even a decade later, did the description take on much more of a multidisciplinary method. In 1987, the World Commission on Environment and Development published a paper, titled "Our Common Future" The article created a wide definition of sustainable development, which states "sustainable development is development that meets the needs of the present without compromising the ability of future generations to meet their own needs." The paper, also repeatedly identified as the *Brundtland Report*, has "since been taken up by almost every international institution, agency and NGO." The *Brundtland Report* became the first manuscript to support sustainable development as a multidisciplinary field, as it clarified that the economy, society, and the environment were crucial in sustainable development. In 1992, the United Nations Conference on Environment and Development developed a program called *Agenda 21*, which supposedly "is the design for sustainability in the 21st century." It is a framework that nations and government severely can obey. Countries that accepted *Agenda 21* are observed by the UN Commission on Sustainable Development that monitors improvement in the enactment of *Agenda 21* and also the Rio Declaration on Environment and Development. Both the *Agenda 21* and the *Brundtland Report* have verified to be broadly used structures that nations, agencies, and organizations use in modern times, yet even though they are accepted, a clear description still is not obvious.

The greatest problem with the concept of sustainable development is the pure quantity of definitions that are presented (Martínez-Graña et al., 2013). A perfect definition of sustainable development has not yet been presented. Although the definition of *Brundtland Report* has extensively been quoted, it does not work in many cases. When reconsidering the definition "sustainable development is development that meets the needs of the present without compromising the ability of future generations to meet their own needs," two main subjects may be noticed. First, the "needs" are not obviously defined. There is a large difference between the needs of an individual residing in an underdeveloped country, in contrary to the needs of someone indwelling in a highly developed country. Furthermore, this definition does not determine any period, as "generations" can only be imprecisely understood. The unbelievable number of definitions existing for sustainable development thus make it a subject that maybe nobody does not have the tendency to follow. An interesting method of inspecting the subjects related to sustainable development is to consider the following quotation that remarks "a combination of uncertainty about what to do, and a feeling of guilt about what is not being done, means that many people seem afraid to expose what they feel is their lack of understanding of sustainable development. Therefore, it is often easier to pretend that it does not need to be addressed." It is obvious that because the definition of sustainable development is ambiguous, peoples and governments deal with it in different ways. Nevertheless, for those that have concurred on a specific definition, the problem of performing emerges.

Countries that consciously attempted to comprehend sustainable development and are eager to make changes at both national and international level confront the problem of performing policies tied to sustainable development. As with *Agenda 21*, the article that gave a framework to countries for sustainable development, there is no "enforcer" of the article. In the other words, countries may pretend to recognize sustainable development, approve to change their policies, but no one in fact imposes them to do so. The *Brundtland Report*, "our inability to promote the common interest in sustainable development is often a product of the relative neglect of economic and social justice within and amongst nations" that is considered as another vital matter in sustainable development, which is something beyond simply sustainable development. The challenge with the enforcement of sustainable development policies is that countries that are in the highest necessity of such policies have no motive to observe them. As the discussion between developed and developing countries is still obviously continued, the developed countries principally are guiding the developing ones the right way to achieve development.

As stated in a report of OECD, "At the institutional level, the interdependent goals of economic growth, social development and environmental protection are managed today by institutions that tend to be independent and fragmented, and that respond to narrow man -dates with closed decision-making bodies. Sustainable development stresses the importance of institutions that are willing to integrate economic, social and environmental objectives at each level of policy development and decision-making."

The age of people is still toughly obvious in the present world. Nonetheless, it is a feature of sustainable development to determine the subjects distinctly that not only a country has with its policies but also the world has with its forces. Sustainable development is consequently a concept that is groundbreaking, so far restricted in the realm of its beholder.

Sustainable development is potentially a revolutionary concept that can transform the way countries act on a national level, and more so on an international level. However, because of its multidisciplinary nature, superlative aims, and flexible understandings, there is no agreement on what a perfect description of the concept may be. Furthermore, its unclear explanation and uncertainty add to the discussions around this concept, as any country may express they are considering sustainable development policies. The problem in the future is firstly to have a succinct definition, and secondly to practically be able to apply it to any country across the world. Because of the potential of sustainable development for more dialogue between the developed and developing countries, each nation should make the essential changes to their own policies in order to prepare for a cleaner, safer, and more efficient environment, economy, and society. What sustainable development mainly calls for is transparency as well as supportive countries that have the tendency to cooperate with each other for the progress of the world. It may be because of this reason that sustainable development is very hard to describe, since each country has its own opinion about what a better world means.

3. Indicator development

Indicator developers use frameworks to provide a mutual language and outlook on the subject and its explanation. This eases indicator development, chiefly when many different actors are involved. The way, in which subjects are formed, gets significant in the explanation and deeper analyses of the results because the frameworks are the assumptions and justifications on which the indicator is based and should be made available to those willing to explain the indicators. Understanding these assumptions and the frameworks is necessary in order to compare and discuss indicators from different institutions because their base frameworks might be different. For the majority of users, however, showing the frameworks themselves, or classes from such frameworks, would only add a useless degree of difficulty that may confuse. Organizations that work on sustainable development should have a share in the definition of the problem, responsive to subjects of proper contribution and representation, and participate in improving science and technology, responsive to subjects of proficiency and quality control. In Clark's opinion, the strongest message to Johannesburg summit was that the research community required playing its historical part in recognizing challenges of sustainability with a higher eagerness to link to other organizations in solving those problems. Such institutions spend most of their time doing pure science or pure politics. They are not likely to be as effective as boundary-spanning institutions that determinedly manage and balance the multiple boundaries within a system tending to be more effective than other institutions in creating information that can affect policy-making. The three criteria of credibility, legitimacy, and salience are key features for describing the effectiveness of SDIs where credibility refers to the scientific and technical adequacy of the measurement system, legitimacy refers to the process of fair dealing with the conflicting values and beliefs of stakeholders, and salience refers to the relevance of the indicator to decision makers. The indicator development process itself is responsible for confirming at least the first two of these criteria. Resources differ noticeably between developed and developing countries. The socioeconomic, environmental, and knowledge contrasts between the two sides may be worsened by such resource distribution. Finding ways to fill this resource gap is necessary for equitable representation, both geographically and in terms of identifying and framing significant subjects. Equitable representation rises the legitimacy and credibility of both the process and the final product. A capability for inducing and applying science and technology is a necessary element of strategies endorsing sustainable development too. Creating acceptable scientific capability and institutional support in developing countries is particularly crucial in order to improve pliability in areas that are vulnerable to the multiple pressures arising from quick, concurrent changes in social and environmental systems.

However, scientific capability itself is not adequate for the aims of making reliable SDIs. Instead, capability making is required, with emphasis on supporting the broader procedures ensuring legitimacy and credibility of the indicator development process. Effective capability making puts emphasis on the main mechanisms of communication, translation, and mediation. Preparing for acceptable communication between stakeholders is necessary, as is confirming that reciprocal understanding is possible. Communication is often prevented by misunderstanding, language differences, and presuppositions about what establishes a convincing discussion. Facilitation or mediation additionally improves transparency of the procedure by considering all viewpoints, and specifying the instructions of behavior and processes as well as defining decision-making standards.

A serious commitment by organizations to handling the restrictions between expertise and decision-making will help take knowledge into action. Forming answerability to main actors across the restriction and using joint outputs to nurture cohesion and commitment to the process are also helpful in emerging capability for sustainable development. Indicator legitimacy and acceptability depend on acknowledgment of the multiplicity of legitimate perspectives. Where there are difficult subjects, the qualities of the decision-making process itself are vital, and procedures planned to open the conversation between stakeholders rather than thinning the authority of science are important to making a wide base of consensus. The role of indicators is to contribute this conversation and decision-making. The value of a specific indicator set differs between users and situations. The users should be able to effect the selection of the indicators that they will have to use. Sometimes this local selection will lead to a loss of comparability as different groups and procedures choose to use different indicators. This can be satisfactory when the core goal of indicators is to endorse effective decision-making. On the other hand, when the core goal is comparability, then more significance should be given to regularization. It is not at all times probable to have both.

In general, sustainability has three main aspects: environmental sustainability, economic sustainability, and social sustainability. Environmental sustainability deals with resource management, conservation of environment, and green analysis, which is innovating and using chemical processes and goods decreasing or even removing the harmful effects to environment and health (Hussain and Keçili, 2019). Recently, green, harmless, and economical substances are applied in order to lessen environment destruction (Mallakpour et al., 2020). Economic sustainability mainly focuses on the optimum use of inputs leading to cost reduction (Hussain and Keçili, 2019). Social sustainability, in general, copes with topics such as quality of life, law, and ethics (Hussain and Keçili, 2019).

4. Environmental sustainability

Environmental sustainable development, nowadays, is vital for the advancement of civilization. Therefore, there is a vast effort for achieving sustainable development. The success of such attempts lies in the hands of scientists trying hard to cross the borders of environmental science and engineering in order to make the planet more environmentally sustainable and livable. Worldwide distresses for sustainability of environment has led to endeavors by scientists and engineers in order to help environmental protection in the future (Hussain and Mishra, 2019).

Meanwhile, science and engineering, yet, have few solutions to the rising distresses about environmental damages, in both developing and developed countries. Therefore, there is an instant necessity for more innovative researches in all aspects of environmental sustainability (Hussain, 2019).

The standard Brundtland definition of sustainable development applies a three-pivot method to precise, in brief, the link between economic activity, quality of life, and the lifetime of ecosystems and natural resources. A country without an effective life support system cannot prosper; lack of helpful social structures and institutions avoid economies from thriving. Additionally, sustainable development has often been understood as social and economic development that should be environmentally sustainable (Millennium Ecosystem Assessment, 2005). In the period since the

Brundtland definition was first published, it has gradually gotten accepted that environmental sustainability has its own competency. Goodland declares that environmental sustainability tries to "improve human welfare by protecting the sources of raw materials used for human needs and ensuring that the sinks for human wastes are not exceeded, in order to prevent harm to humans." This concept definition recognizes that environmental sustainability concerns the putting of restrictions on resource depletion, which is also a principal view of the ecological economics framework of "limits to growth." In addition, Goodland also continues to define environmental sustainability as the burden of restrictions on four major activities influencing the scale of the human economy: "the use of renewable and nonrenewable resources on the source side, and pollution and waste assimilation on the sink side." Moldan, Janouskova, and Hak state that the OECD implement the concept of environmental sustainability within their environmental strategy for the 1st decade of the 21st century, published in 2001 and 2003 The OECD's strategy defined four explicit standards for environmental sustainability: (1) Regeneration—renewable resources shall be used efficiently and their use shall not be permitted to exceed long-term rates of natural regeneration. (2) Substitutability—nonrenewable resources shall be used efficiently and their use limited to levels which can be offset by substitution with renewable resources or other forms of capital. (3) Assimilation—releases of hazardous or polluting substances into the environment shall not exceed their waste assimilative capacity. (4) The avoidance of irreversibility. The OECD used their four explicit standards for environmental sustainability as an instrument of anticipating five interconnected purposes for progressing environmental policies in a sustainable development framework (OECD, 2001):

(1) Preserving ecosystem entirety through the efficient management of natural resources.
(2) Separating of environmental burdens from economic growth.
(3) Improving quality of life.
(4) Increasing global environmental interconnection by enhancing governance and collaboration.
(5) Computing development, mainly using environmental indicators and indices.

The concept of environmental sustainability can be more advanced via the usage of an ecosystem services view, as it strengthens the value relating to nonmonetary ecological qualities and functions, all of which are essential for the OECD's five interconnected purposes to be met. Daily debates "nature's services" to be included of a worldwide life-support system, properties prepared by the geosphere (such as mineral resources), and open space (such as land on the planet's surface, plus the space above and below it). In satisfying the OECD's five purposes for environmental sustainability, human welfare is preserved. On this base, ecosystem services can be measured a vital element of human welfare. Moldan et al. thus deduced that environmental sustainability might be described as the preservation of services of the nature at a proper level.

It necessitates ecosystem services on a local, national, and international scale to be retained in a healthy situation, and by definition necessitates governance systems to have a responsibility of maintenance and regulatory influence on environmental infrastructure.

To present an obvious and objective instrument of computing and representing the environmental sustainability of a nation, it is often useful to use environmental indicators and indices.

There is not unique set of national environmental indicators that is equivalent to the ordinary set of measures, used to estimate economic performance. In economic policy, countries are normally compared on the base of their gross domestic product (GDP) growth and performance. In the environmental field, the most inclusive set of indicators is incorporated within the EPI, which struggles but mostly fails to signify an all-embracing representation, taking the environmental sustainability concept rather than its explicit elements, such as pollution, energy consumption, and soil degradation. A similar absence of inclusiveness can be seen within SDIs, with the subsequent outcome that economic activities assumed in the name of "sustainable development" often last to threaten environmental entirety in a definite area.

Additionally, the subjective process of standardizing and weighting indicators of environmental sustainability is reliably disposed to a high degree of arbitrariness and lack of consideration as the stability and significance of outcomes, decreasing their relevance in terms of policy practice.

In theory, the use of environmental sustainability indices to estimate sustainable development is a necessary means of informing policymakers and general society the relations and trade-offs among its three dimensions. Even though not without their share of critics because of uncertainty as to how well they signify environmental sustainability in practice, environmental indicators and compound indices can be useful instruments for evaluating the circumstances of the environment and observing trends over time, as well as describing circumstances under which resource consumptions are sustainable.

Assessments of sustainability especially environmental sustainability have four common features:

(1) A topic concentration on the connection between human activity and nature;
(2) Direction toward the long-run and an uncertain future;
(3) Normative base in the concept of fairness, between humans of present and future generations as well as between humans and nature; and
(4) Concern for economic efficiency in the distribution of goods and services, as well as their artificial alternatives and counterparts. Taking the full dynamics of environmental sustainability factors as presented in terms of easily explicable indicators is a puzzling mission. The United Nations declares that environmental indices require capturing the next four aspects effectively:
(1) Influences of economic activity on the environment
(2) Impacts of resource productivity on the economy
(3) Effects of environmental degradation on economic productivity
(4) Influences of environmental improvement on society

In spite of obvious unities in the numerous definitions of environmental sustainability, there is not any sets of indicators or a compound index suitably and comprehensively measuring the concept of sustainability in national and international level. Parris and Kates explained three causes for this: (1) the vague nature of sustainable development; (2) the variety of purpose in describing and (3) calculating sustainable development; and confusion about terminology, proper data, and methods of measurement. Then, the variety of indicators and indices needs to quantify any country's environmental sustainability identifications does not lead to a unique, distinct assortment process, and therefore expert judgment has been important.

In addition, some further indices are considered for assessment such as the living planet index, satisfaction with life index, human development index, and sustainable society index. These were rejected for the following reasons:

- Environmental sustainability index: a complex index measuring 21 features of environmental sustainability, but since 2005 it has been substituted by the more comprehensive EPI.
- Barometer of sustainability: comprises all three of the sustainable development dimensions, but the imperfect scope of the environmental feature and lack of a practical approach condensed the EPI a more comprehensive instrument for appraisal.
- Surplus biocapacity index: this index lists countries according to the balance or deficit between their ecological footprint and national biocapacity. Although the surplus or deficit in biocapacity is a central principle for appraisal, an international ranking list proposes nothing further in terms of analytical value.
- Satisfaction with life index: measures subjective welfare across countries relating to their wealth, health, and access to basic education. On the basis that it fails to include any form of ecological focus, the Human Poverty Index is preferable.
- Human development index: although a comprehensive socioeconomic criterion approximately similar to the satisfaction with life index, it also lacks any consideration of environmental subjects and therefore was overlooked in favor of the selected indices.
- Living planet index: emphases explicitly on the subject of biodiversity frameworks and in a worldwide background, therefore lacking adequate scope to measure environmental sustainability subjects in a national background.

5. Sustainable agriculture

Agriculture plays a chief part in sustainable development. Its vital situation as the provider of human sustenance forms the worldwide economy and society's correlation with the natural world. It is hence crucial to attaining a collection of Sustainable Development Goals settled to by the United Nations in 2015 (United Nations, 2015), ranging from eradicating starvation and destitution to enhancing human welfare and decreasing environmental influences (United Nations Economic and Social Council, 2016). Already, over a third of the world's land surface and nearly three-quarters of its freshwater resources are dedicated to agriculture (Dobermann et al., 2013; HLPE, 2013; Pretty et al., 2006). It is both an important cause of global climate change, as a result of land use change and greenhouse gas emissions (Smith et al., 2016), and one of the sectors which is most vulnerable to its influences (Vermeulen et al., 2012). Furthermore, about three-quarters of the poorest people of the world live in rural areas, where agriculture is the core source of employment and income (World Bank, 2007; IFAD, 2011).With growing global population and affluence, the burden on agricultural and natural resources increases. Because of these growing burdens, humans now expect agriculture to provide not only nutritious food but also employment, energy resources, clean water, biodiversity conservation, and more. This condition makes it necessary to recognize and manage the trade-offs between probable profits and harmful effects that may occur as food production interrelates with other features of sustainable agricultural systems (Millenium Ecosystem Assessment, 2005; Tilman et al., 2009; Godfray et al., 2010; Tilman and Clark, 2014). Concepts such as sustainable agricultural intensification (Garnett and Godfray, 2012) and climate-smart agriculture (Lipper et al., 2014) are representing the challenge of attaining several objectives of rising agricultural productivity and rural livelihoods while minimizing negative environmental side effects. As stressed by Garnett and Godfray (2012), sustainable agricultural intensification is not a specific set of tasks but instead provides a conceptual framework for guiding debates on attaining balanced outcomes of intensification. Then, there may be several different methods for sustainable agricultural system. In such systems, suitability may differ according to agro-ecological zone, farming systems, cultural preferences, institutions, and policies. Each of these methods leads to a different collection and/or degree of environment and socioeconomic trade-offs and collaborations that must be renowned and pursued. The successful conversion of the agricultural sector to meet these multiple objectives, thus, necessitates the facility to seek various outcomes, evaluate whether recognized objectives were met, and agree to guide course corrections. In an attempt to make these relations obvious, trade-off analysis for agricultural systems has arose as an increasingly central field of study (Vermeulen et al., 2012; Snapp, 2014).

Initial applications of trade-off analysis in agricultural sustainability appraisals attached biophysical data and models with economic models in order to make a more comprehensive methodology to assessing agricultural sustainability. These initial studies measured the economic, environmental, and health trade-offs of insecticide use. From that time, the use of trade-off analysis to measure agricultural sustainability has increasingly developed as a field of study, growing beyond agronomic and economic products at the field and farm level, to integrate environmental and social products at local and continental scales. A variety of instruments prepare tools to measure the trade-offs and collaborations that come from agricultural intensification (OECD, 2012).

Most people prefer organic foods for which chemical fertilizers and pesticides are not used. Yet, the majority of food consumed in the world is produced by mechanized agriculture, a kind of agriculture where large amounts of crops and livestock are produced through industrialized methods in order to make benefit for property owners. This kind of agriculture depends heavily on several chemicals. It needs a large quantity of fossil fuels and large machines to manage the farms. Although having made it possible to produce large amounts of food, because of negative side effects of mechanized agriculture, there has been a change toward sustainable agriculture.

Nowadays, living at the era of rapid changes in all aspects of life as well as population growth and high development rate, humans encounter more puzzling and complicated problems for environment conservation whose concentration is on establishing new methods to comprehend the consequences of harmful compounds disseminated throughout the environment (Hussain, 2020).

Sustainable agriculture is a kind of agriculture that centers on producing long-run crops and livestock while having least negative side effects on the environment.

This kind of agriculture tends to find a balance between the need for food production and the conservation of the ecological system within the environment. In addition to producing food, there are many overall aims linked to sustainable agriculture, such as preserving water, decreasing the use of chemical fertilizers and pesticides, and considering biodiversity in crops and ecosystem. Sustainable agriculture, moreover, emphasizes on sustaining economic empowerment of farms and aiding farmers progress their skills and quality of life.

Several agriculture approaches can contribute to make agriculture more sustainable. Some of the most popular methods include growing crops that can produce their own nutrients to decrease the use of fertilizers and rotating crops in fields, which minimizes pesticide use because the crops are altering regularly. Another popular method is mixing crops, which decreases the risk of a disease destroying a whole crop and reducing the need for pesticides and herbicides. Sustainable farmers also use water management systems, such as drip irrigation that waste less water.

6. Benefits of sustainable agriculture

There are several advantages for sustainable agriculture, and generally, they can be separated into human health benefits and environmental benefits. In terms of human health, crops grown through sustainable agriculture are preferable for people. Because of the absence of chemical pesticides and fertilizers, people do not consume artificial things. It restricts the risk of people becoming ill from consuming these chemicals. Furthermore, the crops produced through sustainable agriculture can also be more nutritious because the overall crops are healthier and more natural.

Sustainable agriculture has also had positive influences of the environment. One main benefit to the environment is that sustainable agriculture uses 30% less energy per unit of crop yield in comparison to mechanized agriculture. This decreased dependence on fossil fuels leads to the release of less chemicals and pollution into the environment. Pollution of environment may definitely lead to troubles in the health of society and endanger other aspects of sustainability. Air, soil, and water pollution is a critical environmental matter that cause concerns to the humans in general and scientists specifically (Hussain and Keçili, 2019). Sustainable agriculture also benefits the environment by preserving soil quality, decreasing soil degradation and erosion, and saving water. In addition to these benefits, sustainable agriculture also rises biodiversity of the area by providing several organisms with healthy and natural environments to live in.

7. Social sustainability

Social sustainability is the least defined and least understood of the different ways of approaching sustainability and sustainable development. Social sustainability has had noticeably less consideration in public discussion than economic and environmental sustainability.

There are several approaches to sustainability. The first, which suggests a trio of environmental sustainability, economic sustainability, and social sustainability, is the most broadly accepted as a model for addressing sustainability. The concept of "social sustainability" in this approach includes such subjects as social equity, livability, health equity, community development, social capital, social support, human rights, labor rights, place making, social responsibility, social justice, cultural competence, community resilience, and human adaptation.

A second, more recent, approach states that all of the areas of sustainability are social, including ecological, economic, political, and cultural sustainability. These domains of social sustainability are all reliant on the correlation between the social and the natural, with the "ecological domain" defined as human embeddedness in the environment. In these terms, social sustainability includes all human activities (Atkinson, 2003). It is not just related to the centered connection of economics, the environment and the social.

According to the Western Australia Council of Social Services, "Social sustainability occurs when the formal and informal processes; systems; structures; and relationships actively support the capacity of current and future generations to create healthy and livable communities. Socially sustainable communities are equitable, diverse, connected and democratic and provide a good quality of life."

Social Life, a UK-based social institute, focusing on place-based innovation, developed another definition of social sustainability. They defined social sustainability as "a process for creating sustainable, successful places that promote wellbeing, by understanding what people need from the places they live and work. Social sustainability combines design of the physical realm with design of the social world – infrastructure to support social and cultural life, social amenities, and systems for citizen engagement and space for people and places to evolve" (Boulanger, 2004).

Social Life developed a framework for social sustainability (Michele et al., 2015), which has four dimensions: amenities and infrastructure, social and cultural life, voice and influence, and space to grow.

Nobel Laureate Amartya Sen gives the following dimensions for social sustainability (Cherchye and Kuosmanen, 2006):

- Equity—the society delivers equitable opportunities and outcomes for all its members, predominantly the poorest and most vulnerable members of the society
- Diversity—the society endorses and inspires diversity
- Interconnected/social cohesions—the society provides processes, systems, and structures that endorse linkage inside and outside the society at the official, nonofficial, and institutional level
- Quality of life—the society ensures that rudimentary needs are met and raises a good quality of life for all members at the individual, group, and community level
- Democracy and governance—the rudimentary provides democratic processes and open and answerable governance structures
- Maturity—the people accept the responsibility of steady growth and development through wider social traits

In addition, Sustainable Human Development can be considered as development that enhances the abilities of present people without compromising abilities of future generations. In the human development paradigm, environment and natural resources should establish a tool for attaining better standards of living just as income represents a tool for enhancing social expenditure and, finally, welfare.

The different features of social sustainability are often remarked in socially responsible investing (SRI). Social sustainability standards that are frequently used by SRI funds and indexes to rank publicly traded companies comprises society, diversity, employee relations, human rights, product safety, reporting, and governance structure.

8. Economic sustainable development

From the onset of 21st century, the growth of GDP is no longer regarded a central macroeconomic indicator and economic growth is not the major aim of the economy. Since World War II, the emphasis on economic development policy has changed through different paradigms up to the concept of sustainable development. From the Keynesian thought in the period of post-war through mid-1970s, in which economic policy was constructed on strong governmental interferences, to the Monetarist thought in 1980s where creativities to decrease social disparities by integrating deprived groups into the normal economy occurred. In the late 1980s−1990s, the attention on economic development policy moved to a rationalist thought and creativities to advance environmental and general quality of life by employing highly skilled workers and firms began.

The pivotal concepts of the UN identify sustainable development to be assumed as the major purpose of the economy from the new viewpoint. Economic growth, calculated by the rise of GDP, demonstrates the growth of economic activities and is in association with the welfare. A quick economic growth is unlikely to be sustainable. There are countries, specifically developing countries, where there exists high inconsistency between the level of economic activities and the welfare of people—high growth with low development. Development is a wider process than economic growth, as well as welfare cannot be measured only in monetary terms. Economic growth is an external concept, while development is a larger internal one by considering the promotion in standards of living and poverty decrease. Economic growth may cause an enhancement in the standard of living related to a small proportion of the population, while the majority of the population remains poor. It is how the economic growth is distributed among the population that determines the level of development. Economic growth is calculated by the rise of GDP, while economic development is a more complicated process, which necessities more than one indicator. Economic sustainable development investigates all aspects of economic welfare and progress of the society in the present and future. It is a multidimensional concept involving indicators such as economic growth, income distribution, poverty eradicating, and human welfare, all of which are guaranteed to be continued in the future and for the next generations. The first dimension of economic development is economic growth that is calculated by the rise of Gross National Product. However, economic development is a more complicated process, which necessities more than one indicator. Another indicator representing development is income distribution that is how the income earned by society is distributed among the population. The analysis of economic process in many countries indicates that the rapid economic growth has caused serious problems from the viewpoint of sustainable development such as social and local inequality dramatic loss of infrastructure and rural environment, lack of national capital, etc. Sustainable development is a significant improvement of welfare for all residents, not compromising the welfare in the nearest and further future. Economic growth, though, is essential, not adequate condition for the development.

9. Income and wealth distribution

Income distribution is one of the main economic and social challenges countries confront nowadays. The realities indicate that a small proportion of the population make a high percentage of the revenue and it is not equally distributed.

The issue of distribution assumes an "impersonal approach," by dividing labors' part through a market-determined wage. The remaining belongs to the possessors of the capital, as profit by counting also interest. If the market-determined wages could be shown to be "just" on the contribution base, the separation would clearly be just. Nonetheless, no such demonstration has so far been undoubted and definite.

Economics is described as the science of allocation of scarce resources to meet the unlimited needs and desires of the individual members of a given society. Ekins (1997) explicates the need about new directions for economic policy if sustainable development is to be reached. They will be taken in turn under the following heading:

(1) Difference between developed and developing countries
(2) Impartiality in the world economy, and participatory development.

References

Atkinson, T., et al., 2003. Social Indicators: The EU and Social Inclusion. Oxford University Press, Oxford.

Aud, S., Wilkinson-Flicker, S., Kristapovich, P., Rathbun, A., Wang, X., Zhang, J., 2013. The Condition of Education 2013 (NCES 2013-037). U.S. Department of Education, National Center for Education Statistics, Washington, DC. Retrieved [date] from. http://nces.ed.gov/pubsearch.

Biasi, P., Ferrinia, S., Borghesi, S., Rocchi, B., Di Matteo, M., December 2019. Enriching the Italian Genuine Saving with water and soil depletion: national trends and regional differences. Ecol. Indicat. 107, 105573.

Boulanger, P.-M., 2004. Les indicateurs de développement durable: un défi scientifique, un enjeu démocratique. Les séminaires de L'Iddri 12, 1–34.

Cherchye, L., Kuosmanen, T., 2006. Benchmarking sustainable development: a synthetic meta-index approach. In: McGillivray, M., Clarke, M. (Eds.), Understanding Human Well-Being. United Nations University Press, New-York, Paris, Tokyo, pp. 139–169.

da Silva, J., Fernandes, V., Limont, M., Rauen, W.B., 2020. Sustainable development assessment from a capitals perspective: analytical structure and indicator selection criteria. J. Environ. Manag. 260 (2020), 110147.

Dobermann, A., et al., 2013. Solutions for Sustainable Agriculture and Food Systems: Technical Report for the Post-2015 Development Agenda. Thematic Group on Sustainable and Food Systems of the Sustainable Development Solutions Network, New York, USA.

Ekins, P., 1997. The Kuznets curve for the environment and economic growth: examining the evidence. Environ. Plan. 29 (Issue 5), 805−830.

Fernandes, V., Malheiros, T.F., Philippi, J.R., Arlindo and Sampaio, C.A.C., 2012. Metodologia de avaliação estrategica de processo de gestão ambiental municipal. Saúde e Sociedade (USP. Impresso) 21, 128−143. https://doi.org/10.1590/S0104 - 12902012000700011.

Garnett, T., Godfray, C., 2012. Sustainable Intensification in Agriculture. Navigating a Course through Competing Food System Priorities. Food Climate Research Network and the Oxford Martin Programme on the Future of Food. University of Oxford, UK.

Godfray, H.C.J., Beddington, J.R., Crute, I.R., Haddad, L., Lawrence, D., 2010. Food security: the challenge of feeding 9 billion people. Science 327 (5967), 812−818. https://doi.org/10.1126/science.1185383.

Guimaraes, R., Feichas, S., 2009. Challenges in the construction of sustainability indicators. Ambient. Soc. [Online]. 12 (n.2), 307−323. ISSN 1414-753X.

HLPE, 2013. Investing in Smallholder Agriculture for Food Security. A Report by the High Level Panel of Experts on Food Security and Nutrition of the Committee on World Food Security, Rome. Available online at: http://www.fao.org/3/a-i2953e.pdf.

Hussain, C.M., 2019. Handbook of Environmental Materials Management, first ed. Springer International Publishing.

Hussain, C.M., Keçili, R., 2019. Modern Environmental Analysis Techniques for Pollutants, first ed. Elsevier (Chapter 14) Future of environmental analysis.

Hussain, C.M., Mishra, A.K., 2019. Nanotechnology in Environmental Science, vol. 2. John Wiley & Sons.

Hussain, C.M., March 25, 2020. Chapter 18 Future of Environmental Remediation Methods, the Handbook of Environmental Remediation: Classic and Modern Techniques, first ed. Royal Society of Chemistry.

IFAD, 2011. Rural Poverty Report 2011.

Lipper, L., Thornton, P., Campbell, B., et al., 2014. Climate-smart agriculture for food security. Nat. Clim. Change 4, 1068−1072. https://doi.org/10.1038/nclimate2437.

Mallakpour, S., Azadi, E., Hussain, C.M., 2020. Environmentally benign production of cupric oxide nanoparticles and various utilizations of their polymeric hybrids in different technologies. Coord. Chem. Rev. 419, 213378. https://doi.org/10.1016/j.ccr.2020.213378.

Mariann, S., Mária, S., Csete, Tamás, P., 2018. Resilient regions from sustainable development perspective. Eur. J. Sustain. Dev. 7 (1), 395−411.

Martínez-Graña, A.M., Goy y, J.L., Irene De Bustamante Gutiérrez, G., Zazo Cardeña, C., 2013. Characterization of environmental impact on resources, using strategic assessment of environmental impact and management of natural spaces of "Las Batuecas-Sierra de Francia" and "Quilamas" (Salamanca, Spain). Environ. Earth Sci. 71, 39−51 (2014). https://link.springer.com/Godfray. H.C.J.

Meadows, D., 1972. Indicators and Information Systems for Sustainable Development. A Report to the Balaton Group. Published by The sustainability institute. https://edisciplinas.usp.br/pluginfile.php/106023/mod_resource/content/2/texto_6.pdf.

Meadows, D., 1998. A report to the balaton group. Hartland: the sustainability institute. https://donellameadows.org/wp-content/userfiles/IndicatorsInformation.pdf.

Michele, B., Mauthe, S., Allen, G., Shawn, A., John, L., Ping Sun, L., 2015. What determines social capital in a social−ecological system? Insights from a network perspective. Environ. Manag. 55, 392−410.

Millennium Ecosystem Assessment, 2005. Ecosystems and Human Well-Being: Synthesis. Island Press, Washington, DC.

OECD, 2001. OECD Environmental Strategy for the First Decade of the 21st Century Adopted by OECD Environment Ministers, 16 May 2001. OECD, Paris.

OECD, 2003. ECD Guide. Measuring Material Flows and Resource Productivity (Communiqué Adopted by G8 Environment Ministers, Paris, pp. 25–27. Accounting and material flows, 3 June 2003.

OECD, 2012. Meeting of the Environment Policy Committee (EPOC) at Ministerial Level, Meeting of the Environment Policy Committee (EPOC) at Ministerial Level. Available online at: https://www.oecd.org/env/50032165.pdf.

Pinsky, M.L., Worm, B., Fogarty, M.J., Sarmiento, J.L., Levin, S.A., 2015. Marine taxa track local climate velocities. Science 341, 1239–1242.

Pissourios, I.A., 2013. An interdisciplinary study on indicators: a comparative review of quality-of-life, macroeconomic, environmental, welfare and sustainability indicators. Ecol. Indicat. 34, 420–427.

Pretty, J.N., Noble, A.D., Bossio, D., Dixon, J., Hine, R.E., Penningdevries, F.W.T., Morison, J.I.L., 2006. Resource-conserving agriculture increases yields in developing countries. Environ. Sci. Technol. 40 (4), 1114–1119.

Raynaut, C., Zanoni, M., Lana, P.C., 2018. O desenvolvimento sustentavel regional: o que proteger? Quem desenvolver? Desenvolv. Meio Ambiente 47, 275–289. https://doi.org/10.5380/dma.v47i0.62452.

Sartori, S., Latr onico, F., Campos, L.M.S., March 2014. Sustainability and Sustainable Development: A Taxonomy in the Field of Literature. Ambiente & Sociedade. https://www.researchgate.net/publication/270449751.

Smith, A., Snapp, S., Chikowo, R., Thorne, P., Bekunda, M., Glover, J., 2016. Measuring Sustainable Intensification in Smallholder Agrosystems: A Review. Global Food Security. Article (in press).

Snapp, S., 2014. Global Food Security. https://doi.org/10.1016/j.gfs.2016.11.002.

Tilman, D., Clark, M., 2014. Global diets link environmental sustainability and human health. Nature 515, 518–522.

Tilman, D., Socolow, R., Foley, J.A., Hill, J., Larson, E., Lynd, L., Pacala, S., Reilly, J., Searchinger, T., Somerville, C., Williams, R., 2009. Beneficial biofuels—the food, energy, and environment trilemma. Science 325 (5938), 270–271.

UNESCO, 2006. Indicators of Sustainability: Reliable Tools for Decision Making. UNESCO Scope, Policy Briefs. https://unesdoc.unesco.org/ark:/48223/pf0000150005.

United Nations, 1992. Agenda 21, New York: UN. Available online at: https://sustainabledevelopment.un.org/content/documents/Agenda21.pdf.

United Nations, 2015. General Assembly Resolution. September 18. Available at: https://sustainabledevelopment.un.org/post2015/transformingourworld.

United Nations, 2019. General Assembly Resolution. September 18. Available online at: https://sustainabledevelopment.un.org/post2015/transformingourworld.

United Nations Economic and Social Council, 2016. Report of the Inter-Agency and Expert.

United Nations Economic and Social Council, 2017. Report of the Inter-Agency and Expert.

Venturelli, R.C., Galli, A., 2006. Integrated indicators in environmental planning: methodological considerations and applications. Ecol. Indicat. 6, 228–237. https://doi.org/10.1016/j.ecolind.2005.08.023.

Vermeulen, S.J., Campbell, B.M., Ingram, J., 2012. Climate Change and Food Systems, vol. 37, p. 195, 1. www.annualreviews.org.

Waas, T., Huge, J., Block, T., Wright, T., Capistros-Benites, F., Verbruggen, A., 2014. Sustainability assessment and indicators: tools in a decision-making strategy for sustainable development. Sustainability 6, 5512−5534. https://doi.org/10.3390/su6095512.

Wang, J.-J., Jing, Y.-Y., Zhang, C.-F., Zhao, J.-H., 2019. Review on multi-criteria decision analysis aid in sustainable energy decision-making. Renew. Sust. Energ. Rev. 13 (9), 2263−2278.

Wang, H., Prentice, I.C., Ni, J., 2013. Data-based modelling and environmental sensitivity of vegetation in China. Biogeosciences 10, 5817−5830. https://doi.org/10.5194/bg-10-5817-2013.

WCED, 1987. World Commission on Environment and Development. Report Our Common Future. United Nations, New York. https://sustainabledevelopment.un.org/content/documents/5987our-common-future.pdf.

World Bank, 2007. World Development Report (2008). Agriculture for Development. World Bank, Washington, DC. Available online at: https://openknowledge.worldbank.org/handle/10986/5990. License: CC BY 3.0 IGO.

World Conservation Strategy, 2011. Living Resource Conservation for Sustainable Development, World Conservation Strategy. URL. http://data.iucn.org/dbtw-wpd/edocs/WCS-004.pdf.

Important websites

Environment and SafetyEnvironment and Safety. http://www.environmentandsociety.org/.

FAOFAO. http://www.fao.org/.

Millennium AssessmentMillennium Assessment. https://www.millenniumassessment.org/en/index.html.

OCEDOECD. https://www.oecd.org/. OECD. https://www.oecd.org/.

Sustainable DevelopmentSustainable Development. https://sustainabledevelopment.un.org/milestones/wced.

UNUN. https://www.un.org/ecosoc/en/home.

UNUN. https://www.un.org/en/model-united-nations/economic-and-social-council.

UN75UN75. https://un75.online/.

World BankWorld Bank. https://www.worldbank.org/.

To measure the performance of sustainable food supply chain

Pragati Priyadarshinee
Chaitanya Bharathi Institute of Technology, CBIT(A), Hyderabad, Telangana, India

1. Introduction

Sustainability is a major concern for perishable supply chain organization especially for the food industry. The estimated average loss of food product to the consumers is to be 35%, in other words half of the food products to the human is being misused (Riikka Kaipia et al., 2013; Gustavsson et al., 2011; Parfitt et al., 2010). Different supply chain has estimated the average food waste giving some different results, but the studies are in agreement with fruits and vegetables those can be avoided from food wastage and are noteworthy for the food supply chain turnover (Mena et al., 2011; Hanssen and Schakenda, 2011). Sagheer et al. (2009) stated that on competitiveness and value-chain dynamics, there is a dearth of research. It is a major challenge to access the primary data for the agricultural supply chain management. Riikka Kaipia et al. (2013) stated that the success of operation depends upon the shared data in form of data-driven supply chain management. Availability of products in supply chain network is important for speedy delivery. The study revealed improved performance would be achieved with the changes in data sharing and material inflows. A research framework may contain certain factors related to people and society. One factor is definitely related to another factor in the model. The factors related to people can be the management and manufacturer, and the social factors may be the environment, trust, and the food produced (Sagheer et al., 2009).

According to Sachan et al. (2005), the information flow in supply chain management consists of different service distribution, facilities to produce and supply of the essential materials, and it has an integrated approach for delivering products and services to the clients. As per Folinas et al. (2013), value chain can analyze the waste sources in the integrated supply chain management. Value-Stream Mapping can be applied at this stage which is a widely deployment management technique for waste determination in organizational boundaries.

1.1 Objective

All of the research processes include some benefits along with some drawbacks.

The objective of the study is framed in a systematic way in terms of sustainable agricultural supply chain. Like food supply chain, some other related topics were studied to identify the major factors. Reliability of those factors can be checked through

Sustainable Resource Management. https://doi.org/10.1016/B978-0-12-824342-8.00019-5

some independent studies. Thus, the present research area is the outcome of some continuous research process in supply chain management.

The research objectives are framed as follows:

1. To identify the critical success factors for the food industry.
2. To establish the relationship between the factors and the organizational performance.

The remainder is organized as the background theory which gives the literature on different types of sectors. The third section includes the research methodology and model building. Section four gives the result analysis followed by the discussion and conclusion with the future scope and limitations of the study.

2. The background theory

The literature review includes a systematic study of agricultural supply chain management done so far with different operational and analytical models. It includes sustainable food supply chain management, challenges in agricultural supply chain management, and identification of the gaps.

2.1 *Sustainable food supply chain*

As per Li et al. (2014), food distribution is a challenge in production planning. It is critical for the food industry which is the largest manufacturing sector in many of the developing or developed countries. Food industry consumes large amount of natural resources and also always in demand, even if the current system is efficient enough. For the above aspects, sustainable food supply chain has become a global challenge in industries at present. Sustainable supply chain management is a contemporary challenge to grow with the demand supply chain. In supply chain, producer to consumer distribution of goods has a significant role. Food distribution needs to improve with consumer consciousness in sustainable supply chain management.

As per Validi et al. (2014), sustainable development has some common issues of how to ensure procedure involvement for small farmers. The institutional initiatives are also required to satisfy the food safety and quality regulation (Gopal Naik, IIMB). The critical analysis signifies that the food supply chain management is a research field on demand focusing more on case development problems and less integrated methodological approaches for the supply chain optimization (Tsolakis et al., 2014).

Sustainability is a major concern for perishable food supply chain management (Riikka Kaipia et al., 2013). The estimated average loss of food product to the consumers is to be 35%, in other words one-third of the food products to the human is wasted (Gustavsson et al., 2011; Parfitt et al., 2010). Different supply chain has estimated the average food waste giving some different results, past studies agree to the fresh fruits, vegetables, and bakery products for distribution through the supply chain management (Kantor et al., 2007; Griffin et al., 2008; Mena et al., 2011; Hanssen and Schakenda, 2011). Sagheer et al. (2009) stated that on competiveness and value-chain dynamics, there is a dearth of research. It is a major challenge to access the primary data for the agricultural supply chain management. Riikka Kaipia et al. (2013) identified the success rate of operations for the shared data. For ensuring the availability and speeding up the delivery process supply chain is essential part. The study revealed improved performance would be obtained by dataflow and information sharing. Sagheer et al. (2009) gave a framework comprising of social and antisocial elements. The model shows how one action can be influenced by the other.

According to Sachan et al. (2005), supply chain management helps in distributing the materials and services. Supply chain management is shifted to a focused incorporated method (Stevens, 1989; Towill, 1996; Metz, 1998). As per Folinas et al. (2013), value chain can analyze the difference sources of waste in the selected supply chain. Value-stream mapping can be applied at this stage which is a widely deployment management technique for waste determination in organizational boundaries.

2.2 Technology used

For agricultural supply chain to reduce the food wastage, different authors used various techniques for the studies.

Raut et al. (2017) studied upon food wastage and how fresh produce supply chain recycling is possible through three phases of social, ecological, and economic sustainability. The authors used ISM (Interpretive Structural Modeling) for result identification. Kamble et al. (2019) examined the descriptive, predictive, and prescriptive analytics for sustainable agricultural supply chain. Akhtar et al. (2015) studied on data from chief executive officers, managing directors, and senior operations managers of supply chain market.

It is important to identify the relationship among cooperative behavior, relationship management, and alliance performance in sustainable agricultural performance. It investigates how this relationship further influences cooperative behavior (Fu et al., 2018). As per Mangla (2018), the sustainability performance focuses on effective utilization and consumption of natural resources to have the balance on ecological, economic, and societal aspect of agricultural food. It adds demand on business managers that have small profit with more number of customers. Parwez et al. (2016) viewed some important directions of agricultural supply chain. The author identified the issues at various levels of supply chain, agricultural transformation, and information technology roles. Rajabion et al. (2019) presented a model to identify the impact of Information Technology and Systems on farmers' knowledge and business processes on green supply chain management system. Investigated the paradigms of sustainable supply chain management. Yazdani et al. (2019) studied to identify the risk drivers of food and their effects on sustainable agricultural supply chain in connection with circular economy strategy. The most important food drivers are identified with multicriteria decision-making analysis process.

Atul Agarwal et al. (2018) integrated institutional and self-determination theories to integrate the suppliers and markets as noncoercive drivers for colleges as representing a coercive pressure. It leads to a green supply chain management model in which market pressure is mediated by internal impetus, whereas regulatory pressure directly impacts. The model was tested by PLS (Partial Least Square) technique with 60 manufacturing companies in United States.

As per Routroy and Kumar (2016), application of technology is the major part for implementing green practices in manufacturing industries. Technology advancement has a major impact on sustainability. Aboelmaged (2014) divided technology into infrastructure and competence. Technology competence reflects knowledge, skill, and expertise to utilize the infrastructure. Green design, manufacturing processes, and packaging are the applications ICT innovation (Chuang and Yang, 2014).

2.3 Issues and challenges

All of the research processes include some benefits along with some drawbacks. The objective is maintained by adopting a structured and some systematic process. Kassarjan (1997) found a base for validating a research. Other topics were reviewed to form the constructs. Reliability can be ensured by doing the citation analysis and taking the help of independent researchers. Thus, our current research meets the requirement of rigorous research process. Developing countries like India suffer from inadequacies even if the agricultural sector being the primary source of employment. In near future, shortage of natural resources and shortage of supply may arise.

Some more issues are increasing demand for proved sustainable supply chain performance in agriculture sector in a developing country like India. Poor infrastructure, loss of supply chain, and food wastage are the other identified issues that negatively influences supply chain sustainability (Mangla et al., 2014).

2.4 Gap identification

From the literature review, the following research gaps are identified:

1. Most of the studies cover the agricultural supply chain management.
2. Few studies were done on food supply chain management.
3. A robust research model is required to measure the organizational performance.
4. Accurate policy framing is also lagging in the previous studies.
5. More researches are already done on manufacturing and service sectors.

3. Research methodology and model building

Research methodology and framework are the integral part of any research. There are six independent variables and one dependent variable in the model. The six independent variables are Supplier and Customer for joint action (CCJA), Collaborative Green Local Cold Storage (CGLC), Competitive Pressure (CP), Environmental Design Management (EDM), Green Performance Management (GPM), Regulatory Pressure (RP), and one dependent variable Organizational performance (OPR) (Fig. 15.1).

Data collection was done from 317 respondents having knowledge in green revolution or the agricultural food sectors. Questionnaire was formed on 7 constructs and 44 items. The items reduced to 44 after the pilot study. Based on the data research model six hypotheses were framed. Seven-point likert scale was employed for data collection.

Hypothesis development:

H1: Co-operation with Supplier and Customer for joint action (CCJA) has some positive impact on Organizational performance (OPR).
H2: Collaborative Green Local Cold Storage (CGLC) has some positive impact on Organizational performance (OPR).
H3: Competitive Pressure (CP) has some positive impact on Organizational performance (OPR).
H4: Environmental Design Management (EDM) has some positive impact on Organizational performance (OPR).
H5: Green Performance Management (GPM) has some positive impact on Organizational performance (OPR).
H6: Regulatory Pressure (RP) has some positive impact on Organizational performance (OPR).

Survey-based method is used for carrying out the research. It constitutes of EFA, CFA, and SEM. Smart PLS 3.2.2 was used for the analysis.

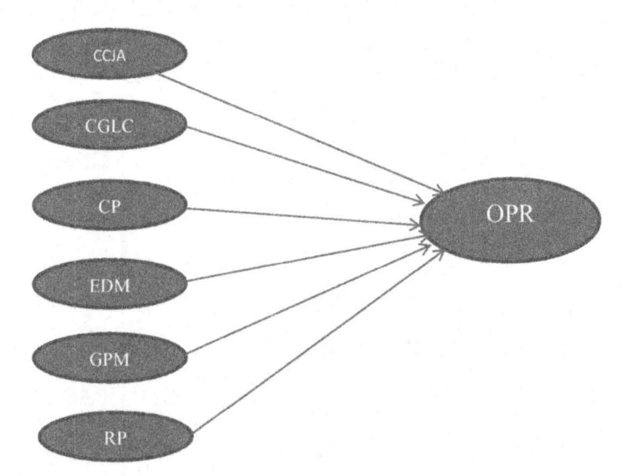

Figure 15.1 Conceptual framework.

4. Result analysis

The survey (EFA, CFA, and SEM) results are predicted below. The constructs calculated for the model are given as shown in Table 15.1.

The above table depicts the composite reliability and cronbach's alpha values of the seven variables; CCJA, CGLC, CP, EDM, GPM, OPR, and RP which satisfies the threshold limit of 0.7 (Table 15.2).

The PLS-SEM path diagram is shown in the figure below (Table 15.3 and Fig. 15.2).

As per the above table, all the hypotheses are accepted. CGLC is having highest effect, GPM is having second higher effect, and EDM is the third high effect on the Organizational Performance (OPR).

Table 15.1 Analysis of constructs of PLS-Model.

Variables	Composite reliability	R square	Cronbach's alpha	Communality	Redundancy
CCJA	0.9529	0	0.9543	0.7716	0
CGLC	0.9505	0	0.9392	0.7333	0
CP	0.8992	0	0.9144	0.6446	0
EDM	0.8595	0	0.8770	0.5161	0
GPM	0.9131	0	0.8913	0.5683	0
OPR	0.9259	0.4768	0.9002	0.7141	0.0003
RP	0.9867	0	0.98	0.9611	0

Table 15.2 Discriminant validity.

		CCJA	EDM	CP	RP	GPM	CGLC	OPR
CCJA	Pearson correlation	**0.7716**						
EDM	Pearson correlation	0.031	**0.5161**					
CP	Pearson correlation	0.081	0.060	**0.6446**				
RP	Pearson correlation	0.161[a]	-0.004	0.164[a]	**0.9611**			
GPM	Pearson correlation	0.001	0.013	0.055	0.023	**0.5683**		
CGLC	Pearson correlation	0.007	0.022	0.001	0.039	0.546[a]	**0.7333**	
OPR	Pearson correlation	0.012	0.018	0.021	0.061	0.545[a]	0.642[a]	**0.7141**

[a]. Correlation is significant at the 0.01 level (two-tailed).

Table 15.3 Hypotheses testing.

Hypothesis no.	Hypothesis	Significance	Results (Supported-Y/N)
H1	Co-operation with Supplier and Customer for joint action (CCJA) positively influences organizational performance (OPR).	0.011	Yes
H2	Collaborative Green Local Cold Storage (CGLC) positively influences organizational performance (OPR).	0.491	Yes
H3	Competitive Pressure (CP) positively influences organizational performance (OPR).	0.017	Yes
H4	Environmental Design Management (EDM) positively influences organizational performance (OPR).	0.034	Yes
H5	Green Performance Management (GPM) positively influences organizational performance (OPR).	0.281	Yes
H6	Regulatory Pressure (RP) positively influences organizational performance (OPR).	0.031	Yes

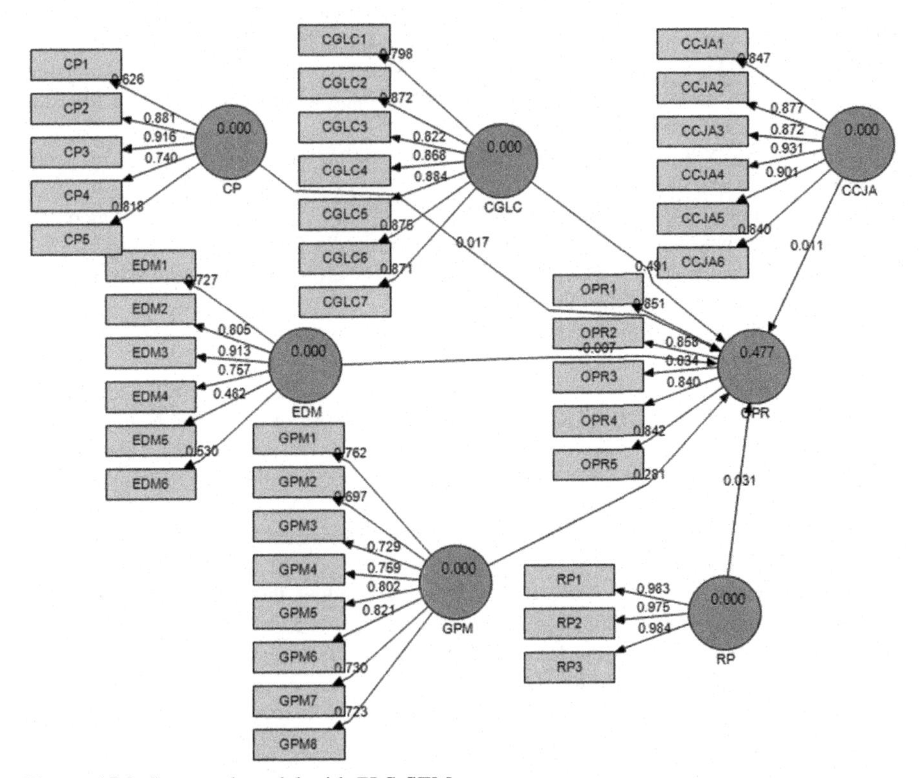

Figure 15.2 Structural model with PLS-SEM.

5. Discussion and conclusion

The study identifies seven critical success factors for the successful food supply chain management. Smart-PLS is used for Structural Equation Modeling (SEM) in the proposed research model. The relationships among the factors were identified through the validity and reliability analysis. The research objectives of the study were framed accordingly. It is identified from the study that CGLC is having highest impact on Organizational Performance.

New managerial policy can be framed from the research output. The study can be compared with the studies on Agricultural Industry. The variables which are missed out in the study can be considered in future for more clarity. The weak points can be improved to sustain themselves in the highly changing market. Regular upgradation is required to stay sustainable. Government should provide the awareness for smart city creation.

References

Aboelmaged, M.G., 2014. Predicting e-readiness at firm-level: an analysis of technological, organizational and environmental (TOE) effects on e-maintenance readiness in manufacturing firms. Int. J. Inf. Manag. 34 (5), 639–651.

Agarwal, A., Giraud-Carrier, F.C., Li, Y., 2018. A mediation model of green supply chain management adoption: the role of internal impetus. Int. J. Prod. Econ. 205, 342–358.

Akhtar, J., Teo, C.L., Lai, L.W., Hassan, N., Idris, A., Aziz, R.A., 2015. Factors affecting delignification of oil palm empty fruit bunch by microwave-assisted dilute acid/alkali pretreatment. BioResources 10 (1), 588–596.

Chuang, K.Y., Ho, Y.S., 2014. A bibliometric analysis on top-cited articles in pain research. Pain Med. 15 (5), 732–744.

Folinas, D., Aidonis, D., Triantafillou, D., Malindretos, G., 2013. Exploring the greening of the food supply chain with lean thinking techniques. Procedia Technology 8, 416–424.

Fu, J., Yu, J., Jiang, C., Cheng, B., 2018. g-C3N4-Based heterostructured photocatalysts. Adv. Energy Mat. 8 (3), 1701503.

Griffin, M., Sobal, J., Lyson, T.A., 2008. An analysis of a community food waste stream. Agric. Hum. Val. 26 (1/2), 67–81.

Gustavsson, J., Cederberg, C., Sonesson, U., van Otterdijk, R., Meybeck, A., 2011. Global Food Losses and Food Waste, Food and Agriculture Organisation of the United Nations. FAO, Rome.

Hanssen, O.J., Schakenda, V., 2011. Nyttbart Matavfall I Norge – Status Og Utviklingstrekk 2010, Østfoldsforskning, Norway.

Kaipia, R., Dukovska-Popovska, I., Loikkanen, L., 2013. Creating sustainable fresh food supply chains through waste reduction. Int. J. Phys. Distrib. Logist. Manag. 43 (3), 262–276.

Kamble, S., Gunasekaran, A., Arha, H., 2019. Understanding the Blockchain technology adoption in supply chains-Indian context. Int. J. Prod. Res. 57 (7), 2009–2033.

Kantor, L.S., Lipton, K., Manchester, A., Oliveira, V., 2007. Estimating and addressing America's food losses. Food Rev. 20 (1), 2.

Kassarjian, J., 1997. The Paradox of leading change. Intl. Inst. Manag. Dev. 40 (12).

Li, D., Wang, X., Chan, H.K., Manzini, R., 2014. Sustainable food supply chain management. Int. J. Prod. Econ. 152 (0), 1–8.

Mangla, S., Kumar, P., Barua, M.K., 2014. An evaluation of attribute for improving the green supply chain performance via DEMATEL method. Intl. J. Mech. Eng. Rob. Res. 1 (1), 30–35.

Mangla, S.K., Luthra, S., Mishra, N., Singh, A., Rana, N.P., Dora, M., Dwivedi, Y., 2018. Barriers to effective circular supply chain management in a developing country context. Prod. Plann. Contr. 29 (6), 551–569.

Mena, C., Adenso-Diaz, B., Yurt, O., 2011. The causes of food waste in the supplier-retailer interface: evidences from the UK and Spain. Resour. Conserv. Recycl. 55 (6), 648–658.

Metz, P.J., 1998. Demystifying Supply Chain Management. Supply Chain Management Review, Winter, pp. 46–55.

Parfitt, J., Barthel, M., Macnaughton, S., 2010. Food waste within food supply chains: quantification and potential for change to 2050. Phil. Trans. Biol. Sci. 365 (1554), 3065–3081.

Parwez, S., 2016. A conceptual model for integration of Indian food supply chains. Global Bus. Rev. 17 (4), 834–850.

Rajabion, L., Shaltooki, A.A., Taghikhah, M., Ghasemi, A., Badfar, A., 2019. Healthcare big data processing mechanisms: the role of cloud computing. Int. J. Inf. Manag. 49, 271–289.

Raut, R., Priyadarshinee, P., Gardas, B., Narkhede, B.E., Nehete, R., 2017. The incident effects of supply chain and cloud computing integration on the business performance- an integrated SEM-ANN approach, Emerald. Benchmark Int. J. 25 (8), 2688–2722.

Routroy, S., Kumar, C.S., 2016. An approach to develop green capability in manufacturing supply chain. Int. J. Process Manag. Benchmark. 6 (1), 1–28.

Sachan, A., Sahay, B.S., Sharma, D., 2005. Developing Indian grain supply chain cost model: a system dynamics approach. Int. J. Prod. Perform. Manag. 54 (3), 187–205.

Sagheer, S., Yadav, S.S., Deshmukh, S.G., 2009. Developing a conceptual framework for assessing competitiveness of India's agrifood chain. Int. J. Emerg. Mark. 4 (2), 137–159.

Stevens, G.C., 1989. Integrating the supply chain. Int. J. Phys. Distrib. Logist. Manag. 19 (8), 3–8.

Towill, D.R., 1996. Industrial dynamics modeling of supply chains. Logist. Inf. Manag. 9 (4), 43–56.

Tsolakis, N.K., Keramydas, C.A., Toka, A.K., Aidonis, D.A., Iakovou, E.T., 2014. Agrifood supply chain management: a comprehensive hierarchical decision-making framework and a critical taxonomy. Biosyst. Eng. 120, 47–64.

Validi, S., Bhattacharya, A., Byrne, P.J., 2014. A case analysis of a sustainable food supply chain distribution system—a multi-objective approach. Int. J. Prod. Econ. 152, 71–87.

Yazdani, M., Kahraman, C., Zarate, P., Onar, S.C., 2019. A fuzzy multi attribute decision framework with integration of QFD and grey relational analysis. Expert Syst. Appl. 115, 474–485.

Sustainable plastic materials management

Juan F. Velasco-Muñoz, José A. Aznar-Sánchez, Belén López-Felices, Daniel García-Arca
Department of Economy and Business, Research Centre CIAIMBITAL, University of Almería, Almería, Spain

1. Introduction

Plastic materials are one of the most successful innovations of the last century (Gu et al., 2017; Kumar et al., 2019). The term plastics covers a wide range of synthetic or semisynthetic materials that are used for an immense number of applications (Andrady and Neal, 2009). It is more appropriate to talk about plastic materials because there is a great diversity of materials under that name, which can be obtained from raw materials of fossil origin (oil, gas, etc.), renewable (cane sugar, starch, vegetable oils, etc.), or mineral (salt). Each of these materials has different characteristics, which make it useful to produce different types of goods (Poonam et al., 2013). Currently, we can find some type of plastic in most of the elements that surround us (Faraj et al., 2020). According to the 2019 annual report of the association of European plastic producers, 359 million tons of plastics were produced worldwide in 2018 (PlasticsEurope, 2019). This production is unevenly distributed worldwide. Asia accounts for 51%, with China standing out with 30% of the world total. The American continent produces 22%, with North America accounting for 18% and Latin America 4%. Africa and the Middle East account for only 7% of the total.

The use of plastic is so widespread due to several properties that make it a unique material and, at the same time, many different materials (Andrady and Neal, 2009; Alam et al., 2018). Plastic is a material with a relatively low density, which makes it a lightweight material. It also has thermal and electrical insulation properties that can, however, be modified if necessary. This material is resistant to the corrosion of many substances, so it is very durable (Kumar et al., 2019). The plastic can be molded into complex shapes and allows the integration of other materials (Iro et al., 2019). Plastics can be modified with fillers, colors, foaming agents, flame retardants, and plasticizers to obtain a specific product for any type of need (Khamtree et al., 2020).

The European Association of Plastics Manufacturers distinguishes between bioplastics, biodegradable plastics, technical plastics, epoxy resins, expanded polystyrene, fluoropolymers, polyolefins, polystyrene, polyurethanes, polyvinyl chloride, thermoplastics, and thermoset plastics (PlasticsEurope, 2020). Bioplastics are manufactured from renewable biological resources (Fojt et al., 2020). An example is the manufacture of ethylene from sugar cane, which is in turn used to make polyethylene.

Sustainable Resource Management. https://doi.org/10.1016/B978-0-12-824342-8.00016-X

Biodegradable plastics can be degraded by microorganisms, which convert them into water, carbon dioxide (or methane), and biomass (Poonam et al., 2013). Engineering plastics include materials such as acrylonitrile butadiene styrene (ABS), used for car bumpers; polycarbonates, used in motorbike helmets; and polyamides (nylons). This group covers harder and more resistant materials, which offer higher performance than standard materials, and are ideal for technical applications (Mošorinski et al., 2020). Epoxy resins are materials whose physical state can change from a low viscosity liquid to a solid with a high melting point. They are often found as a coating for soft drink cans, as a protective coating for beds, shopping carts, or paint to protect the surfaces of boats.

Fluoropolymers are a family of high-performance plastics among which polytetrafluoroethylene stands out, considered the most slippery material in the world. Polyolefins are a family of thermoplastics made from polyethylene and polypropylene, which are produced mainly from oil and natural gas through a polymerization process of ethylene and propylene, respectively (Raab-Obermayr et al., 2017). Polystyrene is a synthetic aromatic thermoplastic polymer made from a liquid petrochemical, which softens with heat and can be converted into semifinished products such as sheets and foils, as well as a wide range of finished articles (Mošorinski et al., 2020). Polyurethane is a resilient, flexible, and durable material, which is used for a wide range of products. Polyvinyl chloride, better known as PVC, was one of the first plastics to be discovered, and is a derivative of salt and oil or gas (Chauhan et al., 2019). It is the third most widely produced synthetic plastic polymer in the world, after polyethylene and polypropylene. Thermoplastics are defined as polymers that can be melted and reshaped almost indefinitely; reacting to temperature so that they melt when heated and harden when cooled. Finally, thermoset plastics are synthetic materials treated in such a way that once heated and formed, they cannot be remelted and change shape (Shieh et al., 2020).

Despite the multiple benefits of plastics use, one of the most threatening and challenging environmental problems globally today is plastics waste management (Thompson et al., 2009; Yu et al., 2018; Li et al., 2020). It is estimated that 300 million tonnes of plastics waste are produced globally per year (Blettler et al., 2017). During the last 70 years, it is estimated that about six billion tonnes of plastic have been produced worldwide, of which only 9% have been recycled and 12% have been incinerated. This implies that 79% of plastic is left untreated in the environment (Tulashie et al., 2020). The accumulation of plastics in the ocean has become a pandemic phenomenon that reaches alarming figures (Al-Thawadi, 2020). It is estimated that more than eight million tons of plastic are discharged into the oceans annually (Jambeck et al., 2015; Santos et al., 2016). Ocean pollution is not only a risk to the survival of aquatic animals but also poses a threat to human health through the food chain, although intake levels and adverse effects on human health are still largely unknown (Klingelhöfer et al., 2020; Maity and Pramanick, 2020).

Among the problems linked to the management of plastic materials, the case of microplastics stands out. These microplastics, which are defined as particles of less than 5 mm, accumulate all over the world, even in the most remote and abandoned parts such as the Arctic snow (Bergmann et al., 2019; Klingelhöfer et al., 2020).

A distinction can be made between primary and secondary microplastics. Primary type microplastics are those that are originally designed with a tiny size to fulfill specific functions as an integral part of cosmetic or textile products, for example (van Wezel et al., 2016; Boucher and Froit, 2017). Secondary microplastics are the result of the degradation process of larger plastics, such as the abrasion of tires (Yu et al., 2018). Microplastic contamination can occur by land to sea transport over river systems. In this case, it has been estimated that the critical points are Southeast Asia and OECD countries due to the contribution of sewage (van Wijnen et al., 2019). On the other hand, the discharge of microplastics to the soil occurs mainly through atmospheric deposition, agricultural application of organic fertilizers, or plastic film padding (Allen et al., 2019; Zhu et al., 2019).

Despite all the above, the management of plastic materials presents opportunities for the development of sustainable models, which contribute to environmental conservation, economic growth, and social welfare. In recent years, innovative plastics have been developed based on biological materials and biodegradable plastics (Agarwal, 2020). A large number of projects have also emerged around directly reusable materials, as well as initiatives for recycling (Gu et al., 2017). These include the use of plastic waste as materials for the construction sector (De Jesus et al., 2018; Tulashie et al., 2020); the development of micromaterials such as microarrays, microreactors, and microfluidics (Alonso-Amigo, 2000; Dow et al., 2018; Qamar et al., 2020); the development of new plastic products for packaging (Carlin, 2019) or for the electronics sector (Klengel et al., 2019); etc.

This study seeks to show the state of the global research on sustainable plastic materials management (SPMM). Answers to the following questions are sought: who are the main agents promoting research on SPMM?; what are the most relevant research lines in this area?; what are the main research gaps? In order to achieve this, the global research conducted on SPMM in the period from 2000 to 2019 will be reviewed. The methodology used will be a qualitative systematic analysis together with a quantitative bibliometric analysis. This study provides an analysis of the global research conducted on SPMM over the last 20 years and can serve as a reference both for researchers and stakeholders interested in this subject.

2. Methodology

To meet the proposed objective, a literature review has been carried out, for which the Scopus database has been selected following previous work on parallel themes (Blettler et al., 2018; Velasco-Muñoz et al., 2018; Aznar-Sánchez et al., 2019). The search parameters were used as follows: TITLE-ABS-KEY (plastic*) and ("sustainable management" or sustainability). These parameters were used in the title, keyword, and summary search fields. The study period selected was from 2000 to 2019, in order to check the contributions to research made during this century. Only documents up to 2019 have been included in order to be able to compare whole year periods (Cossarini et al., 2014). It should be noted that a different search query would give a different result. The sample selection process was carried out in June 2020. The final sample consists of a total of 1447 articles. In addition, a search for articles on "sustainable management" with the same restrictions was conducted to analyze the relative importance of research on SPMM within this overall theme. Once the information was downloaded, processed, and refined, the data were analyzed with reference to number of articles per year, year of publication of the articles, authors of the articles, institutions and countries of affiliation of all authors, the thematic area in which Scopus classifies the articles, the name of the journals in which they have been published and the keywords. The tools used in the data analysis process were Excel (version 2016), SciMAT (v1.1.04), and VOSviewer. Finally, keyword analysis was used to extract the main trends and research needs.

3. Results and discussion

3.1 General evolution on sustainable plastic materials management research

Table 16.1 shows the evolution of the main variables in relation to the research on the use of SPMM in the period 2000−19. The number of articles has increased from 4 in 2000 to 275 in 2019. It should be noted that this line of research has gained importance in recent years, as 61.5% of the articles in the sample have been published in the last 5 years. In order to check whether the growth in the number of articles is due to the general trend in the whole of research into sustainable management, the annual variation in the number of articles published in both lines of research was calculated, taking the first year of the period analyzed as the basis (Fig. 16.1). The average annual growth of articles on sustainable management was 15.1%, while that of articles on SPMM was 24.9%. These data confirm the relevance of the SPMM research line within sustainable management research.

During the period analyzed, a total of 4862 authors have participated in the 1447 articles that make up the sample. This variable has experienced great growth, going from 11 in 2000 to 1062 in 2019. The average number of authors per article has increased from 2.75 to 3.87. In 2000, the four articles analyzed were published in four different journals, while in 2019 the 275 articles in the sample were published in 186 different journals. The average number of articles per journal has remained almost constant, going from 1 to 1.5 at the end of the period analyzed. In total, the 1447 articles that make up the sample have been published in 774 different journals. With respect to the countries which have participated in the preparation of the works, a total of 101 have been counted during the whole period analyzed. The number of countries has increased from 2 in 2000 to 68 in 2019. In terms of citations, the set of documents in the sample has accumulated a total of 21,934 during the whole period. This figure has increased from 1 in 2000, when the first citation in the sample was obtained, to 5483 in 2019. The average number of citations obtained per article has increased from 0.3 in 2000 to 15.1 in 2019.

Table 16.1 Major characteristics of the Sustainable Plastic Materials Management (SPMM) research.

Year	A	AU	J	C	TC	TC/CA
2000	4	11	4	2	1	0.3
2001	1	2	1	1	2	0.6
2002	7	27	7	4	2	0.4
2003	10	21	7	7	7	0.5
2004	12	41	12	10	46	1.7
2005	12	36	12	9	89	3.2
2006	16	41	16	13	100	4.0
2007	23	38	21	12	162	4.8
2008	45	111	41	17	206	4.7
2009	55	142	49	25	283	4.9
2010	57	171	53	25	412	5.4
2011	75	231	69	27	616	6.1
2012	68	228	60	26	790	7.1
2013	81	297	68	33	1171	8.3
2014	91	364	83	38	1413	9.5
2015	104	404	84	43	1877	10.8
2016	140	538	115	44	2262	11.7
2017	143	619	114	50	3110	13.2
2018	228	963	156	50	3902	14.0
2019	275	1062	186	68	5483	15.1

A, The annual number of total articles; *AU*, the annual number of authors; *C*, the annual number of countries; *J*, the annual number of journals; *TC*, the annual number of citations in cumulative articles; *TC/CA*, annual total citation per cumulative article.

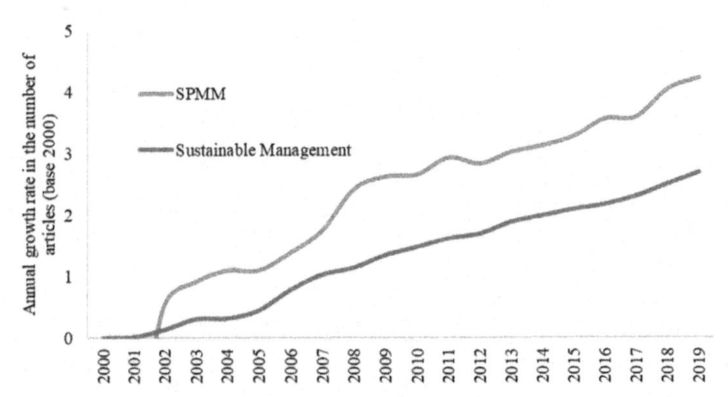

Figure 16.1 Comparative trends in SPMM and Sustainable Management research.

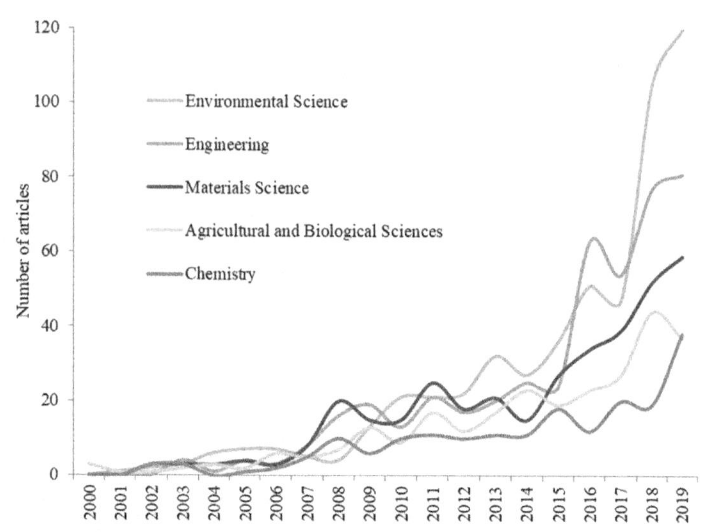

Figure 16.2 Comparative trends of subject categories related to SPMM research.

3.2 Research in SPMM by subject area

Fig. 16.2 shows the evolution of the number of articles published, which have been classified according to the thematic categories established by Scopus. It should be noted that the same study can be classified in more than one category at the same time. The category with the highest number of articles is *Environmental Science*, with 36.6% of the total sample. It is followed by *Engineering* with 31.1%, *Materials Science* with 25.1%, *Agricultural and Biological Sciences* with 18.7%, and *Chemistry* with 13.1%. The rest of the categories are below, representing less than 12% of the total sample studies. The concept of sustainability covers three fundamental areas: environmental, economic, and social. The categories in the economic field (*Economics, Econometrics and Finance*; *and Business, Management, and Accounting*) include 14.1% of the works, while the *Social Sciences* category only has 9.3%. These data show that in research on SPMM, work in the environmental field of sustainability predominates, and makes clear the need to increase the level of knowledge with regard to the other two fields.

3.3 Most relevant journals in SPMM research

Table 16.2 shows the most prolific journals on SPMM in the period 2000–19 and the main characteristics of their articles. As for the countries where these journals are published, three are European (Netherlands, Switzerland, and United Kingdom) and one American (United States). In total, this group of journals has published 261 of the articles in the sample, which represents 18% of the total. In view of these data, we can say that there is a great variety of journals that publish articles on this subject. *Journal of Cleaner Production*, with a total of 64 articles, is the magazine that has

Table 16.2 Major characteristics of the most active journals related to SPMM research.

Journal	A	SJR	H Index	C	TC	TC/A	First A	Last A
Journal of Cleaner Production	64	1.886(Q1)	24	Netherlands	1396	21.8	2010	2019
Sustainability Switzerland	36	0.581(Q2)	9	Switzerland	186	5.2	2015	2019
Resources, Conservation and Recycling	29	2.215(Q1)	18	Netherlands	613	21.1	2005	2019
Waste Management	27	1.634(Q1)	15	United Kingdom	685	25.4	2004	2019
ACS Sustainable Chemistry and Engineering	21	1.766(Q1)	10	United States	380	18.1	2015	2019
Construction and Building Materials	20	1.491(Q1)	12	Netherlands	368	18.4	2012	2019
Waste Management and Research	20	0.650(Q2)	11	United Kingdom	270	13.5	2004	2019
Plastics Engineering	17	0.102(Q4)	2	United States	3	0.2	2009	2019
Horttechnology	14	0.283(Q3)	8	United States	162	11.6	2002	2019
Journal of Applied polymer Science	13	0.541(Q2)	7	United States	120	9.2	2002	2019

A, the annual number of total articles; *C*, country; *First A*, first article of SPMM research by journal; *Last A*, last article; *SJR*, Scimago Journal Ranking; *TC*, the annual number of citations in total articles; *TC/A*, total citation per article.

published more about SPMM. This journal has an H index of 24, a total of 1396 citations, being the largest in the table, and an average of 21.8 citations per article. This journal has a Scimago Journal Rank (SJR) impact factor of 1886 and has been publishing on SPMM since 2010. It is followed by the journal *Sustainability*, which has a total of 36 articles. However, this magazine started publishing on this topic in 2015, 5 years later than the first one. This journal has an H index of 9, a total of 186 citations and an average of 5.2 citations per article, while its SJR impact factor is 0.581. In third place is *Resources, Conservation and Recycling*, with 29 articles. This journal has an H index of 18, a total of 613 citations, and an average of 21.1 citations per article. In addition, this journal has the highest SJR impact factor in the table with 2215. *Waste Management*, which ranks fourth in terms of number of articles with a total of 27, has the highest average number of citations per article with 25.4. In addition, it has the second highest total number of citations with 685, an H index of 15 and an SJR impact factor of 1634. Although the journals *Horttechnology* and *Journal of Applied Polymer Science* have been publishing on SPMM for the longest time (since 2002), they are in ninth and tenth position, with only 14 and 13 articles, respectively.

3.4 Most relevant countries in SPMM research

Table 16.3 shows the most prolific countries on SPMM in the period 2000–19 and the main characteristics of their articles. This group is made up of different countries from all continents except Africa. The United States is the country with the highest number of articles on this subject, with a total of 292. It is also the oldest country, having published since 2000 and still continuing today. It is followed by China with 140 articles, Italy with 106, India with 102, and the United Kingdom with 101. The number

Table 16.3 Major characteristics of the most active countries related to SPMM research.

Country	A	APC	TC	TC/A	H Index	First A	Last A
United States	292	0.890	6556	22.5	39	2000	2019
China	140	0.100	2553	18.2	30	2004	2019
Italy	106	1.758	1538	14.5	26	2004	2019
India	102	0.075	795	7.8	16	2003	2019
United Kingdom	101	1.511	2407	23.8	22	2003	2019
Germany	79	0.950	1065	13.5	18	2006	2019
Australia	77	3.036	1032	13.4	22	2005	2019
Brazil	62	0.294	504	8.1	13	2008	2019
Spain	58	1.232	825	14.2	15	2006	2019
Canada	49	1.304	855	17.4	16	2003	2019

A, the annual number of total articles; *APC*, number of articles per 1mill. inhabitants; *First A*, first article of SPMM research; *Last A*, last article; *TC*, the annual number of citations in total articles; *TC/A*, total citation per article.

of articles per million inhabitants has been calculated to determine the relative weight of each country in the research on SPMM. Based on this variable, the most productive country is Australia with 3036 articles per million inhabitants. It is followed by Italy with 1,758, the United Kingdom with 1,511, Canada with 1,304, and Spain with 1232. However, if we take into account the average number of citations per article, the United Kingdom is in first position with 23.8. Next, we find United States with 22.5, China with 18.2, and Canada with 17.4.

Table 16.4 shows the results of the analysis of the collaborative networks established between the different countries most prolific in SPMM. The average percentage of work done through international collaboration by the group of 10 countries is 37.1%. Canada is the country with the highest percentage of studies carried out through international collaboration, with 51.02%. It is followed by China with 49.29%, the United Kingdom with 48.51%, and Australia with 48.05%. On the opposite side, India with 15.69% and Brazil with 17.74% are the countries with the lowest percentages of international collaboration. Furthermore, the collaboration network has

Table 16.4 Major characteristics in the collaboration of the most active countries related to SPMM research.

Country	IC (%)	NC	Main collaborators	TC/A	
				IC	**NIC**
United States	33.56	45	China, United Kingdom, Australia, Canada, India	22.0	22.7
China	49.29	21	United States, Australia, Canada, Japan, United Kingdom	21.6	15.0
Italy	30.19	22	France, United States, United Kingdom, Spain, Germany	15.3	14.2
India	15.69	15	United States, China, Malaysia, Bangladesh, Brazil	11.4	7.1
United Kingdom	48.51	32	United States, China, Germany, Italy, Netherlands	21.3	26.2
Germany	40.51	21	United Kingdom, China, Finland, Spain, United States	19.9	9.1
Australia	48.05	23	China, United States, Iran, Indonesia, Netherlands	17.2	9.9
Brazil	17.74	12	Canada, Mexico, Australia, Chile, France	9.8	7.8
Spain	36.21	19	Germany, Italy, United Kingdom, United States, Chile	19.8	11.1
Canada	51.02	18	United States, China, Brazil, United Kingdom, Australia	25.7	8.9

IC, international collaborations; *NC*, total number of international collaborators; *NIC*, no international collaborations; *TC/A*, total citation per article.

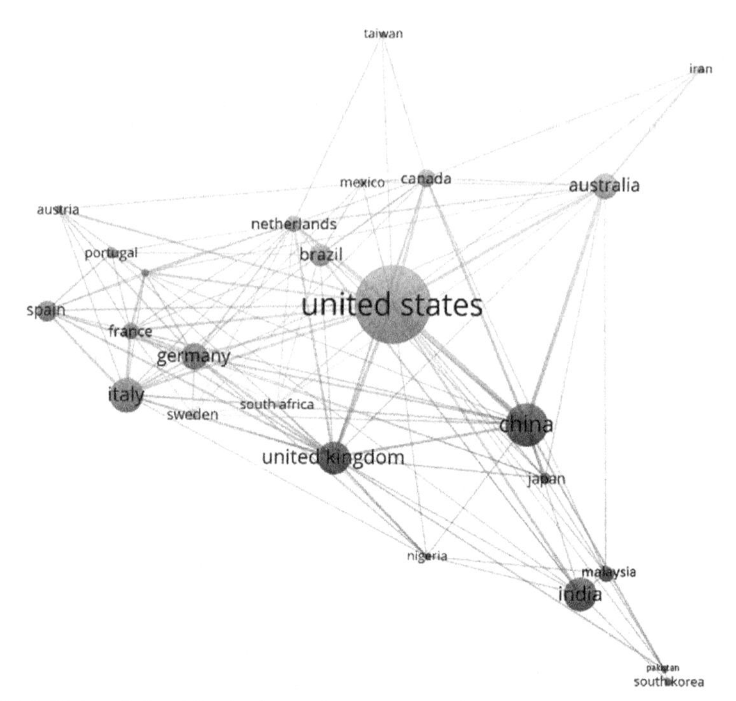

Figure 16.3 Main relationships in the collaboration of the most active countries related to SPMM research.

an average size of 22 countries. The country with the largest collaboration network is the United States, which has 45 different collaborators. It is followed by the United Kingdom with 32, Australia with 23, and Italy with 22. The United States is one of the main partners in all countries except Brazil. In terms of the number of citations, articles made through collaborations have an average of 18.4 citations, while those made without collaboration have an average of 13.2.

The collaborative relationships between the different countries are shown in Fig. 16.3. The size of the circle represents the number of articles from each country and the lines, the relations established between the different countries, the thickness of which depends on the number of contributions. The colors show the groups of collaboration, distinguishing three main clusters. Blue (gray in printed version) is led by the United States in terms of the number of articles and includes Brazil, the Netherlands or Canada among its main collaborators. The red (dark gray in printed version) cluster is led by China and includes the United Kingdom and India among its main collaborators. The green (light gray in printed version) cluster is led by Italy, and also includes Germany, Spain, and France (these colors can be seen in greyscale in the print version).

3.5 Most relevant institutions in SPMM research

Table 16.5 shows the most prolific institutions on SPMM in the period 2000–19 and the main characteristics of their articles. These institutions belong to China, United

Table 16.5 Major characteristics of the most active institutions related to SPMM research.

Institution	C	A	TC	TC/A	H Index	IC (%)	TC/A	
							IC	NIC
Chinese Academy of Sciences	China	20	337	16.9	11	45.0	12.6	20.4
Ministry of Education China	China	17	204	12.0	10	52.9	8.6	15.9
Michigan State University	United States	17	2080	122.4	10	17.7	88.3	129.6
Northwest A&F University	China	16	164	10.3	8	68.8	11.9	6.6
Wageningen University & Research	Netherlands	14	367	26.2	10	64.3	25.9	26.8
Iowa State University	United States	11	100	9.1	5	18.2	32.5	3.9
University of Michigan, Ann Arbor	United States	10	162	16.2	8	40.0	25.8	9.8
University of Western Australia	Australia	10	97	9.7	6	60.0	10.2	9.0
The Royal Institute of Technology KTH	Sweden	9	198	22.0	6	33.3	18.7	0.0
The Ohio State University	United States	9	182	20.2	5	22.2	6.5	24.1
Universidade Estadual Paulista	Brazil	9	57	6.3	5	22.2	3.0	0.0
Università degli Studi di Napoli Federico II	Italy	9	149	16.6	5	22.2	20.5	15.4
Consiglio Nazionale delle Ricerche	Italy	9	230	25.6	5	33.3	43.3	16.7
Universiti Putra Malaysia	Malaysia	9	568	63.1	8	33.3	38.0	75.7

A, the annual number of total articles; *C*, country; *IC*, international collaborations; *NIC*, no international collaborations; *TC*, the annual number of citations in total articles; *TC/A*, total citation per article.

States, Netherlands, Australia, Sweden, Brazil, Italy, and Malaysia. Of these countries, five are among the most prolific (China, United States, Australia, Brazil, and Italy). On the other hand, even though India, United Kingdom, Germany, Spain, and Canada are among the most relevant countries in SPMM research, they do not have any institution among the most prominent. This is due, among other reasons, to the coauthoring relationships established between some institutions and others in the same country. Thus, with a lower number of articles per country, the number of articles from institutions in that country increases.

The group of institutions brings together 11.7% of the articles in the sample and none of them has more than 20 articles. In first place we find the *Chinese Academy of Sciences* with 20 articles, a total of 337 citations, an average of 16.9 citations per article, and an H index of 11. Next, we find the *Ministry of Education China* and *Michigan State University* with 17 articles each. The *Ministry of Education China* has 204 citations in total, an average of 12 citations per article, and an H index of 10. *Michigan State University*, with an H index of 10, is the institution with the highest number of citations, with 2,080, and average citations per article, with 122.4.

With regard to the international collaboration of institutions, the average number of articles carried out through collaboration with other institutions is 38.1%. Only four of them have carried out more than 50% of their published articles on SPMM through international collaboration (*Northwest A&F University*, *Wageningen University and Research*, *University of Western Australia*, and *Ministry of Education China*). The average number of citations in articles written through international collaboration in the group of 14 institutions was 24.7%, and 25.3% for the rest. *Michigan State University* and *Consiglio Nazionale delle Ricerche* are the institutions with the highest number of citations in internationally collaborative articles.

3.6 Most relevant authors in SPMM research

Table 16.6 includes the most productive authors in SPMM research and shows the most outstanding characteristics of their articles. This group of authors is affiliated with 17 different institutions in nine countries. Five of the institutions are from the United States, reaffirming the country's leading position in SPMM research. The number of articles per author is not very high, as only one of the authors has participated in more than five. All these authors began publishing on this subject in 2010, except Manjusri Misra, from the *University of Guelph*, who has published four articles between 2002 and 2016.

Rupinder Singh of the *National Institute of Technical Teachers' Training and Research* is the author with the largest number of articles, with a total of six. However, this author has published the least amount of time on this subject, as his first article was published in 2018. Therefore, he only has a total of 16 citations and an average of 2.7 citations per article. The rest of the authors in the table have five or four articles published on SPMM. If we consider the total number of citations and the average number of citations per article, Manjusri Misra heads the group with 1463 and 365.8, respectively. Also noteworthy is Salit M. Sapuan, from *Universiti Putra* Malaysia, with 375 citations in total and an average of 75 citations per article, and Faris M.

Table 16.6 Major characteristics of the most active authors related to SPMM research.

Author	Articles	Citation	Average citation	H Index	Country	Affiliation	First article	Last article
Singh, Rupinder	6	16	2.7	4	India	National Institute of Technical Teachers' Training and Research	2018	2019
Graves, William R.	5	93	18.6	4	United States	Iowa State University	2013	2016
Grewell, David A.	5	93	18.6	4	United States	North Dakota State University	2013	2016
Koeser, Andrew K.	5	34	6.8	4	United States	Gulf Coast Research and Education Center	2014	2015
McCabe, Kenneth G.	5	93	18.6	4	United States	Iowa State University	2013	2016
Sahajwalla, Veena	5	31	6.2	4	Australia	University of New South Wales	2017	2018
Sapuan, Salit M.	5	375	75.0	5	Malaysia	Universiti Putra Malaysia	2014	2016
Schrader, James A.	5	93	18.6	4	United States	Iowa State University	2013	2016
AL-Oqla, Faris M.	4	354	88.5	4	Jordan	Hashemite University	2014	2016
Bing, Xiaoyun	4	150	37.5	4	Netherlands	Wageningen University and Research	2013	2015

Continued

Table 16.6 Major characteristics of the most active authors related to SPMM research.—cont'd

Author	Articles	Citation	Average citation	H Index	Country	Affiliation	First article	Last article
Bonanomi, Giuliano	4	132	33.0	4	Italy	Università degli Studi di Napoli Federico II	2011	2017
Chai, Qiang	4	39	9.8	3	China	Gansu Agricultural University	2016	2018
Ingrao, Carlo	4	121	30.3	4	Italy	Università degli Studi di Enna "Kore"	2014	2018
Mbohwa, Charles S.	4	102	25.5	3	South Africa	University of Johannesburg	2014	2017
Misra, Manjusri	4	1463	365.8	4	Canada	University of Guelph	2002	2016
Niu, Genhua	4	30	7.5	3	United States	Texas A&M AgriLife Research	2015	2015
Yang, Yiqi Q.	4	46	11.5	3	United States	University of Nebraska–Lincoln	2010	2018
Zhang, Liqun	4	89	22.3	4	China	Beijing University of Chemical Technology	2015	2016
Zhao, Cai	4	39	9.8	3	China	Gansu Provincial Key Laboratory of Aridland Crop Science	2016	2018

AL-Oqla, from *Hashemite University*, with 354 and 88.5, respectively. Salit M. Sapuan also has the highest H index in the group, with 5.

Of the group of authors, six (William R. Graves, Kenneth G. McCabe, Salit M. Sapuan, James A. Schrader, Xiaoyun Bing, and Giuliano Bonanomi) are affiliated with some of the most prolific institutions (Table 16.5). Of particular note are Faris M. AL-Oqla and Charles S. Mbohwa who are among the most prolific authors but whose institutions and countries of affiliation are not among the most prominent. Looking at the collaborative relationships between the authors included in the table, it is notable that four of them (William R. Graves, David A. Grewell, Kenneth G. McCabe, and James A. Schrader) have worked together on all their published articles on SPMM, as they all have five articles, a total of 93 citations and an average of 18.6 citations per article. In addition, three of them are affiliated with *Iowa State University*, while one belongs to *North Dakota State University*, both institutions in the United States. In addition, Salit M. Sapuan and AL-Oqla, Faris M. have collaborated on four of their articles, Qiang Chai and Cai Zhao on another four, and Andrew K. Koeser and Genhua Niu share three articles. None of these authors belong to the same institution but, in the last two cases, their institutions do belong to the same country (China and the United States).

3.7 Keywords analysis in SPMM research

Fig. 6.4 shows the network map of keywords, where the size of the circle varies according to the number of times it has been used and the color represents the group in which the keyword is included, taking into account the number of coappearances. We can observe four different clusters representing the different priority lines of research on SPMM.

The red (dark gray in printed version) cluster has the environment as its central focus and addresses the impact of the use of plastics on the environment. As already mentioned, the pollution of the oceans is of particular concern, so it is to be expected that a broad line of research has been developed in relation to this issue. The results of Ostle et al. (2019) confirm that there has been a significant increase in plastics in the oceans from 1957 to 2016, reaching the North Atlantic and Arctic. Beaumont et al. (2019) show that increases in the amount of plastics in the oceans can have negative impacts on the ecosystem services they provide, particularly in relation to fisheries and aquaculture. Thus, Villarrubia-Gómez et al. (2017) concluded that plastic pollution of marine ecosystems threatens planetary limits due to their irreversibility and global scope. However, there is still much uncertainty regarding the impacts that can be derived from pollution by plastics and microplastics in the oceans, being especially relevant to know the effects that these can generate on human health. In this cluster we also find studies on the environmental impact of new materials from organic materials (Burgueño et al., 2005; Dekiff et al., 2014; Johnston et al., 2018; Kothekar et al., 2018; Schmidt et al., 2019). Spierling et al. (2018), for example, concluded that the use of bio-based plastics could potentially save between 241 and 316 million tons of CO_2 per year. However, more knowledge is still needed about the impact of plastics, how to

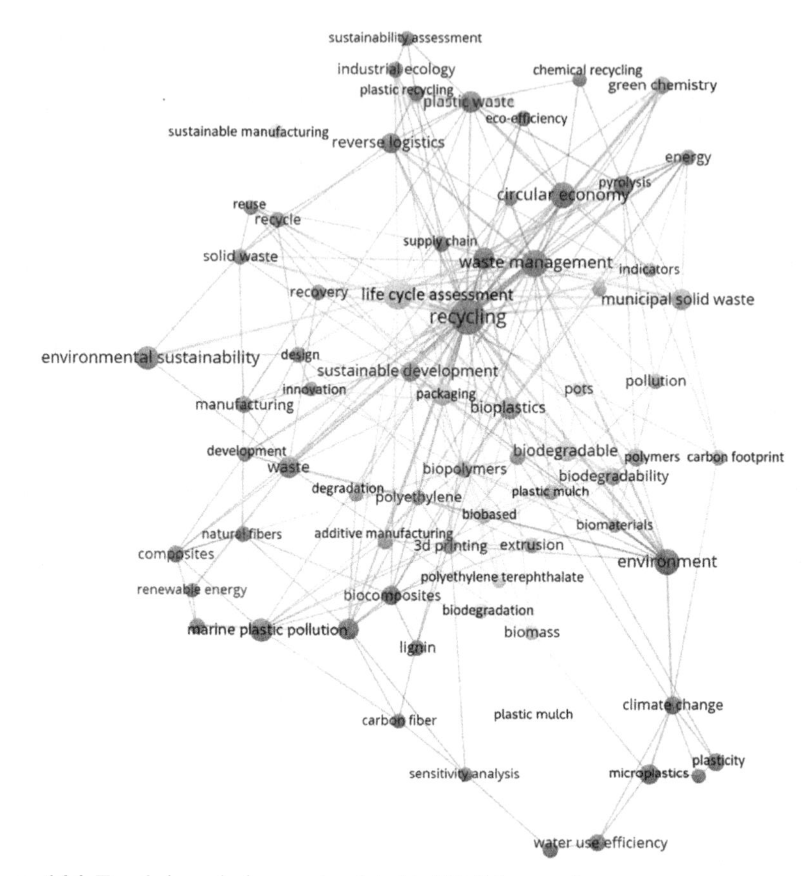

Figure 16.4 Trends in main keywords related to SPMM research.

remedy it, and to what extent new materials are a sustainable alternative for the environment.

The green (gray in printed version) cluster is about the reincorporation of plastic materials into the production process through recycling, reuse, and recovery. Economic growth has been based on a linear model of "use–consumption–disposal" which requires intensive consumption of resources and generates large amounts of waste (Wool et al., 2002; Hahladakis and Iacovidou, 2019). Ensuring sustainability in the future requires replacing this linear model with one based on circularity, in which materials are kept as long as possible in the production cycle. For this reason, a line of research has been developed in relation to techniques for recycling and reusing plastics (Myllytie et al., 2016; Macedo et al., 2018; Dong et al., 2019). Achilias et al. (2018) analyzed two techniques, one mechanical and the other chemical, to recycle polyethylene and polypropylene, and concluded that both processes allow the recovery of at least 90% of the polymers. Dos Santos and Jayaraman (2019) studied the possibility of recycling cross-linkable high-density polyethylene through compression molding, since this material, which is used to manufacture large and rigid products, is not currently recycled. Palombini et al. (2018) analyzed the recovery capacity of small plastic waste on a Brazilian beach and the results showed that around 5% of the concentration mixture of recycled material has mechanical properties suitable for use in new parts. Despite the development of this line of research and efforts in this field, our world is currently only 9% circular (Circle Economy, 2019). This highlights the need to continue making progress in the development of plastic waste management plans, as well as establishing international commitments to guide their implementation.

The purple (mild gray in printed version) cluster focuses on reducing pollution caused by traditional, petroleum-based plastics by using alternatives based on the use of synthetic or organic materials. Brockhaus et al. (2016) conducted a study with product developers in the consumer goods industry, concluding that they are prepared to expand the use of bioplastics in their products but, to do so, it is necessary to ensure the supply of viable materials produced in a sustainable way and that can be recycled. Keziah et al. (2018) used a mixture of glycerol, corn starch, vinegar, and food coloring, while Sharmila et al. (2019) analyzed gelatine, agar, and starch as a basis for the development of new materials. Kamarun et al. (2019) studied the use of natural fibers such as kenaf, sugar palm, or pineapple leaves in the plastic compound industry and concluded that the composition and thermal stability of these fibers are suitable for this purpose. However, further research is still needed in this area to find sustainable materials that have similar characteristics to those commonly used and fulfill their function adequately. It is therefore important to analyze the feasibility of the use of these alternative materials from an economic point of view, as implementing new alternatives may involve very high costs, which may act as a constraint.

In line with the previous cluster, we find that the yellow (light gray in printed version) cluster focuses on the study of the different phases of the value chain through life cycle analysis, with special emphasis on packaging. Most of the packaging of the products which are consumed on a regular basis are made of plastic. It is estimated that, only in the European Union, the packaging sector is the largest generator of plastic waste, with 59% of the total (European Commission, 2018). Therefore, in order to reduce the generation of plastic waste it will be necessary to evaluate the life cycle of the packaging, determining the negative impacts generated in each stage and trying to reduce them, as well as the replacement of traditional plastics by other biodegradable or compostable ones. Tua et al. (2019) analyzed the environmental effectiveness of reusable plastic boxes used for the distribution of fruit and vegetables, concluding that at least three reuses are required for their performance to be greater than that of the single-use box system. Song et al. (2009) concluded that biodegradable packaging is the best alternative for single-use disposable applications, as these materials can be composted. Prathipa et al. (2018) conducted a study on the types of bioplastics suitable for packaging, including polysaccharides, starch, cellulose, polylactic acid, and polyhydroxyalkanoates. In relation to packaging, it is necessary to carry out studies that consider consumers, analyze their characteristics and preferences, as well as how to increase their awareness. Consumers are the final step in the chain and companies aim to satisfy their needs, so if they are proactive in buying recycled products, even if they are more expensive, companies would increase their efforts in this sense. Thus, consumers also play a key role in ensuring that plastic waste is recycled.

4. Conclusions

The aim of this study was to carry out an analysis of the sustainable use of plastic materials worldwide. The results show that the number of articles has followed an increasing trend, particularly in the last 5 years. Furthermore, a comparison of the annual growth rates in the number of articles published on the sustainable management of plastic materials and sustainable management in general has shown that this is an increasingly relevant line of research in this area.

The use of plastic materials is widespread due to their characteristics and multiple uses. This has generated numerous environmental problems derived from the incorrect management of plastic waste at the end of its useful life. The growing concern in this respect has led to extensive research being carried out on this issue. The most promising alternatives are related to increasing the recycling and reuse of plastic materials based on circular economy models, so that they remain in the production process as long as possible. In addition, the replacement of traditional plastics by biodegradable and compostable ones is another relevant alternative, whose implementation could be complementary to the previous one. The implementation of actions aimed at reducing the use of plastic materials requires pilot projects to be carried out at different scales to determine the viability, both technical and economic, of the proposed alternatives. However, the development of these projects may involve a high cost, both in terms of money and time. In this respect, it is also necessary to extend the partnerships at all levels, so that the various research projects in this field can feed back into each other.

Therefore, a key point in moving toward the sustainable management of plastic materials is to raise the general level of awareness in society. To this end, it will be necessary to carry out studies to find out the attitudes and perceptions of the different population groups, to facilitate the implementation of specific action plans aimed at them. On the other hand, the development of agreements at international level, which lay the foundations for conversion toward a more sustainable model of management of plastic materials from both a theoretical and practical point of view, and taking into account society as a whole, would represent a turning point. Finally, research on sustainable plastics management must be developed using a holistic approach that encompasses the economic, social, and environmental fields.

Acknowledgments

This study has been partially supported by the Spanish Ministry of Economy and Competitiveness and the European Regional Development Fund by means of the research project ECO2017-82347-P, by the Research Plan of the University of Almería through a Postdoctoral Contract to Juan F. Velasco Muñoz, and by the Research Plan of the University of Almería through a Gerty Cori Predoctoral Contract to Belén López Felices.

References

Achilias, D., Antonakou, E., Roupakias, C., Megalokonomos, P., Lappas, A., 2018. Recycling techniques of polyolefins from plastic wastes. Glob. NEST J. 10 (1), 114–122. https://doi.org/10.30955/gnj.000468.

Agarwal, S., 2020. Biodegradable polymers: present opportunities and challenges in providing a microplastic-free environment. Macromol. Chem. Phys. 221 (6), 2000017. https://doi.org/10.1002/macp.202000017.

Al-Thawadi, S., 2020. Microplastics and nanoplastics in aquatic environments: challenges and threats to aquatic organisms. Arabian J. Sci. Eng. 45 (6), 4419–4440. https://doi.org/10.1007/s13369-020-04402-z.

Alam, O., Billah, M., Yajie, D., 2018. Characteristics of plastic bags and their potential environmental hazards. Resour. Conserv. Recycl. 132, 121–129. https://doi.org/10.1016/j.resconrec.2018.01.037.

Allen, S., Allen, D., Phoenix, V.R., Le Roux, G., Durántez-Jiménez, P., Simonneau, A., Binet, S., Galop, D., 2019. Atmospheric transport and deposition of microplastics in a remote mountain catchment. Nat. Geosci. 12, 339–344. https://doi.org/10.1038/s41561-019-0335-5.

Alonso-Amigo, M.G., 2000. Polymer microfabrication for microarrays, microreactors and microfluidics. JALA J. Assoc. Lab. Automation 5 (6), 96–101.

Andrady, A.L., Neal, M.A., 2009. Applications and societal benefits of plastics. Applications and societal benefits of plastics. Phil. Trans. R. Soc. B 364, 1977–1984. https://doi.org/10.1098/rstb.2008.0304.

Aznar-Sánchez, J.A., Velasco-Muñoz, J.F., Belmonte-Ureña, L.J., Manzano-Agugliaro, F., 2019. Innovation and technology for sustainable mining activity: a worldwide research assessment. J. Clean. Prod. 221, 38–54. https://doi.org/10.1016/j.jclepro.2019.02.243.

Beaumont, N.J., Aanesen, M., Austen, M.C., Börger, T., Clark, J.R., Cole, M., Hooper, T., Lindeque, P.K., Pascoe, C., Wyles, K.J., 2019. Global ecological, social and economic impacts of marine plastic. Mar. Pollut. Bull. 142, 189–195. https://doi.org/10.1016/j.marpolbul.2019.03.022.

Bergmann, M., Mützel, S., Primpke, S., Tekman, M.B., Trachsel, J., Gerdts, G., 2019. White and wonderful? Microplastics prevail in snow from the Alps to the Arctic. Sci. Adv. 5 (8), eaax1157. https://doi.org/10.1126/sciadv.aax1157.

Blettler, M.C.M., Ulla, M.A., Rabuffetti, A.P., Garello, N., 2017. Plastic pollution in freshwater ecosystems: macro-, meso-, and microplastic debris in a floodplain lake. Environ. Monit. Assess. 189 (11), 581. https://doi.org/10.1007/s10661-017-6305-8.

Blettler, M.C.M., Abrial, E., Khan, F.R., Sivri, N., Espinola, L.A., 2018. Freshwater plastic pollution: recognizing research biases and identifying knowledge gaps. Water Res. 143, 416–424. https://doi.org/10.1016/j.watres.2018.06.015.

Boucher, J., Froit, D., 2017. Primary Microplastics in the Oceans. A Global Evaluation of Sources. IUCN, Gland, Switzerland, p. 43. https://doi.org/10.2305/IUCN.CH.2017.01.en.

Brockhaus, S., Petersen, M., Kersten, W., 2016. A crossroads for bioplastics: exploring product developers' challenges to move beyond petroleum-based plastics. J. Clean. Prod. 127, 84–95. https://doi.org/10.1016/j.jclepro.2016.04.003.

Burgueño, R., Quagliata, M.J., Mehta, G.M., Mohanty, A.K., Misra, M., Drzal, L.T., 2005. Sustainable cellular biocomposites from natural fibers and unsaturated polyester resin for housing panel applications. J. Polym. Environ. 13 (2), 139–149. https://doi.org/10.1007/s10924-005-2945-9.

Carlin, C., 2019. Tokyo pack introduces innovative plastic packaging ideas from asian exhibitors. Plast. Eng. 75 (1), 8–13. https://doi.org/10.1002/peng.20052.

Chauhan, V., Varis, J., Kärki, T., 2019. The potential of reusing technical plastics. Proc. Manuf. 39, 502–508. https://doi.org/10.1016/j.promfg.2020.01.407.

Circle Economy, 2019. The Circularity Gap Report (Accesed 07/29/2020). https://bfc732f7-80e9-4ba1-b429-7f76cf51627b.filesusr.com/ugd/ad6e59_ba1e4d16c64f44fa94fb-d8708eae8e34.pdf.

Cossarini, D.M., MacDonald, B.H., Wells, P.G., 2014. Communicating marine environmental information to decision makers: enablers and barriers to use of publications (grey literature) of the Gulf of Maine Council on the Marine Environment. Ocean Coast Manag. 96, 163–172. https://doi.org/10.1016/j.ocecoaman.2014.05.015.

De Jesus, R.M., Pelaez, E.B., Caneca, M.C., 2018. Experimental study on mechanical behaviour of concrete beams with shredded plastics. Int. J. GEOMATE 14 (42), 71–75. https://doi.org/10.21660/2018.42.7172.

Dekiff, J.H., Remy, D., Klasmeier, J., Fries, E., 2014. Occurrence and spatial distribution of microplastics in sediments from Norderney. Environ. Pollut. 186, 248–256. https://doi.org/10.1016/j.envpol.2013.11.019.

Dong, S.P., Yuan, F., Yang, L.J., Chi, S., Zhong, J.H., Lei, J.X., Bao, L.X., Wang, J.L., 2019. Clean and sustainable biocomposites based on supramolecular interactions induced thermoplasticization of wheat straw powers. J. Clean. Prod. 233, 590–600. https://doi.org/10.1016/j.jclepro.2019.06.101.

Dos Santos, J.N., Jayaraman, R., 2019. Recycling of crosslinked high-density polyethylene through compression molding. J. Appl. Polym. Sci. 136 (43), 48145. https://doi.org/10.1002/app.48145.

Dow, P., Kotz, K., Gruszka, S., Holder, J., Fiering, J., 2018. Acoustic separation in plastic microfluidics for rapid detection of bacteria in blood using engineered bacteriophage. Lab Chip 18 (6), 923–932. https://doi.org/10.1039/C7LC01180F.

European Commission, 2018. Official Journal of the European Union C 433/136. European Parliament Resolution of 13 September 2018 on a European Strategy for Plastics in a Circular Economy (2018/2035(INI)). https://eur-lex.europa.eu/legal-content/EN/TXT/?uri=uriserv:OJ.C_.2019.433.01.0136.01.ENG&toc=OJ:C:2019:433:FULL.

Faraj, R.H., Ali, H.F.H., Sherwani, A.F.H., Hassan, B.R., Karim, H., 2020. Use of recycled plastic in self-compacting concrete: a comprehensive review on fresh and mechanical properties. J. Build. Eng. 30, 101283. https://doi.org/10.1016/j.jobe.2020.101283.

Fojt, J., David, J., Přikryl, R., Řezáčová, V., Kučerík, J., 2020. A critical review of the overlooked challenge of determining micro-bioplastics in soil. Sci. Total Environ. 745, 140975. https://doi.org/10.1016/j.scitotenv.2020.140975.

Gu, F., Guo, J., Zhang, W., Summers, P.A., Hall, P., 2017. From waste plastics to industrial raw materials: a life cycle assessment of mechanical plastic recycling practice based on a real-world case study. Sci. Total Environ. 601–602, 1192–1207. https://doi.org/10.1016/j.scitotenv.2017.05.278.

Hahladakis, J.N., Iacovidou, E., 2019. An overview of the challenges and trade-offs in closing the loop of post-consumer plastic waste (PCPW): focus on recycling. J. Hazard Mater. 380, 120887. https://doi.org/10.1016/j.jhazmat.2019.120887.

Iro, Z.S., Subramani, C., Dash, S.S., 2019. Measurement of electromagnetic properties of plastics and composites using rectangular waveguide. Adv. Intell. Syst. Comput. 846, 381–391. https://doi.org/10.1007/978-981-13-2182-5_36.

Jambeck, J.R., Geyer, R., Wilcox, C., Siegler, T.R., Perryman, M., Andrady, A., Narayan, R., Law, K.L., 2015. Plastic waste inputs from land into the ocean. Science 347 (6223), 768–771. https://doi.org/10.1126/science.1260352.

Johnston, B., Radecka, I., Hill, D., Chiellini, E., Ilieva, V.I., Sikorska, W., Musioł, M., Zieba, M., Marek, A.A., Keddie, D., Mendrek, B., Darbar, S., Adamus, G., Kowalczuk, M., 2018. The microbial production of Polyhydroxyalkanoates from Waste polystyrene fragments attained using oxidative degradation. Polymers 10 (9), 957. https://doi.org/10.3390/polym10090957.

Kamarun, D., Zawawi, E.Z.E., Seth, N.H., Ahmad, N., Karim, S.R.A., 2019. Effect of natural fiber (NF) mix on mechanical strength of NF plastic composites (NFPC). Int. J. Eng. Adv. Technol. 9 (1), 4552–5667. https://doi.org/10.35940/ijeat.A3043.109119.

Keziah, V.S., Gayathri, R., Priya, V.V., 2018. Biodegradable plastic production from corn starch. Drug Invent. Today 10 (7), 1315–1317.

Khamtree, S., Ratanawilai, T., Ratanawilai, S., 2020. The effect of alkaline–silane treatment of rubberwood flour for water absorption and mechanical properties of plastic composites. J. Thermoplast. Compos. Mater. 33 (5), 599–613. https://doi.org/10.1177/0892705718808556.

Klengel, S., Böttge, B., Naumann, F., Mittag, M., Hirsch, U., Ehrich, C., 2019. Microstructure and Thermophysical Characterization of Innovative Plastic Materials for Power Electronics. CIPS 2016 - 9th International Conference on Integrated Power Electronics Systems, ISBN 978-380074171-7.

Klingelhöfer, D., Braun, M., Quarcoo, D., Brüggmann, D., Groneberg, D.A., 2020. Research landscape of a global environmental challenge: Microplastics. Water Res. 170, 115358. https://doi.org/10.1016/j.watres.2019.115358.

Kothekar, S., Shukla, S., Suneetha, V., 2018. A brief study on starch based bio-plastics produced from staple food items. Res. J. Pharm. Technol. 11 (11), 4878–4883. https://doi.org/10.5958/0974-360X.2018.00888.0.

Kumar, S.P., Suman, K.N.S., Ramanjaneyulu, S., 2019. A review on tribological performance characteristics of plastic gears. Int. J. Mech. Eng. Technol. 1, 516–526.

Li, P., Wang, X., Su, M., Zou, X., Duan, L., Zhang, H., 2020. Characteristics of plastic pollution in the environment: a review. Bull. Environ. Contam. Toxicol. 12 March 2020. https://doi.org/10.1007/s00128-020-02820-1.

Macedo, K.R.M., Cestari, S.P., Mendes, L.C., Albitres, G.A.V., Rodrigues, D.C., 2018. Sustainable hybrid composites of recycled polypropylene and construction debris. J. Compos. Mater. 52 (21), 2949–2959. https://doi.org/10.1177/0021998318758367.

Maity, S., Pramanick, K., 2020. Perspectives and challenges of micro/nanoplastics-induced toxicity with special reference to phytotoxicity. Global Change Biol. 26 (6), 3241–3250. https://doi.org/10.1111/gcb.15074.

Mošorinski, P., Prvulovic, S., Josimovic, L., 2020. Determination of the optimal cutting parameters for machining technical plastics. Mater. Tehnol. 54 (1), 11–15. https://doi.org/10.17222/mit.2019.079.

Myllytie, P., Misra, M., Mohanty, A.K., 2016. Carbonized lignin as sustainable filler in biobased poly (trimethylene terephthalate) polymer for injection molding applications. ACS Sustain. Chem. Eng. 4 (1), 102–110. https://doi.org/10.1021/acssuschemeng.5b00796.

Ostle, C., Thompson, R.C., Broughton, D., Gregory, L., Wootton, M., Johns, D.G., 2019. The rise in ocean plastics evidenced from a 60-year time series. Nat. Commun. 10 (1622), 1–6. https://doi.org/10.1038/s41467-019-09506-1.

Palombini, F.L., Demori, R., Cidade, M.K., Kindlein Jr., W., de Jacques, J.J., 2018. Occurrence and recovery of small-sized plastic debris from a Brazilian beach: characterization, recycling, and mechanical analysis. Environ. Sci. Pollut. Res. 25 (26), 26218–26227. https://doi.org/10.1007/s11356-018-2678-7.

PlasticsEurope, 2019. An Analysis of European Latest Plastics Production, Demand and Waste Data. Plastics-The Facts (Accessed 07/29/2020). www.plasticseurope.org.

PlasticsEurope, 2020. Types of Plastics (Accessed 07/29/2020). www.plasticseurope.org.

Prathipa, R., Sivakumar, C., Shanmugasundaram, B., 2018. Biodegradable polymers for sustainable packaging applications. Int. J. Mech. Eng. Technol. 9 (6), 293–303. http://www.iaeme.com/ijmet/issues.asp?JType=IJMET&VType=9&IType=6.

Poonam, K., Rajababu, V., Yogeshwari, J., Patel, H., 2013. Diversity of plastic degrading microorganisms and their appraisal on biodegradable plastic. Appl. Ecol. Environ. Res. 11 (3), 441–449. https://doi.org/10.15666/aeer/1103_441449.

Qamar, A.Z., Asefifeyzabadi, N., Taki, M., Shamsi, M.H., Naphade, S., Ellerby, L.M., 2020. Characterization and application of fluidic properties of trinucleotide repeat sequences by wax-on-plastic microfluidics. J. Mater. Chem. B 8 (4), 743–751. https://doi.org/10.1039/c9tb02208b.

Raab-Obermayr, S., Kossik, A., Gastermann, B., Stopper, M., 2017. Sustainable life cycle management in the thermoset plastics processing industry. Lect. Notes Eng. Comput. Sci. 2228, 862–869.

Santos, R.G., Andrades, R., Fardim, L.M., Martins, A.S., 2016. Marine debris ingestion and Thayer's law – the importance of plastic color. Environ. Pollut. 214, 585–588. https://doi.org/10.1016/j.envpol.2016.04.024.

Sharmila, S., Suganya Rajeswari, E., Singh, N., Kamalambigeswari, R., Kowsalya, E., 2019. Production of bioplastic from biowaste materials and its SEM-EDS report. Int. J. Recent Technol. Eng. 8 (3), 8679–8682. https://doi.org/10.35940/ijrte.C6463.098319.

Shieh, P., Zhang, W., Husted, K.E.L., Kristufek, S.L., Xiong, B., Lundberg, D.J., Lem, J., Veysset, D., Sun, Y., Nelson, K.A., Plata, D.L., Johnson, J.A., 2020. Cleavable comonomers enable degradable, recyclable thermoset plastics. Nature 583 (7817), 542–547. https://doi.org/10.1038/s41586-020-2495-2.

Schmidt, N., Fauvelle, V., Ody, A., Castro-Jiménez, J., Jouanno, J., Changeux, T., Thibaut, T., Sempéré, R., 2019. The amazon river: a major source of organic plastic additives to the tropical North Atlantic? Environ. Sci. Technol. 53 (13), 7513–7521. https://doi.org/10.1021/acs.est.9b01585.

Song, J.H., Murphy, R.J., Narayan, R., Davies, G.B.H., 2009. Biodegradable and compostable alternatives to conventional plastics. Philos. Trans. R. Soc. 364 (1526), 2127–2139. https://doi.org/10.1098/rstb.2008.0289.

Spierling, S., Knüpffer, E., Behnsen, H., Mudersbach, M., Krieg, H., Springer, S., Albrecht, S., Herrmann, C., Endres, H.J., 2018. Bio-based plastics—a review of environmental, social and economic impact assessments. J. Clean. Prod. 185, 476–491. https://doi.org/10.1016/j.jclepro.2018.03.014.

Thompson, R.C., Moore, C.J., vom Saal, F.S., Swan, S.H., 2009. Plastics, the environment and human health: current consensus and future trends. Phil. Trans. R. Soc. B 364 (1526), 2153–2166. https://doi.org/10.1098/rstb.2009.0053.

Tua, C., Biganzoli, L., Grosso, M., Rigamonti, L., 2019. Life cycle assessment of reusable plastic crates (RPCs). Resources 8 (2), 110. https://doi.org/10.3390/resources8020110.

Tulashie, S.K., Boadu, E.K., Kotoka, F., Mensah, D., 2020. Plastic wastes to pavement blocks: a significant alternative way to reducing plastic wastes generation and accumulation in Ghana. Construct. Build. Mater. 241, 118044. https://doi.org/10.1016/j.conbuildmat.2020.118044.

Van Wezel, A., Caris, I., Kools, S.A.E., 2016. Release of primary microplastics from consumer products to wastewater in The Netherlands. Environ. Toxicol. Chem. 35 (7), 1627–1631. https://doi.org/10.1002/etc.3316.

Van Wijnen, J., Ragas, A.M.J., Kroeze, C., 2019. Modelling global river export of microplastics to the marine environment: sources and future trends. Sci. Total Environ. 673, 392–401. https://doi.org/10.1016/j.scitotenv.2019.04.078.

Velasco-Muñoz, J.F., Aznar-Sánchez, J.A., Belmonte-Ureña, L.J., López-Serrano, M.J., 2018. Advances in water use efficiency in agriculture: a bibliometric analysis. Water 10 (4), 377. https://doi.org/10.3390/w10040377.

Villarrubia-Gómez, P., Cornell, S.E., Fabres, J., 2017. Marine plastic pollution as a planetary boundary threat – the drifting piece in the sustainability puzzle. Mar. Pol. 96, 213–220. https://doi.org/10.1016/j.marpol.2017.11.035.

Wool, R.P., Khot, S.N., LaScala, J.J., Bunker, S.P., Lu, J., Thielemans, W., Can, E., Morye, S.S., Williams, G.I., 2002. Affordable composites and plastics from renewable resources: part I: synthesis of monomers and polymers. In: Lankey, R.L., Anastas, P.T. (Eds.), Advancing Sustainability through Green Chemistry and Engineering, ACS Symposium Series, vol. 823. American Chemical Society, Washington, DC, pp. 177–204. https://doi.org/10.1021/bk-2002-0823.ch013.

Yu, Y., Zhou, D., Li, Z., Zhu, C., 2018. Advancement and challenges of microplastic pollution in the aquatic environment: a review. Water Air Soil Pollut. 229, 140. https://doi.org/10.1007/s11270-018-3788-z.

Zhu, F., Zhu, C., Wang, C., Gu, C., 2019. Occurrence and ecological impacts of microplastics in soil systems: a review. Bull. Environ. Contam. Toxicol. 102 (6), 741–749. https://doi.org/10.1007/s00128-019-02623-z.

Websites about topic

Sustainable Plastics. Global Polymer Group. https://www.sustainableplastics.com/.

PlasticsEurope. Association of Plastics Manufacturers. https://www.plasticseurope.org/en.

Plastics Technology. https://www.plastics-technology.com/articles/top-largest-plastic-producing-companies.

Plastics Materials. https://www.plastic-materials.com/.

Plastics Recyclers Europe https://www.plasticsrecyclers.eu/.

The North American Plastics Recyclers Alliance (NAPRA). https://www.plasticsrecyclingalliance.org/.

Other relevant contributions

Hussain, C.M., Mishra, A.K., 2019. Nanotechnology in Environmental Science, 2 Volumes. John Wiley & Sons.

Hussain, C.M., Keçili, R., 2019. Modern Environmental Analysis Techniques for Pollutants, first ed. Elsevier.

Hussain, C.M., 2019. Handbook of Environmental Materials Management, first ed. Springer International Publishing.

Hussain, C.M., March 25, 2020. The Handbook of Environmental Remediation: Classic and Modern Techniques, first ed. Royal Society of Chemistry.

Innovation for new resources development

Milan Majerník [1], *Lucia Bednárová* [2], *Marcela Malindžáková* [2], *Peter Drábik* [1], *Jana Naščáková* [3]

[1]University of Economics in Bratislava, Research Institute of Trade and Sustainable Business, Bratislava, Slovak Republic; [2]Technical University of Košice, Faculty of Mining, Ecology, Process Control and Geotechnologies, Košice, Slovak Republic; [3]University of Economics in Bratislava, Faculty of Business Economy of the University of Economics in Bratislava with the Seat in Košice, Košice, Slovak Republic

1. Introduction

This chapter is devoted to the issue of innovative resources that could significantly contribute to reducing the use of natural resources and critical raw materials in the future.

The use of renewable energy sources brings several positive benefits to society. It increases the security and diversification of energy supplies while reducing the economy's dependence on volatile oil and gas prices. The use of renewable energy sources contributes to reducing greenhouse gas emissions and pollutants. Renewable energy sources contribute to strengthening and diversifying the structure of industry and agriculture (Bednárová et al., 2013). By using them, the principles of sustainable development are fulfilled, which makes them one of the pillars of healthy economic development of society. Increased use of renewable energy sources has an impact on improving the health of the population as well as the environment. In addition, the correct location of renewable energy sources can become a key element in the development of individual regions, which can contribute to achieving better social and economic cohesion in the country. Renewable energy can be produced from several natural sources such as wind, sun, water, geothermal springs, or biomass. Energy from these sources is clean, safe, and inexhaustible. The use of alternative sources is becoming part of all realistic scenarios in support of sustainable development and energy security in the world. Renewable energy sources form a relatively significant part of primary energy sources in many countries and, thanks to the undeniable advantages of their use, they make up an increasing share of total energy production. Alternative fuel is the only solution for mankind's current fuel dilemma (Bednárová and Witek, 2016). And both private and government organizations are throwing millions of dollars at scientists with ideas on how to develop the next "Free and Clean" fuel technology.

Sustainable Resource Management. https://doi.org/10.1016/B978-0-12-824342-8.00005-5

Given the current need to achieve the sustainability of natural resources for future generations, humanity needs to find a way that is both safe and sustainable for it. Recent research has shown alternatives to the use of various organic and inorganic elements. We currently consider very promising developments: hydrogen technologies, microalgae, nanotechnologies, biofuels, heat pump ice storage, biomaterials, etc.

2. Hydrogen

Hydrogen is a high efficiency, low polluting fuel that can be used for transportation, heating, and power generation in places where it is difficult to use electricity or as a CO_2 neutral feedstock for chemical processes. Robert Boyle produced hydrogen gas in 1671 while he was experimenting with iron and acids, but it was not until 1766 that Henry Cavendish recognized it as a distinct element. The element was named hydrogen by the French chemist Antoine Lavoisier. Hydrogen does not exist naturally on Earth. Since it forms covalent compounds with most nonmetallic elements, most of the hydrogen on Earth exists in molecular forms such as water or organic compounds. Combined with oxygen, it is water (H_2O). Combined with carbon, it forms methane (CH_4), coal, and petroleum. It is found in all growing things (biomass).

Hydrogen has the highest energy content of any common fuel by weight, but the lowest energy content by volume.

The number of countries with polices that directly support investment in hydrogen technologies is increasing, along with the number of sectors they target. There are around 50 targets, mandates, and policy incentives in place today that direct support hydrogen, with the majority focused on transport.

It is a high efficiency, low polluting fuel that can be used for transportation, heating, and power generation in places where it is difficult to use electricity. Once hydrogen is produced as molecular hydrogen, the energy present within the molecule can be released, by reacting with oxygen to produce water. This can be achieved by either traditional internal combustion engines or by devices called fuel cells.

The most important primary energy source for hydrogen production currently is natural gas, at 70%, followed by oil, coal, and electricity (as a secondary energy resource). Steam reforming (from natural gas) is the most commonly used method for hydrogen production. To date, only small amounts of hydrogen have been generated from renewable energies, although that amount is set to increase in future. Electrolysis currently accounts for around 5% of global hydrogen production. If hydrogen is extracted from water using a machine called an electrolyzer, which uses an electric current to split H_2O into its constituent parts and renewable or carbon free electricity is used, the gas has a zero-carbon footprint, and is known as green hydrogen.

In daily life, when we refer to hydrogen, we actually refer to H_2 or dihydrogen, the molecule made of two atoms of hydrogen usually in a gaseous form.

2.1 Hydrogen technologies

Hydrogen can be produced from diverse, domestic resources. Currently, most hydrogen is produced from fossil fuels, specifically natural gas. Electricity—from the grid or from renewable sources such as biomass, geothermal, solar, or wind—is also currently used to produce hydrogen. In the longer term, solar energy and biomass can be used more directly to generate hydrogen as new technologies make alternative production methods cost competitive. Hydrogen technologies are technologies that relate to the production and use of hydrogen. Hydrogen technologies are applicable for many uses.

Some hydrogen technologies are carbon neutral and could have a role in preventing climate change and a possible future hydrogen economy. Hydrogen is a chemical widely used in various applications including ammonia production, oil refining, and energy (Badwal et al., 2013). Hydrogen is not a primary energy source, because it is not naturally occurring as a fuel. It is, however, widely regarded as an ideal energy storage medium, due to the ease with which electric power can convert water into its hydrogen and oxygen components through electrolysis and can be converted back to electrical power using a fuel cell. There are a wide number of different types of fuel and electrolysis cells (Badwal, 2014). The potential environmental impact depends primarily on the methods used to generate the hydrogen fuel.

Hydrogen is a clean fuel that, when consumed in a fuel cell, produces only water, electricity, and heat. Hydrogen and fuel cells can play an important role in our national energy strategy, with the potential for use in a broad range of applications, across virtually all sectors—transportation, commercial, industrial, residential, and portable. Hydrogen and fuel cells can provide energy for use in diverse applications, including distributed or combined-heat-and-power; backup power; systems for storing and enabling renewable energy; portable power; auxiliary power for trucks, aircraft, rail, and ships; specialty vehicles such as forklifts; and passenger and freight vehicles including cars, trucks, and buses.

Due to their high efficiency and zero-or near zero-emissions operation, hydrogen and fuel cells have the potential to reduce greenhouse gas emission in many applications. Energy Department–funded analysis has shown that hydrogen and fuel cells have the potential to achieve the following reductions in emissions:

- Light-duty highway vehicles: more than 50% to more than 90% reduction in emissions over today's gasoline vehicles.
- Specialty vehicles: more than 35% reduction in emissions over current diesel and battery-powered lift trucks.
- Transit buses: demonstrated fuel economies of approximately 1.5 times greater than diesel internal combustion engine (ICE) buses and approximately 2 times higher than natural gas ICE buses.
- Auxiliary power units: more than 60% reduction in emissions compared to truck engine idling.
- Combined heat and power systems: 35% to more than 50% reduction in emissions over conventional heat and power sources (with much greater reductions—more than 80%—if biogas or hydrogen from low- or zero-carbon sources is used in the fuel cell).

The greatest challenge for hydrogen production, particularly from renewable resources, is providing hydrogen at lower cost. For transportation fuel cells, hydrogen must be cost-competitive with conventional fuels and technologies on a per-mile basis

(Malindžák et al., 2017). This means that the cost of hydrogen—regardless of the production technology—must be less than $4/gallon gasoline equivalent. To reduce overall hydrogen cost, research is focused on improving the efficiency and lifetime of hydrogen production technologies as well as reducing the cost of capital equipment, operations, and maintenance.

Fuel cells are a promising technology for use as a source of heat and electricity for buildings, and as an electrical power source for electric motors propelling vehicles. Fuel cells operate best on pure hydrogen. But fuels like natural gas, methanol, or even gasoline can be reformed to produce the hydrogen required for fuel cells. Some fuel cells even can be fueled directly with methanol, without using a reformer.

In the future, hydrogen could also join electricity as an important energy carrier. An energy carrier moves and delivers energy in a useable form to consumers. Renewable energy sources, like the sun and wind, cannot produce energy all the time. But they could, for example, produce electric energy and hydrogen, which can be stored until it's needed. Hydrogen can also be transported to locations where it is needed.

The future of hydrogen technology provides extensive and independent hydrogen exploration. It explains the current situation as well as the ways in which hydrogen can help secure the future of clean, secure, and affordable energy and subsequently further develop its potential.

3. Microalgae

Microalgae, commonly known as seaweed, are widely used both in the market and in industry. However, due to their size and specific needs for growth, their mass cultivation is more demanding compared to the cultivation of microalgae and their harvest takes place directly in the ocean. The production of bioenergy in order to meet environment sustainability is one of the most important factors associated with the economic development of the country (Richa et al., 2017).

According to Janet Ranganathan, vice president of science and research at the World Resources Institute, the global population is expected to grow from 20 billion to 9.6 billion by 2050. It is estimated that by the same year, 70% of the population will live in urban areas (Kitamori et al., 2012), which will in turn create increased demand for services and food. An almost two billion increase in population by 2050 will mean that a total of 69% more calories per year need to be produced on Earth (Ranganathan, 2013).

Algae are thought to have been used primarily as food. One example is Nori seaweed, into which sushi is wrapped. Other species of algae are also edible, such as Spirulina and *Palmaria palmata* (also known as Dulse). Dulse is a red alga, commonly available on the market mainly in Ireland and Atlantic Canada. It is eaten raw, fresh, dried, or cooked as spinach. Another alternative is the algae Spirulina, which belongs to a group of bacteria called cyanobacteria. These algae have long been used as a food source in East Africa and pre-Hispanic Mexico. Due to its high content of proteins and other nutrients, it is generally used as a dietary supplement and helps in the treatment of malnutrition.

Chlorella, another popular microalga, has similar nutritional values to Spirulina, but is a unique source of "Chlorella growth factor"—a phytochemical that helps animals and children grow faster. Chlorella is also a source of the cell wall, which has an effect on the breakdown of toxins and heavy metals, such as mercury (the element). The cell wall associates with the toxin and eliminates it from the body. Chlorella is very popular in Japan and is one of the most prescribed supplements in the country.

Porphyra red algae is also harvested and used in several ways. The algae extract industry, otherwise known as phycocolides, requires a constant supply of seaweed, most of which is obtained from offshore power plants. In the tropics, seaweed cultivation has a major impact on the socioeconomy, creating direct jobs for more than 60,000 families in Southeast Asia (mainly the Philippines, Indonesia, Malaysia, and Vietnam), East Africa, and, more recently, India. In particular, the red alga *Kappaphycus alvarezii* is grown, which produces a type of carrageenan, i.e., juice or colloid, which is used as a stabilizer in food, cosmetics, and healthcare.

The method of growing algae on the high seas is simpler and more gentle, and since it does not require traditional agricultural practices, it is not necessary to use fertilizers and pesticides. For the cultivation, open wooden and PVC tanks with polypropylene ropes are used to attach plant seedlings growing in vegetation cycles that range from four to 6 weeks, depending on the fertility of the soil, the season, and the algae strain used. From an ecological point of view, seaweed cultivation provides a substrate and refuge for many species of fish and invertebrates, which play an important role in the fish life cycle. Algae also produce dissolved oxygen, thereby increasing primary productivity in biomass production. Another example of the functional use of algae is described by Antony, T, who in the years 2015–18 carried out the dissertation application of microalgae in the context of industrial design—holistic approaches in the process of design, at the Institute of Design of the Faculty of Architecture STU in Bratislava. He has addressed the issue of sustainable food production, which is currently not energy efficient, is unnecessarily centralized and therefore has a significant share in the production of greenhouse gases. He focused on the possibility of implementing microalgae biotechnology at the level of material objects as well as services, because their most valuable features are multifunctionality and environmentally efficient growth. Spirulina Lamp is a product designed as a sophisticated interior lamp that allows the cultivation of Spirulina in a home or restaurant environment. The basic value of the product lies in the direct consumption of highly nutritious food without the need for intermediaries. Its added value is that the cultivator also uses part of the light to illuminate the interior, making it a full-fledged auxiliary light. The cultivator is the first synthetic lamp of its kind to produce food for humans. The whole device is designed as an interior accessory to the living or dining area. The volume of 25 liters ensures a daily productivity of about 20 g of fresh biomass per day, which covers twice the recommended daily intake of Spirulina for an adult.

4. Biofuels

Biofuels produced from renewable energy sources can be considered as one of the most sustainable alternatives to petroleum-based fuels. Biofuels can also be included as a means for the environmental and economic sustainability of the landscape. Depending on the type of biomass used for production, we can talk about four generations of biofuels. At present, algae (third-generation biofuels) can be considered an ideal energy source for the production of biofuels, as they are characterized by a fast growth rate, they do not occupy agricultural land suitable for growing food. Arias also do not compete with food crops or feed. Because humanity is in dire need of energy for its existence, it is constantly looking for new sources of energy. In recent years, more and more attention has been paid to biofuels produced from biomass. The search for a safe and sustainable source of energy is one of the greatest challenges of our time. The issue of energy security occupies an important place in the security of any state, it is especially important, because only a reliable supply of energy creates the conditions for the harmonious development of a modern economy. Developments in the world associated with the disruption of energy supply on world markets, as well as the global financial and economic crisis, point to the problems arising in energy markets and highlight the issues of energy security. Biofuels are a renewable energy source. We divide them into solid, liquid, and gaseous. Among solid biofuels we include logs, wood chips, briquettes, pellets, sawdust, straw, and hay. Gaseous biofuels are biogas (methane), wood gas, and hydrogen produced by the fission of hydrocarbon biofuels. Liquid biofuels have found application in transport—they are blended into traditional fossil fuels—that is, petrol and diesel. They are mixed in different proportions in different EU countries. Liquid biofuels include bioethanol, MTBE, ETBE, vegetable oils, and biodiesel. Our Slovak biofuel sector produces liquid biofuels, the so-called "First generation" produced from maize, sunflower, and oilseed rape. Elsewhere in the world, first-generation liquid biofuels are also produced from soybeans and palm oil.

"Second-generation" biofuel technologies are currently being developed, which are produced from cellulose, agricultural residues, or even waste. However, second-generation biofuels are not yet commercially produced because they are too expensive. "Third-generation" biofuels that could be produced from algae are also being investigated in laboratories. However, we are probably still decades away from the introduction of second- and third-generation biofuels, as only research and commercial implementation of first-generation liquid biofuels has taken almost 80 years. Biofuels are liquid or gaseous fuels for transport, made from biomass. Biomass is the biodegradable fraction of products, waste and residues from agriculture (including plant and animal substances), forestry and related industries, as well as the biodegradable fraction of industrial and municipal waste.

The following products are considered to be biofuels:

- bioethanol;
- biodiesel: a methyl ester produced from vegetable or animal oil of diesel quality, used as biofuel;
- biogas;
- biomethanol: methanol produced from biomass, used as biofuel;

- biodimethyl ether: dimethyl ether made from biomass, used as biofuel;
- bio-ETBE (ethyl tri-butyl ether): ETBE made from bioethanol. The percentage by volume of bio-ETBE calculated as biofuel is 47%;
- bio-MTBE (methyl tri-butyl ether): a fuel made from biomethanol. The percentage by volume of bio-ETBE calculated as biofuel is 36%;
- synthetic biofuels: synthetic hydrocarbons or mixtures of synthetic hydrocarbons that have been produced from biomass;
- biohydrogen: hydrogen produced from biomass and/or from the biodegradable fraction of waste, used as biofuel;
- pure vegetable oil: oil produced from oily plants by pressing, extraction or similar processes, crude or refined but not chemically modified, provided that its use is compatible with the type of engine concerned and the corresponding emission requirements.

Bioethanol is ethanol produced from biomass and/or from the biodegradable fraction of waste used as biofuel. Biodiesel is a methyl ester made from vegetable or animal oil with diesel quality, used as a biofuel.

Biogas is a fuel gas produced from biomass and/or from a biodegradable fraction of waste, which can be purified to the quality of natural gas, used as biofuel or wood gas. Biomethanol is methanol made from biomass, used as a biofuel. Biodimethyl ether is dimethyl ether made from biomass, used as biofuel. Bio-ETBE (ethyl tri-butyl ether) is ETBE made on the basis of bioethanol. The percentage of bio-ETBE calculated as biofuel is 47%. Bio-MTBE (methyl tri-butyl ether) is a fuel made from biomethanol. The percentage by volume of bio-ETBE, calculated as biofuel, is 36%.

4.1 First-generation biofuels

They are produced by fermentation of sugars and starch, or by transesterification of oils and fats. Generation I biofuels can be prepared from sugar crops (sugar beet, sugar cane) (Fig. 17.1), from starchy crops (potatoes, cereals), and from oil crops and animal fats (rapeseed, soybean, sunflower, oil palm, lard). Nonadvocates of biofuels claim that the use of oilseeds and cereals to produce this generation of biofuels contributes to rising food prices, but the fact is that only 3% of arable land is currently used worldwide for the production of industrial energy crops.

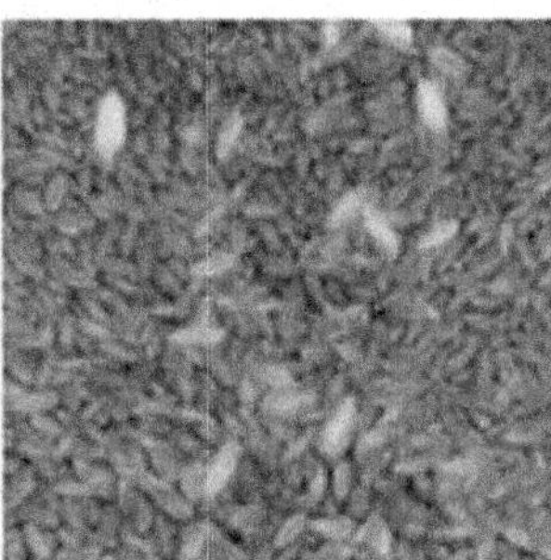

Figure 17.1 Corn, wheat, and sugar cane.
Source: https://commons.wikimedia.org/wiki/File:Corn_grains.JPG.

4.2 Second-generation biofuels

Another variant of sources for biofuels are the so-called non—food cellulose and lignocellulosic raw materials, more precisely, forest waste, leaves, needles, bark, sawdust, straw, hay, maize bark, various agricultural residues (Fig. 17.2). These crops can be processed by various processes and technologies, from hydrolysis, through gasification to cracking. Their advantage is that they do not compete with food crops, they can be grown on lower quality soil and the plus is that in the production of biofuels, the whole crop can be used, not just a part of it. The European Union has agreed that the minimum share of second-generation biofuels in total energy in transport will be 0.2% from 2022, 1% from 2025, and 3.5% in 2030. For first-generation biofuels, a ceiling of 7% on transport energy is set. Currently, second-generation biofuels are in the form of research and gradual introduction into industrial production, they can really affect the transport sector only over the next decade.

4.3 Biofuels third generation

In the future, biofuels made from seaweed or cyanobacteria should also be used. According to scientists, their great advantage will be the high utilization per unit area, which means that, for example, we get as many organic components from the area of two parking spaces as from a soybean field the size of a football field. Algae are the fastest growing plants and are highly productive. In the future, their cultivation in swimming pools or special bioreactors is being considered (Fig. 17.3).

5. Heat pump ice storage

The Bloom Energy Server or Bloom Box is a solid oxide fuel cell power generator made by Bloom Energy, of Sunnyvale, California, that takes a variety of input fuels, including liquid or gaseous hydrocarbons produced from biological sources, to

Figure 17.2 Straw and wood waste.
Source: Straw modified https://upload.wikimedia.org/wikipedia/commons/3/3e/Egg_in_straw_nest.jpg. Wood waste own processing.

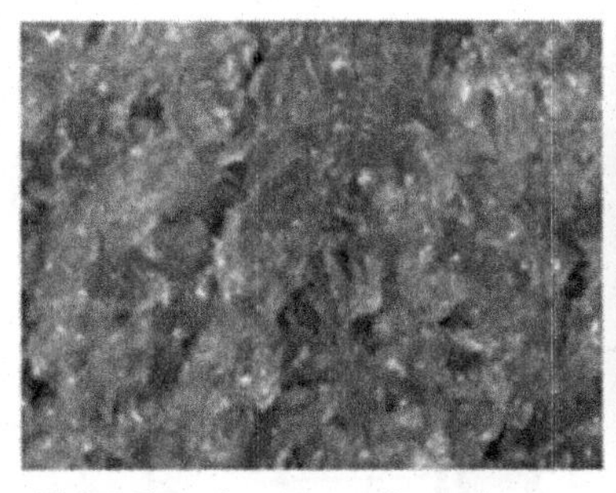
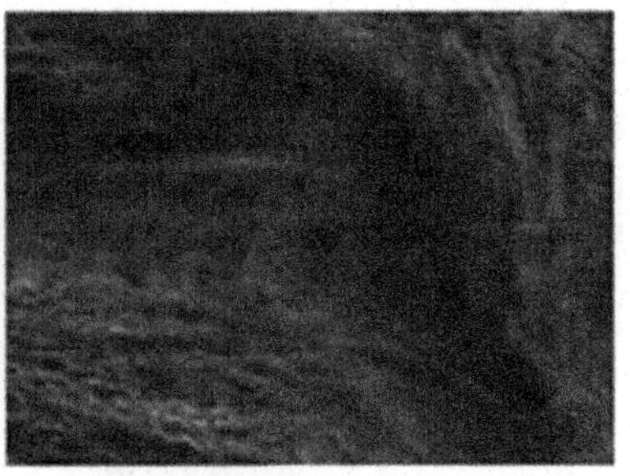

Figure 17.3 Microalgae, cyanobacteria.
Source: https://commons.wikimedia.org/wiki/File:Nodularia_bloom.jpg.

produce electricity at or near the site where it will be used (GLG Expert Contributor" Gerson Lehrman Group.2010). It withstands temperatures of up to 1800°F (980°C). According to the company, a single cell (one 100 × 100 mm plate consisting of three ceramic layers) generates 25 W. The fuel cells have an operational life expectancy of around 10 years; based on predictions on fuel costs, the "break-even" point for those who purchase the device is around 8 years. The cell's technology continues to rely on nonrenewable sources of energy to produce electricity, and because it is not a hydrogen fuel cell, it still produces carbon dioxide (an important greenhouse gas) during operation. In 2011, Bloom stated that 200 servers had been deployed in California for corporations including eBay, Google, Yahoo, and Wal-Mart.

5.1 Bloom Box

It's a collection of fuel cells—skinny batteries—that use oxygen and fuel to create electricity with no emissions.

Fuel cells are the building blocks of the Bloom Box. They are made of sand that is baked into diskette-sized ceramic squares and painted with green and black ink. Each fuel cell has the potential to power one light bulb. The fuel cells are stacked into brick-sized towers sandwiched with metal alloy plates. The fuel cell stacks are housed in a refrigerator-sized unit—the Bloom Box. Oxygen is drawn into one side of the unit, and fuel (fossil fuel, biofuel, or even solar power can be used) is fed into the other side. The two combine within the cell and produce a chemical reaction that creates energy with no burning, no combustion, and no power lines.

Solid oxide fuel cell manufacturer Bloom Energy is entering the commercial hydrogen market by introducing hydrogen-powered fuel cells and electrolyzers that produce renewable hydrogen. Bloom Energy's core technology is based on research done by its founders on using electricity generated by a solar panel to produce fuel and oxygen on planet Mars for NASA. Bloom Energy Servers reversed this process by taking in fuel and air to generate electricity. Bloom is capitalizing on this technology by taking terrestrial renewable power and producing hydrogen using solid oxide electrolyzers (Fig. 17.4). The renewable hydrogen, when produced in this manner, can fuel cars, power resilient AlwaysON Microgrids, or be injected into natural gas pipelines to reduce carbon emissions.

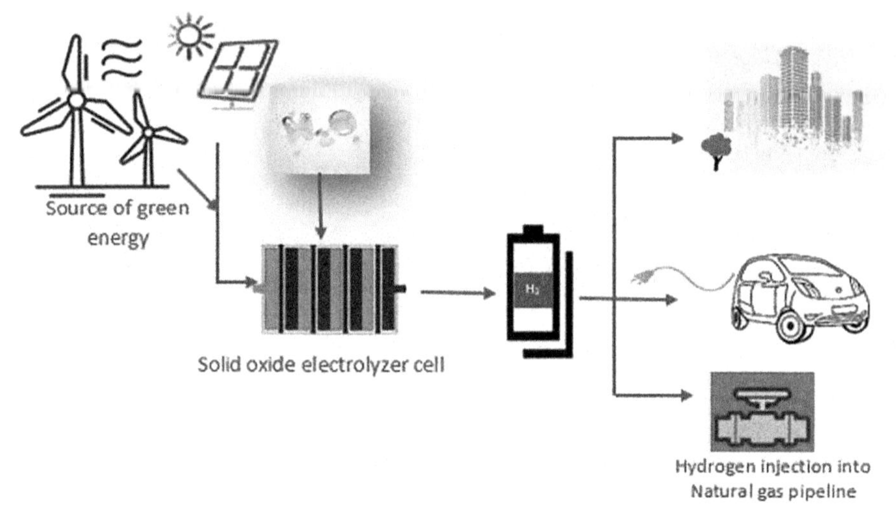

Figure 17.4 Renewable hydrogen production.
Source: Own processing.

5.2 Heat pump ice storage

The rapid increase in many parts of the world of generating capacity by intermittent renewable energy sources, notably wind and solar, has led to a strong incentive to develop energy storage for electricity on a large scale. The extent to which electricity storage can be developed will determine the extent to which those intermittent renewable sources can displace dispatchable sources, taking surplus power on occasions and bridging intermittency gaps. There are questions of scale—power and energy capacity—which are indicated below in particular cases. Also the stored electricity usually needs to be available over days and weeks rather than minutes and hours. Electricity cannot itself be stored on any scale, but it can be converted to other forms of energy which can be stored and later reconverted to electricity on demand. Storage systems for electricity include battery, flywheel, compressed air, and pumped hydro storage. Any systems are limited in the total amount of energy they can store. Their energy capacity is expressed in megawatt-hours (MWh), and the power, or maximum output at a given time, is expressed in megawatts of electric power (MW or MWe). Electricity storage systems may be designed to provide ancillary services to a transmission system including frequency control, and this is the chief role of grid-scale batteries today.

Of course, very effective storage of energy is achieved in fossil fuels and nuclear fuel, before electricity is generated from them. While the focus here is on storage after generation, particularly from intermittent renewable sources, any proper consideration of the question needs also to encompass nuclear fuel for power generation as a more economical option with relatively little materials requirement.

An ice storage system uses a chiller to make ice during off-peak night time hours when energy is cheaper and then melts the ice for peak period cooling needs, effectively shifting the electric load and avoiding higher price energy and demand charges during the day. An ice thermal storage system reduces peak demand, shifts energy usage to nonpeak hours, saves energy, and reduces energy costs. Reduces Peak Demand and Shifts Energy Usage: With less connected horsepower, ice storage can lower peak electrical demand for the HVAC or process cooling system by 50% or more. The math says that a single 2L block of ice is capable of cooling a room down by 11°C (20°F)! Now, there are other things going on that are stopping my room from turning into a refrigerator. The higher the temperature, the more the particles vibrate and faster they move in the material storing energy. The ice storage is an underground water tank that serves as heat source for a heat pump that in turn provides heat for SH and DHW preparation. Heat can be extracted effectively from the ice storage until all the water inside has turned into ice. Since temperature does not change during the phase change from liquid to solid, the source temperature for the heat pump stays constantly around the freezing point. The freezing is interrupted by phases of regeneration, where heat from solar absorbers or from the surrounding soil is introduced into the ice storage. Due to the low temperature in the ice storage, the solar absorbers can be used for regeneration already at low ambient temperature. During the warm summer months, when only little energy is needed for SH and solar heat is abundant, the ice in the ice storage is completely melted and the water heated to around 25°C. Thus, the ice storage is a seasonal storage for solar and ambient heat. The combination of heat pump with seasonal storage and solar absorber could provide a viable alternative to systems with borehole heat exchangers where drilling is prohibited or uneconomical. Since the ice storage is filled solely with water, it could also be installed in water protection areas (Winteler, 2014).

Pumped storage involves pumping water uphill to a reservoir from which it can be released on demand to generate hydroelectricity. The efficiency of the double process is about 70%. Pumped storage comprised 95% of the world's large-scale electricity storage in mid-2016, and 72% of the storage capacity added in 2014. Battery storage, however, is being deployed strongly. Building-scale power storage emerged in 2014 as a defining energy technology trend. This market has grown by 50% year-on-year, with lithium-ion batteries prominent but redox flow cell batteries show promise. Such storage may be to reduce demand on the grid, as back-up, or for price arbitrage.

Pumped storage projects and equipment have a long lifetime—nominally 50 years but potentially more, compared with batteries—8 to 15 years. Pumped hydro storage is best suited for providing peak-load power for a system comprising mostly fossil fuel and/or nuclear generation. It is not so well-suited to filling in for intermittent, unscheduled, and unpredictable generation.

The Thermal Energy Store is the Energy Store associated with the temperature of an object. Electricity cannot itself be stored on any scale, but it can be converted to other forms of energy which can be stored and later reconverted to electricity on demand. Storage systems for electricity include battery, flywheel, compressed air, and pumped hydro storage. Currently the most common type of energy storage is pumped hydroelectric facilities, and we have employed this utility-scale gravity storage technology for the better part of the last century in the United States and around the world. A World Energy Council report in January 2016 projected a significant drop in cost for the majority of energy storage technologies as from 2015 to 2030. Battery technologies showed the greatest reduction in cost, followed by sensible thermal, latent thermal, and supercapacitors. Battery technologies showed a reduction from a range of €100–700/MWh in 2015 to € 50–190/MWh in 2030—a reduction of over 70% in the upper cost limit in the next 15 years. Sodium sulfur, lead acid, and lithium-ion technologies lead the way according to WEC. The report models storage related to both wind and solar plants, assessing the resultant liveliest cost of storage (LCOS) in particular plants. It notes that the load factor and the average discharge time at rated power is an important determinant of the LCOS, with the cycle frequency becoming a secondary parameter. For solar-related storage the application case was daily storage, with 6-h discharge time at rated power. For wind-related storage the application case was for 2-day storage with 24 h discharge at rated power. In the former case, the most competitive storage technology had LCOS of €50–200/MWh. In the latter case, levelized costs were higher and sensitive to the number of discharge cycles per year, and "few technologies appeared attractive."

6. Biomaterials

The term biomaterial can be used to refer to all inanimate materials that are used in medicine for therapy and are directly intended for interaction with the biological system (Ratner et al., 2004). A biomaterial is any substance that has been designed to interact with biological systems for medical purposes—either a therapeutic or a diagnostic one. As a science, biomaterials are about 50 years old. The study of biomaterials is called biomaterials science or biomaterials engineering. Throughout its history, it has seen stable and strong growth, and many companies have invested heavily in developing new products. Biomaterials can be interpreted in various ways, but in general biomaterials are all organic and inorganic, synthetic or natural materials used in medicine to replace or restore soft and hard tissues, in the form of permanent implants in orthopedics, skin surgery, dentistry, etc. (Park and Bronzino, 2002). The development of biomaterials has resulted from the needs of the current high pace of life, bringing with situations in which there are serious accidents or injuries associated with serious damage to a certain part of the organism. The essence of biomaterials can be metals, polymers, ceramics, and composites. The properties of biomaterials are closely related to their application and the functions they have to fulfill in the body after implantation. Depending on how the implant affects living tissue, we divide biomaterials into:

1. **inert**—are characterized by direct contact between the implant and the surrounding bone tissue
2. **bioactive**—they have the ability to form a strong chemical bond with living tissue directly, not through the connective tissue
3. **resorbable**—they serve as a temporary replacement of bones, when the implanted material is gradually absorbed during the renewal of bone tissue.

The condition is that each biomaterial that will be inserted into living tissue is:

- biocompatible—i.e., long-term compatibility of the biomaterial with the surrounding soft tissue of the organism,
- nontoxic—must not cause a change in the tissue reaction that would lead to an allergy,
- noncarcinogenic and corrosion resistant.

According to the interaction with the living system, they can be divided into bioinert and bioactive biomaterials. Bioinert materials can be divided into three basic groups, namely:

- The first generation of biomaterials, which do not affect the biological system and show little or no interaction with the host tissue. The interface between the material and the host tissue is formed by a thin acellular capsule with minimal adhesion and the material is encapsulated in it. These are typically metal implants.
- The second generation of biomaterials are bioactive materials that elicit controlled biological activity at the interface with the host tissue. The main representative of this category is bioactive glass. Within the second generation of biomaterials, bioresorbable materials have also been created, in which, upon interaction with the host tissue, controlled chemical decomposition occurs at the interface until complete resorption of the material.
- The third generation of biomaterials has been designed to stimulate specific cell interactions at the molecular biology level when interacting with host tissue (Hench and Thompson, 2010).

7. Nanotechnologies

We are already talking about the fact that nanotechnology is a scientific discipline that deals with the precise and deliberate manipulation of matter at the atomic level, that is, the size of nanometers. The application of research findings can be found in various fields such as chemistry, biology, physics, or materials science. Thoughts on nanotechnology began sometime in 1959 at a meeting of physicists at the California Institute of Technology. The physicist Richard Feynman described in a discussion entitled "There is still plenty of room at the bottom" a process in which scientists would be able to manipulate and control individual atoms and molecules. However, the concept of nanotechnology did not appear until 10 years later in the study of ultraprecision machining. The era of modern nanotechnology began in 1981 with the advent of the tunneling microscope, which was able to see individual atoms.

Very important question is Why is there a nanotechnology? Answer is not so simple like it seems.

Nanoscience and nanotechnology include the ability to see and control individual atoms and molecules. Everything on earth is made up of atoms. The food we eat, the clothes we wear, the buildings and houses we live in, as well as our own body. But something as small as an atom cannot be seen with the naked eye. In fact, it is impossible to see it even with the microscopes commonly used in schools. Microscopes that can see things the size of nanoparticles have been invented relatively recently. Once scientists had the right tools, such as a scanning tunneling microscope and an atomic force microscope, the age of nanotechnology was born.

Although modern nanosciences and nanotechnologies are completely new, nanomaterials have been used for centuries. Today's scientists and engineers find a wide range of ways to intentionally create nanometer-sized materials and take advantage of their useful properties such as higher strength, lower weight, better light spectrum control, greater chemical reactivity, and the like.

All things around us are made up of atoms, and the properties of these substances depend on how the atoms themselves are arranged in that substance. When we reorganize the atoms into a piece of coal, we can get a diamond. In the field of nanotech, we work with structures that have dimensions of the order of nanometers ($1 \text{ nm} = 10 - 9 \text{ m}$). By precisely assembling them, we can create materials with exactly the required properties. In short, this is the basic principle of how nanotechnologies work.

Some example where we can find a nanotechnology in our basic life:

- electronics,
- electrical engineering (storage media),
- healthcare (drug absorption),
- mechanical engineering (advantageous properties of materials),
- chemical,
- optical (optical filters),
- automotive,
- textile industry (nanofibers for advantageous properties of fabrics).

Scientists are exploring new possibilities and ways in which they can manipulate and modify atoms or molecules. Nanotechnology gives humanity the ability to manipulate atoms and molecules, that is, to select, move, and store them so that things of different shapes and sizes can be made without residues, as is the case today with machining, grinding materials, in addition to creating these materials and objects at microscopic accuracy, at atomic level accuracy. It goes, e.g., for the development of new materials, stronger and lighter than steel, or the application of microscopic nanorobots, which would be able to kill only harmful cells directly in the body, so that they do not damage healthy tissue.

When examining the particles, the researchers came up with an interesting division according to the surface/volume ratio of the particle:

- The surface of the particle is larger than its volume. It also means that there are more atoms on the surface of the particle than inside.
- The particle surface is smaller than the particle volume. It also means that the atoms inside the particle predominates.
- The behavior of the whole particle is determined by the predominant atoms.

Nanoparticles have their special properties and behavior due to the atoms prevailing on their surface.

Classification of nanoproducts in terms of use:

- nano cleaning of various types of surfaces,
- nano protection of various types of surfaces,

- nano thermal insulation,
- nanosilver products (textiles, bottles, pool products),
- nano additives.

If we want to conclude a short excursion on nanotechnology, we must state that it is any technology that is carried out at the level of microscopic particles, i.e., at the nanoscale, for subsequent use in science, technology, industry, and everyday life. It is a combination of knowledge from physics, chemistry, and biology at the level of atoms or molecules to macromolecules, and also the subsequent application of these nanostructures to larger systems, it is the science of mass manipulation at the atomic and molecular level. Nanotechnology brings with it another industrial revolution and also a revolution in science and technology, in materials engineering, medicine, electronics, biotechnology, in information technology, and in other fields and areas.

This chapter is part of the solution of scientific project KEGA 026EU-4/2018 and it was issued with ist support.

Summary

Technological development, innovation and creativity are fundamental determinants of economic prosperity in a globalizing, knowledge-based economy. Economic growth is the result of the development, adoption and creative application of new technologies. Technological transformation is a process that takes place in a creative collaborative partnership of all actors, from creators of innovative development strategies, research organizations, entrepreneurs to citizens. An organization that wants to exist in a competitive environment of global markets must strive to strengthen its position in the market. For this reason, it must innovate and develop new products as well as the way it provides its services. Through innovation, companies try to gain a competitive advantage in the market and provide the customer with something new, to create his hitherto unknown needs. As a result of globalization, organizations face different types of cultures, with different demands on the novelty of products and services provided. Closely related to this is the need to be able to name the differences between existing and desired, the ability to identify the need to change products, services, processes, systems, as well as the behavior of organizations. Changes in the global environment should result in new products, processes and systems that meet user requirements while minimizing resource consumption. Renewable energy sources contribute to the strengthening and diversification of the structure of industry and agriculture, support innovation and the development of information technologies, open up space for new directions and are one of the pillars of building a knowledge economy. Rational management of domestic renewable energy sources is in accordance with the principles of sustainable development, which makes it one of the pillars of healthy economic development of society.

References

Badwal, S.P.S., 2014. Emerging electrochemical energy conversion and storage technologies. Front. Chem. 2 (79) https://doi.org/10.3389/fchem.2014.00079.

Badwal, S.P.S., Giddey, S., Munnings, C., 2013. Hydrogen production via solid electrolytic routes. Wiley Interdiscip. Rev. Energy Environ. 2 (5), 473–487doi. https://doi.org/10.1002/wene.50.

Bednárová, L., Liberko, I., Rovňák, M., 2013. Environmental benchmarking in small and medium sized enterprises and there impact on environment, International Multidisciplinary Scientific GeoConference-SGEM, Albena, Bulgaria, 2013,141-146, ISSN: 1314-2704.

Bednárová, L., Witek, L., et al., 2016. Assessment methods of the influence on environment in the context of eco-design process. In: Majerník, M., et al. (Eds.), Production management and engineering sciences. CRC Press-Taylor & Francis Group, pp. 15–16.

Gerson Lehrman Group, February 22, 2010. Bloom Box: what is it and how does it work? Christ Sci. Mon. Retrieved 2020-09-28.

Hench, L.L., Thompson, I., 2010. J. R. Soc. Interface 7, S379.

Kitamori, K., Manders, T., Dellink, R., Tabeau, A.A., 2012. OECD Environmental Outlook to 2050: The Consequences of Inaction, Revised ed. (OECD Environmental Outlook). OECD. https://doi.org/10.1787/9789264122246-en.

Malindžák, D., Zimon, D., Bednárová, L., et al., 2017. Homogeneous production processes and approaches to their management. Acta montanistica Slovaca 22 (2), 153–160. 2017. https://actamont.tuke.sk/pdf/2017/n2/7malindzak.pdf.

Park, J.B., Bronzino, J.D., 2002. Biomaterials: Principles and Applications. CRC Press, Boca Raton.

Ranganathan, J., 2013. The Global Food Challenge Explained in 18 Graphics. https://www.researchgate.net/publication/288493603_The_Global_Food_Challenge_Explained_in_18_Graphics.

Ratner, B.D., Hoffman, A.S., Schoen, F.J., Lemons, J.E., 2004. Biomaterials Science: An Introduction to Materials in Medicine. Academic Press, New York.

Richa, K., Bhola, R., Gurjar, et al., 2017. https://www.researchgate.net/publication/309204369_Microalgae_An_emerging_source_of_energy_based_bio-products_and_a_solution_for_environmental_issues. (Accessed 2 October 2020).

Winteler, C., et al., 2014. Heat pump, solar Energy and ice storage systems – modelling and seasonal performance. In: 11th IEA Heat Pump Conference 2014. International energy agency SHC, pp. 1–12. May 12–16, 2014, Montréal (Québec) Canada), Retrieved 2020-09-28.

Further reading

Chemické Listy. http://www.chemicke-listy.cz/docs/full/2017_10_614-621.pdf 2020-10-07.

Energy. www.energy.gov/eere/articles/hydrogen-clean-flexible-energy-carrier 2020-09-28.

Energy Storage. https://energystorage.org/why-energy-storage/technologies 2020-10-01.

Mobile Magazine. Bloom Energy Server Unveiled, Bloom Box Not for the Home Just Yet. mobilemag.com. 2010-02-25. Retrieved 2020-09-28.

Mobile Magazine. http://www.mobilemag.com/2010/02/25/bloom-energy-server-unveiled-bloom-box-not-for-the-home-just-yet/#sthash.NitkQsAK.dpuf.

Nanoera.sk. https://nanoera.sk/clanky-o-nanotechnologii/o-nanotechnologii 2020-10-07.
Nanoprom. https://Www.Nanoprom.Sk/Sk/Obsah/Co-Je-To-Nanotechnologia-22.
Slovenské Centrum Dizajnu. www.scd.sk/?designum-1-2020&clanok=vyuzitie-mikrorias-v-kontexte-priemyselneho-dizajnu 2020-10-05.
World Energy. https://www.worldenergy.org/publications/entry/e-storage-shifting-from-cost-to-value-2016.

Development of nanomaterials: valorization of metallic solid waste from copper-based components

18

M.S. El-Eskanadarny[1], *N. Ali*[1], *S.M. Al–Salem*[2]
[1]Energy & Building Research Centre, Kuwait Institute for Scientific Research, Al-Shwaikh Educational Area, State of Kuwait, Kuwait; [2]Environment & Life Sciences Research Centre, Kuwait Institute for Scientific Research, Al-Shwaikh Educational Area, State of Kuwait, Kuwait

1. Introductory remark: on the origin of metal waste and scrap metals

The increase in human evolution can only result in an ever so increase on the dependency of commodities that makes our daily activities easier. This will only increase the reliance on nonrenewable resources to extract fossil fuels to run and operate heavy industries. The ultimate endpoint will unfortunately be the wasted articles and commodities that once were considered to be of high value. Waste materials are defined as a by-product of manmade processes that have no value and require a certain amount of cost and time to handle (Al-Salem et al., 2019a). These waste materials could be of any shape or physical state and have no substantial value on the market when compared to raw or virgin materials. It is therefore sensible to understand that metals and metallic products that are out of service resulting from discarded products or as a by-product of processing lines in industry fall within such a category and could be categorized as "metal waste."

According to recent World Bank data, the highest region in the world that generates waste is East-Asia/Pacific; and metal waste constitutes 4% of the total global average of all waste components (Kaza et al., 2018). This will include waste of various origins and sources that could result from industrial, municipal, and construction and demolition sectors around the globe (Al-Salem, 2019b). The origins of metal waste vary considerably. This is due to the fact that metals have a substantial importance in various applications and could be found in almost all products. Metal waste could originate mainly from electroplating and tanning industry sludge, waste from electrical and electronic equipment (WEEE), arc furnace dust, glass fiber refractory coating, steel plant dust, and metal-containing soil phytoremediation plant (Changming et al., 2018). WEEE alone, a substantial component with an increasing trend worldwide due to electronic appliances and mobile phones increasing dependency, has been a

Sustainable Resource Management. https://doi.org/10.1016/B978-0-12-824342-8.00008-0

subject of various research efforts as of late. This is understandable due to its increasing quantities in the waste stream. Worldwide, WEEE total at some 44.75 million tonnes and Asia alone occupies the major share 41% of total WEEE generated (Ismail and Hanafiah, 2020).

Metal waste could be treated using various techniques and technologies that are well established in industry. Thermal treatment is one example of such which is commonly referred to as *pyrometallurgical* treatment utilizing commonly established techniques such as incineration, melting, and pyrolysis of waste (Al-Salem et al., 2009). Utilizing fourth-state of matter in the form of plasma with high energy density has also proven to be of immense efficiency lately for treating metal containing waste (Changming et al., 2018). Furthermore, special waste such as WEEE contain a significant amount of metals (i.e., base and precious) whereby high-value ones could be easily recovered as an optimal valorization goal. Base metals (e.g., Zn, Sn, Fe, and Al) are commonly leached using mineral acids or ammonia–ammonium media (e.g., H_2SO_4, HNO_3, and HCl) and could be dissolved within such (Hao et al., 2020). On the other hand, precious metals (e.g., Au and Ag) are leached using reagents such as cyanide, thiourea, thiosulfate, and mineral acids (H_2SO_4, HCl, and HNO_3) (Wasserscheid and Keim, 2000). Some metallic waste could be attached to plastic solid waste that could either be disposed of in a landfill site or incinerated. The recent work of Alyousef et al. (2020) has focused on such type of waste which was termed metalized plastic waste (MPW) whereby a thin layer of aluminum (Al) is utilized on top of plastic packaging and end up in the solid waste stream. MPW was proven to be a useful reinforcing material for concrete in their work which could easily be used for construction. On the other hand, low temperature roasting was also proven to be useful for WEEE in the range of operating temperatures between 200 and 325°C (Panda et al., 2020).

2. Major components of metal waste

Increasing the consumption rate of energy generated by fossil fuels and natural resources has rapidly increased. This is due to the desire of societies to have a more convenient and luxurious life. Fig. 18.1 represents the life cycle of the metals, starting from mining and excavation, ending by end users. Traditionally, almost all industries have operated as open systems, transforming the natural resources to products without considering their impact on the environment due to large amounts of energy used (Matsuno et al., 2007). The top part of the figure shows the starting point from ore exploration, estimation of ore reserves, and ore mining through the construction of underground or open-cast mines. The mining and excavations are extremely costly, especially for the underground mines, where the safety and ventilation issues require special constructions and precautions. Most of the cost for metal-end products come from the high cost of mining operations. Mineral dressing operations, which come consequently to the mining, is a costly step requiring large equipment to conduct ore comminutions, flotation, and many other complicated chemical processing

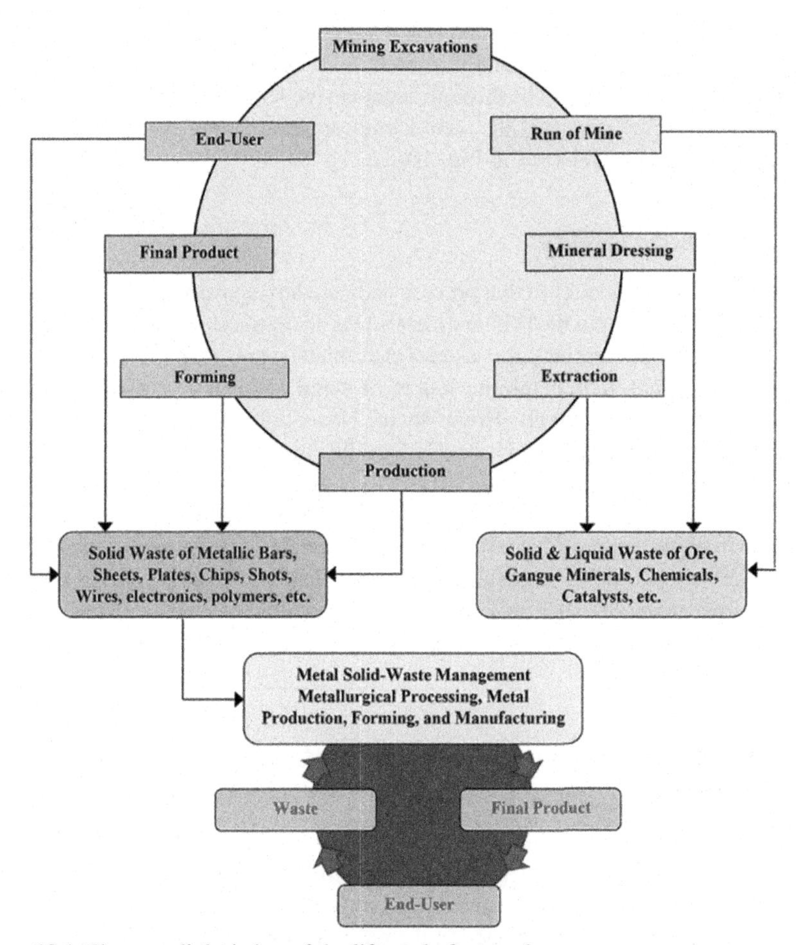

Figure 18.1 The overall depiction of the life cycle for metals.

dedicated to increasing the metal content in the run of mine. The as-dressed ores are then subjected to a more complicated extraction process through a hydro- and/or pyro-metallurgical process applied to obtain the desired metal from the ore. Most of the wastes generated from this process are in liquid form that contain significant toxic chemical compounds. This type of waste material must be separated from the liquid waste through severe refining chemical processing.

Once the metal is liberated, it goes into an intermediate stage so-called production, in which leaching or melting techniques are applied to purify the metal and to increase its purity to be 99.99% or more. The as-produced metal ingots are then formed into the desire shapes of plates, sheets, wires, etc. The as-formed metallic items are then used for manufacturing the desired structure for the end user. In contrast to the run-of-mine to end-user approach, the lower part of Fig. 18.1 presents the way of obtaining high-grade of metal and metal alloys starting from metal-solid waste materials come from different sources such as scraped cars, home electrical devices, workshops, factories, etc. One may say that (SW) metallic materials of solid waste origin are ideal for manufacturing any desired items through inexpensive short approaches. More importantly, metal recycling preserves the natural environment, saving energy, and reducing waste matter, as can be realized in Fig. 18.1.

2.1 Metal recyclability

The term recycling may refer to that process dedicated to recovering the metal from the metallic SW. In contrast to the primary metals that are extracted from mineral dressing and metallurgical processing, the as-recycled metals are called secondary metals (Oguchi et al., 2011). In general, the source of metal SW can be classified into three groups (Chancerel et al., 2009): Home scrap, New scrap, and Obsolete (old) scrap. From the economic point of view, detailed studies and thermodynamics calculations should be undertaken in part to estimate the total amount of metals expected to yield after the recycling process. This is defined as the *recycling efficiency* that can be simply calculated from the output of the recycled metal divided by the input (starting SW). Metals, their alloys, and composites reveal excellent recyclability when compared with any other materials such as ceramics and plastics. Difficulties in the recycling process of metals may come due to the combinations of metals and other materials in the recycled items.

2.2 Potential utilization of recycled metals for advanced technology

Metals are a unique class of materials that they are inherently recyclable. Since they can be successfully recycled repeatedly, the need for natural resources can be minimal. Accordingly, traditional mining, mineral dressing, and extractive processing activities that have a significant hazard impact on the environment may be avoided. Almost all the elements of the periodical table, including metals, semimetals, and rare earth elements, can be efficiently recycled to their pure form. Fig. 18.2 presents some examples of recyclable ferrous, nonferrous metals, and semimetals. Fabrications of advanced metallic materials, such as nanocrystalline metals and alloys, amorphous and metallic glassy alloys, high-entropy alloys, shape memory alloys, and energy storage materials, use costly feedstock materials of pure metal bulk and powders. Almost all these starting metals can be successfully obtained from the recycled metal (secondary source), as recently proposed by El-Eskandarany et al. (2020a, 2020b).

3. Examples of recyclable metals and alloys

The importance of the materials in the transition of our society toward sustainability led us to classify our history into several stages: stone, bronze, iron, porcelain, steel, silicon, and the age of nanotechnology (Fig. 18.3). Since the last few decades,

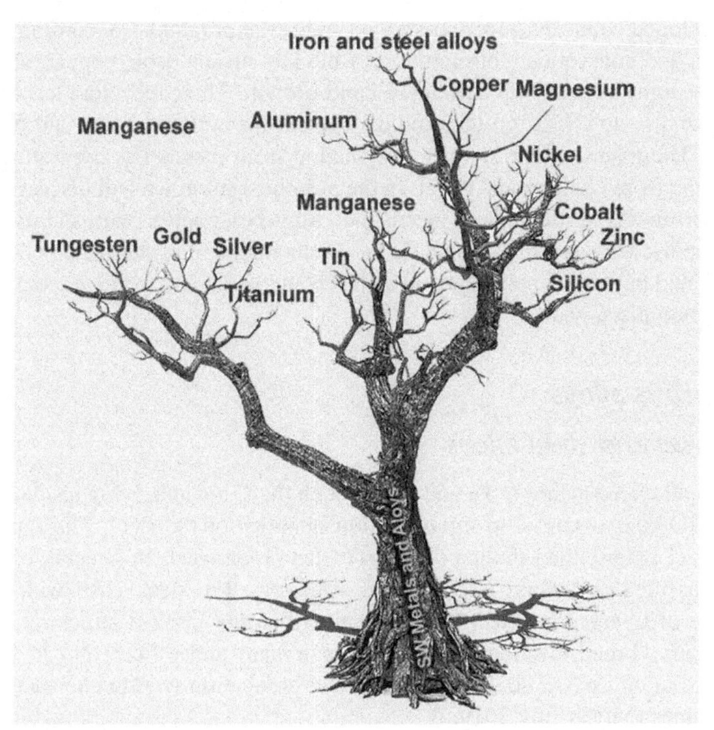

Figure 18.2 Common examples of recyclable ferrous and nonferrous metals.

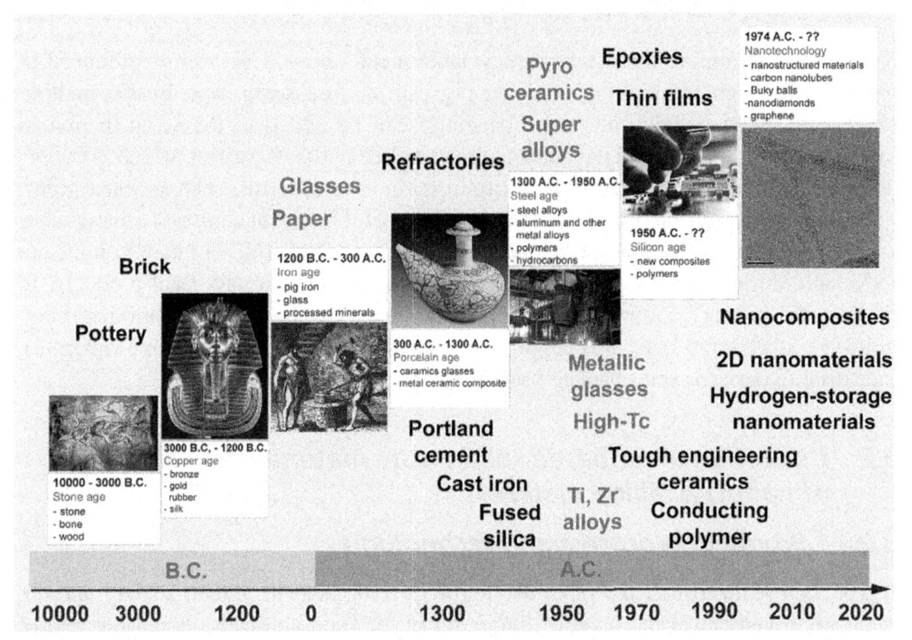

Figure 18.3 The capability of any societies along the human history on developing and instigating of new materials that fit their needs has led to the advancement of their performance and ranking in the worldwide.

many developed countries have started their industrial programs. Accordingly, global production and consequent consumption of the key metals (iron, copper, aluminum, and magnesium) have shown extremely rapid growth. Thus, recycling has become a necessity and is now high on the political agenda for both industrial and developed countries. The management of waste originating from metals has been attractive on an economic front (Ackerman, 1997). In the present section, we will discuss the recycling opportunities of three major metals, i.e., iron (Fe), copper (Cu), and magnesium (Mg), to be used as feedstock materials for producing nanocrystalline powders. Their properties and potential applications in different industrial and medical sectors will be presented and discussed.

3.1 Ferrous alloys

3.1.1 Iron and steel alloys

Steel is an alloy consisting of Fe and C, in which the C content, lying inside the range 0.04%−1.00%, is essential to the execution qualities of the steel. The run of blast furnace steel is unrefined molten steel, so-called crude steel. In general, crude steel can be classified as carbon steel, special steel, or alloy steel; each with different percentage of C and Mn. Steel is the foremost broadly utilized structural metal, in which it finds a broad spectrum of applications in many areas. According to the World Steel Association (WSA, 2020), the crude steel production worldwide reached 156.7 million tonnes (Mt) in July 2019.

3.1.2 Source of iron waste

Steel and cast iron are the essential recyclable metal alloys. Cast iron is produced in foundries through the melting of either pig iron or iron scrap iron. In this melting process, selected metallic alloying element(s) can be added to the scrap in part to improving its corrosion and mechanical property behaviors. Cast iron, which is important recycled source material, may be produced in several forms, such as water pipes, machinery parts, engine blocks, and many other useful items. In contrast to many other scrap materials, steel scraps are unique solid waste, in which 100% of their volume are recyclable through melting, shaping, rolling, and reform into new steel products. In general, the primary sources of steel scrap are cars, construction, and ship-breaking; however, steel scrap may also come from plentiful sources such as home (internal), industrial (external), and obsolete (old) scraps.

3.2 Utilization of Fe-based solid waste materials for fabrication of nanocrystalline materials

3.2.1 Production process and technologies

It has been realized that the rapid economic development in human society and the rapid improvement of the level of living aid in the monotonically increasing rate of personal cars. Accordingly, scrapped vehicles (i.e., end-of-life vehicles) are drastically increasing to create serious problems, which presents a significant challenge in SW treatment and disposal. In The State of Kuwait, as an example, thousands of scrapped cars, which are dumped into scrap yards are considered as hazardous waste since they are subjected to corrosion after some time. Apart from the traditional industrial routes of melting, casting, and shaping processes devoted to producing new steel items from the scrap, the Kuwait Institute of Scientific Research has recently developed an attractive, straightforward approach to fabricate large amounts of uniform Fe nanocrystalline powders. The bulk Fe scrap pieces were firstly cut into smaller pieces (Fig. 18.4) and then chemically activated by sonicating the materials in an acetone bath to remove the oil contamination from their surfaces. The bulk materials are ground into coarse powders (>100 μm in diameter) using a conventional disk mill unit. This primary treatment process is summarized in the flowchart presented in Fig. 18.5.

Figure 18.4 The (A) as-received SW Fe bulk materials after roasting at 250°C, and (B) ultrafine Fe nanocrystalline powders obtained after 200 h of high-energy ball milling. The photos were taken in the Nanotechnology Lab, Energy, and Building Research Center (EBRC) of The Kuwait Institute for Scientific Research (KISR).

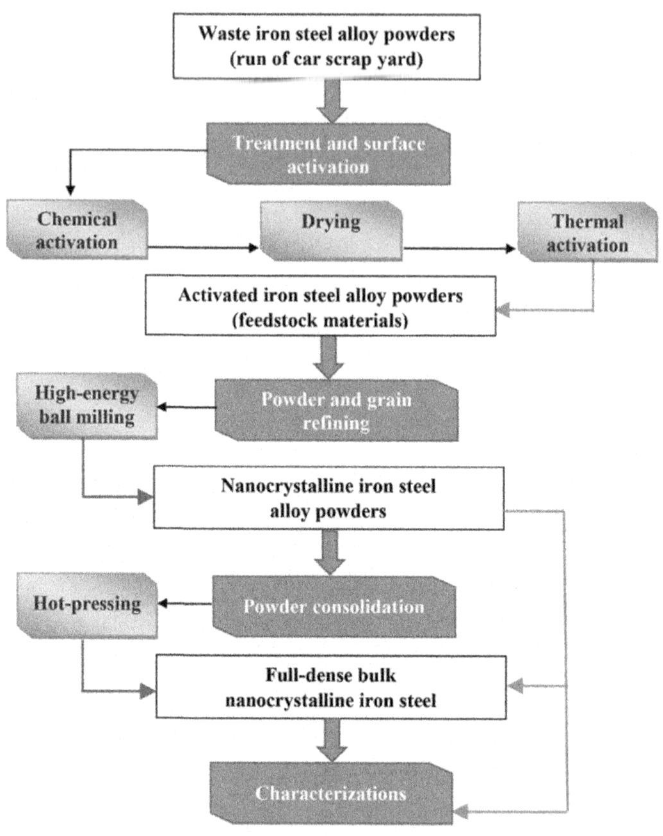

Figure 18.5 Flowchart diagram, describing the experimental sample preparations of Fe SW nanocrystalline powders that were taken place in the Nanotechnology Lab, Energy, and Building Research Centre (EBRC) of The Kuwait Institute for Scientific Research (KISR).

Due to the successive accumulation of the dislocation density, the crystals are disintegrated into subgrains that are initially separated by low-angle grain boundaries. The formation of these subgrains is attributed to the decrease of the atomic level strain. Further ball milling time led to further deformation occurring in the shear bands located in the unstrained parts of the powders which leads to subgrain size reduction so that the orientation of final grains becomes random in crystallographic orientations of the numerous grains and, hence, the direction of slip varies from one grain to another.

3.2.2 *Fabrication of bulk nanocrystalline Fe solid waste powders*

In general, powder metallurgy and technology can be classified into two parts. Firstly, powders production and, secondly, the production of solid objects from the powders, which is called powder consolidation. Now, there are many optional approaches used for preparing of the feedstock powders with controlled nano- and microstructure and for consolidations of the powders into the desired shapes and sizes. Consolidation of nanocrystalline powders, taking the ball-milled Fe powders as example, is an important step for reliable and reproducible determination of the physical/mechanical properties of the fabricated materials and is required for most industrial and structural applications. The end product resulted from a ball milling process usually consists of fine powder particles (~0.5–5 μm in diameter or less). These powders composite of nanocrystalline sizes of about 3–8 nm in diameter. Therefore, they are considered to be ideal feedstock materials that can be used for fabrications and manufacturing of complex and high-performance full dense bulk compacts.

3.2.3 Physical characteristics: magnetic properties of nanocrystalline Fe powders

Herzer (2013) has shown that the grain size of Fe-based nanocrystalline alloys was smaller than the exchange interaction length; thus, the significant magnetic anisotropy is an average over several structural units, which led to a significant decrease in the magnitude. It is well established that Fe-based nanocrystalline alloys reveal very low core loss (W), low coercivity (H_c), and near-zero magnetostrictive coefficient (λ_s) (Kuhnt et al., 2017). All of these attractive magnetic properties make Fe-based nanocrystalline alloys promising core materials used in electronics applications (Gutfleisch et al., 2011).

3.2.4 Potential applications of nanocrystalline Fe-based SW materials

Several advanced Fe-based nanocrystalline and amorphous materials are considered as primary components in today's electronic and electrical engineering industry. As of late, expanding the operating frequency in electronic devices and expanding the demands to save energy and decrease the fabricating costs has gotten to be more vital. This is driven to supplant, in a few cases, the conventional materials with the recently fabricated new family of amorphous and nanocrystalline Fe-based materials.

Nanocrystalline and amorphous alloys of Fe-based ferromagnetic materials, taking $Fe_{80.7}Si_4B_{13}Cu_{2.3}$ as an example, show excellent combinations of magnetic properties (Zhang et al., 2019). They possess exceptionally excellent soft magnetic properties, such as high permeability, high saturation magnetic, and low power. They are used in information handling technology as power devices and magnetic sensors. Selected metals and metalloid materials can be alloyed with the Fe to fabricate advanced magnetic alloys, such as Fe-Cr-Cu-Si-B-Sn powders (Ohta and Chiwata, 2020). This developed alloy exhibit high B_s of 1.55–1.60 T, and core losses of 4800–8000 kW/m^3.

4. Case studies of nonferrous alloys waste

The aim of this research is to obtain high grade of elemental metals, starting from inexpensive feedstock materials. In this section, we will present the results related to the recycling of copper (Cu) and magnesium (Mg) and their potential applications.

4.1 *Utilization of Cu waste for preparations of metallic glassy alloys*

4.1.1 *Introduction and work theme*

The usage of pure copper metal in various applications is restricted by its low strength (Butterworth et al., 1992). Copper can be strengthened successfully by different approaches, including cold working, grain refinement, solid solution hardening, precipitation hardening, dispersion strengthening, etc. Although such modification techniques do lead to a significant increase in Cu's strength, they might also lead to a pronounced reduction in its conductivity (Shijie et al., 2006; Atrens et al., 1997). It has been suggested by Lu et al. (2004) that ultrahigh-strength and high-conductivity Cu can be produced by subjecting the metal to high density of nanoscale twin boundaries. Alloying pure Cu metal with one or more elements (e.g., Ti, Zr, Ni, Al) can lead to significant improvements of mechanical strength and raises the softening temperature (Atrens et al., 1997). More recently, Cu nanoparticles, nanocrystalline Cu-based alloys, and nanocomposites, as well as metallic glassy Cu-based metastable materials, have shown unique and advanced properties (El-Eskandarani et al., 2020). These advanced Cu-based nanocrystalline alloys could have a wide range of useful applications, such as Cu nanoparticles for preparing highly efficient nanocatalysts (Baguc et al., 2018); Cu graphene, Cu graphite, Cu nanodiamonds, and carbon nanotubes for structure, corrosion resistance, thermal and electrical applications (Wang et al., 2019a); $Cu_{50}Ti_{20}Ni_{30}$ (Hendi et al., 2018), and $CuAl_2O_4$ spinel (Wang et al., 2019b) for antibacterial protective coating; Cu nanoparticles for optoelectronic applications (Hendi et al., 2018); Cu nanoparticles for effective photovoltaic applications (Chandrasekaran, 2013); Cu-based metallic glassy alloys for structural corrosion resistance applications (El-Eskanadarny et al., 2001, 2002, 2017); Cu-doped manganese ferrites nanoparticles for high-performance energy storage (Bashir et al., 2017); and finally, Cu-based shape memory alloys (Alaneme and Oketete, 2016).

4.1.2 *Copper-based nanoparticles and metallic glassy alloys*

Cu-based metallic glassy alloys are usually produced when their corresponding melt (liquid phase) fails to crystallize during relatively slow cooling (1 K/s). The field of glassy metals or metallic glasses has seen enormous development during recent years. In general, metallic glasses are new substances with exciting properties, which are of interest not only for basic solid-state physics but also for metallurgy, surface chemistry, and technology. Metallic glasses have properties that are quite different from solid metals, making them promising candidates for technical applications. The following are some of the essential properties that can be found in metallic glasses (El-Eskanadarny, 2020): high mechanical ductility and yield strength, high magnetic permeability, low coercive forces, unusual corrosion resistance, and temperature-independent electrical conductivity.

4.1.3 SW copper

It is well understood that waste is a by-product of our daily activities and social behavior. For hundreds of years, waste materials were viewed as undesired substances that must be dispose of after their primary use. For example, Cu feedstock materials can come from two sources, namely processed and refined run-of-mine ores (primary product) and a recycled end-of-life product (secondary product). Due to the rapid development and the monotonical increase in the level of life, particularly in developing countries, there has been a rapid increase of electric and electronic products. Accordingly, the corresponding Cu SW of wires is dramatically increased. In reference to the drastic consumption of copper, which are used to extract pure Cu metal, there is a significant lack in the production (19,700,000 tons)-to-consumption (25,200,000 tons) weight ratio (Xiao et al., 2016). This ratio is expected to be increased in the next decade due to the rapid growing demands of Cu. Therefore, Cu SW materials, which are made of high purity Cu, have become a very important feedstock material that can be successfully used after refining for producing different types of useful Cu-based advanced alloys and composites. The production of Cu metal from scrap requires much less energy compared with the production from copper ores (Veasey et al., 1993). Accordingly, copper recycling leads to a decrease in the continuous depletion of its natural mineral resources and enables its conservation. Besides, metal-SW recycling has received significant importance over the years, since the extraction of vast quantities of ores by mining approaches produces an enormous amount of pollution and industrial waste (Veasey et al., 1993).

4.1.3.1 Preparations of pure Cu bulk materials starting from Cu SW

Physical examinations and treatment of the as-received Cu SW batches are very important steps to ensure the absence of the SW batches from foreign materials, such as iron-scrape, broken glass pieces, and wood pieces. The batches were chemically treated with diluted HCl solution to remove out all dirties from their surfaces before annealing in a vacuum furnace at 250°C for 6 h. Annealing is an essential process used to ensure evaporating the undesired organic component adhered to the Cu surfaces. The materials were then snipped into smaller pieces and sonicated at room temperature in an acetone bath to remove the oil/fat contamination from their surfaces. Then they were rinsed with ethanol before subjecting to a drying process at 150°C in a vacuum furnace for overnight.

4.1.3.2 Preparations of Zr_2Cu master and amorphous alloys starting from SW Cu and Zr materials

The as-recovered Cu and Zr metals were used as feedstock materials for preparing Zr_2Cu alloy through the arc melt. To ensure the homogeneity in the chemical composition, the molten button was turned over and remelted for 6 times. The chemical analysis of prepared Zr_2Cu indicated the formation of pure master alloy composited of Zr (74.08 wt%), Cu (24.96 wt%), Hf (0.8 wt%), O_2 (0.04 wt%), and C (0.03 wt%). The as-prepared Zr_2Cu was disintegrated into finer pieces (~2.5 cm) and then charged in a lab-type disc mill. The disintegrated powders were classified through a sieving system to collect fine powders of ~50 μm in diameter. Hereafter, the disintegrated powders were introduced to a planetary mill vial (500 mL in volume) made of tool steel alloy vial and well-sealed under He atmosphere, with 75 tool steel balls (11 mm in diameter) inside the glove box. The ball-to-powder weight ratio was kept at a ratio of 20:1. The vial was installed on a planetary-type ball mill, where the ball milling process was carried out at a speed of 250 rpm for 50 h at ambient temperature. The XRD pattern of disintegrated Zr_2Cu alloy powders revealed a long-range ordered structure, as characterized by the sharp Bragg peaks presented in (Fig. 18.6). Formation of amorphous Zr_2Cu phase was confirmed by FE-HRTEM, which manifesting a dense random close-packed structure with mazelike morphology.

The thermal properties, indexed by glass transition temperature (T_g), crystallization temperature (T_x), supercooled liquid region ($\Delta T_x = T_x$-T_g), and the enthalpy change of crystallizations (ΔH_x) of amorphous-Zr_2Cu phase, obtained after 50 h of MD time, were investigated through DSC technique, where its related melting behaviors characterized by melting temperature (T_m), liquids temperature (T_L), and reduced glass transition temperature ($T_{rg} = T_g$/TL) were investigated by differential thermal analysis (DTA) technique. The onset (T_{onset}) and peak temperatures (T_p) of the crystallization peak were measured and found to be 589 and 603°C, respectively, where the ΔH_x (the measured area under the exothermic peak) was −6.88 kJ/mol. It should be emphasized that the broad ΔT_x value (93°C) and its large ΔH_x (−6.88 kJ/mol) implied that binary a-Zr_2Cu system has a GFA. Besides, this metallic glassy system has a good thermal stability, as authenticated by its high values of T_g and T_x.

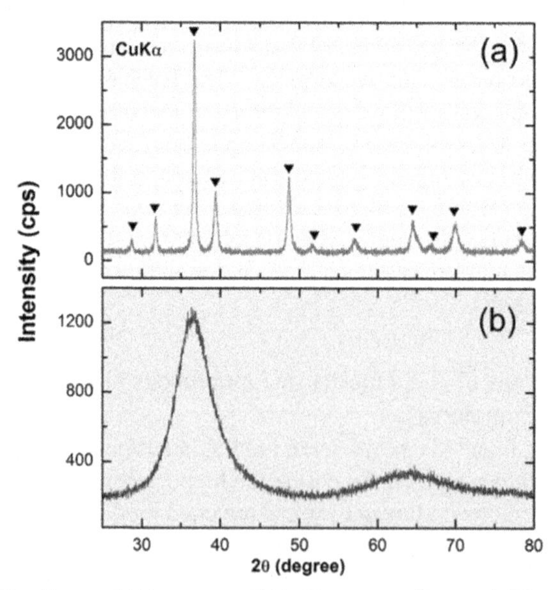

Figure 18.6 XRD patterns of (A) as-arc melt Zr_2Cu master alloy, and (B) as-high energy ball milled Zr_2Cu alloy powders obtained after 50 h of the milling time.

Solid waste copper and zirconium metals were recovered, using induction melting and arc melt techniques, respectively. The as-recovered bulk materials were used as alloying elements to prepare Zr_2Cu master alloy through arc melting. The alloy was then ball milled for 50 h, using a high-energy planetary mill. The as-ball milled powders were obtained after 50 h possessed amorphous structure. This amorphous phase revealed a high-thermal stability and excellent glass forming ability, as implied by its high values of the glass transition (496°C) and crystallization (589°C) temperatures. The broad supercooled liquid region (93°C) and the large ΔH_x (−6.88 kJ/mol) values implied that binary Zr_2Cu system has a good glass-forming ability.

4.2 Synthesizing of high-hydrogen storage nanomaterials starting from Mg waste

Apart from the potential applications of carbon nanotubes and complex hydrides in H_2 storage, the first reports demonstrated the possibility of formation of metal hydride through the chemisorption concept published in 1868 by T. Graham, when he discovered that metallic palladium (Pd) wires were readily charged with H_2 (Graham, 1868). Since then, many metal hydrides, including intermetallic compounds, presented significant capabilities of H_2 absorption in their solid state. Among metal hydrides, magnesium hydride (MgH_2) has been considering as a candidate for H_2 storage materials for producing high-capacity nanomaterials that could be utilized in fuel cell systems (El-Eskandarany, 2019, 2020, Zhang et al., 2020; Song et al., 2017; El-Eskandarany et al., 2016, 2018). The tetragonal phase of β-MgH_2 possesses high volumetric density (6.5 H atoms/cm^3), even higher than pure H_2 gas (0.99 H atoms/cm^3) or liquid H_2 (4.2 H atoms/cm^3) (Seung et al., 2001). The worldwide interest on Mg is owing to its abundance, low cost, lightweight, high gravimetric (7.60 wt.%) and volumetric (110 g/L) hydrogen storage capacities (El-Eskandarny, 2020). Since the early 2000s, many scientific and engineering studies have been carried out to develop the absorption/desorption of hydrogen in metals for fuel cell applications (Lototskyy et al., 2017).

5. Nanofluid fabrication from metallic waste and their potential applications

At the beginning of the 20th century, many researchers showed a large interest into developing enhanced types of fluids, either through fabricating new types or by improving these preexisting ones. Scientists such as Ahuja, Liu et al. and others from Argonne National Laboratory have adopted Maxwell's theoretical work by exploring dispersions containing mm- and μm-scaled particles within conventional fluids (Choi et al., 1991, 1992a, 1992b; Liu et al., 1988; Maxwell, 1881). Their intention was to enhance the thermal conductivity of the hosting fluid by adding solid particles of order of magnitude higher thermal conductivity than that of the liquid. For example, cupper (Cu), aluminum (Al), iron (Fe), and their oxides (i.e., CuO,

Al_2O_3, and Fe_2O_3) have a thermal conductivity of 401, 225.5, 77.5, 76.5, 40, and 6 W/m.K, respectively, whereas the thermal conductivity of pure water is only 0.608 W/m.K (Ali et al., 2018; Han and Fina, 2011; Sezer et al., 2019). Such large deviation in the aforementioned between the two substances showed the promising potential of their approach because when well dispersing these solids in a "basefluid," the mixture net (or effective) thermal conductivity will be somewhere between that of the solid particles and the liquid used, and hence the resulting suspension would have much higher enhanced thermal properties.

Despite this fact, the fabricated suspensions had some limitations especially when dynamically flowing in small passages as they tend to cause clogging or blockage within the small pipelines, due to the natural agglomeration between the dispersed particles. Such obstacle has led Masuda et al. (1993) to propose using ultrafine nanoscaled particles to overcome this problem, after which these types of colloidal where given the name "Nanofluids" by Choi and Eastman (1995) back in the year 1995. One can generally define nanofluids as *suspensions containing homogenously dispersing nanoparticles of less than 100 nm, and preferably of less than 1.0 volumetric concentration (vol. %), in a non-dissolving hosting fluid* (Ali et al., 2019). Furthermore, since the dispersed particles are in the nanoscale, they should in theory cause much higher thermal conductivity enhancement than their mm and μm counterparts. This is attributed to the fact that as the particle size reduces, its exposed surface area greatly increases to its surrounding, causing the effective thermal conductivity to largely increase (Choi and Eastman, 1995; Lee and Choi, 1996).

Nanofluids are fabricated nowadays from a variety of dispersed nanoparticles types, sizes, and shapes into basefluids such as water, ethanol, ethylene glycol, refrigerants, mineral oils, or even a mixture containing more than one of these or other types of liquids (Ali et al., 2018). In order to obtain the optimum output in which a nanofluid was intended for, it is crucial that the user carefully considers the following aspects in their suspension and of the nanoparticle used. This is in addition to potentially having them from various waste sources and components.

1. Nanoparticles: material; shape; size; surface charge; concentration; density; thermal conductivity; and specific heat capacity.
2. Basefluids: type; pH; temperature; molecular charge; density; viscosity; thermal conductivity; and specific heat capacity.
3. Preparation method.
4. Surfactant or particles functionalization (if added or employed).
5. Long- and short-term particles dispersion, chemical, and kinetic stabilities.

Selecting the right combination would help deliver a nanofluid with the favorable thermophysical properties for the targeted or potential application, e.g., solar systems (Bellos et al., 2020; Krishna et al., 2020), air conditioning and refrigeration (AC&R) systems (Akhavan-Behabadi et al., 2015; Babarinde et al., 2018; Coumaressin and Palaniradja, 2014; Hamisa et al., 2020; Wang et al., 2010; Yang et al., 2020a,b), and pool boiling heat transfer (Patra et al., 2017; Yahya et al., 2018).

Since metallic waste can be recycled and rescaled into nanopowder, such source can be very beneficial and an environmentally friendly form of feedstock for fabricating nanofluids. Moreover, when exploring the Elsevier's abstract and citation database, Scopus, for the phrase "Nanofluid from waste metal," it was found that only 29 documents existed from the year 2008 to 2019 (Scopus-Database, 2020). This does not necessarily mean that the area of metallic waste is not of high importance to the field of nanofluids, but is most likely due to the fact that the majority of the scientist working on nanofluid focus on dispersing commercially made particles or grow their own particles within the basefluid. As such, there is a large need for further investigation in this research area. Fig. 18.7 demonstrates the trend in scientific publications covered by the Scopus database.

The following subsection will have a more focus coverage on different aspects related to nanofluids. Subsection 5.1 will demonstrate the different types of nanofluids, whereas Subsection 5.2 will provide the reader with the different production approaches used in producing nanofluids and those suitable for nanoparticles obtained from recycled waste sources. Furthermore, the stability of the suspension and thermophysical properties will be covered in Subsections 5.3 and 5.4, respectively. Finally, Subsection 1.5 will demonstrate the potential applications for such type of advanced fluids along with some of the studies conducted in each.

5.1 Types of nanofluids

As previously mentioned, nanofluids are suspensions containing dispersed nano-sized solid particles in a nondissolving basefluid/s. These particles can be of metallic origin, such as Al, Cu, and Fe; oxides, such as Al_2O_3, CuO, and Fe_2O_3; alloys, such as Ag-Cu, Fe-Ni, and Cu-Zn; metal carbides, such as B_4C, SiC, and ZrC; carbon-based materials, such as carbon nanotube, graphene, diamond; etc. (Ali et al., 2018). In general, scientist have classified nanofluids into two categories. The first category is the commonly known "Nanofluid," where a single type of particles is used in the fabrication process of the suspension. On the contrary, when two or more types of particles are used, the

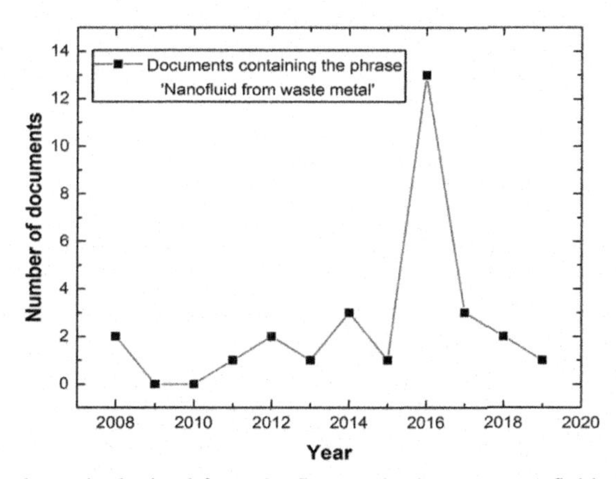

Figure 18.7 Search result obtained from the Scopus database on nanofluids from metal waste illustrating the number of documents published per year on the topic.

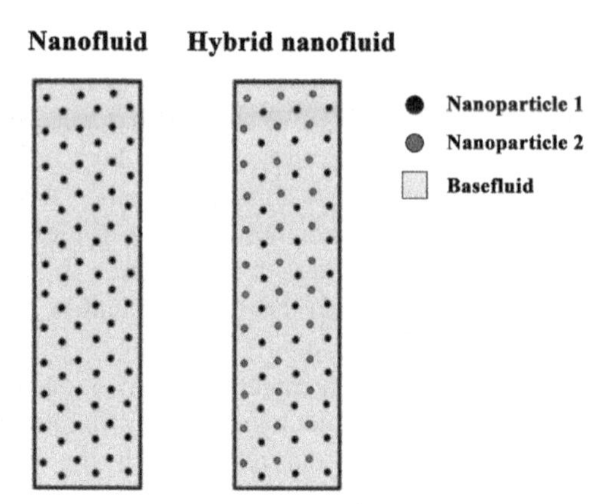

Figure 18.8 An illustration of the two categories of nanofluids, where the left side shows the conventional nanofluid, and the right side shows the hybrid nanofluid.

produced suspension is termed as "Hybrid nanofluid." Fig. 18.8 shows the two categories of nanofluids (i.e., conventional nanofluid and hybrid nanofluid). It is important to note that the first category was initially introduced by Masuda et al. (1993), whereas hybrid nanofluids were studied experimentally for the first time by Jana et al. (2007). Furthermore, there is a large debate between scientist when it comes to composites, as whether their dispersions could be referred to as nanofluids or hybrid nanofluids. Nevertheless, the authors of this work believe that, since the different particles are attached together and moves as a single entity, such as in the case of carbon nanotubes embedded in Cu or the other way around, then it is appropriate to refer to the suspension as nanofluid (Cai et al., 2019; Sundaram et al., 2018). Another thing that was noticed from having reviewed the literature is that the category of the suspension has nothing to do with the number of basefluids used in the production stage but rather is purely related to the number of different types of dispersed nanoparticles (Hamid et al., 2015). Researchers could have, for example, used the terms "Bi-liquid nanofluid" or "Tri-liquid nanofluid" to refer to a nanofluid containing two or three types of basefluids, respectively, and thus deviating from the conventional type of nanofluids (i.e., nanofluids fabricated from a single type of basefluid).

5.2 Production approaches

Nanofluids are formed when solid nanoparticles are dispersed within a basefluid. This mechanism can be mainly accomplished by growing the nano-scaled particles within the hosting liquid or through dispersing preprepared dried nanopowder in the basefluid. The first process is known as the one-step (also known as single-step) nanofluid fabrication approach, whereas the second is referred to as the two-step production method (Almurtaji et al., 2020). When comparing between the two fabrication techniques, the one-step route synthesizes the particles simultaneously within the basefluid while providing a high level of dispersion stability. In the case of metallic waste, the nanofluid can only be prepared with such feedstock though the two-step approach. This is because, as mentioned earlier, the dispersed particles in the one-step method are grown within the basefluid, which is something that cannot be done with recycled metallic waste.

The two-step nanofluid production approach involves an initial estimation of the required nanoparticles concentration, which can be theoretically calculated with respect to the volume percentage (*vol. %*) as per the following:

- For single type of nanoparticles and basefluid (Ali et al., 2018; Almurtaji et al., 2020)

$$vol.\ \% = \frac{V_{np}}{V_{np} + V_{bf}} \times 100 \text{ or } \frac{\left(\frac{m}{\rho}\right)_{np}}{\left(\frac{m}{\rho}\right)_{np} + \left(\frac{m}{\rho}\right)_{bf}} \times 100; \tag{18.1}$$

- For single type of nanoparticles and two types of basefluid (Hamid et al., 2015)

$$vol.\ \% = \frac{\left(\frac{m}{\rho}\right)_{np}}{\left(\frac{m}{\rho}\right)_{np} + \left[\left(\frac{m}{\rho}\right)_{bf1} + \left(\frac{m}{\rho}\right)_{bf2}\right]} \times 100; \tag{18.2}$$

- For two type of nanoparticles and single types of basefluid (Aghahadi et al., 2019; Asadi et al., 2020)

$$vol.\ \% = \frac{\left(\frac{m}{\rho}\right)_{np1} + \left(\frac{m}{\rho}\right)_{np2}}{\left[\left(\frac{m}{\rho}\right)_{np1} + \left(\frac{m}{\rho}\right)_{np2}\right] + \left(\frac{m}{\rho}\right)_{bf}} \times 100; \tag{18.3}$$

- For two type of nanoparticles and two types of basefluid (Kakavandi and Akbari, 2018)

$$vol.\ \% = \frac{\left(\frac{m}{\rho}\right)_{np1} + \left(\frac{m}{\rho}\right)_{np2}}{\left[\left(\frac{m}{\rho}\right)_{np1} + \left(\frac{m}{\rho}\right)_{np2}\right] + \left[\left(\frac{m}{\rho}\right)_{bf1} + \left(\frac{m}{\rho}\right)_{bf2}\right]} \times 100; \tag{18.4}$$

where V, m, ρ, np, $np1$, $np2$, bf, $bf1$, and $bf2$ are the volume, mass, density, nanoparticle, first type of nanoparticle, second type of nanoparticles, basefluid, first type of basefluid, and second type of basefluid, respectively.

5.3 Thermophysical properties

Although the effective thermal conductivity was the main fluid property that derived researchers into fabricating nanofluids, other thermophysical properties of suspensions are still of equal importance when it comes to real-life applications. Properties such as the density, specific heat capacity, and viscosity of the liquid will all change when the nanoparticles get dispersed in the basefluid. While some of these effective properties can be reliably calculated in a theoretically manner (e.g., effective density and effective specific heat capacity), others, such as the effective thermal conductivity and effective viscosity, do not yet have a universal formula. Most of the correlations used today to predict the effective thermal conductivity and effective viscosity of nanofluids are limited to the experimental conditions in which they were developed for. As such, it is important that further work be done into developing a universal equation that can be employed for the two previous properties and for any type of nanofluid. In order to calculate the effective density and effective specific heat capacity of the suspension, the following equations can be reliably used:

$$\rho_{eff} = vol. \times \rho_{np} + (1 - \text{vol.}) \times \rho_{bf} \tag{18.5}$$

$$C_{p_{eff}} = \frac{\rho_{bf} \times (1 - vol.)}{\rho_{nf}} \times C_{p_{bf}} + \frac{\rho_{np} \cdot \text{vol.}}{\rho_{nf}} \times C_{p_{np}} \tag{18.6}$$

where ρ_{eff}, ρ_{np}, ρ_{bf}, $C_{p_{eff}}$, $C_{p_{bf}}$, and $C_{p_{np}}$ are the effective density of the nanofluid, density of the nanoparticles, density of the basefluid, effective specific heat capacity of the nanofluid, specific heat capacity of the basefluid, and specific heat capacity of the nanoparticles, respectively. As mentioned earlier, there are many available formulas in the literature for predicting the effective thermal conductivity and effective viscosity of the suspension (see Ali et al. (2018) and Bakthavatchalam et al. (2020) published works for a list of these correlations) but up to today the optimum method to obtain these two properties is through experimental measurements.

5.4 Potential applications

There are many applications in which nanofluids can play a significant enhancement role in their performance. Examples of such include medical applications (Saleh et al., 2017), liquid fuels combustion enhancement (Wang et al., 2020), air purification systems (Yang et al., 2020a,b), quenching media (Patra et al., 2017; Ramadhani et al., 2019), magnetic sealing (Li et al., 2018), nanolubricants (Asadi et al., 2016; Asadi and Asadi, 2016), heat exchangers working fluid (Almurtaji et al., 2020), air conditioning and refrigeration systems (Kundan and Singh, 2020), gas turbine intercooler (Alsayegh and Ali, 2020), and many more.

References

Ackerman, F., 1997. Why Do We Recycle? Markets, Values and Public Policy. Island Press.

Aghahadi, M.H., Niknejadi, M., Toghraie, D., 2019. An experimental study on the rheological behavior of hybrid Tungsten oxide (WO3)-MWCNTs/engine oil Newtonian nanofluids. J. Mol. Struct. 1197, 497–507. https://doi.org/10.1016/j.molstruc.2019.07.080.

Akhavan-Behabadi, M.A., Sadoughi, M.K., Darzi, M., Fakoor-Pakdaman, M., 2015. Experimental study on heat transfer characteristics of R600a/POE/CuO nano-refrigerant flow condensation. Exp. Therm. Fluid Sci. 66, 46–52. https://doi.org/10.1016/j.expthermflusci.2015.02.027.

Al-Salem, S.M., 2019a. Introduction. Chapter 1 in: Plastics to Energy: Fuel, Chemicals, and Sustainability Implications, first ed. Elsevier, pp. 3–20. https://doi.org/10.1016/B978-0-12-813140-4.00001-7. ISBN: 978-012813140-4;978-012813141-1.

Al-Salem, S.M., 2019b. Major Technologies Implemented for Chemicals and Fuel Recovery. Chapter 2 in: Plastics to Energy: Fuel, Chemicals, and Sustainability Implications, first ed. Elsevier, pp. 21–44. https://doi.org/10.1016/B978-0-12-813140-4.00002-9. ISBN: 978-012813140-4;978-012813141-1.

Al-Salem, S.M., Lettieri, P., Baeyens, J., 2009. Recycling and recovery routes of plastic solid waste (PSW): a review. Waste Manag. 29, 2625–2643.

Alaneme, K., Okotete, E., 2016. Engineering science and technology. Intl. J 19, 1582.

Ali, N., Teixeira, J.A., Addali, A., 2018. A review on nanofluids: fabrication, stability, and thermophysical properties. J. Nanomater. 33. https://doi.org/10.1155/2018/6978130, 2018.

Ali, N., Teixeira, J.A., Addali, A., 2019. Aluminium nanofluids stability: a comparison between the conventional two-step fabrication approach and the controlled sonication bath temperature method. J. Nanomater. 9. https://doi.org/10.1155/2019/3930572, 2019.

Almurtaji, S., Ali, N., Teixeira, J.A., Addali, A., 2020. On the role of nanofluids in thermal-hydraulic performance of heat exchangers—a review. Nanomaterials 10 (4), 734.

Alsayegh, A., Ali, N., 2020. Gas turbine intercoolers: introducing nanofluids—a mini-review. Processes 8 (12), 1572. https://doi.org/10.3390/pr8121572.

Alyousef, R., Mohammadhosseini, H., Tahir, M.M., Alabduljabbar, H., 2020. Green concrete composites production comprising metalized plastic waste fibers and palm oil fuel ash. Mater. Today Proc. https://doi.org/10.1016/j.matpr.2020.04.023 (in press).

Asadi, M., Asadi, A., 2016. Dynamic viscosity of MWCNT/ZnO–engine oil hybrid nanofluid: an experimental investigation and new correlation in different temperatures and solid concentrations. Int. Commun. Heat Mass Tran. 76, 41–45. https://doi.org/10.1016/j.icheatmasstransfer.2016.05.019.

Asadi, A., Asadi, M., Rezaei, M., Siahmargoi, M., Asadi, F., 2016. The effect of temperature and solid concentration on dynamic viscosity of MWCNT/MgO (20–80)–SAE50 hybrid nano-lubricant and proposing a new correlation: an experimental study. Int. Commun. Heat Mass Tran. 78, 48–53. https://doi.org/10.1016/j.icheatmasstransfer.2016.08.021.

Asadi, A., Alarifi, I.M., Foong, L.K., 2020. An experimental study on characterization, stability and dynamic viscosity of CuO-TiO_2/water hybrid nanofluid. J. Mol. Liq. 307, 112987. https://doi.org/10.1016/j.molliq.2020.112987.

Atrens, A., Nairn, J., Fernee, H., FitzGerald, K., Skennerton, G., Olofinjana, A., 1997. Mater. Forum 21, 57.

Babarinde, T.O., Akinlabi, S.A., Madyira, D.M., 2018. Enhancing the performance of vapour compression refrigeration system using nano refrigerants: a review. IOP Conf. Ser. Mater. Sci. Eng. 413, 012068. https://doi.org/10.1088/1757-899x/413/1/012068.

Baguc, I., Ertas, I., Yurderi, M., Bulut, A., 2018. Inorg. Chim. Acta. 483, 431.

Bakthavatchalam, B., Habib, K., Wilfred, C.D., et al., 2020. Comparative evaluation on the thermal properties and stability of MWCNT nanofluid with conventional surfactants and ionic liquid. J. Therm. Anal. Calorim. https://doi.org/10.1007/s10973-020-10374-x.

Bashir, B., Shaheen, W., Asghar, M., Warsi, M., 2017. J. Alloys Compd. 695, 881.

Bellos, E., Tzivanidis, C., Said, Z., 2020. A systematic parametric thermal analysis of nanofluid-based parabolic trough solar collectors. Sust. Energy Technol. Assess. 39, 100714. https://doi.org/10.1016/j.seta.2020.100714.

Butterworth, G.J., Forty, C.B.A., 1992. J. Nucl. Mater. 189, 237.

Cai, X., Wang, Z., Yang, C., Zhou, L., Hu, C., Zhu, W., 2019. Fabrication of CNTs reinforced copper composite powders by electrochemical co-deposition. Integrated Ferroelectrics Int. J. 201 (1), 249–257. https://doi.org/10.1080/10584587.2018.1454767.

Chancerel, P., Rotter, S., 2009. Recycling-oriented characterization of small waste electrical and electronic equipment. Waste Manag. 29, 2336–2352, 5.

Chandrasekaran, S., 2013. Sol. Energy Mater. Sol. Cell. 109, 220.

Changming, D., Chao, S., Gong, X., Ting, W., Xiange, W., 2018. Plasma methods for metals recovery from metal–containing waste. Waste Manag. 77, 373–387.

Choi, S.U.S., Eastman, J.A., 1995. Enhancing thermal conductivity of fluids with nanoparticles. In: Paper Presented at the Conference: 1995 International Mechanical Engineering Congress and Exhibition, San Francisco, CA (United States), 12–17 Nov 1995. Other Information: PBD: Oct 1995. http://www.osti.gov/scitech/servlets/purl/196525.

Choi, U., Tran, T., 1991. Experimental Studies of the Effects of Non-newtonian Surfactant Solutions on the Performance of a Shell-And-Tube Heat Exchanger Recent Developments in Non-newtonian Flows and Industrial Applications, vol. 124. The American Society of Mechanical Engineers, New York, FED, pp. 47–52.

Choi, S.U.S., Cho, Y.I., Kasza, K.E., 1992a. Degradation effects of dilute polymer-solutions on turbulent friction and heat-transfer behavior. J. Non-Newtonian Fluid Mech. 41 (3), 289–307. https://doi.org/10.1016/0377-0257(92)87003-T.

Choi, U., France, D.M., Knodel, B.D., 1992b. Impact of Advanced Fluids on Costs of District Cooling Systems. Argonne National Lab., IL (United States).

Coumaressin, T., Palaniradja, K., 2014. Performance analysis of a refrigeration system using nano fluid. Int. J. Adv. Mech. Eng. 4 (4), 459–470.

El-Eskandarany, M.S., Inoue, A., 2002. Mater. Trans. 43, 608.

El-Eskandarany, M.S., 2020. Mechanical Alloying, Energy Storage, Protective Coatings, and Medical Applications, third ed. Elsevier, Oxford, UK.

El-Eskandarany, M.S., Matsushita, M., Inoue, A., 2001. J. Alloys Compd. 329, 239.

El-Eskandarany, Sherif, M., Shaban, E., Alsairafi, A., 2016. Synergistic dosing effect of TiC/FeCr nanocatalysts on the hydrogenation/dehydrogenation kinetics of nanocrystalline MgH_2 powders. Energy 104, 158–170.

El-Eskandarany, Sherif, M., Al-Nasrallah, E., Banyan, M., Al-Ajmi, F., 2018. Bulk nanocomposite MgH_2/10 wt% (8 Nb_2O_5/2 Ni) solid-hydrogen storage system for fuel cell applications. Int. J. Hydrogen Energy 43, 23382–23396.

El-Eskandarany, Sherif, M., Banyan, M., Al-Ajmi, F., 2019. Environmentally friendly nanocrystalline magnesium hydride decorated with metallic glassy zirconium palladium nanopowders for fuel cell applications. RSC Adv. 9, 27987.

El-Eskandarany, Sherif, M., Ali, N., Sultan Majed, A.-S., 2020. Solid-state conversion of magnesium waste to advanced hydrogen-storage nanopowder particles. Nanomaterials 10, 1037. https://doi.org/10.3390/nano10061037.

El-Eskandarany, Sherif, M., Al-Salem, Majed, S., Ali, N., 2020a. Solid-state conversion of magnesium waste to advanced hydrogen-storage nanopowder particles. Nanomaterials 10, 1037. https://doi.org/10.3390/nano10061037, 2020.

El-Eskandarany, M.S., Al-Salem, Majed, S., Ali, N., 2020b. Top-Down reactive approach for the synthesis of disordered ZrN nanocrystalline bulk material from solid waste. Nanomaterials 10, 1826. https://doi.org/10.3390/nano10091826, 2020.

El-Eskandarany, Sherif, M., Al-Hajji, L., Al-Hazza, A., 2017. Adv. Powder Technol. 28, 814.

Graham, T., 1868. On the relation of hydrogen to palladium. Proc. Roy. Soc. Lond. 17, 212–220.

Gutfleisch, O., Willard, M.A., Brück, E., Chen, C.H., Sankar, S.G., Liu, J.P., 2011. Magnetic materials and devices for the 21st century: stronger, lighter, and more energy efficient. Adv. Mater. 23, 821–842.

Hamid, K.A., Azmi, W.H., Mamat, R., Usri, N.A., Najafi, G., 2015. Investigation of Al2O3 nanofluid viscosity for different water/EG mixture based. Energy Procedia 79, 354–359. https://doi.org/10.1016/j.egypro.2015.11.502.

Hamisa, A.H., Yusof, T.M., Azmi, W.H., Mamat, R., Sharif, M.Z., 2020. The stability of TiO_2/POE nanolubricant for automotive air-conditioning system of hybrid electric vehicles. IOP Conf. Ser. Mater. Sci. Eng. 863, 012050. https://doi.org/10.1088/1757-899x/863/1/012050.

Han, Z., Fina, A., 2011. Thermal conductivity of carbon nanotubes and their polymer nanocomposites: a review. Prog. Polym. Sci. 36 (7), 914–944. https://doi.org/10.1016/j.progpolymsci.2010.11.004.

Hao, J., Wang, Y., Wu, Y., Guo, F., 2020. Metal recovery from waste printed circuit boards: a review for current status and perspectives. Resour. Conserv. Recycl. 157, 104787.

Hendi, A., Rashad, A.M., 2018. Phys. B Condens. Matter 538, 185.

Herzer, G., 2013. Modern soft magnets: amorphous and nanocrystalline materials. Acta Mater. 61, 718–734.

Ismail, H., Hanafiah, M.M., 2020. A review of sustainable e-waste generation and management: present and future perspectives. J. Environ. Manag. 264, 110495.

Jana, S., Salehi-Khojin, A., Zhong, W.H., 2007. Enhancement of fluid thermal conductivity by the addition of single and hybrid nano-additives. Thermochim. Acta 462 (1–2), 45–55. https://doi.org/10.1016/j.tca.2007.06.009.

Kakavandi, A., Akbari, M., 2018. Experimental investigation of thermal conductivity of nanofluids containing of hybrid nanoparticles suspended in binary base fluids and propose a new correlation. Int. J. Heat Mass Tran. 124, 742–751. https://doi.org/10.1016/j.ijheatmasstransfer.2018.03.103.

Kaza, S., Yao, L., Bhada-Tata, P., Van Woerden, F., 2018. What a Waste 2.0: A Global Snapshot of Solid Waste Management to 2050. World Bank Publications. https://doi.org/10.1596/978-1-4648-1329-0.

Krishna, Y., Faizal, M., Saidur, R., Ng, K.C., Aslfattahi, N., 2020. State-of-the-art heat transfer fluids for parabolic trough collector. Int. J. Heat Mass Tran. 152, 119541. https://doi.org/10.1016/j.ijheatmasstransfer.2020.119541.

Kuhnt, M., Marsilius, M., Strache, T., Polak, C., Herzer, G., 2017. Magnetostriction of nanocrystalline (Fe,Co)-Si-B-P-Cu alloys. Scripta Mater. 130, 46–48.

Kundan, L., Singh, K., 2020. Improved performance of a nanorefrigerant-based vapor compression refrigeration system: a new alternative. Proc. IME J. Power Energy. https://doi.org/10.1177/0957650920904553, 0957650920904553.

Lee, S., Choi, S.U.S., 1996. Application of Metallic Nanoparticle Suspensions in Advanced Cooling Systems, vol. 72. American Society of Mechanical Engineers, Materials Division (Publication) MD, pp. 227–234, 10.1007/s10765-017-2218-6.

Li, Z.G., Zhang, J.R., Li, B., Liu, X.B., Yang, F.Y., 2018. Investigation of friction power consumption and the performance of a water turbine seal based on the imbalanced rotation of magnetic nanofluids. In: Paper Presented at the IOP Conference Series: Earth and Environmental Science.

Liu, K., Choi, U., Kasza, K.E., 1988. Measurements of Pressure Drop and Heat Transfer in Turbulent Pipe Flows of Particulate Slurries. Argonne National Lab., IL (USA).

Lototskyy, M.V., Tolj, I., Pickering, L., Sita, C., Barbir, F., Yartys, V., 2017. The use of metal hydrides in fuel cell applications. Prog. Nat. Sci. 27, 3–20.

Lu, L., Shen, Y., Chen, X., Qian, L., Lu, K., 2004. Science 304, 422.

Masuda, H., Ebata, A., Teramae, K., 1993. Alteration of Thermal Conductivity and Viscosity of Liquid by Dispersing Ultra-fine Particles. Alteration of Thermal Conductivity and Viscosity of Liquid by Dispersing Ultra-Fine Particles. Dispersion of Al_2O_3, SiO_2 and TiO_2 Ultra-Fine Particles, pp. 227–233.

Matsuno, Y., Daigo, I., Adachi, Y., 2007. Application of Markov chain model to calculate the average number of times of use of a material in society - an allocation methodology for open-loop recycling - part 2: case study for steel. Int. J. Life Cycle Assess. 12, 34–39.

Maxwell, J.C., 1881. A Treatise on Electricity and Magnetism, 2 ed., vol. 1. Clarendon Press.

Oguchi, M., Murakami, S., Sakanakura, H., Kida, A., Kameya, T., 2011. A preliminary categorization of end of-life electrical and electronic equipment as secondary metal resources. Waste Manag. 31, 2150–2160.

Ohta, M., Chiwata, N., 2020. Development of Fe-based high B_s nanocrystalline alloy powder. J. Magn. Magn Mater. 509 https://doi.org/10.1016/j.jmmm.2020.166838.

Panda, R., Jadhao, P.R., Pant, K.K., Naik, S.N., Bhaskar, T., 2020. Eco-friendly recovery of metals from waste mobile printed circuit boards using low temperature roasting. J. Hazard Mater. 395, 122642.

Patra, N., Gupta, V., Singh, R., Singh, R.S., Ghosh, P., Nayak, A., 2017. An experimental analysis of quenching of continuously heated vertical rod with aqueous Al2O3 nanofluid. Res.-Effi. Technol. 3 (4), 378–384. https://doi.org/10.1016/j.reffit.2017.02.006.

Ramadhani, C.A., Putra, W.N., Rakhman, D., Oktavio, L., Harjanto, S., 2019. A comparative study on commercial grade and laboratory grade of TiO_2 particle in manofluid for quench medium in rapid quenching process. In: Paper Presented at the IOP Conference Series: Materials Science and Engineering.

Saleh, H., Alali, E., Ebaid, A., 2017. Medical applications for the flow of carbon-nanotubes suspended nanofluids in the presence of convective condition using Laplace transform. J. Assoc. Arab Univ. Basic Appl. Sci. 24, 206–212. https://doi.org/10.1016/j.jaubas.2016.12.001.

Scopus-Database, 2020. Nanofluid from Waste Metal Analysis Search Results for the Documents Published from 2008 to 2019. Retrieved August, 2020, from. www.scopus.com.

Seung, M.L., Kay, H.A.,Y.H.L., Gotthard, S., Thomas, F., 2001. A hydrogen storage mechanism in single-walled carbon nanotubes. J. Am. Chem. Soc. 123, 5059–5063.

Sezer, N., Atieh, M.A., Koç, M., 2019. A comprehensive review on synthesis, stability, thermophysical properties, and characterization of nanofluids. Powder Technol. 344, 404–431. https://doi.org/10.1016/j.powtec.2018.12.016.

Shijie, Z., Bingjun, Z., Zhen, Z., 2006. Xin J. J. Rare Earths 24, 385.

Song, J.-zheng, Zhao, Z.-yang, Zhao, X., Fu, R.-dong, Han, S.-min, 2017. Hydrogen storage properties of MgH_2 co-catalyzed by LaH_3 and NbH. Intl. J. Min. Metall. Mat. 24, 1183−1191.

Sundaram, R.M., Sekiguchi, A., Sekiya, M., Yamada, T., Hata, K., 2018. Copper/carbon nanotube composites: research trends and outlook. Royal Soc. Open Sci. 5 (11), 180814. https://doi.org/10.1098/rsos.180814.

Veasey, V.J., Wilson, R.J., Squires, D.M., 1993. The Physical Separation and Recovery of Metals from Wastes. Gordon and Breach Science Publishers, pp. 1−19.

Wang, R., Wu, Q., Wu, Y., 2010. Use of nanoparticles to make mineral oil lubricants feasible for use in a residential air conditioner employing hydro-fluorocarbons refrigerants. Energy Build. 42 (11), 2111−2117. https://doi.org/10.1016/j.enbuild.2010.06.023.

Wang, J., Guo, L.-na, Lin, W.-ming, Chen, J., 2019a. N. Carbon Mater. 34, 161.

Wang, Z., Liang, K., Chan, S.-W., Tang, Y., 2019b. J. Hazard Mater. 371, 5 550.

Wang, X., Zhang, J., Ma, Y., Wang, G., Han, J., Dai, M., Sun, Z.Y., 2020. A comprehensive review on the properties of nanofluid fuel and its additive effects to compression ignition engines. Appl. Surf. Sci. 504 https://doi.org/10.1016/j.apsusc.2019.144581.

Wasserscheid, P., Keim, W., 2000. Ionic liquids-new "solutions" for transition metal catalysis. Angew. Chem. Int. Ed. 39, 3772−3789.

WSA, 2020. World Steel Association. Visited on September 10, 2020. https://www.worldsteel.org.

Xiao, S., Xiong, W., Lijun, W., Ren, Q., 2016. Proc. 4th Intl. Conf. Sust. Energy .Environ. Eng. 500.

Yahya, S.S., Harjanto, S., Putra, W.N., Ramahdita, G., Kresnodrianto, Mahiswara, E.P., 2018. Characterization and observation of water-based nanofluids quench medium with carbon particle content variation. AIP Conf. Proc. 1964 (1), 020006. https://doi.org/10.1063/1.5038288.

Yang, L., Ji, W., Mao, M., Huang, J.-n., 2020a. An updated review on the properties, fabrication and application of hybrid-nanofluids along with their environmental effects. J. Clean. Prod. 257, 120408. https://doi.org/10.1016/j.jclepro.2020.120408.

Yang, L., Jiang, W., Ji, W., Mahian, O., Bazri, S., Sadri, R., Wongwises, S., 2020b. A review of heating/cooling processes using nanomaterials suspended in refrigerants and lubricants. Int. J. Heat Mass Tran. 153, 119611. https://doi.org/10.1016/j.ijheatmasstransfer.2020.119611.

Zhang, C., Li, Y., Duan, Y., Zhang, W., 2019. Preparation and electromagnetic properties of $Fe_{80.7}Si_4B_{13}Cu_{2.3}$ nanocrystalline alloy powders for electromagnetic wave absorbers in X-band. J. Magn. Magn Mater. 497 https://doi.org/10.1016/j.jmmm.2019.165988.

Zhang, X.L., Liu, Y.F., Zhang, X., Hu, J.J., Gao, M.X., Pan, H.G., 2020. Empowering hydrogen storage performance of MgH_2 by nanoengineering and nanocatalysis. Mat. Today Nano 9. https://doi.org/10.1016/j.mtnano.2019.100064.

Further reading

Asadi, A., Pourfattah, F., Miklós Szilágyi, I., Afrand, M., Żyła, G., Seon Ahn, H., Mahian, O., 2019. Effect of sonication characteristics on stability, thermophysical properties, and heat transfer of nanofluids: a comprehensive review. Ultrason. Sonochem. 58, 104701. https://doi.org/10.1016/j.ultsonch.2019.104701.

Drzazga, M., Dzido, G., Lemanowicz, M., Gierczycki, A., 2012. Influence of nonionic surfactant on nanofluid properties. In: Paper Presented at the 14th European Conference on Mixing, Warszawa, Poland. Boston, MA, USA.

EIA, 2019b. Energy Information Administration (EIA). 2019. U.S. Energy-Related Carbon Dioxide Emissions, 2014. U.S. Department of Energy, WashintonWashington, DC. Online at. http://www.eia.gov/environment/emissions/carbon/. (Accessed 31 August 2020).

El-Eskandarany, M.S., 2016. Metallic glassy $Zr_{70}Ni_{20}Pd_{10}$ powders for improving the hydrogenation/dehydrogenation behavior of MgH_2. Sci. Rep. 6, 26936. https://doi.org/10.1038/srep26936.

El-Eskandarany, M.S., 2019. Recent developments in the fabrication, characterization and implementation of MgH_2-based solid-hydrogen materials in the Kuwait Institute for Scientific Research. RSC Adv. 9, 9907−9930.

El-Eskandarany, Sherif, M., 2019. Metallic glassy Ti_2Ni grain-growth inhibitor powder for enhancing the hydrogenation/dehydrogenation kinetics of MgH_2. RSC Adv. 9, 1036.

El-Eskandrany, M., Sherif Al-Azmi, A., 2016. J. Mech. Beh. Biomed. Mat. 56, 183.

Herzer, G., 1997. Nanocrystalline soft magnetic alloys. Handb. Magn. Mater. 10, 415−462.

Ilyas, S.U., Pendyala, R., Marneni, N., 2013. Settling characteristics of alumina nanoparticles in ethanol-water mixtures. & i. Trans Tech Publications. In: Conference, B.E. P.O.f.A., Vol, A.((Eds.), Vol. 372. 2013 2nd International Conference on Advanced Materials Design and Mechanics, ICAMDM. Kuala Lumpur, pp. 143−148, 2013.

Khairul, M.A., Saidur, R., Hossain, A., Alim, M.A., Mahbubul, I.M., 2014. Heat transfer performance of different nanofluids flows in a helically coiled heat exchanger. Adv. Mater. Res. 832, 160−165.

Nair, V., Tailor, P.R., Parekh, A.D., 2016. Nanorefrigerants: a comprehensive review on its past, present and future. Int. J. Refrig. Internationale Du Froid 67, 290−307. https://doi.org/10.1016/j.ijrefrig.2016.01.011.

Pita, F., Castilho, A., 2018. Minerals 8, 517.

Sabiha, M.A., Mostafizur, R.M., Saidur, R., Mekhilef, S., 2016. Experimental investigation on thermo physical properties of single walled carbon nanotube nanofluids. Int. J. Heat Mass Tran. 93, 862−871. https://doi.org/10.1016/j.ijheatmasstransfer.2015.10.071.

Wang, A., Zhao, C., He, A., Men, H., Chang, C., Wang, X., 2016. Composition design of high B_s Fe-based amorphous alloys with good amorphous-forming ability. J. Alloys Compd. 656, 729−734.

World Steel Association, www.worldsteel.org. Visited on September 10, 2020.

Yu, W., Xie, H.Q., 2012. A review on nanofluids: preparation, stability mechanisms, and applications. J. Nanomater. 2012. doi: Artn435873.10.1155/2012/435873.

Future of sustainability and resources management

19

Milan Majerník[1], *Marcela Malindžáková*[2], *Jana Naščáková*[3], *Lucia Bednárová*[2], *Peter Drábik*[1]
[1]University of Economics in Bratislava, Research Institute of Trade and Sustainable Business, Bratislava, Slovak Republic; [2]Technical University of Košice, Faculty of Mining, Ecology, Process Control and Geotechnologies, Košice, Slovak Republic; [3]University of Economics in Bratislava, Faculty of Business Economy of the University of Economics in Bratislava with the Seat in Košice, Košice, Slovak Republic

1. Priorities of social responsibility—from social, economic, and environmental sustainability viewpoint

An enterprise involves a large number of interactions as well as social partnerships that the enterprise cannot ignore because, in addition to business objectives, it is also supposed to meet societal objectives. In the context of company's entrepreneurial activities, a broad responsibility is essential. Social responsibility must take into account not only the social, economic, and environmental aspects but also the ethical aspect. The role of responsible business is to search for the forms of higher quality, morality, because they deal with areas of social action and are not sanctionable by legislative rules. Attention should be drawn to focusing on bringing benefits, for both the enterprise and society, in order to specify the environmental and social consequences. The orientation of social responsibility aims at ensuring that businesses are also beneficial and useful for the community in which they operate and should eliminate the negative consequences of their activities too.

Ecological, social, and economic degradation does not threaten the planet, but humanity and its way of life. CSR is a voluntary and continuous attitude of the company to various problems of society, such as environmental protection, social issues, social employment policy-making, etc. CSR takes into consideration economic, environmental, as well as social aspects. Socially responsible behavior cannot be imposed on businesses where laws are not clearly defined, even though there have already occurred attempts at such approach. In addition to EN ISO 26000:2010, the issue of CSR is dealt with by broader legislation. These are described in the following subsections.

Sustainable Resource Management. https://doi.org/10.1016/B978-0-12-824342-8.00004-3

1.1 Social responsibility in accordance with the requirements of the norm

Social responsibility is specified by EN ISO 26000:2010 Guidance on Social Responsibility, which can be applied to all types of organizations in pursuit of sustainable development. EN ISO 26000:2010 is not a management system area or a certification standard. The main objective is to ensure that organizations implement social and environmental activities in decision-making and ultimately take responsibility for their impact on society and the environment.

Social responsibility must also be introuced into the organizational structure of the company. An essential principle is to take into account legal obligations and requirements. The implementation of CSR also requires the involvement of stakeholders interested in participating in relevant decisions and activities.

The principles of CSR include:

- responsibility—for the impacts on society, the environment, as well as the economy,
  - transparency—of decisions and activities that affect society and the environment,
- ethical behavior—honesty, fairness, and integrity,
  - respect for the interests of the involved parties,
- respect for the rule of law,
- respect for international standards of conduct,
- respect for human rights.

The exercise of social responsibility requires identification, recognition, and involvement of stakeholders. The basis for recognition is cooperation of three relationships, namely relations between the organization and society, between the organization and the stakeholders, and relations between the involved parties themselves (Fig. 19.1). Implementation of these relations may take place differently, mostly in the form of dialogue, either from the side of the organization or vice versa. The priority of the relevant actors' involvement will be

- understanding of the possible consequences of decisions for entities,
- determining the most appropriate way to increase the positive contribution of decisions and activities,
  - increase of transparency, as well as other activities (Fig. 19.1).

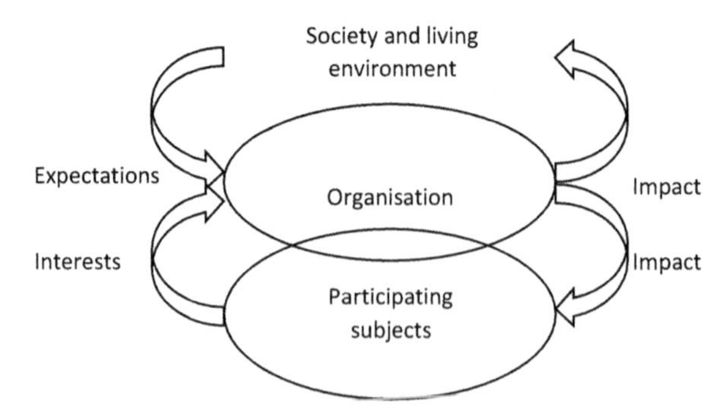

Figure 19.1 Interconnections of three areas. Source: standard EN ISO 26000:2010.

1.2 Key areas of social responsibility

In identifying relevant topics as well as prioritizing, the organization should focus on the following key areas (Fig. 19.2).

A number of social responsibility themes are part of these key areas. Topics related to key areas are presented by different activities and expectations. The dynamics of social responsibility are demonstrated in social and environmental developments. In the future, other topics may be discussed for the realization of social responsibility. The organization's role is to identify and address, for each key area, topics related to decisions and activities or topics which are important to them. It is necessary to take into account short- and long-term objectives for the purpose of assessing the seriousness of the topic. Each organization has to address its own key areas and themes, so the order of issues dealt with in different areas and situations, as well as in contexts, will be different.

1.3 Governance and management

It is one of the most important and key areas in taking responsibility for the impact of decisions and activities in the field of social responsibility. It is a system in which the organization takes and implements decisions aimed at fulfilling its own goals. Elements of social responsibility, processes, as well as mechanisms for their application need to be involved in the decision-making and implementation process. The priority aspect of efficient governance is the management of the organization (standard EN ISO 26000:2010).

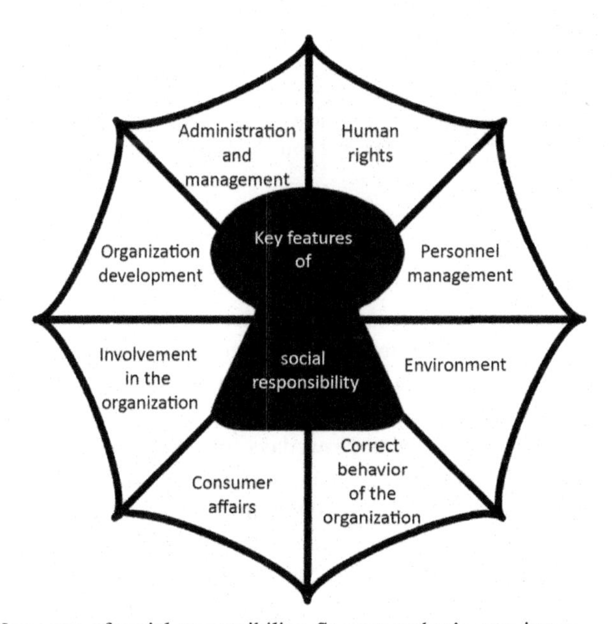

Figure 19.2 Key areas of social responsibility. Source: author's own image.

1.4 Human rights

This sphere includes fundamental rights to which all people are entitled. It covers civil and political rights, economic, social, and cultural rights. Each organization is responsible for respecting human rights, including their impact on society and stakeholders. Human rights are exercised by the organization both within its own premises and in the ranks of its own staff (standard EN ISO 26000:2010).

The field of human rights involves a number of significant topics:

- due diligence,
- situations threatening human rights,
  - exclusion of complicit,
  - dealing with complaints,
  - discrimination and vulnerable groups,
  - civil and political rights,
  - economic, social, and cultural rights,
  - fundamental principles and rights in the workplace.

1.5 Personnel management

This area contains strategies and procedures which are being carried out within the organization as well as the contracted work. The field of personnel management consists of the following (standard EN ISO 26000:2010):

- recruitment and promotion of staff,
- disciplinary procedures,
- complaint handling procedures,
- transfer and relocation of workers,
- termination of employment,
- professional training and skills development,
- health protection, safety, and industrial hygiene,
- strategies and procedures that affect working conditions (working hours, remuneration for the performed work).

1.6 Living environment

Despite the area in which the organization operates, its decisions and activities affect the environment. In reducing the negative impact of an organization on the environment, it is important to adopt an integrated approach that ultimately takes into account both direct and indirect impact of its own decisions and activities on the economy, society, health, and the environment. Nowadays, many ecological problems need to be addressed, such as the depletion of natural resources, pollution, climate change, habitat destruction, species extinction, disruption of ecosystems, as well as the decrease of people's quality of life in cities and also in the countryside. The problems identified from the specific nature of the sector, with a focus on the various factors of the environment (Zimon et al., 2018). The growth of the world's population and consumption represents changes increasing the risk not only to human safety and health but also to the well-being of society too. In this context, it is necessary to identify methods of eliminating and removing the volumes and patterns of both production and consumption that are not sustainable and, at the same time, to ensure that consumption per capita becomes sustainable. The organization is supposed to promote and respect environmental responsibility, preventive approach, environmental risk management, the "polluter principle" applies (EN ISO 26000:2010 standard).

The field of the environment includes:

- protection from pollution,
  - sustainable use of resources,
- climate change mitigation and adaptation,
  - protection of environment, biodiversity, and restoration of natural habitats.

1.7 Correct behavior of the organization

The application of positive outputs in the field of social responsibility requires correct behavior of the organization, due to the use of the organization's relations concerned with other entities. The group of elements of correct behavior consists of (EN ISO 26000:2010 standard):

- noncorrupt behavior,
- responsible involvement in the public sphere,
  - fair economic competition,
  - socially responsible behavior,
- relations with other organizations,
- respect for property rights.

1.8 Consumer affairs

The responsible approach in relation to customers and consumers is demonstrated by the provision of products and services to consumers/customers. The definition of responsibility in this area includes (EN ISO 26000:2010 standard):

- education and provision of accurate information,
  - the use of fair, transparent, and useful marketing information and contractual processes,
  - promotion of sustainable consumption,
  - creation of products and services that are accessible to all consumers/customers.

Responsibility incorporates minimization of the risks arising from the use of products and services through demonstration, production, distribution, information provision, support services, and recall procedures. In addition to these particulars, the areas of consumer affairs include collection and processing of personal data, responsibility for the protection and security of such information, as well as for consumer privacy. The subject matters of consumer affairs are (EN ISO 26000:2010 standard):

- fair marketing, factual and impartial information, fair contractual procedures,
- health protection and safety of consumers,
- sustainable consumption,

- services and support for consumers, handling of claims and conflicts,
- data protection and consumer privacy,
- access to basic services,
- education and information.

Principles that should govern socially responsible practices vis-à-vis consumers include satisfying the basic needs and rights of each person, adequate standard of living, covering adequate food, clothing, housing, as well as continuous improvement of living conditions and availability of basic products and services, including financial funds. Legitimate needs incorporate (EN ISO 26000:2010 standard) safety, information, choice, expression of opinion, redress, education, and healthy environment.

Other needs involve:

- respect for the right to privacy,
  - preventive approach,
- promoting gender equality and empowerment of women,
  - promoting universal design.

1.9 *Commitment to the organization (enterprise) and its development*

An indivisible part of sustainable development is commitment to the organization and development of the organization. The aim of engagement in an organization is to go beyond identification and involvement of stakeholders in relation to the impact of the organization's activities, as well as building and maintaining a relationship with the organization. The development of an organization is a long-term process involving different, sometimes even conflicting interests. The organization needs to be constantly developed in terms of its social, political, economic, and cultural characteristics. The organization depends on the characteristics of the involved social forces. This area covers the following topics (EN ISO 26000:2010 standard):

- commitment to the organization,
  - education and culture,
- job creation and skills development,
- development of and access to technologies,
- wealth and income creation,
  - health,
  - social investment.

Individual topics related to the commitment in organization provide information about the organization's relation to social responsibility, relevant principles, activities, and expectations that are part of the organization. In this context, it is inevitable to talk about transparency, ethical behavior, human rights, the environment, consumers, as well as communication. Commitment in the organization is voluntary and grants organizations/businesses an opportunity to increase their credit on the market, thereby ultimately contributing to implementation and improvement of social responsibility.

Making efforts to move toward long-term prosperity from the viewpoint of social responsibility, it is necessary to specify the priorities of sustainable development and growth.

2. Sustainable development and growth

The concept of sustainable development establishes a balance in which all elements of the global ecosystem are consumed in the same measure. In some form, sustainable development exists among people since the ancient past. Struggling to survive people have to economize rationally and simultaneously they must utilize their property and environment and agriculture elements with great care.

The role of the environment sector in the long term is to focus attention on the following areas: climate change, water supplies, biodiversity protection, protection and sustainable use of natural resources, minimization of production, and waste management, sustainable product production and consumption. All that must be studied in juxtaposition with the policy of circular economy and green growth.

A fundamental change in man's attitude to nature occurred in the 1960s, when, after the peak of rapid post-war economic development in industrial countries, environmental pollution problems came to the fore. Industrial production grew exceptionally fast, increasing not only the amount of production but also consumer waste.

The basic step toward sustainable development is to build and continuously improve legal instruments and institutions, including effective control mechanisms, serving all citizens and society as a whole to help achieve the set objectives. Their action may result in a gradual improvement in the functioning of public administrations, other organizations and institutions (educational institutions play a key role) which would in their turn influence long-term changes in the value orientation and behavior of the population. The value orientation of the population should gradually be changed, sustainable production and consumption patterns should be preferred, and at the same time consumer lifestyles should be economically and morally disadvantaged. Indicators of the condition of the environment (burdens and threats to the countryside) would be improved as a result of application of the new philosophy of economic and social development of the society. Environmental indicators for sustainable development contain:

1. mining of mineral resources in mil.m^3,
2. area susceptible to geodynamic processes in ha,
3. the funds spent on the protection of the geological environment,
4. annual groundwater and surface water abstractions—the total annual gross volume of groundwater and surface water collected for use as a percentage of the total average annual volume of available water resources,
5. household water consumption per capita—amount of water consumed per person in households in L/dwelling/day,
6. groundwater reserves—quantity of groundwater reserves in m^3,
7. concentration of fecal coliform microorganisms in freshwater—the proportion of freshwater sources contaminated by fecal coliforms above the permissible level in %,

8. total surface water pollution—proportion of flows classified in Classes IV and V in at least one indicator,
9. wastewater treatment—share of treated wastewater from the total wastewater in %,
10. the proportion of treated watercourses in % of the total length of watercourses,
11. volume of wastewater discharged into watercourses in m^3,
12. the quantity of pollutants discharged into watercourses in m^3,
13. funds spent on the purification of water resources,
14. quality of drinking water—% of noncompliant samples from the total number of samples,
15. change in land use—share of land use change per unit of time in %,
16. total investments allocated to nature conservation,
17. greenhouse gas emissions—amount of greenhouse gases, carbon oxides, methane, NO_x in gigagrams (Gg),
18. sulfur dioxide emissions—the amount of sulfur dioxide emitted into the air in tons,
19. emissions of nitrogen oxides—amount of nitrogen oxides emitted into the air in tons,
20. emissions of heavy metals—the amount of heavy metals emitted into the air in tons,
21. concentration of pollutants in urbanized areas—air pollution index,
22. expenditure on reducing air pollution—share of investment allocated to air protection in SK,
23. generation of industrial and municipal solid waste—production of industrial and municipal waste per inhabitant in t/1 inhab.,
24. generation of hazardous wastes—amount of hazardous waste produced per inhabitant in t/1 inhab.,
25. production of radioactive waste—amount of produced radioactive waste,
26. amount of disposed-of waste per capita—amount of disposed-of waste per inhabitant in t/1 inhab.,
27. recycling and reuse of waste—the amount of waste used and recycled from the total generated waste in %,
28. disposal of municipal waste—amount of disposed-of municipal waste per inhabitant in t/1 inhab.,
29. the area of landfills for waste from the cadastral area,
30. land contaminated with hazardous waste—area of hazardous waste landfills in ha.

Basic principles of environmental protection:

- precautionary vigilance,
- prevention,
- liability of the polluter (paid by the polluter),
- high level of environment protection,
- redress of damage at the source,
- sustainable development,
- essentials of environmental law.

The following strategic objectives of sustainable development (www.11) are specified in the framework of the approach to the long-term environmental priority:

1. Improving the health condition of the population and health care, improving lifestyles.
2. Development of an integrated model of agriculture.
3. Restructuring, modernization, and recovery of the manufacturing sector.
4. Improvement of transport and technical infrastructure, development of tourism.

5. Restructuring and modernization of the banking sector.
6. Reducing energy and raw material intensity and increasing the efficiency of the Slovak economy.
7. Reducing the share of the use of nonrenewable natural resources along with the rational use of renewable resources.
8. Reducing the environmental burden on the environment.
9. Mitigating the effects of global climate change, ozone depletion, and natural disasters.
10. Improving the quality of the environment in regions.

Quality and environmental management is one of the central concepts in the field of management. **Total** Quality and Environmental Management **(TQEM)** is the cutting edge of approaches in implementing multiprocess sustainability strategies. TQEM is "an approach of continuous improving of the quality of processes and products intervening in the environment with the participation of all organisation's levels and functions." TQEM has proved to be one of the most useful and most widely implemented frameworks for the establishment of an environmental management system, as it provides companies with a framework for clarifying the ecological impact of products and processes. It helps clarify several misconceptions concerning the relation between quality and environment protection (Malindžáková, 2018):

- environmental performance of products—in addition to price and quality—is becoming the most significant competitive factor,
- public opinion increasingly favors businesses that produce their products environmentally friendly and, conversely, discriminates against businesses that neglect environmental protection in the form of consumer disinterest. This has an impact on the creation of enterprise's image, forming part of its social responsibility,
- the market sector of environmental technologies is increasingly expanding, which is also linked with ever-tightening environmental protection rules. Demand is growing not only for air protection facilities, water treatment, waste disposal but also for low-waste technologies, alternative sources of energy production, etc.

Our businesses, if they want to establish themselves on foreign markets in the future, must take these facts into account at present. All these arguments are in favor of the implementation of environmentally oriented company management, the essence of which is integration of environmental protection into the system of corporate objectives and adoption of an appropriate strategy for the development of new products for their achievement.

The environment and its protection have increasingly been the subject of international negotiations and conferences. The first conference on the living environment was organized in Stockholm Resilience Centre (1972). It was gradually followed by other international events. In December 1983, the UN General Assembly appointed a World Commission on Environment and Development. An important result of its work was the definition of the concept of sustainable development and the definition of its 16 basic principles, which should be obligatory for governments of all countries (Malindžáková, 2013).

1. Principle of promoting the development of human resources.	**9.** The principle of cultural and social integrity.
2. Ecological principle.	**10.** The principle of nonviolence.
3. The principle of self-regulatory and self-supporting developments.	**11.** Principle of emancipation and participation.
4. Effectiveness principle.	**12.** Principle of solidarity.
5. Principle of reasonable sufficiency.	**13.** Principle of subsidiarity.
6. Principle of precautionary vigilance and foresight.	**14.** Principle of acceptable errors.
7. The principle of respect for the needs and rights of future generations.	**15.** Principle of optimization.
8. The principle of intragenerational, intergenerational, and global equality of rights of the people on the Earth.	**16.** Principle of socially, ethically, and environmentally favorable management, decision-making, and behavior.

The concepts of sustainability and sustainable development came into use in the early 1970s, especially in the context of the knowledge that uncontrollable growth of any type (of population, production, consumption, pollution, etc.) is not sustainable in the environment of existing finite resources. Milestones in the general introduction and development of the sustainable development concept included in the first place the report *Our Common Future* (Brundtlandtová, 1987). In June 1992, 20 years after the Stockholm Conference, the United Nations Conference on Environment and Development was held in Rio de Janeiro (1992), which became known as the RIO SUMMIT. At this conference, the fifth day of June was declared the Day of Environment. In 1996, 132 indicators of sustainable development were approved, which were divided into four groups:

- social (38),
- economic (23),
- environmental (56),
- institutional (15).

2.1 Sustainable development—Agenda 2030

For the purpose of achieving sustainable development, a set of global priorities, entitled *Agenda 2030*, has been worked out, following the direction of the United Nations. The process of globalization is characterized by the creation of global structures in different areas of social reality, as well as in different fields of human activity, aimed at increasing the internal complexity of these structures by gradually linking the subglobal structures. The dynamic development of individual subglobalization processes is interlinked, each subprocess has its own consequences for the others and is influenced by them. Key principles under Agenda 2030 were released in

Figure 19.3 Global goals of sustainable development.

2015 in the document *Transforming Our World: An Agenda 2030 for Sustainable Development.* The main motto of this document is "transformation, integration, and universality." The 2030 Agenda consists of 17 global Sustainable Development Goals, which are further developed into 169 partial goals. The subject matter of this document is to focus on the structural political, economic, and social transformation of individual countries of the world in response to the threats that humanity is currently facing. The crucial point here is interconnection between the three pillars of sustainable development, namely economic, social, and environmental (Fig. 19.3).

Agenda 2030 is not legally binding. The aim of this agenda is to focus on sustainable development and implement this sustainable development in national policies, strategies, and plans so as to contribute to the achievement of global objectives. The concept of implementing Agenda 2030 consists of three main commitments (www.12):

- Agenda 2030 (2015) for sustainable development,
- Addis Abeba Action Programme on Financing for Development (2016),
- the Paris Agreement to the United Nations Framework Convention on Climate Change (2014).

Another objective of their implementation is to eliminate poverty in all forms and dimensions, as well as to eradicate extreme poverty. These are the greatest global challenges and simultaneously a necessary requirement for sustainable development. They are aimed at ensuring human rights for all, achieving gender equality, and empowering women and girls.

3. Ecological growth

Scientists who maintain that economic growth at some point begins to bring more negatives than positives are proposing the so-called *"green (ecological) growth"* as a solution.

According to the report of the Organisation for Economic Co-operation and Development (OECD), entitled *Towards green growth*, green growth not only promotes economic growth and development but also protects natural resources so that we can continue using the resources and environmental services on which our social well-being and quality of life depend. It is therefore necessary to promote investment and innovation, which will become the basis for sustainable economic growth and allow new economic opportunities to emerge. Return to the original model would not, according to the OECD, be unreasonable and ultimately unsustainable, with risks to economic growth and development posed by the costs and constraints associated with human capital. It could lead to shortage of water and other natural resources, increased air and water pollution, climate change, and biodiversity loss. These consequences would be irreversible, so strategies need to be adopted to achieve ecological, green growth. According to the report, green growth has the potential to solve economic and environmental problems, and open up access to new sources of growth through the following channels:

Productivity—incentives to promote efficiency in the use of reserves and natural resources—increase productivity, reduce waste and energy consumption, make alternative sources available for their efficient use.

Innovation—innovation opportunities stimulated by political measures and framework conditions that will enable environmental problems to be tackled in new ways.

New markets—creating new markets by stimulating demand for green technologies, goods. and services, creating the potential for new jobs.

Confidence—building investor confidence through better predictability and stability of measures by which governments will address environmental problems.

Stability—stable macroeconomic conditions—for example, by reviewing composition and efficiency as well as public spending and increasing revenues through pollution charges. The risk of the following negative effects on growth may drop:

- **lack of resources** that increases investment costs, for example, when capital-intensive infrastructure is needed due to depletion of water supplies or a decline in water quality (e.g., desalination facilities). In this respect, the loss of natural capital may be higher than the gains in economic activity, which would hinder the direction toward sustainable growth;
- **imbalances in natural systems** that increase the risk of serious, sudden, and potentially irreversible consequences, leaving major damage, for example, to some fish population and may also be reflected in biodiversity as a result of persistent climate change. Attempts at identification of a possible boundary show that in some cases, such as climate change, the global nitrogen cycle or biodiversity loss, that boundary has already been crossed (s. 32 Stockholm Resilience Centre—Planetary boundaries).

Proponents of the "green-growth" ideology believe that if we remove the negative side of economic growth (i.e., increasing consumption of nonrenewable natural resources, increasing deforestation, rising CO_2 emissions, etc.), we will gradually stop burdening the environment, but "technological growth" will be maintained. However, according to Albert Bartlett (2012), the Professor of Biophysics, this "green growth" approach is a part of the problem rather than a solution to the challenges of economic growth. This is also why a number of economists and politicians see further growth as the only (or simplest) way to address the problems caused by growth (economic growth, emissions growth, population, consumption of natural resources, energy sources, debt growth). Based on their research, Baek and Kim (2013) argue that only economic growth and rising living standards will allow us to make our new technologies and energy sources cleaner, thus reducing harmful emissions. In their view, it is inevitable to grow, fighting against negative climate change, not try to turn back time. Baranzini et al. (2013) in their study claim that we only consume energy when we need it, because we will not create this need artificially by improper use of energy. As an example, they point out that, just as people cannot gain weight if they do not receive more food than what they need for their survival, the economy cannot grow if it does not have enough energy to maintain this growth.

Green growth may be slightly less unsustainable than *the current economic* "industrial" growth, based mainly on the consumption of nonrenewable energy sources, but this is not sufficient to make the environment sustainable, says Bartlett (2012). According to him, the only sustainable growth is "zero growth" (stagnation) although, as mentioned in the previous chapter, the functioning of our socioeconomic system is based on stable, exponential growth. Bartlett also points out that the world is in a phase of "ecological blowout" and therefore it is necessary and desirable to ensure a decrease of consumption in all directions and areas of life on Earth.

4. Zero growth—economic stagnation

In 2001, the "Degrowth" movement established in France, supporting the environmental theory that the economy does not need to grow. The first conference on the subject of economic degrowth was held in 2008 in Paris, and the issue of this alternative path has currently become a subject of academic research. In this context, new notions arise—*culture of transition, simple life, postgrowth,* or *economics of collapse*. Supporters of the movement criticize especially the economic system based on both stable economic growth and gdp growth, which requires ever-increasing production and consumption. Tim Jackson, a British environmental economist and professor of sustainable development, points out that overall consumption of raw materials and energy has still been growing despite efficient production and technological progress. Consumption grows faster than efficiency. Promoters of degrowth seek to remove the link between the political, technological, educational, and information systems and short-term economic interests.

One of the pioneers of the movement is a Frenchman François Schneider, an industrial ecology researcher who contributed to establishing of the Research Institute for a Sustainable Europe in Vienna. He is currently based in Barcelona, where he cofounded the Research and Degrowth group, which is dedicated to economic degrowth research. Members of the movement are advocating decrease in production and consumption because, in their view, overconsumption makes the basis of long-term environmental problems and social inequality. They maintain that reducing consumption need not be linked to individuals' suffering and a drop in living standards. A sense of happiness and contentment can be perceived on the basis of collaboration, lower consumption, greater interest in art, music, family, culture, and the community in which people live. At individual level, zero growth can be achieved by voluntary simplification of life. According to Schneider, the goal is a "sustainable society" which respects ecosystems through adequate production/consumption/recycling, and a society that guarantees fairness among people in the use of available natural resources.

"The path from economic growth to zero growth (stagnation) is not easy. Economic growth has changed from something we strive for into something that keeps us entrapped, because otherwise there is a risk of collapse. Nevertheless, there is a risk of collapse, too, if we don't stop economic growth," explains Johanis (Johanis Nada—an environmental economist, environmentalist, and researcher) (www.4).

5. Circular economy

The circular economy is a sustainable development strategy that establishes functional and healthy relations between the nature and human society. By perfectly closing the flows of materials in long cycles, it opposes our current linear system, where raw materials are converted into products, sold, and, at the end of their life cycles, incinerated or landfilled.

The circular economy and its basic principles are based on the idea that all product and material flows should be reinvolved in their cycles after their use, where they will become reresources for new products and services again. This means that the waste as such will no longer exist. While substitution of primary materials by secondary ones may offer part of the solution, recycling does not represent a definitive and simultaneously attractive solution, as its processes are energy-intensive and generally mean the degradation of materials—all of which lead to an increase in demand for original materials.

The circular economy goes beyond recycling as it is based on a restoring industrial system, leading to the end of waste. Recycling can be thought of as the outer cover of the entire circular economy, requiring greater energy consumption than the inner cover of the circular economy, which means, above all, repair/modification, reuse, or treatment.

The aim is not only to create an improved life cycle and use of the product itself but also to minimize energy consumption (Lacy and Rutqist, 2016).

6. Low-carbon economy

Just as the expressions "Bronze Age" and "Iron Age" are used to describe history, we might call the present the "carbon period" (Rifkin, 2011).

The global environmental crisis stems largely from a global energy system based on fossil energy sources, argues Jeffrey D. Sachs (2014), a professor of sustainable development at Columbia University in New York. About 80% of all primary energy in the

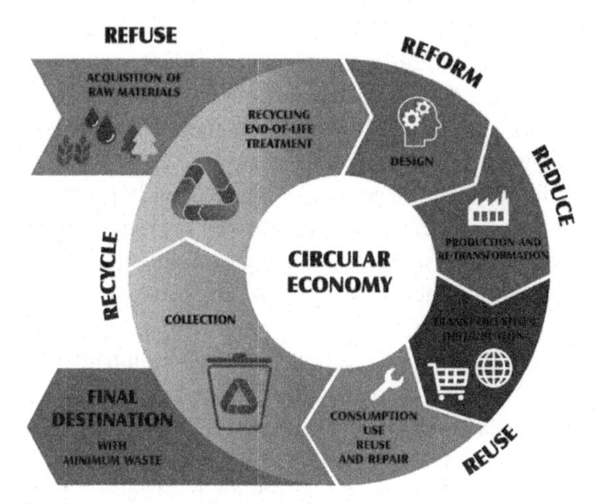

Figure 19.4 Circular economy.

world comes from coal, oil, and gas. When these resources are being burned, they release CO_2, which changes climatic conditions on Earth, gradually warming the Earth. The physical foundations of these processes have been known for over a 100 years. However, several oil companies also spend sums of money on creating ambiguities where scientists have already arrived at a clear agreement: If we want to save the Earth, ensure social well-being and quality of life for future generations, we must shift our focus onto a new, low-carbon energy system.

According to Sachs (2014), this transformation consists of three parts:

1. Improved energy efficiency—to achieve the same level of well-being consuming less energy than before (e.g., to construct buildings which will take advantage of sunlight and natural air circulation and require less energy for heating and air conditioning).
2. Use solar, wind, water, and geothermal energy and other forms which are not based on fossil fuels as much as possible and, where necessary, utilize fossil fuels only to a limited extent. There are technologies which use these alternatives safely, at an affordable price, and on a scale large enough to replace almost all the coal and much of the oil we consume today. Only natural gas, the purest burning fossil fuel, remained an important source of energy.
3. Capture CO_2 emissions from power plants before they escape into the atmosphere. The entrapped CO_2 would then be pumped underground or below the seabed, where it would be safely stored for a long time. The method of carbon capture and storage (CCS) is already used on a small scale and, if it turns out to be appropriate for widespread use, coal-dependent countries (China, India, and the United States) could continue to exploit their stocks.

Low-carbon energy strategies could also include an increasing tax on CO_2 emissions, efforts to research and develop low-carbon technologies, the transition to electric vehicles, and regulations leading to the phasing-out of all coal-burning power plants, with the exception of the ones employing CCS. Sachs (2014) sees a problem in that in the countries where powerful lobby groups advocate the established coal or oil interests, politicians mostly are not interested in promoting the inevitability of low-carbon energy.

In 2009, the then British Prime Minister Gordon Brown announced the start of the Green Revolution, a scale comparable to the Industrial Revolution. The Green Revolution should lead to a new ecological age in which carbon footprint will become an essential benchmark for any economic activity. As the main motive is to reduce greenhouse gas emissions, the term "low-carbon economy" *has entered political rhetoric.*

A low-carbon economy is a new approach that gradually pushes out the concept of sustainable development to date. It is a vision of an economy in which considerations of combating climate change, saving, efficiency and rationality in the use of natural resources, as well as environmental protection, are perceived through the optics of economy.

The low-carbon economy offers solutions to combat climate change and environmental decline, and it can also be a tool for economic recovery and the creation of new business opportunities and jobs. This concept overlaps with the knowledge economy and also includes answers to questions concerning stability and independence from fossil energy sources. The low-carbon economy has become the subject of political agenda, the Europe 2020 Strategy, and the European Union's policies for decades to come. The Union sets binding targets for reducing greenhouse gas emissions, reducing energy consumption, and replacing traditional, fossil energy sources with renewable sources. The European Commission developed a "Roadmap for moving to a competitive low-carbon economy in 2050" and incorporated the same considerations into energy and transport strategies and plans. The new regional policy will include the investment priority "Transition to a low-carbon economy in all sectors" and each member state is expected to allocate adequate money from euro funds onto it.

According to Mojžiš (2012), the true contribution of measures for reducing the carbon footprint depends on what will be defined by the European Union as low carbon. The gas lobby is trying to push natural gas through as a low-carbon fuel to secure access to European funds. In turn, the coal lobby sees an advantage in burning technology and in storing of the produced carbon, not in fuel. Therefore, it is pushing for the adoption of CCS, for low-carbon technologies. Nuclear energy, which is being presented as a source with huge amounts of energy and minimum greenhouse gas emissions, is also often characterized as low carbon. If we wanted to be consistent, we should seek to exclude all processes dependent on nonrenewable energy sources. However, this would assume deep reflection and making changes related to the social division of labor, the degree of needs satiation, and consumer behavior– Mojžiš (2012).

The first industrial revolution brought about civilizational changes in the 19th century. As a result of the introduction of technologies powered by steam (train, printing machines), the first printing materials, books, and magazines occurred, literacy grew, and factories were established that produced a large number of goods. In the Second Industrial Revolution, people replaced their horses with cars, which logically was related to greater demand for oil. In the United States, which became the world's largest oil producer, the web of motorways began to spread within a few years. Another big change was brought about by the invention of telephone, radio, and television. Similar radical changes to the ones triggered by the first and second industrial revolutions should accompany a third industrial revolution, which will bring about a change of focus onto new energy sources and communication technologies, argues Rifkin (2011).

Opponents of the Third Industrial Revolution argue that alternative energy sources are not sufficient to propel the world economy and cover our current energy consumption. Geothermal energy represents a great potential of almost unused green energy as yet. Between 2005 and 2010, geothermal energy use increased by 20%. Although, according to research, 39 countries have the potential to meet their energy needs merely through geothermal energy, only 9 of them have so far started using it. Candidates for geothermal energy use within the European Union are Italy, France, Germany, Austria, Hungary, Poland, and Slovakia. The future, according to Rifkin (2011), also lies in small alternative energy productions that any household can have. According to Rifkin's surveys, in the European Union up to 40% of the roofs surface and 15% of the buildings surface are suitable for the location of photovoltaic systems. At the same time, renewable energy repositories will have to be built and, for the time being, hydrogen repositories seem to be an appropriate option.

7. Implementation of social responsibility in the organization's activities

For the purpose of integration of social responsibility, organizations may have special techniques, created to introduce new procedures into their own decision-making processes as well as activities, they may also have effective systems for communication or internal evaluation. On the other hand, organizations may have sophisticated corporate governance systems or other aspects of social responsibility. The implementation of social responsibility into praxis derives from the existing strategies and systems. Some activities are carried out in a new way, or organizations/companies have created techniques to implement new procedures. Following the existing procedures or newly established procedures, it is necessary to identify links and evaluations between the basic characteristics of the organization and CSR. It follows that it is necessary to identify themes of social responsibility within each key area. Within the possibilities, the evaluation shall include

- type and size of the organization, the purpose and nature of its activities,
  - places in which the organization operates, including evaluation, existence of a legal framework in relation to social responsibility, assessment of social, environmental, and economic characteristics,
  - information on the organization's social responsibility activities from past periods,
  - characteristics of the workforce or employees,
- work conditions and social protection,
  - social dialog,
  - safety and health protection at work,
- roles, visions, values, principles, and code of conduct,
- interests of internal and external stakeholders, including social responsibility,
  - structure and nature of decision-making in the organization,
  - personal development and on-the-job training.

7.1 Notion of corporate social responsibility

Understanding of the notion of corporate social responsibility necessarily involves orientation at the following three areas:

1. **Due diligence**—specifies a complex active process aimed at identification of the current and potential negative social, environmental, and economic impact oriented toward minimizing or eliminating of the given impact. Elements of the due diligence process depend on the size and conditions of the organization:
- the means to assess the potential impact of existing and proposed activities on the strategic objectives,
- the means to integrate key areas of social responsibility into the organization's activities,
- the means to monitor performance in the course of time, according to priority and procedures,
- action to address the negative impact of relevant decisions and activities.
2. **Determining the context and importance of key areas for the organization**—the organization's priorities need to include familiarization with all key areas related to the organization's activities.

The organization is supposed to identify the following tasks as possible:

- process a list of all custom activities,
- identify stakeholders,
- specify the activities of the organization itself as well as the activities of organizations within the sphere of influence,
- identify realistic key areas so that the organization or the subjects within the sphere of influence or the value chain are able to carry out the relevant activities,
- examine the possible impacts of the organization's decisions and activities on stakeholders as well as sustainable development,
- explore ways in which stakeholders can influence the plans, decisions, and activities of the concerned organization in the context of social responsibility,
  - specify topics of social responsibility, relating the day-to-day activities of the organization, which are up-to-date or occasional in specific circumstances.
3. **Organization's sphere of influence**

In assessing the organization's sphere of impact, account should be taken of

- ownership, administration, and management,
- economic relations,
- legal and political status,
- public opinion.

Applying the organization's impact onto other entities, it is important to increase the positive or minimize the negative impact on sustainable development or to achieve both objectives simultaneously. The application of impacts can be carried out in the following ways:

- determine contractual provisions or incentives,
- make public statements of the organization,
- participation of the community, political representatives, or other stakeholders,
- take investment decisions,
- share knowledge and information,
- implement joint projects,
  - collaborate with the media,

- promote best practices,
- establish cooperation/partnerships with organizations and other entities.

4. Prioritizing topics

From the viewpoint of integrating social responsibility into both the organization's activities and its day-to-day practices, it is necessary to identify priorities and ensure that they are respected. The priority specification ought to include a requirement for the involvement of stakeholders. The order of priorities can be modified over time. The following aspects should be taken into consideration when determining the priority themes of an organization:

- the organization's current activities in accordance with legislation, international standards as well as international standards of conduct, best practices,
- the impact of a particular topic on the organization's ability to meet substantial objectives,
  - the impact of the action on the sources necessary for implementation,
  - the time perspective for achieving the desired results,
- financial implications if the topic will not be addressed as a matter of priority,
- simplicity, clarity, and speed of implementation in order to increase awareness and motivation in the field of social responsibility within the organization.

The aim of social responsibility is to contribute to sustainable development, social prosperity, and health. Social responsibility thus becomes one of the important influences on the performance of the organization. More than ever before, the performance of an organization in relation to the social environment in which it operates and its impact on the nature of the environment becomes a critical part of assessing its overall performance and its ability to operate effectively. It reflects the growing recognition of the need to ensure a healthy ecosystem, social equality, and governance in the organization, contribution to quality of life. With increasing frequency organizations are being exposed to critical views of stakeholders, including clients and consumers (prosumer ≡ producer and consumer), workers, community members, NGOs, financiers, investors, firms, and other commercial entities.

Perception of social responsibility exercised by the organization may affect

- general name/position of the organization,
- the ability to attract and retain workers, and also customers, clients, or users,
- maintaining employees' morale and productivity,
- the views of investors, sponsors, and financial institutions and their relations to the government, media, suppliers, partners, customers, and companies in which they operate.

The standard EN ISO 26000:2010 is intended to help organizations wishing to achieve mutual trust between stakeholders, improving, respecting, and applying social responsibility. Its origination and development were preceded by the following standards:

- international standard SA8000:2008—social responsibility,
- AA 1000 AccountAbility:2008—standard on dialogue with interest groups, audit, reporting,
- EN ISO 14001:2015—Environmental Management Systems. Requirements.
- EMAS:2009—Environmental Management and Audit System,
- OHSAS 18001:2008—Occupational Health and Safety Assessment Series.

TQM tools and philosophy can be used to improve environmental efficiency/performance by eliminating waste or reducing its impact. The system thus created is called TQEM. Comprehensive environmental quality management is the cutting edge of approaches in implementing sustainability strategies for managed processes. TQEM is "an approach to continuous improving of the quality of processes and products intervening in the environment by participating in all levels and functions in the organization." TQEM can be considered one of the most widely implemented frameworks for the establishment of an environmental management system because it provides companies with a framework to clarify the environmental impact of products and processes (Benková et al., 2007).

For the design and implementation of such a system, Albert Cherns offers the following checklist (Benková et al., 2007):

1. Compatibility—the process of designing a sociotechnical system should be aligned with its objective. In case of success of this system, it is necessary to give workers more freedom to make decisions and implement their knowledge.
2. Minimization of critical technical conditions—only important requirements should be specified when designing the system.
3. Disagreement management—disagreements cannot always be eliminated, but can be managed, evaluated until the source of these disagreements is identified (CE diagnostic principle, cause-effect). The source of disagreements may be a quality defect or an equipment failure.
4. Adaptability—the organization should be sensitive toward changes in the environment. People ought to be aware of the importance of protecting the environment and adapting to new requirements.
5. Effective position boundary—marginal conditions must be specified in each organization. Grouping of similar products in production brings about a reduction in production time and a reduction in the resources needed to transport materials between departments.
6. Information flows—the system should be designed in such a way that it provides information to those who process it further.
7. Feedback consistent with objectives—a system promoting rewards and humanity should be designed to clearly demonstrate the model behavior.
8. Proposal as iteration process—no system can be perfect. Even environmental assumptions are changing, not just quality requirements due to technological and user paradigms.

8. Industry and waste

In the past, waste separation was an unknown term for people. We are living in times when more and more people become aware not only of the need and necessity to minimize waste generation but also of the treatment and recovery of the waste they generate, which surrounds us. Waste separation is important in the material recovery process, which has the following advantages:

- lower equipment failure rate or lower maintenance costs,
- better calorific value of selected burnable components,
- improvement of the quality of the environment,
- lower fees for residents.

Industry produces a lot of industrial waste through its technologies. These so-called by-products of production pollute and devalue the environment in the near as well as distant surroundings. For many reasons, the term "industrial waste" is not considered to be entirely correct and appropriate. In principle, it is a secondary raw material which, even under the current conditions, is already used or it is a secondary raw material which, with the current knowledge of science and technology, is not yet efficiently (considering costs and utility) useable for the production of new products.

Following the production processes, input materials are processed in larger volumes and subsequently amounts of waste substances from the production process increase. Individual types of waste include solid, liquid, and gaseous waste products from all industries, as well as wastes of every kind. A more sophisticated and efficient use of such wastes is accompanied, on the one hand, by a difficult situation concerning raw materials, on the other hand, by a greater threat to the environment from different wastes. At present, significant emphasis is placed on waste management, waste minimization, as well as the use of secondary raw materials. Homogeneous production processes and approaches to their management can also affect in significant way a conception of produce waste. Waste production in the field of technology is currently on an increasing trend. Municipal and hazardous waste amount has also been on an upward trend in recent years.

According to data from www.13, the comparison of the amount of municipal waste generated in the Slovak Republic in 2009 was 339 kg/person and increased to 348 kg/person in 2016. Out of the selected countries, unfortunately, the most communal waste (approximately 50%–80%) ends up in landfills in the Czech Republic, Poland, and Slovakia. Recycling is the most effective in Cyprus (66%), followed by Austria (59%), Slovenia (58%), and the Netherlands (53%). In these countries, around 50% of waste ends up in incinerators. As a result, more economically advanced countries are both burning and recycling more waste.

9. Sustainable scientific and technological progress and implementation of innovation technologies in energetics

While there are gradual changes in Europe toward a "green" and sustainable future, the United States still believes in unfettered economic growth, unlimited production, and consumption of traditional energy sources, according to Rifkin (2011). Japan, India, and the Middle East take a similar stance to the United States. Authoritarian governments and oligarchs who have become rich in oil or have controlled countries for decades reject the idea of alternative energy sources. This is because oil and natural gas are so-called "elite energy sources" which cannot be found everywhere and which are not accessible to everyone. Oligarchs, sheikhs, politicians, and various interest groups are not keen on changes in resources and energy consumption, because they acquired their wealth and power, thanks to the production and consumption of traditional energy sources. A similar situation exists in the nuclear industry. Most countries have halted their nuclear programs after the Chernobyl disaster, but the nuclear industry is now coming to the forefront as the best solution to climate change, global warming, and scarcity of fossil energy sources. Nuclear lobby groups are persuading governments of individual countries that they do not need to introduce changes concerning alternative, renewable energy sources because of relatively affordable and clean nuclear energy.

"Many people still do not understand that the era of fossil fuels is coming to an end and that we need to change the way we think. Several civilizations in history have reached the point at which they were forced to change. Some of them did, others didn't. Compared to the past, our situation differs in that the probability of temperature rises on Earth is increasing and can trigger the onset of mass mortality of animals and plants, with which comes the real possibility of a nationwide extinction of our race," claims Jeremy Rifkin, the president of The Foundation on Economic Trends, creator of the EU Strategy for Measures to Ensure Long-Term Energy Sustainability and author of *The Third Industrial Revolution*. In this book, Rifkin not only draws attention to current environmental problems but also points to possible optimistic prospects for the future—provided that the way in which energy is obtained will be reevaluated and changed.

Just as in the past, in addition to the geological and technical information needed to calculate the time of depletion of stocks, it is necessary to consider economic factors that could, to some extent, prolong or even shorten the life of energy reserves. The continuation or resuming of the current economic crisis may lead to a reduction in energy demand and subsequently to an increase in the time of possible extracting energy raw materials. Recovering from the economic crisis can also cause a significant increase in energy consumption, as in 2010, when the world energy consumption increased by 6.3% (Heinberg, 2011; BP Statistical Review of World Energy, 2011).

The British Petroleum annually report on fossil reserves worldwide. Reserves are considered to be resources which are economically viable from the viewpoint of geological and technological circumstances. The life of reserves is expressed as the ratio of reserve to production for a given year. Production is understood as the amount extracted in a given year which also characterizes global consumption.

Estimates of the durability of resource stocks mostly refer to the current trend in production and consumption and the current population of the Earth. The declared lifetime applies only if the production and consumption of the resource remains at its current level. However, the global trend in energy consumption is characterized by steady growth in production and consumption on all continents. Despite the differences between developed and developing countries, forecasters expect an increase in fossil energy consumption in all areas of the world. This is related not only to the growth of the Earth's population but also to the increasing consumption and income of the population, and to the rate of growth of the economy. The durability of resource stocks is crucial to the future development of consumption. Mining, production, and consumption of resource stocks, in turn, depend on what the prices of resources are on world markets.

Calculations and estimates concerning the period of depletion of resource stocks hide many pitfalls. Apart from the fact that the declared lifetime applies only if the production and consumption of a given resource remains at the current level, a mere increase in energy consumption suffices to shorten the life of the stocks and unexpected new discoveries of resources may in turn extend the life of the stocks. However, it is clear that the nonrenewable energy sources do not have endless reserves.

In his studies and analyses, Bartlett (2006) operates with the following formula to calculate the period of depletion of stocks:

$$T_{EE} = \frac{1}{r} \ln\left(\frac{rR_0}{L_0} + 1\right)$$

where

T_{EE}—lifetime of nonrenewable resource R, consumption of which grows exponentially.

R_0—resource quantity at time $t = 0$, demonstrable resource reserves (inventories) in a given year.

r—growth rate (g/100)—growth rate as an average for each year.

L_0—resource consumption at time $t = 0$, resource consumption in a given year.

However, the calculated number of years of stocks durability will only be valid at the current level of production and zero increase in fossil energy consumption.

10. Implementation of innovation technologies and macroeconomic feedback

Some natural energy sources are nonreproducible and many of the renewable energy sources are potentially exhaustible. Due to resource constraints and exhaustibility, the theory of unlimited economic growth is problematic, and for that reason even sustainable development—green growth or zero decline in production—may not be easily feasible. Neoclassical literature focused on economic growth and natural resources examines under what conditions a continued growth is possible, or at least nondecreasing consumption and usefulness.

A hallmark of technological change since the Industrial Revolution is the change of the use of renewable sourcing instead of nonrenewable sources, especially fossil fuels. However, the use of new technologies based on fossil fuels has increased the productivity of sectors, such as agriculture, forestry, and fishery, and these have not improved or even maintained the productive capacity of the ecosystem that produces crops, wood, and fish. There is a gradual deterioration of land, deforestation, and overfishing. Pollution from the use of fossil fuels, mining, production, and overconsumption continue to impair the ecosystem's production capacity (Stern, 2004).

Judson et al. (1999) observed, using a regression curve on a large set of data, relationships for energy consumption in each of a number of energy-intensive spheres, such as industry, construction, transport, agriculture, households, and others. They tracked various impacts of innovation and new technologies, revealing increasing energy consumption over time in the household and agricultural sectors and, conversely, declining consumption in industry and construction. Technical innovations tend to bring more energy-intensive appliances to households and, on the contrary, energy-efficient technologies to industry.

Based on their research, Khazzoom (1980) and Brookes (1990) assume that energy-saving innovations can ultimately result in even greater energy consumption if the money saved is spent on other goods and services that depend on energy consumption in their production. The actual production of energy services required by the producer or consumers needs energy. Innovation that reduces the amount of energy needed to produce an energy service unit reduces the real price of energy services. This results in an increased demand for energy services and therefore for energy itself. The lower price of energy will also affect income, which will increase the demand for all goods in the economy and therefore also for the energy needed to produce them, thus bringing the so-called feedback effect.

In his scientific study, Howarth (1997) arrived at a conclusion that the feedback effect (macroeconomic feedback) is always lower than the initial reduction in energy consumption caused by innovation, so that improving energy efficiency actually reduces overall energy consumption. In addition to substitution, technological changes are an important factor mitigating the scarcity of resources in the standard growth model. He says that even if substitution opportunities are limited, sustainability is still possible precisely through technological change.

According to Stern (2004), arguments advocating technological change as a solution to problems would be more persuasive if it were clear that technological change is not the same as substitution. There is more than one type of inputs substitution and also more than one reason why substitution may be limited. In substitution, we can use compensation from a category of similar production inputs—for example, between different fuels—or from a category of different types of inputs—for example, between energy and machinery. A distinction should also be made between substitution at microlevel (in one production process or in one enterprise) and at macrolevel (in the economy as a whole). It is clear that some types of substitution can be made in a certain country, but not worldwide.

Although neoclassical models consider technological change and substitution to be two separate phenomena, other approaches do not. The neo-Ricardian approach assumes that only one method of production is available at a certain time and any change in this method is a change in technology. On the contrary, the neoclassical approach presupposes that at one point in time there simultaneously exist countless effective methods. There is a change in technology when new, more effective methods are discovered. In a sense, however, these new methods represent a replacement of knowledge by other factors of production, and there will also still be *thermodynamic* limitations on the extent of the reduction of energy and material flows. The first law of thermodynamics describes the principle of preservation of matter. In order to obtain a given material output, larger or equal quantities of matter must enter the manufacturing process together with residues such as pollutants or waste. Therefore, for any production process which produces material outputs, there always are minimum requirements on input material. The second thermodynamic law implies that a minimum amount of energy required to carry out the transformation of matter is necessary. Therefore, limits must be set for energy for the substitution of other production factors. All economic processes need energy, although some activities do not require direct material processing—which only applies to the microlevel. At macrolevel, all economic processes require indirect material consumption, either to support labor or capital production (Ayres and Kneese, 1996).

It is also important whether new technology moves in the "right" direction. If natural resources are not valued correctly due to market failure (a common phenomenon that is the main subject matter of the study of traditional environmental economics), then there will be insufficient motivation to develop technologies that reduce resource and energy consumption. Instead, technological change will not lead to smaller, but to greater use of resources. Another reason for alleviating technological optimism, according to Stern (2004), is that new technologies are often double-edged swords in terms of their overall impact on natural capital. New technologies that alleviate the scarcity of nonrenewable resources can produce more waste or other harmful types of waste and therefore have a greater impact on renewable natural capital and ecosystem services.

On the basis of the above statements, it can be concluded that scientific and technological progress, the development and implementation of innovative technologies may also have negative effects in addition to the undeniable advantages and benefits. Not reducing the importance of benefits, it is at the same time necessary to draw attention to possible negatives, in order to reduce them as much as possible, because their complete elimination is impossible due to macroeconomic feedback.

11. Conclusion

According to previous findings, the implementation of innovative technologies in the framework of scientific and technological progress may in some cases be linked to the effects of so-called macroeconomic feedback, not addressing problems related to the preference for stable economic growth and the associated increasing consumption of natural and energy resources, the increase in population, and material consumption, as well as environmental and other related problems. However, there is also the so-called "new lifestyle change" which should aim to reduce unjustified consumption in order to develop another type of economy (preferably zero or green growth) reflecting real needs and not the needs artificially induced by marketing communication tools. Lifestyle change could benefit the future from the perspective of a gradual reduction in the consumption of natural and energy resources, unjustified material consumption, and an improvement in the quality of life of the population without macroeconomic feedback.

It is also necessary to promote the efficient and justified exploitation of the potential of renewable energy sources in the regions and to apply the principles of the so-called "smart energetic," which include self-sufficiency, local control, sustainability, support for the local economy, and efficiency.

Making efforts to ensure the social well-being and quality of life of the population, in compliance with application of the proposed solutions, the economy does not have to be based only on stable economic growth and other related aspects (increasing consumption of natural and energy resources) which may be the cause of the aforementioned environmental, economic, demographic, and social problems.

"The chapter is part of the solution of scientific project KEGA 026EU-4/2018 and KEGA 017TUKE-4/2019 **it was issued with ist support."**

References

Ayres, R.U., Kneese, A.V., 1996. Production, consumption, and externalities. Am. Econ. Rev. 59 (3), 282–297. Pub ID: 103-352-001.

Baek, J., Kim, H.S., 2013. Is economic growth good or bad for the environment? Empirical evidence from Korea. Energy Econ. 36, 744–749. Dostupné na: http://www.sciencedirect.com/science/article/pii/S0140988312003180.

Baranzini, A., Weber, S., Bareit, M., Mathys, N.A., 2013. The causal relationship between energy use and economic growth in Switzerland. Energy Econ. 36, 464–470. Dostupné na: http://www.sciencedirect.com/science/article/pii/S0140988312002290?np=y.

Bartlett, A., 2012. Arithmetic, Population and Energy. Odborná prednáška amerického profesora Alberta Bartletta. Priatelia zeme. Dostupné na: http://www.priateliazeme.sk/cepa/sk/informacie/temy/922-aritmetika-populacia-a-energia.

Bartlett, A., 2006. A depletion protocol for non-renewable natural resources: Australia as an example. In: Natural Resources Research, vol. 15, pp. 151–164. https://doi.org/10.1007/s11053-006-9018-1 (3).

Benková, M., Floreková, Ľ., Bogdanovská, G., 2007. Systémy riadenia kvality. Elfa, Košice.

British Petrol, 2011. BP Statistical Review of World Energy. Dostupné na: http://www.bp.com/liveassets/bp_internet/globalbp/globalbp_uk_english/reports_and_publications/statistical_energy_review_2011/STAGING/local_assets/pdf/coal_section_2011.pdf.

Brookes, L., 1990. The greenhouse effect: the fallacies in the energy efficiency solution. In: Energy Policy, vol. 18, pp. 199–201.

Brundtlandtová, G.H., 1987. Our Common Future: Report of the World Commission on Environment and Development, 1st. Oxford University Press, Oxford, pp. 55–155.

Climate Change, 2014. Impacts, Adaptation, and Vulnerability, Summary for Policymakers. Dostupné na: http://ipcc-wg2.gov/AR5/images/uploads/WG2AR5_SPM_FINAL.pdf.

General Assembly, 2015. Transforming Our World: the 2030 Agenda for Sustainable Development.

Heinberg, R., 2011. The End of Growth: Adapting to Our New Economic Reality, vol. 336. New Society Publishers, Gabriola Island, Kanada, ISBN 978-0-86571-695-7.

Howarth, R.B., 1997. Energy efficiency and economic growth. Contemp. Econ. Pol. 25, 1–9.

Judson, R.A., Schmalensee, R., Stoker, T.M., 1999. Economic development and the structure of demand for commercial energy. Energy J. 20 (2).

Lacy, P., Rutqvist, J., 2016. Waste to Wealth: The Circular Economy Advantage. Springer, ISBN 1137530707, pp. 2016–2264.

Mojžiš, M., 2012. Nízky uhlík. Dostupné na: http://blog.jetotak.sk/kriticka-ekonomia/2012/08/20/miroslav-mojzis-nizky-uhlik/.

Malindžák, D., Zimon, D., Bednárová, L., et al., 2017. Homogeneous production processes and approaches to their management. Acta montanistica Slovaca 22 (2), 153–160. https://actamont.tuke.sk/pdf/2017/n2/7malindzak.pdf.

Malindžáková, M., 2013. Štandardy Pre Environment, 1st. Technická Univerzita v Košiciach, Košice, pp. 5–15.

Malindžáková, M., 2018. Kvalita logistických systémov. Edičné stredisko F BERG, Košice, ISBN 978-80-553-2664-1.

Norma EN ISO 26000:2010, n.d. Guidance on social responsibility.

Rifkin, J., 2011. The Third Industrial Revolution, How Lateral Power Is Transforming Energy, the Economy and the World, vol. 291. Palgrave Macmillan, New York.

Sachs, J.D., 2014. Karbónová loby zabíja planétu. Dostupné na: http://komentare.hnonline.sk/komentare-167/karbonova-loby-zabija-planetu-603568.

Stern, D.I., 2004. Economic growth and energy. In: Encyklopedia of Energy, vol. 2.

Stockholm Resilience Centre, 1972. About Stockholm Resilience Centre. Dostupné na: http://www.stockholmresilience.org/aboutus.4.aeea46911a3127427980003326.html.

www.4, n.d. http://pravdu.cz/ekologie/nerust-alternativni-ekonomicka-cesta.

www.11, n.d. www.enviroportal.sk.

www.12, n.d. https://www.tcu.sk/aktuality/tyzden-celozivotneho-ucenia-podporuje-agendu-2030.

www.13, n.d. https://www.europarl.europa.eu/news/sk/headlines/society/20180328STO00751/odpadove-hospodarstvo-v-eu-fakty-a-cisla.

Zimon, D., Gajewska, T., Malindzakova, M., 2018. Implementing the requirements of ISO 9001 and improvement logistics processes in SMES which operate in the textile industry. Autex Res. J. 18 (4), 392–397. https://doi.org/10.1515/aut-2018-0020.

Further reading

Abulfotuh, F., 2007. Energy efficiency and renewable technologies: the way to sustainable energy future. In: Desalination the International Journal on the Science and Technology of Desalting and Water Purification, vol. 209. Published by Elsevier B.V., pp. 275–282. ISBN: 0011-9164.

Carson, R.T., 2010. The environmental kuznets curve: seeking empirical regularity and theoretical structure. Oxford Journals Rev. Environ. Econ. Pol. 4 (1), 3–23. Published online December 22, 2009, online ISSN 1750-6824.

Daly, H.E., Cobb JR., J.B., 1989. For the Common good: Redirecting the Economy toward Community, the Environment, and a Sustainable Future. Beacon Press, Boston.

Donner, S., 2007. Emissions Intensity: Declining for Decades. Dostupné na: http://simondonner.blogspot.sk/2007/05/emissions-intensity-declining-for.html.

Frankel, J., 2014. Omyl-o-oligarchii. Dostupné na: http://openiazoch.zoznam.sk/cl/144373/Omyl-o-oligarchii.

Gajdoš, J., Naščáková, J., Andrejovský, A., Ručinský, R., 2011. Ener supply - dôvody zapojenia sa do projektu. In: Nekonferenčný recenzovaný zborník v rámci riešenia projektov VEGA, KEGA, APVV, ENER SUPPLY. Vydavateľstvo EKONÓM, Bratislava, ISBN 978-80-225-3207-5, pp. 33–40.

Garrett, T.J., 2012. The Physics of Long-run Global Economic Growth. Dostupné na: http://www.inscc.utah.edu/~tgarrett/Economics/Economics.html.

Grubb, M., 1993. The Earth Summit Agreements: A Guide and Assessment. Royal Institute of International Affairs Series. Energy and Environmental Programme. Earthscan, London.

Hegyi, L., 2012. Návrh modelu vzťahov prírodných limitov a trvaloudržateľného rozvoja spoločnosti. Diplomová práca, Technická univerzita v Košiciach, Strojnícka fakulta, Katedra environmentalistiky.

Islam, S.M.N., Clarke, M., 2002. The relationship between economic development and social welfare: a new adjusted GDP measure of welfare. In: Social Indicators Research, pp. 201–216.

Kenny, C., 2005. Does development make you happy? Subjective wellbeing and economic growth in developing countries. In: Social Indicators Research, Roč, vol. 73, pp. 199–219, 29–57.

Khazzoom, D.J., 1980. Economic implications of mandated efficiency standards for household appliances. The Energy Journal 1 (4), 21–39.

Klinec, I., 2005a. Zelené myslenie, zelená budúcnosť. Alternatívne ekonomické a sociálne teórie podporujúce smerovnaie k udržateľnému rozvoju. Olomouc, vol. 258. Univerzita Palackého Olomouc, Česká republika (s).

Klinec, I., 2005b. Alternatívne ekonomické teórie podporujúce smerovanie k trvalo udržatelnému rozvoju. In: Sborník z projektu „ K uržitelnému rozvoji Ceské republiky: vytvárení podmínek". Univerzita Karlova v Praze, Centrum pro otázky životního prostředí, Praha, pp. 52–119. Svazek 2: Teoretická východiska, souvislosti, instituce.

Lequiller, F., 2005. Is GDP a satisfactory measure of growth?. In: The OECD Observer, 246/247, pp. 30–31.

Lisý, J., 1999. Výkonnosť ekonomiky a ekonomický rast. Iura Edition, Bratislava.

Mazanec, P., 2011. Odklon od jaderné energie zvyšuje poptávku po dřevu. Praha. Dostupné na: http://ekolist.cz/cz/zpravodajstvi/zpravy/odklon-od-jaderne-energie-zvysuje-poptavku-po-drevu.

Meadows, D., et al., 1972. The Limits to Growth. Universe Books, New York.

Musil, P., 2009. Globální energetický problém a hospodářská politika: se zaměřením na obnovitelné zdroje. C.H. Beck, Praha.

Naščáková, J., Pčolinská, L., Gajdoš, J., 2011a. Kvalita života a jej dimenzie. In: Nekonferenčný recenzovaný zborník čiastkové výsledky riešených projektov „VEGA, KEGA, APVV, ENER SUPPLY". Ekonomická univerzita v Bratislave Podnikovohospodárska fakulta v Košiciach, Košice, ISBN 978-80-225-3207-5, pp. 56–65.

Naščáková, J., Pčolinská, L., Gajdoš, J., 2011b. Vybrané spôsoby merania ekonomického rastu a spoločenského blahobytu. In: Nekonferenčný recenzovaný zborník v rámci riešenia projektov VEGA, KEGA, APVV, ENER SUPPLY. Vydavateľstvo EKONÓM, Bratislava, ISBN 978-80-225-3207-5, pp. 66–70.

Naščáková, J., Pudło, P., 2011. Meranie kvality života z pohľadu ekonomicko-politického prostredníctvom ukazovateľov hrubého domáceho produktu, indexu udržateľného hospodárskeho blahobytu a reálneho pokroku. In: Nekonferenčný recenzovaný zborník v rámci riešenia projektov VEGA, KEGA, APVV, ENER SUPPLY. Vydavateľstvo EKONÓM, Bratislava, ISBN 978-80-225-3207-5, pp. 71–78.

Nátr, L., 1998. Rostliny, lidé a trvale udržitelný život člověka na Zemi. Skripta. Nakladatelství Karolinum, Praha, ISBN 80-7184-681-3.

OECD, 2018. Global Outlook on Financing for Sustainable Development 2019. Time to Face the Challenge.

Pecho, J., 2012. Zmena klímy - globálny problém s lokálnymi dopadmi. In: Klimatická zmena a lokálny rozvoj – výzva pre samosprávy: Príručka pre samosprávy. Karpatský rozvojový institut, Košice.

Rockström, J., et al., 2009. Planetary boundaries: exploring the safe operating space for humanity. In: Ecology and Society, vol. 14, p. 32 (2), Dostupné na: http://www.ecologyandsociety.org/vol14/iss2/art32.

Rowe, J., 2009. The Cult of GDP/Gross Domestic Product Ignores Wealth Generated by the Commons/. Dostupné na: http://onthecommons.org/cult-gdp-0.

Simon, J.L., 2006. Největší bohatství, vol. 666. Centrum pro studium demokracie a kultury, Brno. ISBN: 80-7325-082-9.

Steard, J.G., Steard, W.E., 2012. Manažment pre malú planétu - Prečo je dôležité meniť stratégie neobmedzeného rastu na stratégie udržateľnosti. Vydavateľstvo: Eastone, ISBN 9788081092169.

Správa OECD, 2011. Towards Green Growth, ISBN 978-92-64-094970.

Vlčková, E., 2014. S uhlím vypouštíme džina z lahve, říká švédský ekolog. Lidové noviny, Praha. Dostupné na: http://www.lidovky.cz/s-uhlim-vypoustime-dzina-z-lahve-rika-svedsky-ekolog-pcs-/ln_veda.asp?c=A100510_114257_ln_veda_ev.

Willquist, K., 2012. Vodík: zdroj zelené energie v budoucnosti? Dostupné na: http://www.scienceinschool.org/2012/issue22/hydrogen/czech.

Index

Note: 'Page numbers followed by "*f*" indicate figures and "*t*" indicate tables.'

T

Printed in the United States
by Baker & Taylor Publisher Services